Analytic Combinatorics in Several Variables

Mathematicians have found it useful to enumerate all sorts of things arising in discrete mathematics: elements of finite groups, configurations of ones and zeros, graphs of various sorts; the list is endless. Analytic combinatorics uses analytic techniques to do the counting: generating functions are defined and their coefficients are then estimated via complex contour integrals. This book is the result of nearly fifteen years of work on developing analytic machinery to recover, as effectively as possible, asymptotics of the coefficients of a multivariate generating function. It is the first book to describe many of the results and techniques necessary to estimate coefficients of generating functions in more than one variable.

Aimed at graduate students and researchers in enumerative combinatorics, the book contains all the necessary background, including a review of the uses of generating functions in combinatorial enumeration as well as chapters devoted to saddle point analysis, Groebner bases, Laurent series and amoebas, and a smattering of differential and algebraic topology. All software along with other ancillary material can be located via the book website, www.cs.auckland.ac.nz/~mcw/Research/mvGF/asymultseq/ACSVbook/.

ROBIN PEMANTLE is a Professor in the Department of Mathematics at the University of Pennsylvania.

MARK C. WILSON is a Senior Lecturer in the Department of Computer Science at the University of Auckland.

CAMBRIDGE STUDIES IN ADVANCED MATHEMATICS

All the titles listed below can be obtained from good booksellers or from Cambridge University Press.

For a complete series listing visit: www.cambridge.org/mathematics.

Already published

99 D. Goldfeld *Automorphic forms and L-functions for the group GL(n,R)*
100 M. B. Marcus & J. Rosen *Markov processes, Gaussian processes, and local times*
101 P. Gille & T. Szamuely *Central simple algebras and Galois cohomology*
102 J. Bertoin *Random fragmentation and coagulation processes*
103 E. Frenkel *Langlands correspondence for loop groups*
104 A. Ambrosetti & A. Malchiodi *Nonlinear analysis and semilinear elliptic problems*
105 T. Tao & V. H. Vu *Additive combinatorics*
106 E. B. Davies *Linear operators and their spectra*
107 K. Kodaira *Complex analysis*
108 T. Ceccherini-Silberstein, F. Scarabotti & F. Tolli *Harmonic analysis on finite groups*
109 H. Geiges *An introduction to contact topology*
110 J. Faraut *Analysis on Lie groups: An introduction*
111 E. Park *Complex topological K-theory*
112 D. W. Stroock *Partial differential equations for probabilists*
113 A. Kirillov, Jr *An introduction to Lie groups and Lie algebras*
114 F. Gesztesy et al. *Soliton equations and their algebro-geometric solutions, II*
115 E. de Faria & W. de Melo *Mathematical tools for one-dimensional dynamics*
116 D. Applebaum *Lévy processes and stochastic calculus (2nd Edition)*
117 T. Szamuely *Galois groups and fundamental groups*
118 G. W. Anderson, A. Guionnet & O. Zeitouni *An introduction to random matrices*
119 C. Perez-Garcia & W. H. Schikhof *Locally convex spaces over non-Archimedean valued fields*
120 P. K. Friz & N. B. Victoir *Multidimensional stochastic processes as rough paths*
121 T. Ceccherini-Silberstein, F. Scarabotti & F. Tolli *Representation theory of the symmetric groups*
122 S. Kalikow & R. McCutcheon *An outline of ergodic theory*
123 G. F. Lawler & V. Limic *Random walk: A modern introduction*
124 K. Lux & H. Pahlings *Representations of groups*
125 K. S. Kedlaya *p-adic differential equations*
126 R. Beals & R. Wong *Special functions*
127 E. de Faria & W. de Melo *Mathematical aspects of quantum field theory*
128 A. Terras *Zeta functions of graphs*
129 D. Goldfeld & J. Hundley *Automorphic representations and L-functions for the general linear group, I*
130 D. Goldfeld & J. Hundley *Automorphic representations and L-functions for the general linear group, II*
131 D. A. Craven *The theory of fusion systems*
132 J. Väänänen *Models and games*
133 G. Malle & D. Testerman *Linear algebraic groups and finite groups of Lie type*
134 P. Li *Geometric analysis*
135 F. Maggi *Sets of finite perimeter and geometric variational problems*
136 M. Brodmann & R. Y. Sharp *Local cohomology (2nd Edition)*
137 C. Muscalu & W. Schlag *Classical and multilinear harmonic analysis, I*
138 C. Muscalu & W. Schlag *Classical and multilinear harmonic analysis, II*
139 B. Helffer *Spectral theory and its applications*

Analytic Combinatorics in Several Variables

ROBIN PEMANTLE

The University of Pennsylvania

MARK C. WILSON

University of Auckland

CAMBRIDGE
UNIVERSITY PRESS

CAMBRIDGE UNIVERSITY PRESS
Cambridge, New York, Melbourne, Madrid, Cape Town,
Singapore, São Paulo, Delhi, Mexico City

Cambridge University Press
32 Avenue of the Americas, New York, NY 10013-2473, USA

www.cambridge.org
Information on this title: www.cambridge.org/9781107031579

First published 2013

Printed in the United States of America

A catalog record for this publication is available from the British Library.

Library of Congress Cataloging in Publication data

Pemantle, Robin.
Analytic combinatorics in several variables / Robin Pemantle, The University of Pennsylvania,
Mark C. Wilson, University of Auckland.
pages cm – (Cambridge studies in advanced mathematics ; 140)
Includes bibliographical references and index.
ISBN 978-1-107-03157-9 (hardback)
1. Combinatorial enumeration problems. 2. Functions of several complex
variables. I. Wilson, Mark C. (Mark Curtis), 1967– II. Title.
QA164.8.P46 2013
511′.6–dc23 2012049386

ISBN 978-1-107-03157-9 Hardback

To the memory of Philippe Flajolet, on whose shoulders stands all of the work herein.

Contents

PART II: MATHEMATICAL BACKGROUND

PART III: MULTIVARIATE ENUMERATION

Preface

The term "analytic combinatorics" refers to the use of complex analytic methods to solve problems in combinatorial enumeration. Its chief objects of study are generating functions (Flajolet and Sedgewick, 2009, page vii). Generating functions have been used for enumeration for more than a hundred years, going back to Hardy and, arguably, to Euler. Their systematic study began in the 1950s (Hayman, 1956). Much of the impetus for analytic combinatorics comes from the theory of algorithms, arising, for example, in the work of Knuth (2006). The recent, seminal work by Flajolet and Sedgewick (2009) describes the rich univariate theory with literally hundreds of applications.

The multivariate theory, as recently as the mid-1990s, was still in its infancy. Techniques for *deriving* multivariate generating functions have been well understood, sometimes paralleling the univariate theory and sometimes achieving surprising depth (Fayolle, Iasnogorodski, and Malyshev, 1999). Analytic methods for recovering coefficients of generating functions once the functions have been derived have, however, been sorely lacking. A small body of analytic work goes back to the early 1980s (Bender and Richmond, 1983); however, even by 1995, of 100+ pages in the *Handbook of Combinatorics* devoted to asymptotic enumeration (Odlyzko, 1995), multivariate asymptotics received fewer than six.

This book is the result of work spanning nearly fifteen years. Our aim has been to develop analytic machinery to recover, as effectively as possible, asymptotics of the coefficients of a multivariate generating function. Both authors feel drawn to this area of study because it combines many areas of modern mathematics. Functions of one or more complex variables are essential, but also algebraic topology in the Russian style, stratified Morse theory, computational algebraic methods, saddle-point integration, and of course the basics of combinatorial enumeration. The many applications of this work in areas such as bioinformatics, queueing theory, and statistical mechanics are not surprising when we realize how widespread is the use of generating functions in applied combinatorics and probability.

The purpose of this book is to pass on what we have learned, so that others may learn it and use it before we forget it. The present form of the book grew out of graduate-level mathematics courses that developed, along with the theory, at the University of Wisconsin, Ohio State University, and the University of Pennsylvania. The course was intended to be accessible to students in their second year of graduate study. Because of the eclectic nature of the required background, this presents something of a challenge. One may count on students having seen calculus on manifolds by the end of a year of graduate studies, in addition to some complex variable theory. One may also assume some willingness to do some outside reading. However, some of the more specialized areas on which multivariate analytic combinatorics must draw are not easy to get from books. This includes topics such as the theory of amoebas (Gel'fand, Kapranov, and Zelevinsky, 1994) and the Leray-Petrovsky-Gårding theory of inverse Fourier transforms. Other topics such as saddle-point integration and stratified Morse theory exist in books but require being summarized to avoid a semester-long detour.

We have dealt with these problems by summarizing a great amount of background material. Part I contains the combinatorial background and will be known to students who have taken a graduate-level course in combinatorial enumeration. Part II contains mathematical background from outside of combinatorics. The topics in Part II are central to the understanding and execution of the techniques of analytic combinatorics in several variables. Part III contains the theory, all of which is new since the turn of the millennium and only parts of which exist in published form. Finally, there are appendices, almost equal in total size to Part II, which include necessary results from algebraic and differential topology. Some students will have seen these, but for the rest, the inclusion of these topics will make the present book self-contained rather than one that can only be read in a library.

We hope to recruit further researchers into this field, which still has many interesting challenges to offer, and this explains the rather comprehensive nature of the book. However, we are aware that some readers will be more focused on applications and seek the solution of a given problem. The book is structured so that after reading Chapter 1, it should be possible to skip to Part III and pick up supporting material as required from previous chapters. A list of publications using the multivariate methods described in this book can be found on our website: `www.cs.auckland.ac.nz/~mcw/Research/mvGF/asymultseq/ACSVbook/`.

The mathematical development of the theory belongs mostly to the two authors, but there are a number of individuals whose help was greatly instrumental in moving the theory forward. The complex analysts at the University of Wisconsin-Madison, Steve Wainger, Jean-Pierre Rosay, and Andreas Seeger, helped the authors (then rather junior researchers) to grapple with the problem in its earliest incarnation. A similar role was played several years later by Jeff McNeal. Perhaps the greatest thanks are due to Yuliy Baryshnikov, who translated the Leray-Petrovsky theory and the work of Atiyah-Bott-Gårding into terms the authors could understand and coauthored several articles. Frank Sottile provided help with algebra on many occasions; Persi Diaconis arranged for a graduate course while the

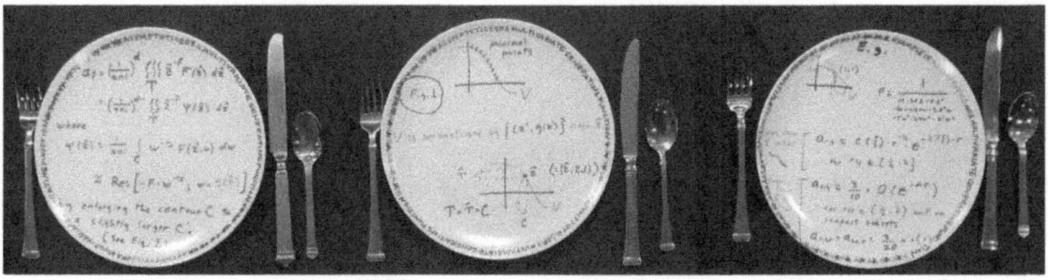

Figure 0.1 Customized "asymptotics of a multivariable generating function" dinner plates.

first author visited Stanford in 2000; Richard Stanley answered our numerous miscellaneous queries. Thanks are also due to our other coauthors on articles related to this project, listed on the project website linked from the book website. Alex Raichev and Torin Greenwood helped substantially with proofreading and with computer algebra implementations of some parts of the book. Thanks also to valuable proofreading contributions from Lily Yen. All software can be located via the book website.

On a more personal level, the first author would like to thank his wife, Diana Mutz, for encouraging him to follow this unusual project wherever it took him, even if it meant abandoning a still productive vein of problems in probability theory. The sentiment in the probability theory community may be otherwise, but the many connections of this work to other areas of mathematics have been a source of satisfaction to the authors. The first author would also like to thank his children, Walden, Maria, and Simi, for their participation in the project via the Make-A-Plate company (see Figure 0.1).

The second author thanks his wife Golbon Zakeri, children Yusef and Yahya, and mother-in-law Shahin Sabetghadam for their help in carving out time for him to work on this project, sometimes at substantial inconvenience to themselves. He hopes they will agree that the result is worth it.

PART I

Combinatorial Enumeration

1

Introduction

1.1 Arrays of Numbers

The main subject of this book is an array of numbers

$$\{a_{r_1,\ldots,r_d} : r_1, \ldots, r_d \in \mathbb{N}\}.$$

This is usually written as $\{a_{\boldsymbol{r}} : \boldsymbol{r} \in \mathbb{N}^d\}$, where as usual $\mathbb{N} = \{0, 1, 2, \ldots\}$. The numbers $a_{\boldsymbol{r}}$ may be integers, real numbers, or even complex numbers. We always use d to denote the dimension of the array. The variables r, s, and t are reserved as synonyms for r_1, r_2, and r_3, respectively, to avoid subscripts in examples of dimensions up to three.

The numbers $a_{\boldsymbol{r}}$ usually come with a story – a reason they are interesting. Often they count a class of objects parametrized by $\boldsymbol{r}$. For example, it could be that $a_{\boldsymbol{r}}$ is the multinomial coefficient $a_{\boldsymbol{r}} := \binom{|\boldsymbol{r}|}{r_1 \cdots r_d}$, with $|\boldsymbol{r}| := \sum_{j=1}^{d} r_j$, in which case $a_{\boldsymbol{r}}$ counts sequences with r_1 1's, r_2 2's, and so forth up to r_d occurrences of the symbol d. Another frequent source of these arrays is in probability theory. Here, the numbers $a_{\boldsymbol{r}} \in [0, 1]$ are probabilities of events parametrized by $\boldsymbol{r}$. For example, a_{rs} might be the probability that a simple random walk of r steps ends at the integer point s.

How might one understand an array of numbers? There might be a simple, explicit formula. The multinomial coefficients, for example, are given by ratios of factorials. As Stanley[1] (1997) points out in the introduction, a formula of this brevity seldom exists; when it does, we don't need fancy techniques to describe the array. Often, if a formula exists at all, it will not be in closed form but will have a summation in it. As Stanley says, "There are actually formulas in the literature (nameless here forevermore) for certain counting functions whose evaluation requires listing all of the objects being counted! Such a 'formula' is completely worthless." (Example 1.1.4 page 2) Less egregious are the formulae containing functions that are rare or complicated and whose properties are not

[1] Much of the presentation in this first section is heavily influenced by Stanley – see the notes to this chapter.

immediately familiar to us. It is not clear how much good it does to have this kind of formula.

Another way of describing arrays of numbers is via recursions. The simplest recursions are finite linear recursions, such as the recursion

$$a_{r,s} = a_{r-1,s} + a_{r,s-1}$$

for the binomial coefficients. A recursion for $a_{\boldsymbol{r}}$ in terms of values $\{a_{\boldsymbol{s}} : \boldsymbol{s} < \boldsymbol{r}\}$ whose indices precede $\boldsymbol{r}$ in the coordinate-wise partial order may be pretty unwieldy, perhaps requiring evaluation of a complicated function of all $a_{\boldsymbol{s}}$ with $\boldsymbol{s} < \boldsymbol{r}$. However, if the recursion is of bounded complexity, such as a linear recursion $a_{\boldsymbol{r}} = \sum_{\boldsymbol{j} \in F} c_{\boldsymbol{j}} a_{\boldsymbol{r}-\boldsymbol{j}}$ for some finite set $\{c_{\boldsymbol{j}} : \boldsymbol{j} \in F\}$ of constants, then the recursion gives a polynomial time algorithm for computing $a_{\boldsymbol{r}}$. Still, even in this case, the estimation of $a_{\boldsymbol{r}}$ is not at all straightforward. Thus, although we look for recursions to help us understand number arrays, recursions rarely provide definitive descriptions.

A third way of understanding an array of numbers is via an estimate. If one uses Stirling's formula

$$n! \sim \frac{n^n}{e^n}\sqrt{2\pi n},$$

one obtains an estimate for binomial coefficients

$$a_{r,s} \sim \left(\frac{r+s}{r}\right)^r \left(\frac{r+s}{s}\right)^s \sqrt{\frac{r+s}{2\pi r s}} \tag{1.1.1}$$

and a similar estimate for multinomial coefficients. If number theoretic properties of $a_{\boldsymbol{r}}$ are required, then we are better off sticking with the formula $(r+s)!/(r!\,s!)$, but when the approximate size of $a_{\boldsymbol{r}}$ is paramount, then the estimate (1.1.1) is better.

A fourth way to understand an array of numbers is to give its generating function. The *generating function* for the array $\{a_{\boldsymbol{r}}\}$ is the series $F(\boldsymbol{z}) := \sum_{\boldsymbol{r}} a_{\boldsymbol{r}} \boldsymbol{z}^{\boldsymbol{r}}$. Here $\boldsymbol{z}$ is a d-dimensional vector of indeterminates $(z_1, \ldots, z_d)$, and $\boldsymbol{z}^{\boldsymbol{r}}$ denotes the monomial $z_1^{r_1} \cdots z_d^{r_d}$. In our running example of multinomial coefficients, the generating function

$$F(\boldsymbol{z}) = \sum_{\boldsymbol{r}} \binom{|\boldsymbol{r}|}{r_1 \ \cdots \ r_d} z_1^{r_1} \cdots z_d^{r_d}$$

is written more compactly as $1/(1 - r_1 - \cdots - r_d)$. Stanley calls the generating function "the most useful but the most difficult to understand" method for describing a sequence or array.

One reason a generating function is useful is that the algebraic form of the function is intimately related to recursions for $a_{\boldsymbol{r}}$ and combinatorial decompositions for the objects enumerated by $a_{\boldsymbol{r}}$. Another reason is that estimates (and exact formulae if they exist) may be extracted from a generating function. In other words, formulae, recursions, and estimates all ensue once a generating function is known.

1.2 Generating Functions and Asymptotics

We employ the usual asymptotic notation, as follows. If f, g are real valued functions, then the statement "$f = O(g)$" is shorthand for the statement "$\limsup_{x \to x_0} |f(x)|/|g(x)| < \infty$." It must be made clear at which value, x_0, the limit is taken; if f and g depend on parameters other than x, it must also be made clear which is the variable being taken to the limit. Most commonly, $x_0 = +\infty$; in the statement $a_n = O(g(n))$, the limit is always taken at infinity. The statement "$f = o(g)$" is shorthand for $f(x)/g(x) \to 0$, again with the limiting value of x specified. Lastly, the statement "$f \sim g$" means $f/g \to 1$ and is equivalent to "$f = (1 + o(1)) \cdot g$" or "$f - g = o(g)$"; again, the variable and its limiting value must be specified. Two more useful notations are $f = \Omega(g)$, which just means $g = O(f)$, and $f = \Theta(g)$, which means that both $f = O(g)$ and $g = O(f)$ are satisfied. An ***asymptotic expansion***

$$f \sim \sum_{j=0}^{\infty} g_j$$

for a function f in terms of a sequence $\{g_j : j \in \mathbb{N}\}$ satisfying $g_{j+1} = o(g_j)$ is said to hold if for every $M \geq 1$, $f - \sum_{j=0}^{M-1} g_j = O(g_M)$. This is equivalent to $f - \sum_{j=0}^{M-1} g_j = o(g_{M-1})$. Often we slightly extend the notion of an asymptotic series by saying that $f \sim \sum_{j=0}^{\infty} a_n g_n$ even when some a_n vanish, as long as $f - \sum_{n=0}^{M-1} a_n g_n = O(g_M)$ and infinitely many of the $\{a_n\}$ do not vanish.

A function f is said to be ***rapidly decreasing*** if $f(n) = O(n^{-K})$ for every $K > 0$, ***exponentially decaying*** if $f(n) = O(e^{-\gamma n})$ for some $\gamma > 0$, and ***super-exponentially decaying*** if $f(n) = O(e^{-\gamma n})$ for every $\gamma > 0$.

Example 1.2.1 Let $f \in C^\infty(\mathbb{R})$ be a smooth real function defined on a neighborhood of zero. Thus it has a Taylor expansion whose n^{th} coefficient is $c_n := f^{(n)}(0)/n!$. If f is not analytic, then this expansion may not converge to f (e.g., if $f(x) = e^{-1/x^2}$ then $c_n \equiv 0$) and may even diverge for all nonzero x, but we always have Taylor's remainder theorem:

$$f(x) = \sum_{n=0}^{M-1} c_n x^n + c_M \xi^M$$

for some $\xi \in [0, M]$. This proves that

$$f \sim \sum_n c_n x^n$$

is always an asymptotic expansion for f near zero. ■

All these notations hold in the multivariate case as well, except that if the limit value of z is infinity, then a statement such as $f(z) = O(g(z))$ must also specify how z approaches the limit. Our chief concern is with the asymptotics of $a_{\mathbf{r}}$ as $\mathbf{r} \to \infty$ in a given direction. More specifically, by a ***direction***, we mean an element of $(d-1)$-dimensional projective space whose class contains a d-tuple of positive real numbers. Often we parametrize positive projective vectors by the corresponding unit vector $\hat{\mathbf{r}} := \mathbf{r}/|\mathbf{r}|$. It turns out that a typical asymptotic formula for $a_{\mathbf{r}}$ is $a_{\mathbf{r}} \sim C|\mathbf{r}|^{\alpha}\mathbf{z}^{-\mathbf{r}}$, where $|\mathbf{r}|$ is the sum of the coordinates of $\mathbf{r}$, and the d-tuple $\mathbf{z}$ and the multiplicative constant C depend on $\mathbf{r}$ only through $\hat{\mathbf{r}}$. In hindsight, formulae such as these make it natural to consider $\mathbf{r}$ projectively and take $\mathbf{r}$ to infinity in prescribed directions. In its original context, the above quote from Stanley referred chiefly to univariate arrays, i.e., the case $d = 1$. As is seen in Chapter 3, it is indeed true that the generating function $f(z)$ for a univariate sequence $\{a_n : n \in \mathbb{N}\}$ leads, almost automatically, to asymptotic estimates for a_n as $n \to \infty$. [Another notational aside: we use $f(z)$ and a_n instead of $F(z)$ and a_r in one variable to coincide with notation in the univariate literature.]

To estimate a_n when f is known, begin with Cauchy's integral formula:

$$a_n = \frac{1}{2\pi i}\int z^{-n-1} f(z)\, dz\,. \tag{1.2.1}$$

The integral is a complex contour integral on a contour encircling the origin, and one may apply complex analytic methods to estimate the integral. The necessary knowledge of residues and contour shifting may be found in an introductory complex variables text such as Conway (1978) or Berenstein and Gay (1991), although one obtains a better idea of univariate saddle point integration from Henrici (1988) or Henrici (1991).

The situation for multivariate arrays is nothing like the situation for univariate arrays. In 1974, when Bender published his review article (Bender, 1974) on asymptotic enumeration, the asymptotics of multivariate generating functions was largely a gap in the literature. Bender's concluding section urges research in this area:

> Practically nothing is known about asymptotics for recursions in two variables even when a generating function is available. Techniques for obtaining asymptotics from bivariate generating functions would be quite useful. (page 512)

In the 1980s and 1990s, a small body of results was developed by Bender, Richmond, Gao, and others, giving the first partial answers to questions of asymptotics of generating functions in the multivariate setting. The first article to concentrate on extracting asymptotics from multivariate generating functions was Bender (1973), already published at the time of Bender's survey, but the seminal work is Bender and Richmond (1983). The hypothesis is that F has a singularity of the form $A/(z_d - g(\mathbf{x}))^q$ on the graph of a smooth function g, for some real exponent q, where $\mathbf{x}$ denotes $(z_1, \ldots, z_{d-1})$. They show, under appropriate further hypotheses on F, that the probability measure μ_n one obtains by renormalizing $\{a_{\boldsymbol{r}} : r_d = n\}$ to sum to 1 converges to a multivariate normal when appropriately rescaled. Their method, which we call the ***GF-sequence method***, is to break the d-dimensional array $\{a_{\boldsymbol{r}}\}$ into a sequence of $(d-1)$-dimensional slices and consider the sequence of $(d-1)$-variate generating functions

$$f_n(\boldsymbol{x}) = \sum_{\boldsymbol{r}:r_d=n} a_{\boldsymbol{r}} \boldsymbol{z}^{\boldsymbol{r}} .$$

They show that, asymptotically as $n \to \infty$,

$$f_n(\boldsymbol{x}) \sim C_n g(\boldsymbol{x}) h(\boldsymbol{x})^n \tag{1.2.2}$$

and that sequences of generating functions obeying (1.2.2) satisfy a central limit theorem and a local central limit theorem.

These results always produce Gaussian (central limit) behavior. The applicability of the entire GF-sequence method is limited to the single, although important, case where the coefficients $a_{\boldsymbol{r}}$ are nonnegative and possess a Gaussian limit. The work of Bender and Richmond (1983) has been greatly expanded upon, but always in a similar framework. For example, it has been extended to matrix recursions (Bender, Richmond, and Williamson, 1983), and the applicability has been extended from algebraic to algebraico-logarithmic singularities of the form $F \sim (z_d - g(\boldsymbol{x}))^q \log^\alpha(1/(z_d - g(\boldsymbol{x})))$ (Gao and Richmond, 1992). The difficult step is always deducing asymptotics from the hypotheses $f_n \sim C_n \, g \cdot h^n$. Thus some publications in this stream refer to such an assumption in their titles (Bender and Richmond, 1999), and the term "quasi-power" has been coined for such a sequence $\{f_n\}$.

1.3 New Multivariate Methods

The research presented in this book grew out of several problems encountered by the first author concerning bivariate and trivariate arrays of probabilities. One might have thought, based on the situation for univariate generating functions, that results would exist, well known and neatly packaged, that gave asymptotic estimates for the probabilities in question. At that time, the most recent and complete reference on asymptotic enumeration was Odlyzko's 1995 survey (Odlyzko, 1995). Only six of its more than 100 pages are devoted to multivariate asymptotics, mainly to the GF-sequence results of Bender et al. Odlyzko's section on multivariate methods closes with a call for further work in this area. Evidently, in the multivariate case, a general asymptotic formula or method was not known, even for the simplest imaginable class, namely rational functions. This stands in stark contrast to the univariate theory of rational functions, which is trivial (see Chapter 3). The relative difficulty of the problem in higher dimensions is perhaps unexpected. The connections to other areas of mathematics such as Morse theory are, however, quite intriguing, and these, more than anything else, have caused us to pursue this line of research long after the urgency of the original motivating problems had faded.

Odlyzko (1995) describes why he believes multivariate coefficient estimation to be difficult. First, the singularities are no longer isolated, but form $(d-1)$-dimensional hypersurfaces. Thus, he points out, "Even rational multivariate functions are not easy to deal with." Second, the multivariate analogue of the one-dimensional residue theorem is the considerably more difficult theory of Leray (1959). This theory was later fleshed out by Aĭzenberg and Yuzhakov (1983), who spent a few pages in their Section 23 on generating functions and combinatorial sums. Further progress in using multivariate residues to evaluate coefficients of generating functions was made by Bertozzi and McKenna (1993), although at the time of Odlyzko's survey, none of the works based on multivariate residues such as Lichtin (1991) and Bertozzi and McKenna (1993) had resulted in any kind of systematic application of these methods to enumeration.

The focus of this book is a recent vein of research, begun in Pemantle and Wilson (2002) and continued in Pemantle and Wilson (2004); Lladser (2003); Wilson (2005); Lladser (2006); Raichev and Wilson (2008); Pemantle and Wilson (2008); DeVries (2010); Pemantle and Wilson (2010), and Raichev and Wilson (2012b), as well as several others (Baryshnikov and Pemantle, 2011; DeVries, van der Hoeven, and Pemantle, 2012). This research extends ideas that are present to some degree in Lichtin (1991) and Bertozzi and McKenna (1993), using complex methods that are genuinely multivariate to evaluate coefficients via the multivariate Cauchy formula

$$a_r = \left(\frac{1}{2\pi i}\right)^d \int_T z^{-r-1} F(z)\, dz\,. \tag{1.3.1}$$

By avoiding symmetry-breaking decompositions such as $F = \sum f_n(z_1, \ldots, z_{d-1}) z_d^n$, one hopes the methods will be more universally applicable and the formulae more canonical. In particular, the results of Bender et al. and the results of Bertozzi and McKenna (1993) are seen to be two instances of a more general result estimating the Cauchy integral via topological reductions of the cycle of integration. These topological reductions, although not fully automatic, are algorithmically decidable in large classes of cases. An ultimate goal, stated in Pemantle and Wilson (2002) and Pemantle and Wilson (2004), is to develop software to automate all of the computation.

We can by no means say that the majority of multivariate generating functions fall prey to these new techniques. The class of functions to which the methods described in this book may be applied is larger than the class of rational functions, but similar in spirit: the function must have singularities, and the dominant singularity must be a pole. This translates to the requirement that the function be meromorphic in a neighborhood of a certain polydisk (see the remark following Pemantle and Wilson [2008, Theorem 3.16] for exact hypotheses), which means that it has a representation, at least locally, as a quotient of analytic functions. Nevertheless, as illustrated in Pemantle and Wilson (2008) and in the present book, meromorphic functions cover a good number of combinatorially interesting examples.

Throughout these notes, we reserve the variable names

$$F = \frac{G}{H} = \sum_{r} a_r z^r$$

for the meromorphic function F expressed (locally) as the quotient of analytic functions G and H. We assume this representation to be in lowest terms. What this means about the common zeros of G and H will be clearer once stratifications have been discussed. The variety $\{z : H(z) = 0\}$ at which the denominator H vanishes is called the ***singular variety*** and is denoted by $\mathcal{V}$. We now describe the method briefly (more details are provided in Chapter 8).

(i) Use the multidimensional Cauchy integral (1.3.1) to express a_r as an integral over a d-dimensional torus T in $\mathbb{C}^d$.

(ii) Observe that T may be replaced by any cycle homologous to $[T]$ in $H_d(\mathcal{M})$, where $\mathcal{M}$ is the domain of holomorphy of the integrand.

(iii) Deform the cycle to lower the modulus of the integrand as much as possible; use Morse theoretic methods to characterize the minimax cycle in terms of *critical points*.

(iv) Use algebraic methods to find the critical points; these are points of $\mathcal{V}$ that depend on the direction $\hat{r}$ of the asymptotics and are saddle points for the magnitude of the integrand.

(v) Use topological methods to locate one or more *contributing* critical points z_j and replace the integral over T by an integral over *quasi-local* cycles $C(z_j)$ near each z_j.

(vi) Evaluate the integral over each $C(z_j)$ by a combination of residue and saddle point techniques.

When successful, this approach leads to an asymptotic representation of the coefficients a_r of the following sort. The set of directions r is partitioned into finitely many cones K. On the interior of each cone, there is a continuously varying set $\mathtt{contrib}(r) \subseteq \mathcal{V}$ that depends on r only through the projective vector $\hat{r}$ and formulae $\{\Phi_z : z \in \mathtt{contrib}\}$ that involve r and $z(\hat{r})$. Uniformly, as r varies over compact projective subsets of such a cone,

$$\begin{aligned} a_r &\sim \frac{1}{(2\pi i)^d} \int_{[T]} z^{r-1} F(z)\, dz \\ &= \frac{1}{(2\pi i)^d} \sum_{z \in \mathtt{contrib}} z^{-r-1} F(z)\, dz \\ &\sim \sum_{z \in \mathtt{contrib}} \Phi_z(r) . \end{aligned} \tag{1.3.2}$$

The first line of this is steps (i) and (ii). In the second line, the set `contrib` is a subset of the set `critical` of critical points in step (iii). The set `critical` is easy to compute (see step [iv]), whereas determining membership in the subset `contrib` can be challenging (see step [v]). The explicit formulae $\Phi_z(\boldsymbol{r})$ in the last line are computed in step (vi), sometimes relatively easily (Chapter 9) and sometimes with more difficulty (Chapter 10 and especially Chapter 11).

1.4 Outline of the Remaining Chapters

The book is divided into three parts, the third of which is the heart of the subject: deriving asymptotics in the multivariate setting once a generating function is known. Nevertheless, some discussion is required of how generating functions are obtained, what meaning can be read into them, what are the chief motivating examples and applications, and what did we know how to do before the recent spate of research described in Part III. Another reason to include these topics is to make the book into a somewhat self-contained reference. A third is that in obtaining asymptotics, one must sometimes return to the derivation for a new form of the generating function, turning an intractable generating function into a tractable one by changing variables, re-indexing, aggregating, and so forth. Consequently, the first three chapters comprising Part I form a crash course in analytic combinatorics. Chapter 2 explains generating functions and their uses, introducing formal power series, their relation to combinatorial enumeration, and the combinatorial interpretation of rational, algebraic, and transcendental operations on power series. Chapter 3 is a review of univariate asymptotics. Much of this material serves as mathematical background for the multivariate case. Although some excellent sources are available in the univariate case, for example, Wilf (2006), van Lint and Wilson (2001), and Flajolet and Sedgewick (2009), none of these is concerned with providing the brief yet reasonably complete summary of analytic techniques that we provide here. It seems almost certain that someone trying to understand the main subject of these notes will profit from a review of the essentials of univariate asymptotics.

Carrying out the multivariate analyses described in Part III requires a fair amount of mathematical background. Most of this is at the level of graduate coursework, ideally already known by practicing mathematicians but in reality forgotten, never learned, or not learned in sufficient depth. The required background is composed of small to medium-sized chunks taken from many areas: undergraduate complex analysis, calculus on manifolds, saddle point integration (both univariate and multivariate), algebraic topology, computational algebra, and Morse theory. Many of these background topics would be a full semester's course to learn from scratch, which of course is too much material to include here, but we also want to avoid the scenario in which a reference library is required each time a reader picks up this book. Accordingly, we have included substantial background material.

This background material is separated into two pieces. The first piece is the four chapters that comprise Part II. This contains material that we feel should be read or skimmed before the central topics are tackled. The topics in Part II have been sufficiently pared down so that it is possible to learn them from scratch if necessary. Chapters 4 and 5 describe how to asymptotically evaluate saddle point integrals in one and several variables, respectively. Intimate familiarity with these is needed for the analyses in Part III to make much sense. Most of the results in these chapters can be found in a reference such as Bleistein and Handelsman (1986). The treatment here differs from the usual sources in that Fourier and Laplace type integrals are treated as instances of a single complex-phase case. Working in the analytic category, analytic techniques (contour deformation) are used whenever possible, after which comparisons are given to the corresponding C^∞ approach (which uses integration by parts in place of contour deformation).

The last two chapters in Part II concern algebraic geometry. Chapter 6 covers techniques in computational algebra, such as may be found in Cox, Little, and O'Shea (2005) and other books on Gröbner bases. It is possible to skip this, if one wants to understand the theory and does not care about computation; however, few users of analytic combinatorics live in a world where computation does not matter. Chapter 7 concerns domains of convergence of power series and Laurent series. The notion of an amoeba from Gel'fand, Kapranov, and Zelevinsky (1994) is introduced, a relation given to convergence of Laurent series. This chapter is perhaps the least necessary of the four chapters in Part II, providing motivation and conceptual explanation more than it provides tools for computing or proving theorems. Nevertheless, because it contains exactly those pieces of several complex variable theory needed to make sense out of multivariate Cauchy integrals, it is an efficient vehicle for ensuring that the reader understands the analytic aspects of multivariate power series.

The remaining background material is relegated to the appendices, of which there are three. Each of these contains a reduction of a semester's worth of material. It is not expected that the reader will go through these in advance; rather they serve as references so that frequent library visits will not be needed. Appendix A presents for beginners all relevant knowledge about calculus on manifolds and algebraic topology. Manifolds and tangent and cotangent vectors are defined, differential forms in $\mathbb{R}^n$ are constructed from scratch, and integration of forms is defined. Next, complex differential forms are defined. Finally, the essentials of algebraic topology are reviewed: chain complexes, homology and cohomology, relative homology, Stokes' Theorem, and some important exact sequences. Appendix B summarizes classical Morse theory – roughly the first few chapters of Milnor's classic text (1963). Appendix C then introduces the notion of stratified spaces and describes the stratified version of Morse theory as developed by Goresky and MacPherson (1988). Part II and the appendices have a second function. Some of the results used in Part III are often quoted in the literature from sources that do not provide a proof. On more than one occasion, when organizing the material in this book, we found that a purported reference to a proof led ultimately to nothing. Beyond serving as a mini-reference library, therefore, the background sections provide some key proofs and corrected citations to eliminate ghost references and the misquoting of existing results.

The heart of this book, Part III, is devoted to new results in the asymptotic analysis of multivariate generating functions. The first chapter, Chapter 8, is a conceptual guide and overview. The techniques presented in the subsequent chapters can be carried out blindly – in fact, it would be fair to say that they were in the original research articles by Pemantle and Wilson (2002); and Pemantle and Wilson (2004). The purpose of Chapter 8 is to unify the disparate results of the subsequent chapters and exhibit hidden mathematical structure behind the analyses. The chapter begins with a section on the exponential rate function and its relation to convex duals. Next, the key concepts from Morse theory are lifted from the appendices to give meaning and direction to the search for deformations of chains of integration. Some general observations follow regarding critical points and their computation, after which these are related back via so-called minimal points to the exponential rate function. The chapter closes with a discussion of certain optimized chains of integrations (quasi-local relative cycles) and the residue and saddle point integrals to which they lead. The chapter finally arrives at (8.6.3), which is the template (1.3.2) without any specification of the set `contrib` or the formulae Φ_z.

Having reduced the computation of a_r to saddle point integrals with computable parameters, plugging in results on saddle point integration yields theorems for the end user. These break into several types. Chapter 9 discusses the case where the critical point z is a smooth point of $\mathcal{V}$. This case is simpler than the general case in several respects. The residues are simple, so multivariate residue theory is not needed. Also, in the smooth setting, only regular Morse theory is needed, not stratified Morse theory. Chapter 10 discusses the case where $\mathcal{V}$ is the intersection of smooth hypersurfaces near z and simple multivariate residues are needed. Both of these cases are reasonably well understood. A final case that arises for rational functions is a singularity with nontrivial monodromy. In this case our knowledge is limited, but some known results are discussed in Chapter 11. This chapter is not quite as self-contained as the preceding ones; in particular, some results from Baryshnikov and Pemantle (2011) are quoted without proof. This is because the technical background for these analyses exceeds even the relatively large space we have allotted for background. The article by Baryshnikov and Pemantle (2011), which is self-contained, already reduces by a significant factor the body of work presented in the celebrated work of Atiyah, Bott, and Gårding (1970), and further reduction is only possible by quoting key results. In Chapter 12, a number of examples are worked whose analyses follow the theory in Chapters 9–11. Finally, Chapter 13 is devoted to further topics. These include higher order asymptotics, algebraic generating functions, diagonals, and a number of open problems.

Notes

The viewpoint in Section 1.1 is borrowed from the introduction to Stanley (1997). The two very different motivating problems alluded to in Section 1.3 were the hitting time generating function from Larsen and Lyons (1999) and the Aztec Diamond placement probability generating function from Jockusch, Propp, and Shor (1998). The first versions of the six-step program at the end of Section 1.3 that were used to obtain multivariate asymptotics involved expanding a torus of integration until it was near a minimal critical point and then doing some surgery to isolate the main contribution to the integral as the integral of a univariate residue over a complementary $(d-1)$-dimensional chain. This was carried out in Pemantle and Wilson (2002); and Pemantle and Wilson (2004). This method was brought to the attention of the authors by several analysts at Wisconsin, among them S. Wainger, J.-P. Rosay, and A. Seeger. Although their names do not appear in any bibliographic citations associated with this project, they are acknowledged in these early publications and should be credited with useful contributions to this enterprise. The second author maintains an ongoing database of work in this area, residing at the project website linked from the book website.

Exercises

1.1 (asymptotic expansions need not converge)
Find an asymptotic expansion $f \sim \sum_{j=0}^{\infty} g_j$ for a function f as $x \downarrow 0$ such that $\sum_{j=0}^{\infty} g_j(x)$ is not convergent for any $x > 0$. Conversely, suppose that $f(x) = \sum_{j=0}^{\infty} g_j(x)$ for $x > 0$ and $g_{j+1} = o(g_j)$ as $x \downarrow 0$; does it follow that $\sum_{j=0}^{\infty} g_j$ is an asymptotic expansion of f?

1.2 (bivariate asymptotics)
Prove or give a counterexample: if g is a continuous function and for each λ we have $a_{rs} = g(\lambda) + O(r+s)^{-1}$ as $r, s \to \infty$ with $r/s \to \lambda$, then

$$a_{rs} \sim g(r/s) \text{ as } r, s \to \infty$$

as λ varies over a compact interval in $\mathbb{R}^+$.

1.3 (Laplace transform asymptotics)
Let A be a smooth real function in a neighborhood of zero and define its Laplace transform by

$$\hat{A}(\tau) := \int_0^{\infty} e^{-\tau x} A(x)\, dx .$$

Let us note that

$$\int_0^{\infty} x^n e^{-\tau x}\, dx = n!\tau^{-n-1}. \tag{1.4.1}$$

Writing $A(x) = \sum_n c_n x^n$ where $c_n = A^{(n)}(0)/n!$ as in Example 1.2.1 and integrating term by term using (1.4.1) suggests the series

$$\sum_n A^{(n)}(0)\tau^{-n-1} \tag{1.4.2}$$

as a possible asymptotic expansion for $\hat{A}$. Although the term-by-term integration is completely unjustified, show that the series (1.4.2) is a valid asymptotic expansion of $\hat{A}$ in decreasing powers of τ as $\tau \to \infty$.

2
Generating Functions

This chapter gives a crash course on generating functions and enumeration. For a more lengthy introduction, we recommend Wilf (2006). Chapter 14 of van Lint and Wilson (2001) also provides a fairly concise but readable treatment. Proofs for facts about formal power series may be found in Stanley (1997, Section 1.1). A comprehensive treatment of the relation between power series operations and corresponding combinatorial constructions on finite sets is the encyclopedic reference by Goulden and Jackson (2004). Chapters 1–3 of Flajolet and Sedgewick (2009) contain a very nice treatment as well. Bender and Williamson (1991) and Kauers and Paule (2011).

Throughout the book, but particularly in this chapter, the notation $[n]$ denotes the set $\{1, \ldots, n\}$.

2.1 Formal Power Series

From an algebraic viewpoint, the ring of formal power series is obtained by imposing a particular ring structure on the set of complex arrays of numbers. However, we usually think of them in functional notation as follows. Let $z_1, \ldots, z_d$ be indeterminates, and consider the set of formal expressions of the form $\sum_{\boldsymbol{r}} f_{\boldsymbol{r}} \boldsymbol{z}^{\boldsymbol{r}}$, which we denote by $\mathbb{C}[[z_1, \ldots, z_d]]$. Addition is defined by $(f+g)_{\boldsymbol{r}} = f_{\boldsymbol{r}} + g_{\boldsymbol{r}}$, and multiplication is defined by convolution: $(f \cdot g)_{\boldsymbol{r}} = \sum_{\boldsymbol{s}} f_{\boldsymbol{s}} g_{\boldsymbol{r}-\boldsymbol{s}}$. The sum in this convolution is always finite, so there is no question of convergence. Each array $\{f_{\boldsymbol{r}} : \boldsymbol{r} \in \mathbb{N}^d\}$ corresponds to an element of $\mathbb{C}[[z_1, \ldots, z_d]]$, called its ***generating function***. The simplest-looking element z_i corresponds to the array δ_i having a 1 in position 1 in dimension i and 0's elsewhere. A common notation for $f_{\boldsymbol{r}}$ is $[\boldsymbol{z}^{\boldsymbol{r}}] f$, read as "the $\boldsymbol{z}^{\boldsymbol{r}}$-coefficient of f."

The additive identity in $\mathbb{C}[[z_1, \ldots, z_d]]$ is the zero series, and the multiplicative identity is the series with a 1 in position $(0, \ldots, 0)$ and 0 elsewhere. It is an easy exercise to see that f has a multiplicative inverse if and only if $f_{\boldsymbol{0}} \neq 0$. Thus $\mathbb{C}[[z_1, \ldots, z_d]]$ is a *local ring*, meaning there is a unique maximal ideal, $\mathfrak{m}$, the set of non-units. Local rings come equipped with a notion of convergence, namely $f_n \to f$ if and only if $f_n - f$ is eventually in $\mathfrak{m}^k$ for every k. An easier way to say this is that for all $\boldsymbol{r}$ there is an $N(\boldsymbol{r})$ such that $(f_n)_{\boldsymbol{r}} = f_{\boldsymbol{r}}$ for $n \geq N(\boldsymbol{r})$. An open ***polydisk*** centered at $\boldsymbol{z} \in \mathbb{C}^d$ and with polyradius $\boldsymbol{b} \in (\mathbb{R}^+)^d$ is the set

$$\{\boldsymbol{y} \in \mathbb{C}^d : |y_j - z_j| < b_j \quad \text{for } 1 \leq j \leq d\}.$$

Let $\mathcal{N}$ be an open polydisk centered at the origin in $\mathbb{C}^d$, i.e., a set $\{\boldsymbol{z} : |z_i| < b_i, 1 \leq i \leq d\}$. Suppose that $f, g \in \mathbb{C}[[z_1, \ldots, z_d]]$ are absolutely convergent on $\mathcal{N}$, i.e., $\sum_{\boldsymbol{r}} |f_{\boldsymbol{r}}||w_1|^{r_1} \cdots |w_d|^{r_d} < \infty$ when all $|w_i| < b_i$, and similarly for g. Then $f + g$ and $f \cdot g$ are absolutely convergent on $\mathcal{N}$ as well, and the sum and product in the ring of formal power series are the same as in the ring of analytic functions in $\mathcal{N}$. Because a finite intersection of neighborhoods of the origin is a neighborhood of the origin, the subset of $\mathbb{C}[[z_1, \ldots, z_d]]$ of series that converge in some neighborhood of the origin is a subring. This is called the ***ring of germs of analytic functions***. It is not all of $\mathbb{C}[[z_1, \ldots, z_d]]$, and it is denoted by $\mathbb{C}\{z_1, \ldots, z_d\}$.

That is, there are some formal power series that fail to converge anywhere (except at the origin), and for these it will not work to apply analytic methods. One can, however, consider a different mapping from arrays to formal power series, for example, $\sum_{\boldsymbol{r}} f_{\boldsymbol{r}} \boldsymbol{z}^{\boldsymbol{r}} / g(\boldsymbol{r})$ for a judiciously chosen g. A good choice is often to let $g(\boldsymbol{r})$ be a product of some or all of the quantities $r_i!$; a generating function normalized by factorials is called an ***exponential generating function***. Not only may this normalization cause the power series to converge, but the behavior of exponential generating functions under convolution has an important combinatorial interpretation. Several examples of this are given in Section 2.5.

One can go backwards as well. If $f = g/h$, where g and h are convergent on a neighborhood of the origin and $h \notin \mathfrak{m}$, then f is analytic on a neighborhood of the origin; in fact, it is analytic where g and h are and where h is nonzero, and its Taylor series is equal to g/h in $\mathbb{C}[[z_1, \ldots, z_d]]$. Similarly, one may define formal differentiation by

$$\frac{\partial}{\partial z_j} f = \sum_{\boldsymbol{r}} r_j f_{\boldsymbol{r}} \boldsymbol{z}^{\boldsymbol{r} - \delta_j},$$

and this will agree with analytic differentiation on the domain of convergence of F.

The interior $\mathcal{D}$ of the domain on which the formal power series F converges is the union of open polydisks. In particular, it is the union of tori and is hence characterized by its intersection $\mathcal{D}_{\mathbb{R}}$ with $\mathbb{R}^d$. The set $\mathcal{D}$ is in fact pseudoconvex, meaning that the set $\operatorname{Re}\log \mathcal{D}$ defined by $(x_1, \ldots, x_d) \in \operatorname{Re}\log \mathcal{D}$ if and only if $(e^{x_1}, \ldots, e^{x_d}) \in \mathcal{D}$ is a convex order ideal (i.e., a set closed under $\leq$ in the coordinate-wise partial order on $\mathbb{R}^d$). See Hörmander (1990, Section 2.5) for these and other basic facts about functions of several complex variables. Just as we use r, s, and t for r_1, r_2, and r_3, to make examples more readable, we use x, y, and z for z_1, z_2, and z_3.

2.2 Rational Operations on Generating Functions

A d-variate ***combinatorial class*** is a set $\mathcal{A}$, which is the disjoint union of finite sets $\{\mathcal{A}_{\boldsymbol{r}} : \boldsymbol{r} \in \mathbb{N}^d\}$ in some natural way. In this section, $F = \sum_{\boldsymbol{r}} a_{\boldsymbol{r}} \boldsymbol{z}^{\boldsymbol{r}}$ will "generate" a combinatorial class $\mathcal{A}$, i.e., $|\mathcal{A}_{\boldsymbol{r}}| = a_{\boldsymbol{r}}$ for all $\boldsymbol{r}$. We also say that F "counts $\mathcal{A}$ by ϕ," where ϕ is the map taking $x \in \mathcal{A}$ to the $\boldsymbol{r}$ for which $x \in \mathcal{A}_{\boldsymbol{r}}$. Arithmetical operations in the ring of formal power series were defined to correspond to existing operations on analytic power series. It is instructive to find interpretations for these operations on the combinatorial level. Here follows a list of set-theoretic interpretations for rational operations. The combinatorial wealth of these interpretations explains why there are so many rational generating functions in combinatorics.

Equality: Bijection

It goes almost without saying that equality between two generating functions F and G corresponds to bijective correspondence between the classes they generate: $|\mathcal{A}_{\boldsymbol{r}}| = |\mathcal{B}_{\boldsymbol{r}}|$ for all $\boldsymbol{r}$.

Multiplication by z_j: Re-indexing

In the univariate case, the function $zF(z)$ generates the sequence $0, a_0, a_1, a_2, \ldots$. Similarly, in the multivariate case, $z_j F(\boldsymbol{z})$ generates $\{b_{\boldsymbol{r}}\}$ where $b_{\boldsymbol{r}} = a_{\boldsymbol{r}-\boldsymbol{\delta}_j}$, which is defined to be zero if any coordinate is negative.

Re-indexing the other direction is more complicated. In the univariate case, the sequence $a_1, a_2, \ldots$ is generated by the function $(f - f(0))/z$. In the multivariate case, the sequence $\{a_{\boldsymbol{r}+\boldsymbol{\delta}_j}\}$ is generated by $(F - F(z_1, \ldots, z_{j-1}, 0, z_{j+1}, \ldots, z_d))/z_j$.

Sums: Disjoint Unions

If F generates a class $\mathcal{A}$ and G generates a class $\mathcal{B}$, then $F + G$ generates the class C, where $C_{\boldsymbol{r}}$ is the disjoint union of $\mathcal{A}_{\boldsymbol{r}}$ and $\mathcal{B}_{\boldsymbol{r}}$. The interpretations of equality, multiplication by z_j, and sums on the combinatorial level are pretty simple, but already one may find examples that are not completely trivial.

Example 2.2.1 (binary sequences with no repeated 1's) Let $\mathcal{A}_n$ be the set of sequences of 0's and 1's of length n that do not begin with 1 and have no two consecutive 1's. Each such sequence ends either in 0 or in 01. The sequence that remains can be any sequence in $\mathcal{A}_{n-1}$ or $\mathcal{A}_{n-2}$, respectively. Thus stripping off the last one or two symbols respectively yields a bijective correspondence between $\mathcal{A}_n$ and the disjoint union $\mathcal{A}_{n-1} \cup \mathcal{A}_{n-2}$. At the generating function level, we see that $F(z) = zF(z) + z^2F(z)$ – well, almost! If we take $\mathcal{A}_n$ to be empty for $n < 0$, the correspondence still works for $n = 1$, but it fails for $n = 0$. Thus, actually,

$$zF(z) + z^2F(z) = F(z) - 1.$$

Via operations in the formal power series ring, we may rearrange to isolate the 1 and then divide by $1 - z - z^2$ to obtain

$$F(z) = \frac{1}{1 - z - z^2}.$$

■

Example 2.2.2 (binomial coefficients) Let $\mathcal{A}_{r,s}$ be the set of colorings of the set $[r+s] := \{1, \ldots, r+s\}$ for which r elements are red and s elements are green. Decomposing according to the color of the last element, $\mathcal{A}_{r+s}$ is in bijective correspondence with the disjoint union of $\mathcal{A}_{r-1,s}$ and $\mathcal{A}_{r,s-1}$. This is a combinatorial interpretation of the identity $\binom{r+s}{r} = \binom{r+s-1}{r} + \binom{r+s-1}{r-1}$ and holds as long as $r + s > 0$. It follows that

$$F(x, y) - 1 = xF(x, y) + yF(x, y),$$

and solving for F gives

$$F(x, y) = \frac{1}{1 - x - y}.$$

■

Products: Convolutions

If F generates the class $\mathcal{A}$ and G generates the class $\mathcal{B}$, then FG generates the class C defined by letting $C_{\boldsymbol{r}}$ be the disjoint union of cartesian products $\mathcal{A}_{\boldsymbol{s}} \times \mathcal{B}_{\boldsymbol{r}-\boldsymbol{s}}$ over all $\boldsymbol{s} \le \boldsymbol{r}$. This is the canonical definition of a product in any category of graded objects.

Students of probability theory will recognize it as a convolution. Suppose that F and G have nonnegative coefficients. Suppose furthermore that $F(\mathbf{1}) = G(\mathbf{1}) = 1$, where $\mathbf{1}$ is the d-vector of 1's (i.e., the coefficients of each sum to 1). Then F is the ***probability generating function*** for a probability distribution on $\mathbb{N}^d$ that gives mass $a_{\boldsymbol{r}}$ to the point $\boldsymbol{r}$, and G is likewise a probability generating function. The product FG of the power series generates the convolution of the distributions: the distribution of the sum of independent picks from the two given distributions. Thus the study of sums of independent, identically distributed random variables taking values in $\mathbb{N}^d$ is equivalent to the study of powers of such a generating function F. The laws of large numbers in probability theory may be derived via generating function analyses, whereas the central limit theorem is always proved essentially this way. In Chapter 9, versions of these laws are proved for coefficients of generating functions far more general than powers of probability generating functions.

A useful trick with products is as follows.

Example 2.2.3 (enumerating partial sums) Let $F(z)$ enumerate the class $\mathcal{A}$, and let $G(z) = 1/(1 - z)$ enumerate a class $\mathcal{B}$ with $|\mathcal{B}_n| = 1$ for all n. Then FG

enumerates the class C, where C_n is the disjoint union $\biguplus_{j=0}^{n} \mathcal{A}_j$. Consequently, the generating function for the partial sums $\sum_{j=0}^{n} a_j$ is $F(z)/(1-z)$. ■

The Operation $1/(1-F)$*: Finite Sequences*

Let $\mathcal{B}$ be a combinatorial class, and let $\mathcal{A}$ be the class of finite sequences of elements of $\mathcal{B}$, graded by total weight, meaning that the sequence $(x_1, \ldots, x_k)$ belongs to $\mathcal{A}_{\mathbf{r}}$ if $x_j \in \mathcal{B}_{\mathbf{s}^{(j)}}$ for $1 \le j \le k$ and $\sum_{j=1}^{k} \mathbf{s}^{(j)} = \mathbf{r}$. Then $\mathcal{A}$ is the disjoint union of the empty sequence, the class of singleton sequences, the class of sequences of length 2, and so forth, and summing the generating functions gives $F = 1 + G + G^2 + \cdots$. Provided that $\mathcal{B}$ has no elements of weight zero ($\mathcal{B}_0 = \emptyset$), this converges in the ring of formal power series and is equal to $1/(1-G)$. If G grows no faster than exponentially, then both sides of this equation converge analytically in a neighborhood of the origin and are equal to $1/(1-G)$.

A simple example of this is to count the binary strings of Example 2.2.1 by the number of 0's and the number of 1's, rather than by total length. Any such sequence may be uniquely decomposed into a finite sequence of the blocks 0 and 01. Letting these have weights $(1, 0)$ and $(1, 1)$, respectively, the class $\mathcal{B}$ of blocks has generating function $x + xy$. The generating function for $\mathcal{A}$ is therefore $F(x, y) = 1/(1 - x - xy)$. We may collapse this to a univariate function by using weights 1 and 2 instead of $(1, 0)$ and $(1, 1)$, recovering the generating function $1/(1 - z - z^2)$.

Example 2.2.4 (prefix codes) Let T be a finite rooted binary tree (every vertex has either 0 or 2 children) whose vertices are identified with finite sequences of 0's and 1's. Any sequence of 0's and 1's may be decomposed into blocks by repeatedly stripping off the initial segment i.e., a leaf of T. The decomposition is unique; it may end with a partial block, i.e., an internal node of T.

Here is a derivation of the generating function F counting all binary sequences by their length and the number of blocks. Let $B(x)$ be the univariate generating function counting leaves of T by depth. The generating function for blocks by length and number of blocks is $yB(x)$ because each block has number of blocks equal to 1. The generating function for binary sequences with no incomplete blocks is therefore $1/(1 - yB(x))$. Allowing incomplete blocks, each sequence uniquely decomposes into a maximal sequence of complete blocks, followed by a (possibly empty) incomplete block. Letting $C(x)$ count incomplete blocks by length, we see that $1 + y(C(x) - 1)$ counts incomplete blocks by length and number of blocks, and therefore

$$F(x, y) = \frac{1 + y(C(x) - 1)}{1 - yB(x)} .$$

■

Lattice paths yield a large and well-studied class of examples.

Example 2.2.5 (lattice paths) Let E be a finite subset of $\mathbb{N}^d$ not containing $\mathbf{0}$ and let $\mathcal{A}$ be the class of finite sequences $(\mathbf{0} = x_0, x_1, \ldots, x_k)$ of elements of $\mathbb{N}^d$ with $x_j - x_{j-1} \in E$ for $1 \leq j \leq k$. We call these *paths with steps in* E.

Let $B(z) = \sum_{r \in E} z^r$ generate E by step size. Then $1/(1 - B(z))$ counts paths with steps in E by ending location. This includes examples we have already seen. Multinomial coefficients count paths ending at $\boldsymbol{r}$ with steps in the standard basis directions $e_1, \ldots, e_d$; the generating function $1/(1 - \sum_{j=1}^{d} z_j)$ follows from the generating function $\sum_{j=1}^{d} z_j$ for E. ■

Example 2.2.6 (Delannoy numbers) Let $\mathcal{A}_r$ be the lattice paths from the origin to $\boldsymbol{r}$ in $\mathbb{Z}^2$ using only steps that go north, east, or northeast to the next lattice point. The numbers $a_r := |\mathcal{A}_r|$ are called ***Delannoy numbers*** (Comtet, 1974, Exercise I.21). The generating function $x + y + xy$ for E leads to the Delannoy generating function

$$F(x, y) = \frac{1}{1 - x - y - xy}.$$

■

One final example comes from Corteel, Louchard, and Pemantle (2004).

Example 2.2.7 (no gaps of size 2) Let $\mathcal{B}_n$ be the class of subsets of $[n]$ where no two consecutive members are absent. It is easy to count $\mathcal{B}_n$ by mapping bijectively to Example 2.2.1. However, in Corteel, Louchard, and Pemantle (2004), an estimate was required on the number of such sets that were mapped into other such sets by a random permutation. To compute this (actually, to compute the second moment of this random variable), it sufficed to count the pairs $(S, T) \in \mathcal{B}_n^2$ by n, $|S|$, $|T|$ and $|S \cap T|$.

A four-variable generating function $F(x, y, z, w)$ may be derived by investigating what may happen between consecutive elements of $S \cap T$. Identify $(S, T) \in \mathcal{B}_n^2$ with a sequence α in the set $\{(0, 0), (0, 1), (1, 0), (1, 1)\}^n$, where a $(1, 1)$ in position j denotes an element of $S \cap T$, a $(1, 0)$ denotes an element of $S \setminus T$, and so forth. If j and $j + r$ are positions of consecutive occurrences of $(1, 1)$, then the possibilities for the string $\alpha_{j+1} \cdots \alpha_{j+r}$ are as follows:

(i) $\alpha_{j+1} = (1, 1)$: the only possibility is $r = 1$.
(ii) $\alpha_{j+1} = (0, 0)$: the only possibility is $r = 2$ and $\alpha = ((0, 0), (1, 1))$.
(iii) $\alpha_{j+1} = (1, 0)$: then $r \geq 2$ may be arbitrary and α alternates between $(1, 0)$ and $(0, 1)$ until the final $(1, 1)$.
(iv) $\alpha_{j+1} = (0, 1)$: then $r \geq 2$ may be arbitrary and α alternates between $(0, 1)$ and $(1, 0)$ until the final $(1, 1)$.

In the first case, the generating function $G_1(x, y, z, w)$ for blocks by the four weights is just $xyzw$. In the second case, $G_2(x, y, z, w) = x^2yzw$. In the third case, one may write the block as either $((1, 0))$ or $((1, 0), (0, 1))$, followed by zero or more alternations of length two; decomposing this way shows the generating function to be $G_3(x, y, z, w) = xyzw\frac{xy + x^2yz}{1 - x^2yz}$. Similarly, we see that $G_4(x, y, z, w) = xyzw\frac{xz + x^2yz}{1 - x^2yz}$. Summing these gives a block generating function of

$$G(x, y, z, w) = xyzw\frac{(1 + x)(1 - x^2yz) + xy + xz + 2x^2yz}{1 - x^2yz}.$$

Finally, we use the $1/(1 - G)$ formula. Stringing together blocks of the four types gives all legal sequences of any length that end in $(1, 1)$; thus this class has generating function $1/(1 - G)$. These correspond to pairs $(S, T) \in \mathcal{B}_n^2$ with $n \in S \cap T$ and are in bijective correspondence (via deletion of the element n) to all pairs in $\mathcal{B}_{n-1}^2$, except that when $n = 0$, it is not possible to delete n. The bijection reduces the weight of each (S, T) by $(1, 1, 1, 1)$. Thus

$$F(x, y, z, w) = \left(\frac{1}{1 - G(x, y, z, w)} - 1\right) / (xyzw)$$

$$= \frac{(1 + x)(1 - x^2yz) + xy + xz + 2x^2yz}{1 - x^2yz - xyzw[(1 + x)(1 - x^2yz) + xy + xz + 2x^2yz]}.$$

■

Transfer Matrices: Restricted Transitions

Suppose we want to count words (i.e., finite sequences) in an alphabet V, but only some consecutive pairs are allowed. Let E be the set of allowed pairs. Allowed words of length n are equivalent to paths of length n in the directed graph (V, E). To count these by length, let $\boldsymbol{M}$ be the incidence matrix of (V, E), i.e., the square matrix indexed by V, with $M_{vw} = 1$ if $(v, w) \in E$ and $M_{vw} = 0$ otherwise. The number of allowed paths of length n from v to w is $(\boldsymbol{M}^n)_{vw}$. If we wish to count paths by length, we must sum $(z\boldsymbol{M})^n$ over n. Thus the generating function counting finite paths from v to w by their length is

$$F(z) = \sum_{n=0}^{\infty}((z\boldsymbol{M})^n)_{vw} = \left[(\boldsymbol{I} - z\boldsymbol{M})^{-1}\right]_{vw}.$$

This formula is quite versatile. To count all allowed paths by length, we may sum in v and w; a convenient way to notate this is trace$((\boldsymbol{I} - z\boldsymbol{M})^{-1}\boldsymbol{J})$, where $\boldsymbol{J}$ is the $|V| \times |V|$ square matrix of 1's. Alternatively, we may count by features other than length. The most general way to count is by the number of each type of transition: enumerate $E = \{e_1, \ldots, e_k\}$ and let $\tilde{M}_{vw} = z_k$ if $e_k = (v, w)$ and $\tilde{M}_{vw} = 0$ if $(v, w) \notin E$; then $[(\boldsymbol{I} - \tilde{\boldsymbol{M}})^{-1}]_{vw}$ counts paths from v to w by transitions and trace$((\boldsymbol{I} - \tilde{\boldsymbol{M}})^{-1}\boldsymbol{J})$ counts all paths by transitions.

Example 2.2.8 (binary strings revisited) The transfer matrix method may be used to count the paths of Example 2.2.1. Let $V = \{0, 1\}$ and $E = \{(0, 0), (0, 1), (1, 0)\}$

contain all directed edges except (1, 1). Then

$$\boldsymbol{M} = \begin{bmatrix} 1 & 1 \\ 1 & 0 \end{bmatrix}$$

so

$$\mathbf{Q} := (\boldsymbol{I} - z\boldsymbol{M})^{-1} = \frac{1}{1 - z - z^2}\begin{bmatrix} 1 & z \\ z & 1 - z \end{bmatrix}.$$

The paths from 0 to 0 having n transitions, $n \geq 0$, are in one-to-one correspondence, via stripping off the last 0, to the words in Example 2.2.1 of length n. Thus the generating function is the (0, 0)-entry of $\mathbf{Q}$, namely $1/(1 - z - z^2)$. ■

Composition: Block Substitution

Let F be a d-variate generating function and $G_1, \ldots, G_d$ be d generating functions in any number of variables, all with vanishing constant terms. One may define the formal composition $F \circ (G_1, \ldots, G_d)$ as a limit in the formal power series ring:

$$F \circ (G_1, \ldots, G_d) := \lim_{n \to \infty} \sum_{|\boldsymbol{r}| \leq n} a_{\boldsymbol{r}} \mathbf{G}^{\boldsymbol{r}} . \tag{2.2.1}$$

The degree of any monomial in $\mathbf{G}^{\boldsymbol{r}} := G_1^{r_1} \cdots G_d^{r_d}$ is at least $|\boldsymbol{r}| := \sum_{j=1}^{d} r_j$ by the assumption that $G_j(\mathbf{0}) = 0$ for all j; hence the $\boldsymbol{z}^{\boldsymbol{r}}$-coefficient of the sum does not change once $n > |\boldsymbol{r}|$, and the limit exists in the formal power series ring. Even if some $G_j(0) \neq 0$, it may still happen that the sum converges in the ring of analytic functions, meaning that the infinitely many contributions to all coefficients are absolutely summable.

A slightly unwieldy abstract combinatorial interpretation of this is given in Sections 2.2.20–2.2.22 of Goulden and Jackson (2004). Let $\mathcal{A}, \mathcal{B}_1, \ldots, \mathcal{B}_d$ be the classes generated respectively by $F, G_1, \ldots, G_d$. The class corresponding to $F \circ (G_1, \ldots, G_d)$ is obtained as a disjoint union over elements $x \in \mathcal{A}$ of d-tuples $(C_1, \ldots, C_d)$, where C_i is a sequence of length r_i of elements of $\mathcal{B}_i$. The weight of such a d-tuple is the sum of the weights of the C_i, which are in turn the sum of the weights of all r_i elements of $\mathcal{B}_i$. The following examples should clarify this.

Example 2.2.9 (queries) Queries from a database have integer computation times associated with them. There are b_k queries of size k, $k \geq 1$. The protocol does not allow two long queries in a row, where long is defined as of size greater than some number M. How many query sequences are there of total time n?

The sequences of queries are bijectively equivalent to the composition $\mathcal{A} \circ (\mathcal{B}_1, \mathcal{B}_2)$, where $\mathcal{A}$ is the class from Example 2.2.8, counted by numbers of 0's and 1's, and $\mathcal{B}_1$ and $\mathcal{B}_2$ are respectively the short queries and the long queries, counted by computation time. Thus the function $F(G_1, G_2)$ counts queries by time, where $F(x, y) = 1/(1 - x - xy)$, $G_1(z) = \sum_{k=1}^{M} b_k z^k$ and $G_2(z) = \sum_{k>M} b_k z^k$. ■

The following example may seem a natural candidate for the transfer matrix method, but it is simpler to analyze it as from the viewpoint of compositions.

Example 2.2.10 (Smirnov words) Let $\mathcal{A}$ be the class of ***Smirnov words***, i.e., words in the alphabet $[d]$ with no consecutive repetition of any symbol allowed. Of course $|\mathcal{A}_n| = d \cdot (d-1)^{n-1}$, but suppose we wish to count differently.

Let F count Smirnov words by number of occurrences of each symbol, and let G count the class $\mathcal{B}$ of all words on the alphabet $[d]$, also by number of occurrences of each symbol. Starting with $x \in \mathcal{A}$ and substituting an arbitrary nonzero string of the symbol j for each occurrence of j in x produces each element of $\mathcal{B}$ in a unique way. The generating function for a nonempty string of j's is $\dfrac{z_j}{1-z_j}$, whence

$$G(\mathbf{z}) = F\left(\frac{z_1}{1-z_1}, \ldots, \frac{z_d}{1-z_d}\right).$$

Solve for F by setting $y_j = z_j/(1+z_j)$ to obtain

$$F(\mathbf{y}) = G\left(\frac{y_1}{1+y_1}, \ldots, \frac{y_d}{1+y_d}\right)$$

and use $G(\mathbf{z}) = 1/(1 - \sum_{j=1}^{d} z_j)$ to get

$$F(\mathbf{z}) = \frac{1}{1 - \sum_{j=1}^{d} \frac{z_j}{1+z_j}}.$$

■

One subject in probability theory, namely the study of branching processes, is almost always dealt with by means of generating functions.

Example 2.2.11 (Galton-Watson process) Let $f(z)$ be a probability generating function, i.e., $f(z) = \sum_{n=0}^{\infty} p_n z^n$ with $p_n \geq 0$ and $\sum_{n=0}^{\infty} p_n = 1$. A ***branching process*** with offspring distribution f is a random family tree with one progenitor in generation 0 and each individual in each generation having a random number of children; these numbers of children born to the individuals in a generation are independent, and each is equal to n with probability p_n. The random number of individuals in generation n is denoted Z_n. What is the probability, $p_{n,k}$, that $Z_n = k$?

We compute the probability generating function for Z_n inductively as follows. The probability generating function for Z_1 is just f. Suppose we know the probability generating function $g_n := \sum_k p_{n,k} z^k$ for Z_n. Interpret this as saying that there is a total mass of $p_{n,k}$ configurations with $Z_n = k$. In a configuration with $Z_n = k$, the next generation is composed of a sequence of k families, each independently having size j with probability p_j. The probability generating function for such a sequence is f^k, whence $g_{n+1} = g_n \circ f$. Inductively then, $g_n = f^{(n)} := f \circ \cdots \circ f$, a total of n times. Observe that, unless $p_0 = 0$ (no extinction), this composition is not defined in the formal power series ring, but because all functions involved are convergent on the unit disk, the compositions are well defined analytically. ■

Example 2.2.12 (Branching random walk) Associate to each particle in a branching process a real number, which we interpret as the displacement in one dimension between its position and that of its parent. If these are independent of each other and of the branching, and are identically distributed, then one has the classical branching random walk. A question that has been asked several times in the literature, for example, in Kesten (1978), is: beginning with a single particle, say at position 1, does there exist a line of descent that remains to the right of the origin for all time?

To analyze this, modify the process so that X denotes the number of particles ever to hit the origin. Let us examine this in the simplest case, where the branching process is deterministic binary splitting ($p_2 = 1$) and the displacement distribution is a random walk that moves one unit to the right with probability $p < 1/2$ and one unit to the left with probability $1 - p$. If we modify the process so that particles stop moving or reproducing when they hit the origin, then an infinite line of descent to the right of the origin is equivalent to infinitely many particles reaching the origin.

To analyze the process, therefore, we let X be the number of particles ever to hit the origin (still begin with a single particle at 1). Let ϕ be the probability generating function for X:

$$\phi(z) = \sum_{n=2}^{\infty} a_n z^n \quad \text{where } a_n := \mathbb{P}(X = n).$$

If the initial condition is changed to a single particle at position 2, then the number of particles ever to reach the origin will have probability generating function $\phi \circ \phi$. To see this, apply the analysis of the previous example, noting that the number of particles ever to reach 1 before any ancestor has reached 1, together with their collections of descendants who ever reach 0, form two generations of a branching process with offspring distribution the same as X.

Each of the two children in the first generation is located at 0 with probability $1 - p$ and at 2 with probability p, so the probability generating function for the contribution to X of each child is $(1 - p)z + p\phi(\phi(z))$. The two contributions are independent and so their sum is a convolution, whose probability generating function is therefore the square of this. Thus we have the identity

$$\phi(z) = [(1 - p)z + p\phi(\phi(z))]^2 . \tag{2.2.2}$$

Although this does not produce an explicit formula for ϕ, it is possible from this to derive asymptotics for $\phi(t)$ as $t \uparrow 1$, allowing us to use so-called *Tauberian theorems* to recover asymptotic information about a_n. ■

2.3 Algebraic Generating Functions

After rational functions, most people consider algebraic functions to be the next simplest class. These arise frequently in combinatorics as well. One reason, having to do with recursions satisfied by algebraic functions, is taken up in the next section. Another reason is that when a combinatorial class solves a convolution equation, its generating function solves an algebraic equation. A famous univariate example of this is as follows.

Example 2.3.1 (binary trees and Catalan numbers) Let $\mathcal{A}$ be the class of finite, rooted, sub-binary trees. This class is defined recursively as follows: the empty tree (no vertices) is in $\mathcal{A}$; every element of $\mathcal{A}$ with $n \geq 1$ vertices has a root that has a left and a right subtree; the possible ordered pairs (L, R) of left and right subtrees are just all ordered pairs of previously defined trees, the cardinalities of which sum to $n - 1$. Let $\mathcal{A}_n$ denote the subclass of binary trees with n vertices. The cardinality of $\mathcal{A}_n$ is the nth ***Catalan number***, usually denoted C_n; binary trees are one of dozens of classes counted by the Catalan numbers. A (by no means exhaustive) list of 66 of these is given in Stanley (1999, Problem 6.19).

The recursion implies that for $n \geq 1$, the set $\mathcal{A}_n$ is in bijection with the disjoint union $\mathcal{B}_n := \biguplus_{k=0}^{n-1} \mathcal{A}_k \times \mathcal{A}_{n-1-k}$. This is a re-indexed convolution of $\mathcal{A}$ with itself. At the level of generating functions, we see that the re-indexed convolution has generating function $zF(z)^2$. Taking into account what happens when $n = 0$ yields

$$F(z) - 1 = zF(z)^2 . \tag{2.3.1}$$

This ought to have a unique formal power series solution because any solution to this obeys the defining recursion and the initial condition for the class of binary trees. To solve (2.3.1) in the ring of formal power series, let us first find solutions in the ring of germs of analytic functions, because we know more operations there. The quadratic formula yields two solutions:

$$F(z) = \frac{1 \pm \sqrt{1 - 4z}}{2z}.$$

If either of these two functions is analytic near zero, then the series must have coefficients C_n, because this is the unique solution to the convolution equation represented by $F = 1 + zF^2$. Because the denominator vanishes at zero, division can be valid only if the numerator also vanishes at zero. Choosing the negative root yields such a numerator, and one may check by rationalizing the denominator that the resulting function is analytic near zero:

$$F(z) = \frac{1 - \sqrt{1 - 4z}}{2z} = \frac{2}{1 + \sqrt{1 - 4z}},$$

which is analytic in the disk $|z| < 1/4$. ■

The Kernel Method

The rest of this section is devoted to the ***kernel method***, one of the most prolific sources of algebraic generating functions. The kernel method is a means of producing a generating function for an array $\{a_{\mathbf{r}}\}$ satisfying a linear recurrence

$$a_{\mathbf{r}} = \sum_{\mathbf{s}\in E} c_{\mathbf{s}} a_{\mathbf{r}-\mathbf{s}} \tag{2.3.2}$$

for some constants $\{c_{\mathbf{s}} : \mathbf{s} \in E\}$, except when $\mathbf{r}$ is in the ***boundary condition***, which is made precise later. The set E is a finite subset of $\mathbb{Z}^d$ but not necessarily of $\mathbb{N}^d$. Indeed, if $E \subseteq \mathbb{N}^d$, then Example 2.2.5 generalizes easily to show that $F(\mathbf{z}) = \sum_{\mathbf{r}} a_{\mathbf{r}} \mathbf{z}^{\mathbf{r}}$ is rational. There is one further condition on E: its convex hull must not intersect the negative orthant $\{\mathbf{r} : \mathbf{r} \le \mathbf{0}\}$. This ensures that the recursion is well founded (see Lemma 2.3.3).

The kernel method is of interest to the present study because it often produces generating functions that, even though they are not rational, satisfy the meromorphicity assumptions that allow us to compute their asymptotics. It is shown in Bousquet-Mélou and Petkovšek (2000) that the complexity of F increases with the number of coordinates in which points of E are allowed to take negative values. Just as allowing no negative coordinates in E causes F to be rational, it turns out that allowing only one negative coordinate in E causes F to be algebraic. This is shown in Bousquet-Mélou and Petkovšek (2000), along with counterexamples when E contains points with two different negative coordinates. The remainder of this section draws heavily on Bousquet-Mélou and Petkovšek (2000). We begin, though, with an example.

Example 2.3.2 (A random walk problem) Two players move their tokens toward the finish square, flipping a fair coin each time to see who moves forward one square. At present the distances to the finish are $1 + r$ and $1 + r + s$. If the second player passes the first player, the second player wins; if the first player reaches the finish square, the first player wins; if both players are on the square immediately preceding the finish square, then it is a draw. What is the probability of a draw?

Let a_{rs} be the probability of a draw, starting with initial positions $1 + r$ and $1 + r + s$. Conditioning on which player moves first, one finds the recursion

$$a_{rs} = \frac{a_{r,s-1} + a_{r-1,s+1}}{2},$$

which is valid for all $(r, s) \ge (0, 0)$ except for $(0, 0)$, provided that we define a_{rs} to be zero when at least one coordinate is negative. The relation $a_{rs} - (1/2)a_{r,s-1} - (1/2)a_{r-1,s+1} = 0$ suggests we multiply the generating function $F(x, y) := \sum a_{rs} x^r y^s$ by $1 - (1/2)y - (1/2)(x/y)$. To clear denominators, we multiply by $2y$: define $Q(x, y) = 2y - y^2 - x$ and compute $Q \cdot F$. We see that the coefficients of this vanish with two exceptions: the $x^0 y^1$ coefficient corresponds to $2a_{0,0} - a_{0,-1} - a_{-1,1}$ which is equal to 2, not 0, because the recursion does not hold at $(0, 0)$ (a_{00} is set equal to 1); the $y^0 x^j$ coefficients do not vanish for $j \geq 1$ because, due to clearing the denominator, these correspond to $2a_{j,-1} - a_{j,-2} - a_{j-1,0}$. This expression is nonzero because, by definition, only the third term is nonzero, but the value of the expression is not given by prescribed boundary conditions. That is,

$$Q(x, y)F(x, y) = 2y - h(x) \tag{2.3.3}$$

where $h(x) = \sum_{j \geq 1} a_{j-1,0} x^j = xF(x, 0)$ will not be known until we solve for F.

This generating function is in fact a simpler variant of the one derived in Larsen and Lyons (1999) for the waiting time until the two players collide, which is needed in the analysis of a sorting algorithm. Their solution is to observe that there is an analytic curve in a neighborhood of the origin on which Q vanishes. Solving $Q = 0$ for y in fact yields two solutions, one of which, $y = \xi(x) := 1 - \sqrt{1 - x}$, vanishes at the origin. Because ξ has a positive radius of convergence, we have, at the level of formal power series, that $Q(x, \xi(x)) = 0$, and substituting $\xi(x)$ for y in (2.3.3) gives

$$0 = Q(x, \xi(x))F(x, \xi(x)) = 2\xi(x) - h(x).$$

Thus $h(x) = 2\xi(x)$ and

$$F(x, y) = 2\frac{y - \xi(x)}{Q(x, y)} = \frac{2}{1 + \sqrt{1 - x} - y}.$$

■

A General Explanation of the Kernel Method

Let $\boldsymbol{p}$ be the coordinate wise infimum of points in $E \cup \{\boldsymbol{0}\}$, i.e., the greatest element of $\mathbb{Z}^d$ such that $\boldsymbol{p} \leq \boldsymbol{s}$ for every $\boldsymbol{s} \in E \cup \{\boldsymbol{0}\}$. Let

$$Q(\boldsymbol{z}) := \boldsymbol{z}^{-\boldsymbol{p}} \left(1 - \sum_{\boldsymbol{s} \in E} c_{\boldsymbol{s}} \boldsymbol{z}^{\boldsymbol{s}} \right),$$

where the normalization by $\boldsymbol{z}^{-\boldsymbol{p}}$ guarantees that Q is a polynomial but not divisible by any z_j. We assume $\boldsymbol{p} \neq \boldsymbol{0}$, because we already understand in that case how the recursion leads to a rational generating function. The ***boundary value locations*** are any $B \subseteq \mathbb{N}^d$ closed under $\leq$. In the examples that follow, B will always be the singleton $\{\boldsymbol{0}\}$. The boundary values are a set of values $\{b_{\boldsymbol{r}} : \boldsymbol{r} \in B\}$. We study the initial value problem with initial conditions

$$a_{\boldsymbol{r}} = b_{\boldsymbol{r}} \text{ for all } \boldsymbol{r} \in B \tag{2.3.4}$$

and with the recursion (2.3.2) assumed to hold for all $\boldsymbol{r} \in \mathbb{N}^d \setminus B$, and with the convention that summands with $\boldsymbol{r} - \boldsymbol{s} \notin \mathbb{N}^d$ are zero. Thus the data for the problem are E, Q, B, and $\{b_{\boldsymbol{r}} : \boldsymbol{r} \in B\}$. Figure 2.1 shows an example of this with $E = \{(2, -1), (-1, 2)\}$, with B taken to be the y-axis; the most natural way to depict these polynomials is via a ***Newton diagram***, namely the set of vector exponents in $\mathbb{N}^d$.

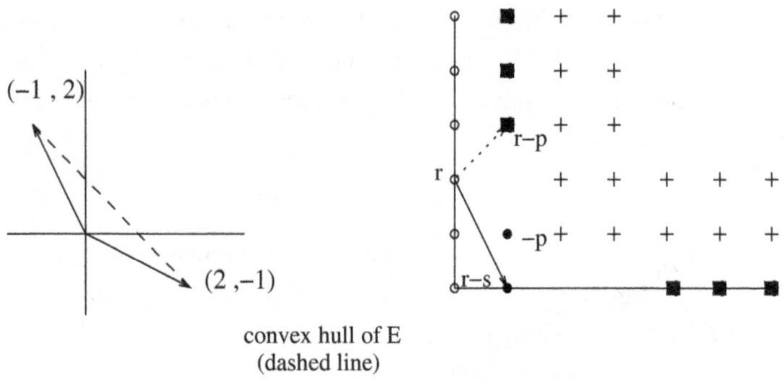

Figure 2.1 The set E and a Newton diagram of K and U.

Let Z denote the set $\mathbb{N}^d \setminus B$ and let $F_Z := \sum_{r \in Z} a_r z^r$ be the generating function for those values for which the recursion (2.3.2) holds. It is more convenient to work with F_Z and then recover F from F_Z via $F = F_Z + F_B$, where

$$F_B = \sum_{r \in B} b_r z^r .$$

To apply the kernel method, one examines the product QF_Z. There are two kinds of contribution to QF_Z. First, for every pair $(\boldsymbol{r}, \boldsymbol{s})$ with $\boldsymbol{s} \in E, \boldsymbol{r} \in Z$, and $\boldsymbol{r} - \boldsymbol{s} \in B$, there is a term $c_s b_{r-s} z^{r-p}$ coming from the difference between the coefficient of z^{r-p} in QF_Z and the coefficient in QF, which vanishes. Let

$$K(z) := \sum_{r \in Z, s \in E, r-s \in B} c_s b_{r-s} z^{r-p}$$

denote the sum of these terms. The "K" stands for "known," because the coefficients of K are determined by the boundary conditions, which are known. The example in Figure 2.1 has terms of K in the first two rows and columns of Z. Second, for every pair $(\boldsymbol{r}, \boldsymbol{s})$ with $\boldsymbol{s} \in E, \boldsymbol{r} - \boldsymbol{s} \in Z$, and $\boldsymbol{r} \notin Z$, there is a term $-c_s a_{r-s} z^{r-p}$ coming from the fact that the recursion does not hold at $\boldsymbol{r}$. Let

$$U(z) = \sum_{r-s \in Z, s \in E, r \notin Z} c_s a_{r-s} z^{r-p}$$

denote these terms. The "U" stands for "unknown," because these coefficients are not explicitly determined from the boundary conditions. In the example from Figure 2.1, U has one row and one column of terms; the value of $\boldsymbol{r}$ leading to the xy^3-term of U is pictured.

Lemma 2.3.3 (Bousquet-Mélou and Petkovšek, 2000, Theorem 5) *Let E be a finite subset of $\mathbb{N}^d$ whose convex hull does not intersect the negative orthant. Let $\{c_s : \boldsymbol{s} \in E\}$ be constants, let $\boldsymbol{p}$ be the coordinate wise infimum of E as above, let*

$B \subseteq \mathbb{N}^d$ be closed under $\leq$, let $\{b_s : s \in B\}$ be constants, and let

$$K(z) := \sum_{r \in Z, s \in E, r-s \in B} c_s b_{r-s} z^{r-p}.$$

Then there is a unique set of values $\{a_r : r \in Z\}$ such that (2.3.2) *holds for all $r \in Z$. Consequently, there is a unique pair of formal power series F_Z and U such that*

$$QF_Z = K - U.$$

Furthermore, if K is analytic in a neighborhood of the origin, then so are U and F_Z.

Proof The convex hull of E and the closed negative orthant are disjoint convex polyhedra so there is a hyperplane that separates them and meets neither. The normal vector may be perturbed slightly to obtain a rational vector $\boldsymbol{v}$ such that $\boldsymbol{v} \cdot \boldsymbol{s} > 0$ for all $\boldsymbol{s} \in E$ and $\boldsymbol{v} \cdot \boldsymbol{s} < 0$ for all $\boldsymbol{s} \neq \boldsymbol{0}$ in the negative orthant. The vector $\boldsymbol{v}$ must have positive coordinates. Clearing denominators, we may assume $\boldsymbol{v}$ is an integer. Linearly order $\mathbb{N}^d$ by the value of the dot product with $\boldsymbol{v}$, breaking ties arbitrarily, to produce a well-ordering, $\preceq$, of $\mathbb{N}^d$ and hence of Z.

Now proceed by induction on $\boldsymbol{r}$ with respect to $\preceq$. Fix $\boldsymbol{m} \in Z$. If $\boldsymbol{s} \in E$ and $\boldsymbol{r} \prec \boldsymbol{m}$, then $\boldsymbol{r} - \boldsymbol{s} \prec \boldsymbol{m}$. Consequently, the validity of (2.3.2) for all $\boldsymbol{r} \prec \boldsymbol{m}$ depends only on values $a_{\boldsymbol{r}}$ with $\boldsymbol{r} \prec \boldsymbol{m}$. Assume for induction that there is a unique set of values of $\{a_{\boldsymbol{r}} : \boldsymbol{r} \prec \boldsymbol{m}\}$ such that (2.3.2) holds for $\boldsymbol{r} \prec \boldsymbol{m}$. Imposing (2.3.2) for $\boldsymbol{r} = \boldsymbol{m}$ then uniquely specifies $a_{\boldsymbol{m}}$, completing the induction.

To show convergence, let $\gamma' = \log \sum_{s \in E} |c_s|$. By analyticity of K, we may choose $\gamma' \geq \gamma$ for which $|b_{\boldsymbol{r}}| \leq \exp(\gamma\, \boldsymbol{r} \cdot \boldsymbol{v})$. With this as the base step, it follows by induction that this holds for $a_{\boldsymbol{r}}$ in place of $b_{\boldsymbol{r}}$:

$$\begin{aligned} |a_{\boldsymbol{r}}| &\leq \left(\sum_{s \in E} |c_s| \right) \sup_{s \in E} |a_{\boldsymbol{r}-\boldsymbol{s}}| \\ &\leq e^{\gamma} \sup_{\boldsymbol{m} \cdot \boldsymbol{v} < \boldsymbol{r} \cdot \boldsymbol{v}} |a_{\boldsymbol{m}}| \\ &\leq e^{\gamma} e^{\gamma\, (\boldsymbol{r} \cdot \boldsymbol{v} - 1)} \\ &= e^{\gamma\, \boldsymbol{r} \cdot \boldsymbol{v}}, \end{aligned}$$

establishing an exponential bound on $|a_{\boldsymbol{r}}|$ and hence analyticity of F near the origin. From this, analyticity of F_Z and U follow. □

The previous lemma is based on a formal power series approach. Another way of thinking about this is that F_Z is trying to be the power series K/Q, but because Q vanishes at the origin, one must subtract some terms from K to cancel whatever factor of Q vanishes at the origin. The kernel method turns this intuition into a precise statement.

Theorem 2.3.4 (Bousquet-Mélou and Petkovšek, 2000, Theorem 13) *Let $d \geq 2$ be arbitrary, and suppose the boundary locations B are of the form $\{\mathbf{r} : \mathbf{r} \not\geq \mathbf{s}\}$ for some $\mathbf{s} \in \mathbb{N}^d$. If $p_1, \ldots, p_{d-1} \geq 0 > p_d$ and the boundary generating function $K(z)$ is algebraic, then F is algebraic.*

Proof Suppose $\mathbf{r} \notin Z$ and $\mathbf{r} - \mathbf{s} \in Z$ with $\mathbf{s} \in E$. We know that $\mathbf{r} - \mathbf{s}' \notin Z$, where $\mathbf{s}'_j = \mathbf{s}_j$ for all $j \leq d-1$ and $\mathbf{s}'_d = 0$; this is because the complement of Z is closed under coordinate-wise $\leq$ and the first $d-1$ coordinates of any point in E are nonnegative. Thus $s_d - p_d < r_d \leq s_d$. It follows that U is $x_d^{s_d+1}$ a polynomial of degree at most $p_d - 1$ in x_d.

The polynomial Q is equal to $z_d^{p_d} - \sum_{s \in E} c_s z_d^{p_d} \mathbf{z}^{\mathbf{s}}$. It is convenient to regard this as a polynomial in z_d over the field of algebraic functions of $z_1, \ldots, z_{d-1}$. The degree of Q in z_d is at least p_d. Let $\{\xi_i(z_1, \ldots, z_{d-1})\}$ be the roots of this polynomial. At least p_d of these, when counted with multiplicities, satisfy $\xi_i(\mathbf{0}) = 0$: this follows from the fact that $(0, \ldots, 0, j) \notin E$ for any negative j, whence the polynomial $Q(0, \ldots, 0, z_d)$ has multiplicity p_d at 0.

If the p_d such roots of Q are distinct, then the equation $QF_Z = K - U$ evaluated at each ξ_i leads to p_d equations

$$U(\xi_i) = K(\xi_i).$$

The Lagrange interpolation formula (Pólya and Szegő, 1998, Section V1.9) produces a polynomial P given its values $y_1, \ldots, y_k$ at any k points $x_1, \ldots, x_k$:

$$P(x) = \sum_{j=1}^{n} y_j \prod_{i \neq j} \frac{x - x_i}{x_j - x_i}. \tag{2.3.5}$$

Over any field of characteristic zero, and in particular over the algebraic functions of $x_1, \ldots, x_d$, this is the unique polynomial of degree at most $k - 1$ passing through the k points. Taking $k = p_d, x_i = \xi_i$, and $y_i = K(\xi_i)$ shows that U is given by (2.3.5). Thus U is a rational function of algebraic functions and is therefore algebraic. Finally, if the ξ_i are not distinct, one has instead the p_d equations:

$$U(\xi_i) = K(\xi_i),\ U'(\xi_i) = K'(\xi_i),\ \ldots,\ U^{(m_i-1)}(\xi_i) = k^{(m_i-1)}(\xi_i),$$

where m_i is the multiplicity of the root ξ_i. One may replace the Lagrange interpolation formula by the Hermite interpolation formula (Isaacson and Keller, 1994, Section 6.1, Problem 10), which again gives U as a rational function of each $K(\xi_i)$ and its derivatives. □

Specializing further to $d = 2$ and $B = \{\mathbf{0}\}$ gives the following explicit formula for F.

Corollary 2.3.5 (Bousquet-Mélou and Petkovšek, 2000, Equation [24]) *Suppose further that $d = 2$, $\mathbf{p} = (0, -p)$, and $B = \{\mathbf{0}\}$ with boundary value $b_{\mathbf{0}} = 1$. There will be exactly p formal power series $\xi_1, \ldots, \xi_p$ such that $\xi_j(0) = 0$ and*

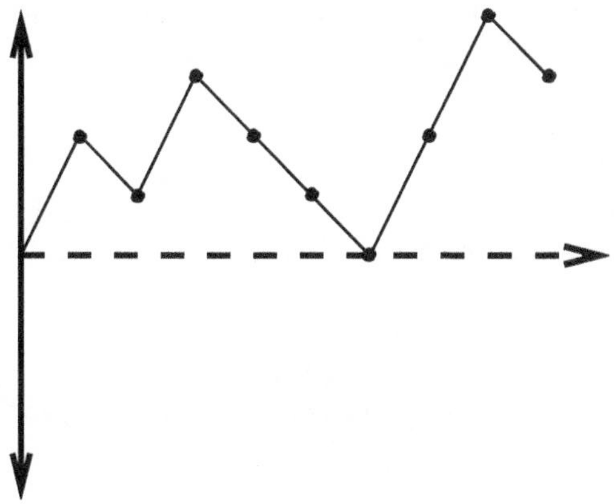

Figure 2.2 A generalized Dyck path of length nine with $E = \{(1, 2), (1, -1)\}$.

$Q(x, \xi_j(x)) = 0$, and we may write $Q(x, y) = -C(x)\prod_{j=1}^{p}(y - \xi_j(x))\prod_{j=1}^{P}(y - \rho_j(x))$ for some r and $\rho_1, \ldots, \rho_P$. The generating function F_Z will then be given by

$$F_Z(x, y) = \frac{K(x, y) - U(x, y)}{Q(x, y)} = \frac{\prod_{j=1}^{p}(y - \xi_j(x))}{Q(x, y)} = \frac{1}{-C(x)\prod_{j=1}^{r}(y - \rho_j(x))}.$$

Proof Work in the ring $\mathbb{C}[[x]][y]$ of polynomials in y with coefficients in the local ring of power series in x converging in a neighborhood of zero. The asserted factorization of Q follows from its vanishing to order p at $y = 0$ and having degree $p + r$ (as a polynomial in y with coefficients in $\mathbb{C}[[x]]$). By definition, $K(x, y) = y^p$. Recalling that the degree in y of $U(x, y)$ is at most $p - 1$, it follows that the degree of $K(x, y) - U(x, y)$ in y is exactly p. If we know p factors $(y - a_j)$ and the leading coefficient C of a polynomial of degree p, then the polynomial is completely determined: it must be $C\prod_{j=1}^{p}(y - a_j)$. Because $K - U$ vanishes on $y = \xi_j(x)$ in a neighborhood of zero for all j, it is divisible by $\prod_{j=1}^{p}(y - \xi_j(x))$. The leading coefficient of $K - U$ is the same as the leading coefficient of K, namely 1. Therefore, $K - U = \prod_{j=1}^{p}(y - \xi_j(x))$, which establishes the conclusion of the corollary. □

Dyck, Motzkin, Schröder, and Generalized Dyck Paths

Let E be a set $\{(r_1, s_1), \ldots, (r_k, s_k)\}$ of integer vectors with $r_j > 0$ for all j and $\min_j s_j = -p < 0 < \max_j s_j = P$. The ***generalized Dyck paths*** from $(0, 0)$ to (r, s) with increments in E are the paths that never go below the horizontal axis. Figure 2.2 shows on example.

Let $F(x, y) = \sum_{r,s} a_{rs}x^r y^s$ generate the number, a_{rs}, of generalized Dyck paths to the point (r, s). In the notation of the previous discussion, we have $\boldsymbol{q} = (0, 0)$, $F = F_q$, $Q(x, y) = y^p(1 - \sum_i x^{r_i} y^{s_i})$, and $C(x) = \sum_{i:s_i=P} x^{r_i}$. The

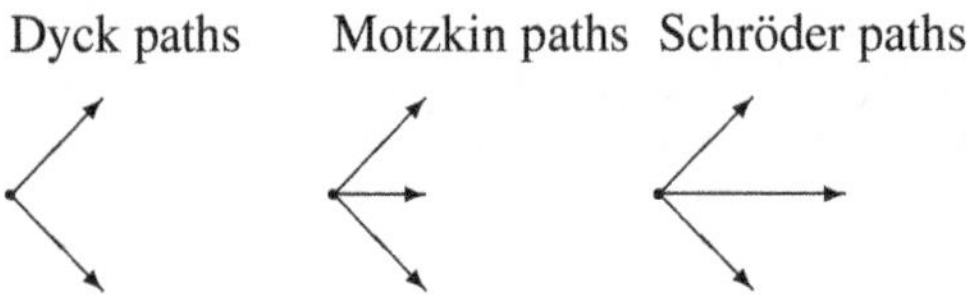

Figure 2.3 Legal steps for three types of paths.

special case $p = P = 1$, i.e., vertical displacement of at most 1 per step, occurs often in classical examples.

Proposition 2.3.6 *Let E, p, P, $C(x)$ be as above and suppose that $p = P = 1$. Then the generating function for generalized Dyck paths with steps from E is given by*

$$F(x, y) = \frac{\xi(x)}{a(x) - C(x)\xi(x)y}$$

where $a(x) = \sum_{i:s_i=-1} x^{r_i}$.

Proof Here, Q is quadratic in y, and we may simplify the formula for F as follows. The product $\xi\rho$ equals $a(x)/C(x)$ where $a(x) = \sum_{i:s_i=-1} x^{r_i}$, and hence

$$F(x, y) = \frac{\xi(x)/a(x)}{1 - [C(x)\xi(x)/a(x)]y}.$$

□

We now discuss the three standard examples from Bousquet-Mélou and Petkovšek (2000), depicted in Figure 2.3.

Dyck Paths. When $E = \{(1, 1), (1, -1)\}$ we have the original Dyck paths. We have $p = 1 = P$ and $Q(x, y) = y - xy^2 - x$. Here $C(x) = x$, and $Q(x, y) = -x(y - \xi(x))(y - \rho(x))$, where $\xi(x) = (1 - \sqrt{1 - 4x^2})/(2x)$ and $\rho(x) = (1 + \sqrt{1 - 4x^2})/(2x)$ is the algebraic conjugate of ξ. Note that ρ is a formal Laurent series and $\rho\xi = 1$.

Thus, following the preceding discussion,

$$F(x, y) = \frac{1}{-x(y - \rho(x))} = \frac{\xi(x)/x}{1 - y\xi(x)}.$$

Setting $y = 0$ recovers the fact that the Dyck paths coming back to the x-axis at $(2n, 0)$ are counted by the Catalan number C_n.

Motzkin Paths. Let $E = \{(1, 1), (1, 0), (1, -1)\}$. In this case, the generalized Dyck paths are known as ***Motzkin paths***. Again, $Q(x, y) = y - xy^2 - x - xy$. Now ρ and ξ are given by $(1 - x \pm \sqrt{1 - 2x - 3x^2})/(2x)$ and

$$F(x, y) = \frac{\xi(x)/x}{1 - y\xi(x)} = \frac{2}{1 - x + \sqrt{1 - 2x - 3x^2} - 2xy}.$$

Schröder Paths. Here $E = \{(1,1),(2,0),(1,-1)\}$. We have $C(x) = x$, $Q(x,y) = y - xy^2 - x^2y - x$, and ρ and ξ are given by $(1 - x^2 \pm \sqrt{1 - 6x^2 + x^4})/(2x)$. Again,

$$F(x,y) = \frac{\xi(x)/x}{1 - y\xi(x)} = \frac{2}{1 + \sqrt{1 - 6x^2 + x^4} - x^2 - 2xy}.$$

2.4 Generating Functions

The more explicitly a generating function is described, the better are the prospects for getting information out of it. This includes not only asymptotic estimation, but also proving bijections and, in general, relating the class being counted to other combinatorial classes. Rational generating functions are easy to work because they are specified by finite data: both numerator and denominator are a finite sum of monomials with integer exponents and (usually) integer coefficients.

We have seen that some common and very natural combinatorial operations take us from the class of rational functions to the larger class of algebraic generating functions. These also have canonical representations: if f is algebraic, then there is a minimal polynomial P for which $P(f) = 0$; f may be specified by writing down the coefficients of P, which are themselves polynomials and therefore finitely specified. In Section 6.1, we discuss techniques in computational algebra that allow one to manipulate algebraic functions by performing manipulations directly on the minimal polynomials. This makes the class of algebraic generating functions quite nice to work with. There are, however, common combinatorial operations that take us out of the class of algebraic functions, and this drives us to consider one further step in the hierarchy. A more complete discussion of this hierarchy for univariate functions may be found in Stanley (1999, Chapter 6). We give a brief summary here, first for univariate functions and then for multivariate functions.

Univariate D-Finite Functions

Some notation, taken from Stanley (1999), will come in handy. We have already been using $\mathbb{C}[z]$ and $\mathbb{C}[[z]]$ to denote, respectively, the polynomials and formal power series over $\mathbb{C}$. To discuss algebraic and D-finite functions, it is more convenient to work over a field. Denote by $\mathbb{C}(z)$ the field of fractions of $\mathbb{C}[z]$, which are just the rational functions. Denote by $\mathbb{C}((z))$ the field of fractions of $\mathbb{C}[[z]]$; because $\mathbb{C}[[z]]$ has the unique maximal ideal $\langle z \rangle$, the fraction field $\mathbb{C}((z))$ coincides with the ring of Laurent polynomial series $\mathbb{C}[[z]][1/z]$. The ring of algebraic formal power series is defined to be the set $\mathbb{C}_{\text{alg}}[[z]]$ of elements of $\mathbb{C}[[z]]$ that are algebraic over $\mathbb{C}(z)$. In other words, clearing denominators, $f \in \mathbb{C}_{\text{alg}}[[z]]$ if and only if f is a formal power series and

$$\sum_{j=0}^{m} P_j f^j = 0$$

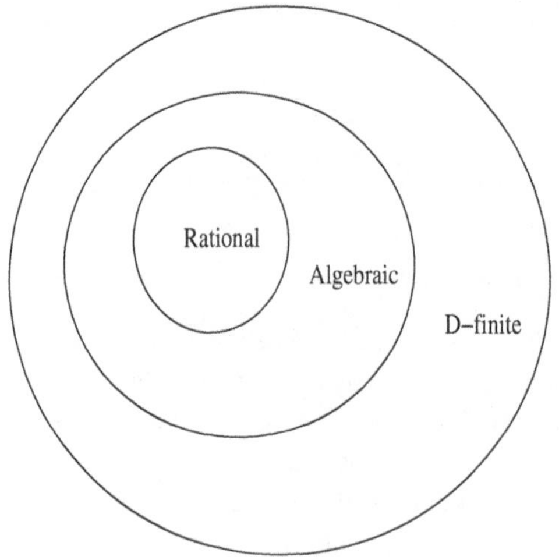

Figure 2.4 Some classes of generating functions.

for some m and some $P_0, \ldots, P_m \in \mathbb{C}[z]$. Equivalently, f is algebraic if and only if the powers $1, f, f^2, \ldots$ span a finite dimensional vector space in $\mathbb{C}((z))$ over $\mathbb{C}(z)$. Given a formal power series $f = \sum_{n=0}^{\infty} a_n z^n$, the formal derivative f' is defined, as one expects, to be $\sum_{n=0}^{\infty}(n+1)a_{n+1}z^n$. If f is analytic in an open neighborhood $\mathcal{N}$ of zero, then f' is analytic on $\mathcal{N}$ as well. We have not discussed a combinatorial interpretation for differentiation, but it is clear that differentiation will arise from the operation "multiply the nth term by n." Shortly, we will generalize this operation to *polynomial recursion*.

Definition 2.4.1 *A formal power series $f \in \mathbb{C}[[z]]$ is* ***D-finite*** *if and only if there is an integer m and polynomials $P, P_0, \ldots, P_m \in \mathbb{C}[z]$ with $P_m \neq 0$ such that*

$$P + P_0 f + P_1 f' + \cdots + P_m f^{(m)} = 0\,. \tag{2.4.1}$$

Equivalently, f is D-finite if and only if f and its derivatives span a finite dimensional vector space in $\mathbb{C}((z))$ over $\mathbb{C}(z)$.

Remark The natural definition of a D-finite series is one whose derivatives span a finite dimensional space over the polynomials. Vector spaces over fields are simpler than modules over rings, so we phrase this in terms of vector space dimension over the rational functions. This requires that the rational functions act on $\mathbb{C}[[z]]$, which requires us to extend $\mathbb{C}[[z]]$ to $\mathbb{C}((z))$. The same phenomenon will occur in the multivariate setting in Definition 2.4.6, the only difference being that the fraction field of $\mathbb{C}[[z]]$ is not a finitely generated extension of $\mathbb{C}[[z]]$.

Variants of this definition are discussed in Stanley (1999, Proposition 6.4.1). The Venn diagram depicting the hierarchy in Figure 2.4 is justified by the following proposition.

Proposition 2.4.2 (Stanley, 1999, Theorem 6.4.6) *If f is algebraic, then f is D-finite.*

Proof By definition, because f is algebraic, there is a polynomial $P = \sum_{j=0}^{m} P_j y^j$ in $\mathbb{C}[z, y]$ for which $P(z, f) = 0$. Implicit differentiation yields

$$f' = -\frac{\partial P(z, y)/\partial z|_{y=f}}{\partial P(z, y)/\partial y|_{y=f}}. \tag{2.4.2}$$

To justify this formally, assume P to be of minimal degree in y; then $\partial P(z, y)/\partial y$ is of lesser degree in y, so it is nonzero when evaluated at $y = f$, and taking the derivative of the equation $P(z, f) = 0$ shows that $\partial P(z, y)/\partial z + f' \partial P(z, y)/\partial y$ vanishes when evaluated at $y = f$, justifying (2.4.2).

We have shown that f' is in the field of fractions of the ring $\mathbb{C}[z][f]$. The quotient rule now implies that the derivative of any element of $\mathbb{C}(z, f)$ is again in $\mathbb{C}(z, f)$. By induction, all derivatives of f are in $\mathbb{C}(z, f)$. However, this is a finite extension of $\mathbb{C}(z)$ for any algebraic f. Thus f and its derivatives span a finite vector space over $\mathbb{C}(z)$, finishing the proof. □

Recall that a series $f = \sum_{n=0}^{\infty} a_n z^n$ is rational if and only if the sequence $\{a_n : n \geq 0\}$ satisfies a linear recurrence with constant coefficients. There is no such quick characterization of coefficients of algebraic generating functions, but there is for D-finite functions, which is another reason the D-finite generating functions are a natural class.

Definition 2.4.3 (P-recursiveness [1 variable]) *A sequence $\{a_n : n \geq 0\}$ is said to be* ***P-recursive*** *(short for "polynomially recursive") if there exist polynomials $P_0, \ldots, P_m$ with $P_m \neq 0$ such that*

$$P_m(n)a_{n+m} + P_{m-1}(n)a_{n+m-1} + \cdots + P_0(n)a_n = 0 \tag{2.4.3}$$

for all $n \geq 0$.

Example 2.4.4 Let $a_n = 1/n!$. Then $(n + 1)a_{n+1} - a_n = 0$ for all $n \geq 0$, so $\{a_n : n \geq 0\}$ is P-recursive. The generating function for $(n + 1)a_{n+1}$ is f', implying $f' - f = 0$ and showing that f is D-finite. This differential equation has solution $f(z) = a_0 e^{-z}$. ■

The connection between P-recursion and D-finiteness illustrated in the previous example is generalized by the following result, which is stated as Stanley (1999, Proposition 6.4.3) and attributed to Comtet (1964). The proof consists of matching up coefficients.

Theorem 2.4.5 (D-finite ⇔ P-recursive) *A sequence $\{a_n : n \geq 0\}$ is P-recursive if and only if its generating function $f = \sum_{n=0}^{\infty} a_n z^n$ is D-finite.*

Proof First suppose f is D-finite. Let $\{P_k : 0 \leq k \leq m\}$ be as in (2.4.1), and let $b_{k,j}$ denote the x^j-coefficient of P_k. The x^{n-k} coefficient of $f^{(j)}$ is equal to $(n - k + j)_j a_{n-k+j}$, where $(u)_j := u(u - 1) \cdots (u - j + 1)$ denotes the falling factorial. Equating the coefficient of x^n to zero in the left-hand side of (2.4.1) gives

$$\sum_{j=0}^{m} \sum_{k} b_{k,j}(n-k+j)_j a_{n-k+j} = 0 .$$

This is a linear equation in $\{a_{n+j} : j \in [a, b]\}$ for some finite interval $[a, b]$ whose coefficients are polynomials in n of degrees at most m. It does not collapse to $0 = 0$ because for any j such that $b_{m,j} \neq 0$, the coefficient of a_{n-j+m} is $b_{m,j} n^m + O(n^{m-1}) \neq 0$. We may re-index so that $a = 0$.

Conversely, suppose that (2.4.3) is satisfied. The polynomials $\{(n+j)_j : j \geq 0\}$ form a basis for $\mathbb{C}[n]$ that is triangular with respect to the basis $\{n^j\}$, hence each P_k is a finite linear combination $\sum c_{k,j}(n+j)_j$. Plugging this into (2.4.3) yields

$$\sum_{k=0}^{m} \sum_{j=0}^{\deg P_k} c_{k,j}(n+j)_j a_{n+k} = 0 .$$

The rules for differentiating formal power series extend to formal Laurent series $C((x))$, in which $(n+j)_j a_{n+k}$ is just the x^n coefficient of $(f \cdot x^{j-k})^{(j)}$. Thus

$$\sum_{k=0}^{m} \sum_{j=0}^{\deg P_k} c_{k,j}(f \cdot x^{j-k})^{(j)} = 0$$

in $\mathbb{C}((x))$. Using the product rule, this becomes a nontrivial linear ODE in f with coefficients in $\mathbb{C}[x][x^{-1}]$, and multiplying through by a sufficiently high power of x gives a relation of the form (2.4.1). □

Multivariate D-Finite Functions

Let $\mathbb{C}[\boldsymbol{z}]$, $\mathbb{C}(\boldsymbol{z})$ and $\mathbb{C}[[\boldsymbol{z}]]$ denote, respectively, the polynomials, rational functions, and formal power series in d variables, $z_1, \ldots, z_d$. Formal partial differentiation is defined in the obvious way: if $f = \sum_{\boldsymbol{r}} a_{\boldsymbol{r}} \boldsymbol{z}^{\boldsymbol{r}}$, then $\partial f / \partial z_j := \sum_{\boldsymbol{r}} (r_j + 1) a_{\boldsymbol{r}+\delta_j} \boldsymbol{z}^{\boldsymbol{r}}$, where δ_j has a 1 in position j and 0's elsewhere. Generalizing the univariate definition, a power series $f \in \mathbb{C}[[\boldsymbol{z}]]$ is said to be D-finite if f and all its iterated partial derivatives generate a finite dimensional vector space over $\mathbb{C}(\boldsymbol{z})$; here one must interpret $\mathbb{C}[[\boldsymbol{z}]]$ as embedded in the ring (actually a field) of formal quotients of elements of $\mathbb{C}[[\boldsymbol{z}]]$ by polynomials.

The correct analogue of P-recursiveness in the multivariate case is not so obvious. The following definition is Definition 3.2 in Lipshitz (1989). Note that the definition is recursive in the dimension, d.

Definition 2.4.6 (P-recursiveness [d variables]) *Suppose P-recursiveness has been defined for arrays of dimension $d-1$. Then the array $\{a_{\boldsymbol{r}} : \boldsymbol{r} \in \mathbb{N}^d\}$ is said to be P-recursive if there is some positive integer k such that the following two conditions hold.*

(i) *For each* $j \in [d]$ *there are polynomials* $\{P_\nu^j : \nu \in [k]^d\}$, *not all vanishing, such that*

$$\sum_{\nu \in [k]^d} P_\nu^j(r_j) a_{\mathbf{r}-\nu} = 0$$

as long as $\mathbf{r} \geq k \cdot \mathbf{1}$ *coordinate-wise.*

(ii) *All the* $(d-1)$*-variate arrays obtained from* $\{a_\mathbf{r}\}$ *by holding one of the* d *indices fixed at a value less than* k *are P-recursive.*

The following result extends the connection between P-recursiveness and D-finiteness to the d-variate setting. The proof, which we omit, may be found in Lipshitz (1989).

Theorem 2.4.7 *The array* $\{a_\mathbf{r} : \mathbf{r} \in \mathbb{N}^d\}$ *is P-recursive if and only if the generating function* $f(\mathbf{z}) := \sum_\mathbf{r} a_\mathbf{r} \mathbf{z}^\mathbf{r}$ *is D-finite.*

Diagonals

D-finite generating functions are finitely specifiable and the arrays they generate satisfy nice recursions. There is one more reason to expand our horizons to this class of function, namely its closure properties. The classes of rational and algebraic generating functions are both closed under the ring operations. It will come as no surprise that the D-finite functions are closed under these as well.

Theorem 2.4.8 *If* f *and* g *are D-finite, then so are* $f+g$ *and* fg.

Proof Let V be the vector subspace of $\mathbb{C}((x))$ spanned over $\mathbb{C}(x)$ by $f+g$ and all its derivatives. Clearly V is contained in the sum of subspaces $V_f + V_g$ spanned by the derivatives of f and g, respectively; hence V is finite dimensional and in fact the dimension is bounded by $\dim(V_f) + \dim(V_g)$.

Let $f^{(\mathbf{r})}$ denote the partial derivative of f taken r_i times with respect to z_i for each $1 \leq i \leq d$. The products $f^{(\mathbf{r})} g^{(\mathbf{s})}$ span a finite dimensional space, in fact a space of dimension at most $\dim(V_f) \cdot \dim(V_g)$. By the product rule, every derivative $(fg)^\mathbf{r}$ is in this space. □

A more interesting and nontrivial closure property has to do with diagonals. The *diagonal* of the bivariate power series $F(x,y) := \sum_{r,s=0}^\infty a_{rs} x^r y^s$ is the univariate series $\operatorname{diag} F(z) := \sum_{n=0}^\infty a_{n,n} z^n$. This may be generalized to any number of variables.

Definition 2.4.9 (diagonal of a formal power series) *Let* $F(\mathbf{z}) := \sum_\mathbf{r} a_\mathbf{r} \mathbf{z}^\mathbf{r}$ *be a formal power series in* d *variables. The* ***diagonal*** *of* F *corresponding to the surjection* $\pi : [d] \twoheadrightarrow [t]$, *where* $1 \leq t \leq d$, *is the* t*-variate series* $F_\pi(\mathbf{z}) := \sum_\mathbf{s} a_{\pi^{-1}\mathbf{s}} \mathbf{z}^\mathbf{s}$. *The* ***elementary diagonal*** *is the diagonal corresponding to the surjection sending* 1 *to* 2 *and each other element of* $[d]$ *to itself.*

When $t = 1$, *the resulting univariate series is called the* **complete diagonal** *and denoted* diag F.

Remark Every diagonal can be formed by iterating the elementary diagonal and permutation of variables. A more relaxed definition of diagonal is provided in Section 13.2. Sometimes one speaks of the generating functions $\sum_n a_{n\alpha} z^n$, where α is a fixed direction, as diagonals. When α has integer coordinates, then this can be expressed in terms of the complete diagonal.

A result appearing in Hautus and Klarner (1971) and credited to Furstenberg (1967) is that the diagonal of a bivariate rational power series is always algebraic. This result, although it does not solve the general bivariate asymptotic problem, is handy when one is only interested in the main diagonal. This occurs more often than we might at first think. For example, the Lagrange inversion formula can be described in terms of diagonals – see Section 12.3. However, it cannot be used iteratively, because the result does not apply to algebraic generating functions. Second, a rational function in more than two variables need not have algebraic diagonals. See Section 13.1 for more details.

Because the Hautus-Klarner-Furstenberg diagonal extraction method is constructive, it is computationally useful, and we present a proof and an example shortly.

In 1988, Lipshitz proved that D-finite series are closed under taking diagonals.

Theorem 2.4.10 (Lipshitz, 1988) *Any diagonal of a D-finite series is D-finite.*

This is the main result of Lipshitz (1988) and the proof of this, though not long, is a little too much to reproduce here. This is, in some sense, the final word on the hierarchy in Figure 2.4. A consequence is that the ***Hadamard product*** of two d-variate D-finite power series is again D-finite. The Hadamard product of $\sum_r a_r \mathbf{z}^r$ with $\sum_r b_r \mathbf{z}^r$ is simply the function $\sum_r a_r b_r \mathbf{z}^r$. The Hadamard product of F and G is a generalized diagonal of $F(\mathbf{x})G(\mathbf{y})$, whence closure under Hadamard products follows from closure under generalized diagonals. Lest the class of D-finite functions seem too good to be true, we should point out that it is not closed under composition. It is shown in Stanley (1999) that $F \circ G$ is D-finite if F is D-finite and G is algebraic, but $G \circ F$ need not be D-finite in this case.

We owe a proof that the diagonal of a bivariate rational series is algebraic. Stanley (1999) provided a formal power series proof, but it relies on some results we have not developed on the algebraic closure of $\mathbb{C}[[x]]$ (Puiseux's Theorem). The following analytic proof has the advantage of providing computational information.

Theorem 2.4.11 (Hautus-Klarner-Furstenberg diagonal extraction) *Let* F *be a rational power series in two variables. Then* diag F *is algebraic.*

Proof Write

$$F(x, y) = \frac{P(x, y)}{Q(x, y)},$$

where P and Q are coprime polynomials and $Q(0, 0) \neq 0$. Let $h = \text{diag}\, F$. Note that F converges in a neighborhood of the origin. Hence when $|y|$ is sufficiently small, the function $F(z, y/z)$ is absolutely convergent for z in some annulus $A(y)$. Treating y as a constant, we view $F(z, y/z)$ as a Laurent series in z inside the annulus $A(y)$; the constant term C of this series is equal to $h(y)$. Thus if we are able to evaluate this constant term as a function of y, we will have the function whose power series in a neighborhood of 0 is h.

By Cauchy's integral formula, $h(y)$ is equal to

$$\frac{1}{2\pi i} \int_{\mathbb{C}} \frac{P(z, y/z)}{zQ(z, y/z)}\, dz,$$

where $\mathbb{C}$ is any circle in the annulus of convergence $A(y)$. By the Residue Theorem,

$$h(y) = \sum \text{Res}\left(\frac{P(z, y/z)}{zQ(z, y/z)}; \alpha \right),$$

where the sum is over residues at poles α inside the inner circle of the annulus $A(y)$. The residues are all algebraic functions of y, so we have represented h as the sum of algebraic functions of y. □

Remark When computing, one needs to know which poles are inside the circle. They are precisely those that converge to zero as $y \to 0$.

Example 2.4.12 (Delannoy numbers continued) Recall that $F(z, w) = 1/(1 - z - w - zw)$, so

$$z^{-1} F(z, y/z) = \frac{1}{z - z^2 - y - yz}.$$

The poles of this are at

$$z = \frac{1 - y}{2} \pm \frac{1}{2}\sqrt{1 - 6y + y^2}.$$

Let α_1 denote the root going to zero with y, i.e., the one with the minus sign, and let α_2 denote the other root. Because $z^{-1}F(z, y/z) = -1/[(z - \alpha_1)(z - \alpha_2)]$, the residue at $z = \alpha_1$ is just $-1/(\alpha_1 - \alpha_2)$ which is simply $(1 - 6y + y^2)^{-1/2}$. Thus

$$h(y) = \frac{1}{\sqrt{1 - 6y + y^2}}. \tag{2.4.4}$$

■

We close the section on D-finite functions with some good news and some bad news concerning diagonal extraction. The good news is that some recent work of Chyzak and Salvy (1998) has made the extraction of diagonals effective. That is, one may feed a rational function, F, into a black box, along with a specification of which diagonal to compute, and the output gives the polynomial coefficients of a differential equation witnessing the D-finiteness of the diagonal. The bad news is that although the method may be adapted to other directions with rational slopes by means of the substitution $F(x^p, y^q)$, the complexity of the computation increases with p and q, so there is no way to take limits, no uniformity, and hence no way to use this method to obtain asymptotics that are truly bivariate (allowing both indices to vary simultaneously). There are other difficulties too – for much more on this topic, see Section 13.1.

2.5 Exponentiation: Set Partitions

Let exp denote the power series $\sum_{n=0}^{\infty} z^n/n!$. The exponential e^F of a formal power series in any number of variables may be defined as $\exp \circ F$; this is well defined as long as $F(\mathbf{0}) = 0$, with formal composition defined by (2.2.1). The map $F \mapsto \exp(F) - 1$ on the space of functions with $F(\mathbf{0}) = 0$ is inverted by the map $F \mapsto \log(1 + F) := \sum_{n=1}^{\infty}(-1)^{n-1} F^n/n$. Exponentiation turns out to have a very useful combinatorial interpretation.

Let $\mathcal{B}$ be a combinatorial class with $b_n := |\mathcal{B}_n|$. Define a *$\mathcal{B}$-partition* of $[n]$ to be a set of pairs $\{(S_\alpha, G_\alpha) : \alpha \in I\}$, where the collection $\{S_\alpha : \alpha \in I\}$ is a partition of the set $[n]$, and each G_α is an element of $\mathcal{B}_j$ for $j = |S_\alpha|$. Define the class $\exp(\mathcal{B})$ to be the class of $\mathcal{B}$-partitions enumerated by n, i.e., $\exp(\mathcal{B})_n$ is the class of $\mathcal{B}$-partitions of $[n]$.

Example 2.5.1 Take $\mathcal{B}$ to be the class of connected graphs with labeled vertices, enumerated by number of vertices. Given $S \subseteq [n]$ and a graph G with $|S|$ vertices, labeled $1, \dots, |S|$, let $\langle S, G \rangle$ denote the graph G with each label j replaced by s_j, where $S = \{s_1 < \cdots < s_{|S|}\}$. Replacing each pair (S_α, G_α) by $\langle S, G \rangle$, we have an interpretation of $\exp(\mathcal{B})_n$ as a collection of connected graphs whose labels are $1, \dots, n$, each used exactly once. In other words, the exponential of the class of labeled connected graphs is the class of all labeled graphs. ■

The use of the word "exponential" for the combinatorial operation described previously above is justified by the following theorem.

Theorem 2.5.2 (exponential formula) *Let $g(z)$ be the exponential generating function for the class $\mathcal{B}$, i.e.,* $g(z) = \sum_{n=0}^{\infty} \frac{b_n}{n!} z^n$. *Then* $\exp(g(z))$ *is the exponential generating function for* $\exp(\mathcal{B})$, *and* $\exp(y\, g(z))$ *is the* semi-exponential generating function *whose $y^k z^n$-coefficient is $1/n!$ times the number of elements of* $\exp(\mathcal{B})_n$ *with $|I| = k$.*

Proof Let $\mathcal{A}$ denote the bivariate class $\exp(\mathcal{B})$ broken down by $|I|$, so that $a_{nk} := |\mathcal{A}_{nk}|$ counts elements of $\exp(\mathcal{B})_n$ with $|I| = k$. Because every set of size k may be listed in $k!$ different orders, we see that $a_{nk} = u_{nk}/k!$, where u_{nk} counts sequences of pairs, $((S_j, G_j), \dots, (S_k, G_k))$ of partitions of $[n]$ and associated elements of $\mathcal{B}$.

Just as $g(z)$ is the exponential generating function for the class $\mathcal{B}$, it is also the ordinary generating function for the class $\mathcal{B}$ re-weighted so that each element of $\mathcal{B}_k$ counts with weight $1/k!$. Therefore, $g(z)^k$ counts sequences of length k with the weight of a sequence given by $w(G_1, \dots, G_k) := \prod_{j=1}^k (1/|G_j|!)$. The reason it is useful to count these sequences by total weight is that every such sequence appears exactly $\binom{n}{|G_1| \cdots |G_k|}$ times as the sequence of second coordinates of sequences of pairs counted by u_{nk}. This multinomial coefficient is exactly $n!\, w(G_1, \dots, G_k)$, so we see that

$$u_{nk} = n![y^k z^n](y\, g(z))^k.$$

The relation between u_{nk} and a_{nk} yields

$$\frac{a_{nk}}{n!} = [y^k z^n] \frac{(y\, g(z))^k}{k!}.$$

Because $(y\, g(z))^k$ has a y^j-coefficient only when $j = k$, we may sum the right-hand side:

$$\frac{a_{nk}}{n!} = [y^k z^n] \sum_{k=0}^{\infty} \frac{(y\, g(z))^k}{k!} = \exp(y\, g(z)).$$

Thus $\exp(y\, g(z))$ is the semi-exponential generating function for $\{a_{nk}\}$ (the ordinary generating function for $a_{nk}/n!$). This proves the second claim. The first follows from setting $y = 1$. □

To use the exponential formula, one needs exponential generating functions to input.

Example 2.5.3 (egf for permutations) The number of permutations of $[n]$ is $n!$, so the exponential generating function for permutations is

$$f(z) = \sum_{n=0}^{\infty} \frac{n!}{n!} z^n = \frac{1}{1-z}. \tag{2.5.1}$$

Subtracting 1, the generating function for non-empty permutations is $z/(1-z)$. ■

Example 2.5.4 (egf for cycles) The fraction of all permutations of size n that consist of a single n-cycle is $1/n$. This follows, for instance, by computing recursively the probability that $\pi^k(1) = 1$ given $\pi^j(1) \neq 1$ for all $j \leq k-1$. The exponential generating function for non-empty n-cycles is therefore

$$g(z) = \sum_{n=1}^{\infty} \frac{1}{n} z^n = \log\left(\frac{1}{1-z}\right). \tag{2.5.2}$$

■

Example 2.5.5 (permutations by number of cycles) A permutation is the commuting product of its cycles. Thus the class $\mathcal{A}$ of permutations is the exponential of the class $\mathcal{B}$ of non-empty cycles. By the exponential formula, the relation $f = \exp(g(z))$ holds between them, which agrees with (2.5.1) and (2.5.2): $1/(1-z) = \exp(\log(1/(1-z)))$. Enumerating permutations by cycle, we get the exponential generating function

$$f(z) = \exp\left(y \log \frac{1}{1-z}\right).$$

We may write this compactly as $\frac{1}{(1-z)^y}$, although this has no content because we have no definition of the y-power of a series other than exponentiation of y times the logarithm. ■

The number of permutations of $[n]$ with k cycles is called a ***Stirling number of the first kind*** and is denoted variously as $\left[{n \atop k} \right]$, $(-1)^k s(n,k)$, $c(n,k)$ as well as other notations.

Example 2.5.6 (set partitions) The Stirling numbers of the second kind, denoted $\left\{ {n \atop k} \right\}$, count partitions of $[n]$ into k nonempty sets. To count partitions $\{S_\alpha : \alpha \in I\}$, set $|\mathcal{B}_n| = 1$ for each $n \geq 1$. The generating function for $\mathcal{B}$ is $\exp(z) - 1$, so the generating function for partitions by number of sets is

$$\exp(y(e^z - 1)).$$

In particular, the exponential generating function for all partitions is

$$e^{e^z - 1}.$$

■

Example 2.5.7 (partitions into ordered sets) Just as permutations are the exponential of the class of cycles, the exponential of the class of non-empty permutations is the class of partitions into ordered sets, i.e., collections of sequences $\{(x_{11}, \ldots, x_{1n_1}), \ldots, (x_{k1}, \ldots, x_{kn_k})\}$ where each element of $[n]$ appears exactly once as some $x_{i,j}$. Thus the semi-exponential generating function $F(y,z)$ for partitions into ordered sets by total size number and number of sets is given by exponentiating again

$$F(y,z) = \exp\left(y \frac{z}{1-z}\right). \tag{2.5.3}$$

■

Example 2.5.8 (involutions) An involution is a permutation whose square is the identity. Equivalently, all its cycles must be of length 1 or 2. Thus the class $\mathcal{A}$ of involutions is the exponential of the class $\mathcal{B}$ of cycles of length 1 or 2, enumerated by length. There is just one of each length, so the exponential generating function g of the class $\mathcal{B}$ class is $z + \frac{z^2}{2}$. Hence the exponential generating function f for the class of involutions is $\exp(z + z^2/2)$. ■

Example 2.5.9 (2-regular graphs) A 2-regular graph is a simple graph (no loops or multiple edges) in which every vertex has degree 2. A labeled 2-regular graph is the union of labeled, undirected cycles, whence the class of labeled 2-regular graphs is the exponential of the class of labeled undirected cycles. Let $\mathcal{A}$ denote this class. We do not allow parallel edges, so the cycles must have length at least 3. What is the number $a_n := |\mathcal{A}_n|$ of labeled 2-regular graphs on n vertices?

Every undirected cycle of length $n \geq 3$ corresponds to two directed cycles. Counting a permutation π as having weight $w = 2^{-N(\pi)}$ where $N(\pi)$ is the number of cycles, and letting p be the proportion of permutations having no short cycles (cycles of length less than 3), we see that $a_n = n!\, p\overline{w}$, where $\overline{w}$ is the average of w over permutations having no short cycles. It is known that $p = \Theta(1)$ and $N(\pi) \sim \log n$ for all but a vanishing proportion of permutations, so it would seem likely that $a_n/n! = \Theta(2^{-\log n}) = \Theta(n^{-\log 2})$. This gives a rigorous lower bound: by convexity of 2^{-x}, the average of 2^{-N} over permutations with no short cycles is at least $2^{-\overline{N}}$. It takes a generating function, however, to correct this to a sharp estimate.

Let $u(z)$ be the exponential generating function for undirected cycles of length at least 3. By (2.5.2),

$$u(z) = \frac{1}{2}\left(\log\frac{1}{1-z} - z - \frac{z^2}{2}\right).$$

Applying the exponential formula shows that the exponential generating function for labeled 2-regular graphs is

$$\frac{e^{-\frac{1}{2}z-\frac{1}{4}z^2}}{\sqrt{1-z}}.$$

Methods in Chapter 3 convert this quickly into a good estimate. ■

Notes

The transfer matrix method is very old. Two classical combinatorics texts in which it is discussed are Stanley (1997, Section 4.7) and Goulden and Jackson (2004, Chapter 2). The discussion of the kernel method borrows liberally from Bousquet-Mélou and Petkovšek (2000). The method itself, which appears to have been rediscovered several times, has been taken much further; see, for example Flatto and McKean (1977), Flatto and Hahn (1984), and Fayolle, Iasnogorodski, and Malyshev (1999) for some applications involving nontrivial amounts of number theory. The discussion of the exponential formula is inspired by Wilf (2006). There, the origin of the exponential formula is attributed to the doctoral work of Riddell (Riddell and Uhlenbeck, 1953), becoming greatly expanded by Bender and Goldman (1970/1971) and Foata and Schützenberger (1970).

Most of the proofs in Section 2.4 are taken from Stanley (1999). An earlier definition of P-recursiveness appeared in the literature, but it was discarded because of its failure to be equivalent to D-finiteness; counterexamples are given in Lipshitz (1989). Lipshitz's Theorem replaced two earlier proofs with gaps, found in Gessel (1981) and Zeilberger (1982). It also solved a problem of Stanley (1980, Question 4e).

Algorithms for finding a differential equation satisfied by an algebraic function go back, apparently, at least to Abel, and work is still continuing – see Bostan, Chyzak, Salvy, Lecerf, and Schost (2007) as a starting point.

One might consider a still larger class of generating function, namely the ***differentially algebraic*** functions, defined to be those that satisfy an equation

$$P(z, f, f', \ldots, f^{(m)}) = 0$$

for some $m > 0$ and some polynomial P. The question of possible behaviors of the coefficient sequence of such a function is wide open; some of the few known results are in Rubel (1983); and Rubel (1992).

In the other direction, the theory of hypergeometric sequences is well developed. These are the univariate formal power series satisfying a *first-order* linear recurrence with polynomial coefficients (alternatively, the ratio of successive coefficients a_{n+1}/a_n is some fixed rational function of n) and hence correspond to a special subclass of generating functions. A substantial algorithmic theory exists, well described at an elementary level in the book Petkovšek, Wilf, and Zeilberger (1996). There is a multivariate theory also – we recommend starting with recent works by Abramov and Petkovšek.

With regard to effective computing within each of these classes, a great deal is known about algebraic computations. Some of this is discussed in Section 6.1. For D-finite functions, there has been substantial recent progress, a very brief discussion of which is given in Section 6.3. Two good references are Chyzak and Salvy (1998) and Saito, Sturmfels, and Takayama (2000); see also Chyzak, Mishna, and Salvy (2005) for computing with symmetric D-finite functions. Hypergeometric functions are easier to deal with, and more precise results are possible. Several packages for commonly used computer algebra systems are available (we do not list them here because of the risk of giving outdated references). Little or nothing is known regarding effective computability in the class of differentially algebraic functions.

Exercises

2.1 (Counting domino tilings)

A *domino* or *dimer* is a union of two unit squares along a common edge. Let a_{nk} be the number of ways of placing k non-overlapping dominoes on a $2 \times n$ grid. Find the generating function for these numbers.

2.2 (Counting almost binary trees)

Define a class of "d-ary until the end" trees, by altering the definition in Example 2.3.1 so that each vertex must have at most d children *unless* all the children are leaves, in which case an arbitrary number is permitted. Adapt the argument from

Example 2.3.1 to compute a generating function G that counts these trees by the number of vertices.

2.3 (algebraic to D-finite conversion)
From the defining algebraic equation for the Catalan number generating function $C(z) = \sum_n C_n z^n$, derive a first-order linear differential equation with polynomial coefficients for C, and thence a first-order linear recurrence for C_n. Use this to deduce the explicit formula for C_n in terms of factorials.

2.4 (not a diagonal)
Let $F(x, y) = \sum_{r,s} a_{rs} x^r y^s$ be a bivariate formal power series. What is the difference between diag F and $F(x, x)$? Give a combinatorial interpretation of the latter.

2.5 ("diagonal" of slope 2)
Obtain the generating function for the next-simplest slice, $a_{2i,i}$, of the Delannoy numbers and compare the algebraic complexity to the diagonal generating function (2.4.4).

2.6 (generating function)
Let $p_0 = 1$ and define $\{p_N : N \geq 1\}$ recursively by

$$p_N = \frac{1}{3N+1} \sum_{j=2}^{N} p_{j-2} p_{N-j}$$

(the sum is empty when $N = 1$). This generating function from Limic and Pemantle (2004) gives the probability that a genome in a certain model due to Kaufmann and Levin cannot be improved by changing one allele. Find a differential equation satisfied by the generating function $f(z) := \sum_{N=0}^{\infty} p_N z^N$. Then use Maple (or its equivalent) to solve this Riccati equation explicitly in terms of Bessel functions. Among the solutions, find the only one i.e., analytic in a neighborhood of the origin.

2.7 (joint GF of left-right and right-left maxima)
A *left-to-right maximum* (respectively right-to-left maximum) of a permutation π of $1, \ldots, n$ is a position i for which $\pi_i > \pi_j$ for all $j < i$ (respectively all $j > i$). Derive an explicit formula for the bivariate (semi-exponential) generating function that enumerates permutations by length and number of left-to-right maxima. Then derive the trivariate generating function that also counts right-to-left maxima.

3
Univariate Asymptotics

Throughout this chapter, $f(z) = \sum_{n=0}^{\infty} a_n z^n$ is a univariate generating function for the sequence $\{a_n\}$.

3.1 Rational Functions: An Explicit Formula

For rational functions in one variable, estimation is not needed because there is an explicit formula for a_n. Some special cases form the basis for this. The first is obvious, and the second is easy to check by induction on k.

$$\text{If } f(z) = \frac{1}{1 - z/\rho} \quad \text{then } a_n = \rho^{-n} \tag{3.1.1}$$

$$\text{If } f(z) = \frac{1}{(1-z/\rho)^k} \quad \text{then } a_n = \binom{n+k-1}{k-1} \rho^{-n} \tag{3.1.2}$$

Now let $f = p(z)/q(z)$ be a rational function i.e., analytic at $z = 0$. We may and shall assume without loss of generality that p and q are relatively prime polynomials and that $q(0) = 1$.

Case 1: Distinct Roots. Suppose the roots $\rho_1, \ldots, \rho_t$ of q are distinct. Then

$$q(z) = \prod_{j=1}^{t} \left(1 - \frac{z}{\rho_j}\right).$$

Let $q_j(z) := q(z)/(1 - z/\rho_j)$. The ideal generated by all the q_j is all of $\mathbb{C}[z]$, hence any polynomial p may be written as $\sum p_j q_j$ for some polynomials p_j. This proves the ***partial fraction expansion***

$$f = \frac{p}{q} = \sum_{j=1}^{t} \frac{p_j q_j}{q} = \sum_{j=1}^{t} \frac{p_j}{1 - z/\rho_j}.$$

This may be written in the canonical form

$$f(z) = p_0(z) + \sum_{j=1}^{t} \frac{c_j}{1 - z/\rho_j},$$

where $\{c_j\}$ are constants that will shortly be evaluated (see (3.1.8)) as

$$c_j = -\frac{p(\rho_j)}{\rho_j \, q'(\rho_j)} \, . \tag{3.1.3}$$

By (3.1.1), for $n > \deg(p_0)$ we have $a_n = \sum_{j=1}^{t} c_j \rho_j^{-n}$. The leading term of this approximation is

$$a_n \sim -\frac{p(\rho_*)}{\rho_* \, q'(\rho_*)} \, \rho_*^{-n}$$

where ρ_* is the root of minimum modulus. If there are several roots of minimal modulus, $\rho_1, \ldots, \rho_\nu$, then the leading term is

$$a_n \sim \sum_{j=1}^{\nu} -\frac{p(\rho_j)}{\rho_j \, q'(\rho_j)} \rho_j^{-n} = |\rho_j|^{-n} \left(\sum_{j=1}^{\nu} -\frac{p(\rho_j)}{\rho_j \, q'(\rho_j)} \omega_j^n \right)$$

where $\omega_j = \rho_j / |\rho_j|$ is on the unit circle. In the special case in which the minimum modulus poles of f are on the unit circle, then if $\{a_n\}$ are real, ω_j are necessarily roots of unity, and hence the sequence $\{a_n\}$ is eventually periodic; conversely, every eventually periodic sequence can be obtained in this way.

When there is just one root of minimum modulus, the second term is exponentially smaller than the first:

$$a_n = c_* \rho_*^{-n} \left(1 + O\left(\left| \frac{\rho_*}{\rho_\dagger} \right|^n \right) \right)$$

where $\rho_\dagger$ is the next smallest root, and there is a polynomial correction if ρ_* is a multiple root. Although exponentially good estimates are the best one normally hopes for, it can be a nontrivial task to determine how close $|\rho_\dagger|$ is to ρ_* or even which root has the least modulus. To see how to make these determinations automatically, consult Gourdon and Salvy (1996).

Case 2: Repeated Roots. Let the root ρ_j have multiplicity m_j and let $q_j(z)$ now denote $q(z)/(1 - z/\rho_j)^{m_j}$. The same algebraic argument as before shows there is a partial fraction expansion, one canonical form of which is

$$f(z) = p_0(z) + \sum_{j=1}^{t} \frac{p_j(z)}{(1 - z/\rho_j)^{m_j}} ,$$

with p_0 a polynomial and p_j polynomials of degree at most $m_j - 1$ and not vanishing at ρ_j. One can further break down $p_j/(1 - z/\rho_j)^{m_j}$ as a sum $\sum_{i=0}^{m_j - 1} c_{ji}/(1 - z/\rho_j)^i$. By (3.1.2),

$$a_n = \sum_{j=1}^{t} \sum_{i=0}^{m_j} c_{ji} \binom{n+i-1}{i-1} \rho_j^{-n} .$$

The binomial coefficients, viewed as functions of n, are polynomials of degree $i-1$. The leading term(s) in this sum are the ones that minimize $|\rho_j|$ and, among those, maximize i. If there is only one maximum multiplicity root of minimum modulus, let the root be denoted ρ_* and the multiplicity i_*. The leading term becomes

$$a_n \sim c_* \, \rho_*^{-n} \binom{n+i_*-1}{i_*-1}, \tag{3.1.4}$$

where it will be shown shortly that

$$c_* = -\frac{p(\rho_*)}{\rho_* \, q_*(\rho_*)} . \tag{3.1.5}$$

Meromorphic Functions of One Variable

Let us compare the formal power series solution, which is complete but specialized, to an analytic solution. First note that the radius of convergence, R, of the power series for f is equal to the minimum modulus of a singularity for f, which we have denoted ρ_*. For any power series with radius of convergence R, a preliminary estimate is obtained by integrating $z^{-n-1} f(z)$ over a circle of radius $R - \varepsilon$:

$$\begin{aligned} |a_n| &= \left| \frac{1}{2\pi i} \int z^{-n-1} f(z) \, dz \right| \\ &\le (R-\varepsilon)^{-n} \sup_{|z|=R-\varepsilon} |f(z)| \end{aligned}$$

and therefore

$$\limsup_n \frac{1}{n} \log |a_n| \le -\log(R-\varepsilon)$$

for any $\varepsilon > 0$. If f is integrable on the circle of radius R, then the integrand is $O(R^{-n})$ and the contour has length $2\pi R$, so the estimate improves to

$$a_n = O(R^{-n}).$$

In the other direction, because there is a singularity of modulus R, we know the series does not converge for $|z| > R$, so $\log a_n \ge n(-\log R - \varepsilon)$ infinitely often. This establishes

$$\limsup \frac{1}{n} \log |a_n| = -\log R . \tag{3.1.6}$$

Thus we get the correct exponential rate, at least for the limsup, with no work at all.

Next, use Cauchy's integral formula to write

$$a_n = \frac{1}{2\pi i} \int_C \frac{dz}{z^{n+1}} f(z),$$

where C is any contour enclosing the origin and contained in the domain of convergence of f. Let C be a circle of radius $r < \rho_*$, and let C' be a circle of radius $R > \rho_*$. Assume that ρ_* is the only root of q of minimum modulus and that the moduli of other roots are greater than R. We then have, by the residue theorem,

$$\int_C \frac{dz}{z^{n+1}} f(z) - \int_{C'} \frac{dz}{z^{n+1}} f(z) = -2\pi i \operatorname{Res}(z^{-n-1} f(z); \rho_*). \tag{3.1.7}$$

It is conceivable the reader may have gotten this far but never seen residues, in which case the text by Conway (1978) is recommended, and it can meanwhile be taken on faith that the residue at a simple pole and pole of order $k > 1$, respectively, are defined by

$$\operatorname{Res}(g; r) = \lim_{z \to r} (z - r) g \,,$$

$$\operatorname{Res}(g; r) = \frac{1}{(k-1)!} \frac{d^{k-1}}{dz^{k-1}} ((z-r)^k g) \,.$$

If $f = p/q$ has a simple pole at ρ_* then the residue is just $\rho_*^{-n-1} p(\rho_*)/q'(\rho_*)$. The integral over C' is bounded by $2\pi R^{-n} \sup_{|z|=R} |f(z)|$ and is therefore exponentially smaller than the residue. Thus the leading term asymptotic for a_n is

$$a_n = -\rho_*^{-n} \frac{p(\rho_*)}{\rho_* q'(\rho_*)} + O(R^{-n}). \tag{3.1.8}$$

In fact, we may send C' to infinity, thus picking up all the terms. This will be a sum of terms $-\rho_i^{-n-1} p(\rho_i)/q'(\rho_i)$. This makes good on the promise to prove (3.1.3), showing also that even when one can work everything out algebraically, an analytic approach may still add something.

If there is more than one root of minimum modulus, one may simply sum the contributions. If the root ρ_j appears with multiplicity m_j, then the residue at ρ_j comes out to be

$$-\binom{n + m_j - 1}{m_j - 1} \rho_j^{-n} \frac{p(\rho_j)}{\rho_j q_j(\rho_j)} + O\left(n^{d-2} \rho_j^{-n}\right) .$$

The lower order terms are easy to compute, although the expressions are longer because of the higher order derivatives required.

Aside from providing a shortcut to the constants, the analytic approach has the advantage of generality. The partial fractions approach required that $f = p/q$ be a rational function. The residue computation gives an answer whenever f is ***meromorphic*** in a disk of radius greater than $|\rho_*|$, meaning that it is the quotient of analytic functions. In other words, if $f(z) = p(z)/q(z)$ with p and q analytic on a disk of radius R and q vanishing at some point a inside the disk, then the previous estimates are still valid: $a_n = a^{-n} p(a)/q'(a) + O(R - \varepsilon)^{-n}$ if a is a simple pole and $a_n = c n^{m-1} a^{-n} + O(n^{m-2} a^{-n})$ if q has a root of multiplicity $m > 1$ at a.

3.2 Saddle Point Methods

One of the crowning achievements of elementary complex analysis is development of techniques to evaluate integrals by deforming the contour of integration. Much of this can be grouped together as "saddle point methods" aimed at discovering the best deformation. In several variables, topology comes into play, and this forms the content of Chapter 8, which is at the heart of this book. To prepare for this, and because it is useful in itself, a tutorial in univariate saddle point integration is given in this section.

When the modulus of an integrand falls steeply on either side of its maximum, most of the contribution to the integral comes from a small interval about the maximum. If that were so, then multiplying the integrand by the length of the interval where the modulus is near its maximum (or doing something slightly more fancy) would give an easy estimate. Most contours, however, do not have this property. To see this, note first that this estimate cannot hold if the contour can be deformed so as to decrease the maximum modulus of the integrand, because then the integral would be less than the claimed estimate. Let γ be a contour and denote the logarithm of the integrand by I. At a point z_0 where the modulus of the integrand is maximized, $\text{Re}\{I'\}$ vanishes along γ. Generically, $\text{Im}\{I'(z_0)\}$ will not vanish along γ. By the Cauchy-Riemann equations, $\text{Re}\{I'\}$ in the direction perpendicular to the contour is equal to $\text{Im}\{I'\}$ along γ. When this does not vanish, γ may be locally perturbed, fixing the endpoints but pushing the center in the direction of increasing $\text{Re}\{I\}$, thereby decreasing the maximum modulus $|e^I|$ of the integrand on the contour. In other words, if the modulus of the integrand is maximized on γ at z_0 and this maximum cannot be reduced by perturbing γ, then both the real and imaginary part of I' must vanish at z_0; hence z_0 is a critical point for I.

The univariate saddle point method consists of the following steps.

(i) Locate the zeros of I' (a discrete set of points).
(ii) Determine whether the contour can be deformed to minimize $\text{Re}\{I\}$ at such a point.
(iii) Estimate the integral via a Taylor series development of the integrand.

In Chapter 4, we see that for families parametrized by λ of integrals

$$\int A(z)\exp(-\lambda\phi(z))$$

(into which form the Cauchy integral may be put), one may often get away with approximating the critical point $z_0(\lambda)$ by the critical point z_0 for ϕ, ignoring A and removing the dependence of z_0 on λ. This approximation is often good enough to provide a complete asymptotic approximate expansion of the integral.

Here, we consider cases where this does not work but where we can deal directly with $z_0(\lambda)$. For the second step not to fail, either f must be entire, or the saddle point (where I_n' vanishes) must be in the interior of the domain of convergence of f. This is not asking too much, and the method is widely applicable. We are not sure when Cauchy's formula was first combined with contour integration methods to estimate power series coefficients. One seminal work by Hayman (1956) defines a broad class of functions, called ***admissible functions***, for which the saddle point method can be shown to work and the Gaussian approximation mechanized. The title of Hayman's publication refers to the fact that when one takes $f(z) = e^z$, one recovers Stirling's approximation to $n!$.

Examples of Hayman's Method

The next few examples apply Hayman's methods to univariate generating functions derived in Chapter 2. They all rely on the estimate

$$\int_\gamma A(z)\exp(-\lambda\phi(z))\,dz \sim A(z_0)\sqrt{\frac{2\pi}{\phi''(z_0)\lambda}}\exp(-\lambda\phi(z_0)), \qquad (3.2.1)$$

which holds when A and ϕ are smooth and $\mathrm{Re}\{\phi\}$ is minimized in the interior of γ at a point z_0 where ϕ'' does not vanish. (In the notation, A is for "amplitude" and ϕ is for "phase.") This is generalized and proved in Theorem 4.1.1. However, to show that it is elementary, a direct verification is provided for the first example.

Example 3.2.1 (ordered-set partitions: an isolated essential singularity) Evaluate (2.5.3) at $y = 1$ to obtain the exponential generating function

$$f(z) = \exp\left(\frac{z}{1-z}\right)$$

for the number a_n of partitions of $[n]$ into ordered sets. By Cauchy's formula (1.2.1),

$$\frac{a_n}{n!} = \frac{1}{2\pi i}\int z^{-n-1}\exp\left(\frac{z}{1-z}\right)dz\,.$$

We show that

$$a_n \sim n!\sqrt{\frac{1}{4\pi e}}n^{-3/4}\exp(2\sqrt{n}).$$

To estimate this, we find the critical point z_n. Denote the logarithm of the integrand by $I_n(z) := -(n+1)\log z + \dfrac{z}{1-z}$ and compute the derivative

$$I_n' = \frac{-n-1}{z} + \frac{1}{(1-z)^2}\,.$$

The root closest to the origin is $1-\beta_n$, where $\beta_n = n^{-1/2} + O(n^{-1})$. The only singularities of the integrand are at $z = 0, 1$, so we may expand the contour to a circle passing through $1-\beta_n$. It is a little more convenient to work in the Riemann sphere (there is no singularity at infinity), so that we may deform the contour into the line $z = 1 - \beta_n + it$, $t \in \mathbb{R}$. Denote this contour by γ.

The hope now is that the integral is well approximated by integrating the degree-two Taylor approximation of I_n. Specifically, we hope that (cf. (3.2.1) with $\lambda = n + 1$ and $\phi = -I_n(1 - \beta_n)$)

$$\frac{1}{2\pi i}\int_{-\infty}^{\infty} \exp(I_n(1 - \beta_n + it))\,(i\,dt) \tag{3.2.2}$$

$$\approx \frac{1}{2\pi}\int_{-\infty}^{\infty} \exp[I_n(1 - \beta_n) + \frac{1}{2}I_n''(1 - \beta_n)(it)^2]\,dt \tag{3.2.3}$$

$$= \sqrt{\frac{1}{2\pi I_n''(1 - \beta_n)}}\exp(I_n(1 - \beta_n)). \tag{3.2.4}$$

This hope is easily verified as follows. We compute

$$\begin{aligned} I_n''(1 - \beta_n) &= \frac{n+1}{(1 - \beta_n)^2} + \frac{2}{\beta_n^3} \\ &= (2 + o(1))n^{3/2}. \end{aligned}$$

This tells us that the main contribution to (3.2.2) should come from the region where $|t|$ is not much larger than $n^{-3/4}$. Accordingly, we pick a cutoff a little greater than that, say $L = 2n^{-3/4}\log n$, and break the integrals (3.2.2) and (3.2.3) into two parts, $|t| \le L$ and $|t| > L$. Up to the cutoff the two integrals are close, and past the cutoff they are both small.

More precisely, we define

$$M_1 := \int_{|t|\ge L} \left|\exp\left[I_n(1 - \beta_n) + \frac{1}{2}I_n''(1 - \beta_n)(it)^2\right]\right| dt$$

$$M_2 := \int_{|t|\ge n^{-1/2}} |\exp(I_n(1 - \beta_n + it))|\,dt$$

$$M_3 := \int_{n^{-1/2} > |t| \ge L} |\exp(I_n(1 - \beta_n + it))|\,dt$$

$$M_4 := \int_{|t|<L} \left|\exp[I_n(1 - \beta_n) + \frac{1}{2}I_n''(1 - \beta_n)(it)^2] - \exp(I_n(1 - \beta_n + it))\right| dt$$

so that M_1 is the integral in (3.2.3) beyond L, whereas M_2 and M_3 together are the integral in (3.2.2) beyond L, and M_4 is the difference between (3.2.2) and (3.2.3) on $[-L, L]$. Letting $M := \exp(I_n(1 - \beta_n))$, we will prove an upper bound of $M \cdot \exp(-c(\log n)^2)$ on M_1, M_2, and M_3 and will show also that $M_4 = o(Mn^{-3/4})$. Because $Mn^{-3/4}$ is the order of magnitude of the integral in (3.2.4), this suffices to show that a_n is asymptotic to the quantity in (3.2.4).

The bound on M_1 is a standard Gaussian tail estimate. The standard deviation is $I''(1-\beta_n)^{-1/2} = (2^{-1/2} + o(1))n^{-3/4}$. The region $|t| \geq L$ is therefore $(\sqrt{8} + o(1)) \log n$ standard deviations in the tail. A two-sided tail of t standard deviations is $O(e^{-t^2/2})$ times the whole integral, which proves the asserted upper bound on M_1 for any $c < 8$.

The contribution to (3.2.2) when $|t| > L$ has been broken into two parts, M_2 and M_3. To bound M_3, we pull out the factor of M:

$$M_3 \leq M \int_{L<|t|<n^{-1/2}} \exp\left(\mathrm{Re}\left\{I_n(1-\beta_n+it) - I_n(1-\beta_n)\right\}\right) dt\,.$$

The real part of $-(n+1)\log(1-\beta_n+it)$ is maximized at $t = 0$, whence

$$\mathrm{Re}\left\{I_n(1-\beta_n+it) - I_n(1-\beta_n)\right\} \leq \mathrm{Re}\left\{\frac{1-\beta_n+it}{\beta_n - it} - \frac{1-\beta_n}{\beta_n}\right\}.$$

Thus

$$\begin{aligned} M_3 &\leq M \cdot \int_{L<|t|<n^{-1/2}} \exp\left(\mathrm{Re}\left\{\frac{1-\beta_n+it}{\beta_n-it} - \frac{1-\beta_n}{\beta_n}\right\}\right) dt \\ &= M \cdot \int_{L<|t|<n^{-1/2}} \exp\left(\frac{-t^2}{\beta_n^3 + \beta_n t^2}\right) \\ &\leq M \cdot \int_{L<|t|<n^{-1/2}} \exp\left(-\frac{t^2}{2\beta_n^3}\right) dt \end{aligned}$$

because the β_n^3 term is the greatest term in the denominator when $t < n^{-1/2}$. Plugging in $\beta_n \sim n^{-1/2}$ and $L = 2n^{-3/4}\log n$ proves the desired upper bound on M_3 for any $c < 2$.

To bound M_2, observe first that when $|t| \geq n^{-1/2}$, the exponent $-t^2/(\beta_n^3 + \beta_n t^2)$ decreases to $-\beta_n^{-1} \sim -\sqrt{n}$. This is small, but integrating it over the unbounded region $[n^{-1/2}, \infty]$ requires some further damping. We obtain this from the $-(n+1)\log z$ term that we previously ignored:

$$\begin{aligned} \frac{|\exp(I_n(1-\beta_n+it))|}{\exp(I_n(1-\beta_n))} &\leq \frac{|1-\beta_n|^n}{|1-\beta_n+it|^n} \exp\left(\mathrm{Re}\left\{\frac{1-\beta_n+it}{\beta_n-it} - \frac{1-\beta_n}{\beta_n}\right\}\right) \\ &\leq (1+t^2)^{-n/2} \exp\left(-(1+o(1))n^{1/2}\right). \end{aligned}$$

Here, the bound of $(1+t^2)^{-n/2}$ on the z^{-n-1} term follows from the fact that $1-\beta_n < 1$ and that $|x/(x+it)|$ is increasing in $x \geq 0$. Integrating from $t = n^{-1/2}$ to ∞, the factor $(1+t^2)^{-n/2}$ integrates to $o(1)$. This implies that

$$\frac{M_2}{M} \leq \exp\left(-(1+o(1))n^{-1/2}\right)$$

which is certainly $o(\exp -c(\log n)^2)$.

Finally, for M_4, we use the Taylor approximation

$$\left| I_n(1-\beta_n+it) - I_n(1-\beta_n) + \frac{1}{2}t^2 I_n''(1-\beta_n) \right| \le \frac{1}{6}t^3 \sup_{|s|\le L} |I_n'''(s)|.$$

Differentiating $I' = -(n_1)/z + 1/(1-z)^2$ twice, we find that $I'''(z) \sim 6/(1-z)^4$ near $z=1$ and hence that the right-hand side is bounded by $(1+o(1))t^3n^2 = (8+o(1))n^{-1/4}\log^3 n$. Whenever $f > 0$ and $g \le \varepsilon f$, it follows that $\int g \le \varepsilon \int f$. In particular, because the integrand of (3.2.3) is everywhere positive, this implies that $M_4 \le cn^{-1/4}\log^3 n$ times the common value of the integrals in (3.2.3) and (3.2.4), as desired.

Having established that (3.2.4) is the leading term, we now compute it. Using the formula for $I_n''(z)$ and the formula

$$\beta_n = n^{-1/2} - \frac{1}{2}n^{-1} + O(n^{-3/2}),$$

we get

$$\begin{aligned}
&\sqrt{\frac{1}{2\pi I_n''(1-\beta_n)}}\exp(I_n(1-\beta_n)) \\
&= (1+o(1))\sqrt{\frac{1}{4\pi n^{3/2}}}\exp\left(-(n+1)\log(1-\beta_n) - 1 + \frac{1}{\beta_n}\right) \\
&= (1+o(1))\sqrt{\frac{1}{4\pi n^{3/2}}}\exp\Bigg(- (n+1)(-n^{-1/2} + O(n^{-3/2})) \\
&\qquad\qquad -1 + n^{1/2} + \frac{1}{2} + O(n^{-1/2})\Bigg) \\
&= (1+o(1))\sqrt{\frac{1}{4\pi e}}n^{-3/4}\exp(2\sqrt{n}).
\end{aligned}$$

Note that computing the full asymptotic development is almost as easy. The cutoff is calibrated so that the remainder after k terms of the Taylor expansion is always small on $|t| \le L$, and essentially the same computation suffices to derive an asymptotic series. ■

Example 3.2.2 (involutions: an entire function) Let $f(z) = \exp(z + z^2/2)$ be the exponential generating function for the number a_n of involutions in the permutations group S_n, as in Example 2.5.8. This is an entire function, so we apply Hayman's method. Denote

$$I_n(z) = \log(f(z)z^{-n-1}) = z + \frac{z^2}{2} - (n+1)\log z.$$

Setting the derivative equal to zero gives the quadratic $z^2 + z - (n+1) = 0$. The roots are $-\frac{1}{2} \pm \sqrt{n + \frac{5}{4}}$. The coefficients a_n are positive, whereas $\exp(I_n(z))$ alternates in sign near the negative root so a_n cannot be approximated by the integrand near the negative root, and we therefore try taking $z_0 = \sqrt{n + \frac{5}{4}} - \frac{1}{2}$.

Let γ be the circle centered at the origin through z_0. It is easy to verify that real part of I_n on γ is maximized at z_0. The estimate

$$[z^n]f(z) = \frac{1}{2\pi i} \int_\gamma \exp(I_n(z_n))\, dz \sim \exp(I_n(z_0)) \sqrt{\frac{1}{2\pi \; I_n''(z_0)}}$$

is justified the same way as in the previous example. Using the approximations

$$\begin{aligned}
z_0 &= n^{1/2} - \frac{1}{2} + \frac{5}{8} n^{-1/2} + O(n^{-3/2}) \\
\frac{z_0^2}{2} &= \frac{1}{2} n - \frac{1}{2} n^{1/2} + \frac{3}{4} + o(1) \\
\log(z_0) &= \frac{1}{2} \log n - \frac{1}{2} n^{-1/2} + \frac{1}{2} n^{-1} + O(n^{-3/2}) \\
I_n(z_0) &= -\frac{1}{2} n \log n + \frac{1}{2} n + n^{1/2} - \frac{1}{2} \log n - \frac{1}{4} + o(1) \\
I_n''(z_0) &= 2 + o(1)
\end{aligned}$$

and using Stirling's formula to approximate $n!$, we find that

$$\begin{aligned}
a_n &= n!\,[z^n]f(z) \\
&\sim \exp(n \log n - n) \sqrt{2\pi n}\, \exp(I_n(z_0)) \sqrt{\frac{1}{2\pi \; I_n''(z_0)}} \\
&\sim \exp(n \log n - n)\, n^{1/2} \exp\left(-\frac{1}{2} n \log n + \frac{1}{2} n^{1/2} - \frac{1}{2} \log n - \frac{1}{2} \right) 2^{-1/2} \\
&= \exp\left(\frac{1}{2} n \log n - \frac{1}{2} n + n^{1/2} - \frac{1}{2} \log 2 - \frac{1}{4} + o(1) \right).
\end{aligned}$$

Evidently, a_n is near $\sqrt{n!}$. Pulling out the $\sqrt{n!}$ yields the slightly more transparent

$$a_n \sim \sqrt{n!} e^{\sqrt{n}} (8\pi e n)^{-1/4}.$$

■

3.3 Circle Methods

When f has a branch singularity, for example, a logarithm or non-integral power, there is often no way to maneuver the contour through a saddle. As we have seen when deriving the crude estimate (3.1.6), pushing the contour to the boundary of the disk of convergence will give some improvement. This leads to ***circle methods***, such as Darboux's Theorem. The following preliminary estimate is well known to harmonic analysts.

Lemma 3.3.1 *Suppose a complex-valued function f on the circle γ of radius R is k times continuously differentiable for some integer $k \geq 0$. Then*

$$\int_\gamma z^{-n-1} f(z)\, dz = O(n^{-k} R^{-n})$$

as $n \to \infty$.

Proof Replacing f by $f(z/R)$, we may assume without loss of generality that $R = 1$. Integrating by parts,

$$\int_\gamma z^{-n} f(z)\, dz = \int_\gamma \frac{1}{n-1} z^{1-n} f'(z)\, dz,$$

with the $\int_\gamma \frac{z^{1-n}}{1-n} f(z)\, dz$ term dropping out because γ has no boundary. By induction on k,

$$\int_\gamma z^{-n} f(z)\, dz = \frac{1}{k!\binom{n-1}{k}} \int_\gamma z^{k-n} f^{(k)}(z)\, dz.$$

Because $f^{(k)}$ is continuous, it is bounded on the unit circle, so this last integral is bounded independently of n, and the lemma follows from $k!\binom{n-1}{k} \sim n^k$. □

The following version of Darboux's Theorem may be found in Henrici (1991, Theorem 11.10b).

Theorem 3.3.2 (Darboux) *Suppose that $f(z) = (1 - z/R)^\alpha \psi(z)$ for some $R > 0$, some $\alpha \notin \mathbb{N}$, and some function ψ with radius of convergence greater than R. Denote the coefficients of ψ expanded about R by $\psi(z) = \sum_{n=0}^{\infty} b_n (R - z)^n$. Then the coefficients $\{a_n\}$ of f have asymptotic expansion*

$$a_n \sim R^{-n} \sum_{k=0}^{\infty} c_k n^{-\alpha-1-k}.$$

The coefficients c_k are given by explicit linear combinations of $b_0, \ldots, b_k$, and the leading term is

$$a_n \sim \frac{\psi(R)}{\Gamma(-\alpha)} n^{-\alpha-1} R^{-n}.$$

Proof Again, assume without loss of generality that $R = 1$. Begin by recalling some elementary facts about the power series $(1 - z)^\alpha$. Its coefficients are the formal binomial coefficients $(-1)^n \binom{\alpha}{n}$ defined by

$$\binom{x}{n} := \prod_{j=1}^{n} \frac{x - j + 1}{j}.$$

As a function of n, these coefficients are asymptotically approximated by

$$(-1)^n \binom{\alpha}{n} \sim \frac{1}{\Gamma(-\alpha)} n^{-\alpha-1}. \qquad (3.3.1)$$

Furthermore, an asymptotic series for $\binom{\alpha}{n}$ in terms of decreasing powers $n^{-\alpha-1-k}$ is known. Therefore, a triangular linear map converts an asymptotic series $a_n \sim \sum_{k=0}^{\infty} c'_k (-1)^n \binom{\alpha+k}{n}$ into a series $a_n \sim \sum_{k=0}^{\infty} c_k n^{-\alpha-1-k}$, with $c_0 = c'_0/\Gamma(-\alpha)$.

To find the series in c'_k, let m be an integer greater than $\text{Re}\{-\alpha\}^+$, and let ψ_m be the mth remainder term in the Taylor series for ψ:

$$(1-z)^m \psi_m(z) = \psi(z) - \sum_{k=0}^{m-1} b_k (1-z)^k.$$

Multiplying by $(1-z)^\alpha$ yields

$$f(z) - \sum_{k=0}^{m-1} b_k (1-z)^{\alpha+k} = (1-z)^{\alpha+m} \psi_m(z)$$

on the open unit disk, and hence as formal power series. Taking the z^n-coefficient on both sides yields

$$a_n - \sum_{k=0}^{m-1} b_k (-1)^n \binom{\alpha+k}{n} = [z^n](1-z)^{\alpha+m} \psi_m(z). \qquad (3.3.2)$$

This proves the desired expansion, provided that the right-hand side of (3.3.2) is $o(n^{-\alpha-m})$.

By assumption, $\alpha + m \geq 0$, so the function $(1-z)^{\alpha+m}\psi_m$ is $\lfloor \alpha + m \rfloor$ times continuously differentiable on the unit circle, which implies that the right-hand side of (3.3.2) is $O(n^{-\lfloor \alpha+m \rfloor})$. This is not quite small enough, but replacing m by $m+1$ adds a term known to be $O(n^{-\alpha-m-1})$ to the left-hand side of (3.3.2) while reducing the right-hand side to $O(n^{-\lfloor \alpha+m+1 \rfloor})$, which is good enough. □

Example 3.3.3 (2-regular graphs: an algebraic singularity) Let $f(z) = e^{-z/2-z^2/4}/\sqrt{1-z}$ be the exponential generating function for 2-regular graphs that was derived in Example 2.5.9. Apply Darboux's Theorem with $R = 1, \alpha = -1/2$, and $\psi = \exp(-z/2 - z^2/4)$. Then $\psi(0) = e^{-3/4}$, $\Gamma(-\alpha) = \sqrt{\pi}$ and the number a_n of 2-regular graphs on n labeled vertices is estimated by

$$a_n \sim n! \frac{e^{-3/4}}{\sqrt{\pi n}}.$$

■

3.4 Transfer Theorems

A closer look at the proof of Darboux's Theorem shows that one can do better. Analyticity of $f/(R-z)^\alpha$ beyond the disk of radius R was used only to provide a development of f in decreasing powers of $R - z$. Also, a sharper estimate than Lemma 3.3.1 will allow us to obtain a sufficiently good estimate on the remainder term without replacing m by $m+1$, which is crucial if the hypotheses are weakened to a finite asymptotic expansion.

There are various results along these lines, our favorite among which are the ***transfer theorems*** of Flajolet and Odlyzko (1990). Their idea was that the estimates could be improved to $a_n = O(n^{-\alpha-1})$ for the coefficients of *any* power series $f(z)$ i.e., $O(1-z)^\alpha$. They reduced the necessary domain of analyticity of f as far as possible (to a neighborhood of the unit disk in the slit plane), and while they were at it they generalized the scope beyond powers to other branch singularities.

To state the main theorem of Flajolet and Odlyzko (1990), we define `alg-log` to be the class of functions that are a product of a power of $R-z$, a power of $\log(1/(R-z))$, and a power of $\log\log(1/(R-z))$. Analogously to (3.3.1), one begins with a description of asymptotics for all functions in the class `alg-log`. We refer to Flajolet and Odlyzko (1990) for the proof of the following lemma.

Proposition 3.4.1 (Flajolet and Odlyzko, 1990, Theorem 3B) *Let α, γ, and δ be any complex numbers other than nonnegative integers, and let*

$$f(z) = (1-z)^\alpha \left(\frac{1}{z}\log\frac{1}{1-z}\right)^\gamma \left(\frac{1}{z}\log\left(\frac{1}{z}\log\frac{1}{1-z}\right)\right)^\delta .$$

Then the Taylor coefficients $\{a_n\}$ of f satisfy

$$a_n \sim \frac{n^{-\alpha-1}}{\Gamma(-\alpha)}(\log n)^\gamma (\log\log n)^\delta .$$

Remark When α, γ, or δ is a nonnegative integer, different formulae hold. For example, for the case $\alpha \in \mathbb{N}$, $\gamma \notin \mathbb{N}$, $\delta = 0$, the estimate

$$a_n \sim Cn^{-\alpha-1}(\log n)^{\gamma-1} \tag{3.4.1}$$

is known: the coincidence of α with a nonnegative integer causes an extra log in the denominator.

Given a positive real R and an $\varepsilon \in (0, \pi/2)$, the so-called ***Camembert-shaped region***,

$$\{z : |z| < R+\varepsilon, z \neq R, |\arg(z-R)| \geq \pi/2 - \varepsilon\},$$

denoted $\Delta(R, \varepsilon)$, is shown in Figure 3.1.

Theorem 3.4.2 (transfer Theorem) *Let $f(z) = \sum_{n=0}^\infty a_n z^n$ be analytic in a Camembert-shaped region $\Delta(R, \varepsilon)$. If $g(z) = \sum_{n=0}^\infty b_n z^n \in$ `alg-log`, then the following hold.*

(i) $f(z) = O(g(z)) \Rightarrow a_n = O(b_n)$;
(ii) $f(z) = o(g(z)) \Rightarrow a_n = o(b_n)$;
(iii) $f(z) \sim g(z) \Rightarrow a_n \sim b_n$

In particular, when $f(z) \sim C(R-z)^\alpha$, this result subsumes Theorem 3.3.2.

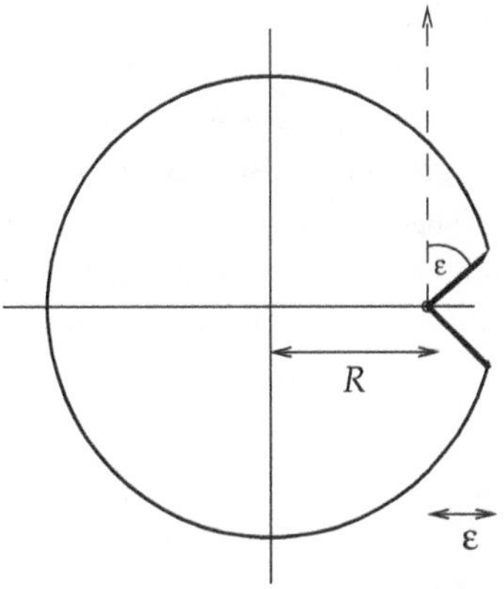

Figure 3.1 A Camembert-shaped region.

So as not to devote too much space to computation, we only prove this theorem with the class `alg-log` replaced by the class of powers $(1-z)^\alpha$; this result still greatly improves Darboux's Theorem.

Proof for $g(z) = (1-z)^\alpha$ Assume without loss of generality that $R = 1$. Note that for $g = (1-z)^\alpha$, the nth coefficient is of order $n^{-\alpha-1}$. Next, note that the assumption that $f(z) = O((1-z)^\alpha)$ near $z = 1$ implies (using only continuity, not analyticity) that for some K, $|f(z)| \le K|1-z|^\alpha$ everywhere on $\Delta(R, \varepsilon)$. The contour of integration in Cauchy's formula will be a contour γ constructed as the union of four pieces (see Figure 3.2). Let γ_1 be the circular arc parameterized by $1 + n^{-1}e^{it}$ for $\xi \le t \le 2\pi - \xi$. Let γ_2 be the line segment between $1 + n^{-1}e^{i\xi}$ and the number β of modulus $1 + \eta$ and $\arg(\beta - 1) = \xi$. Let γ_3 be the arc on the circle of radius $1 + \eta$ running between β and $\overline{\beta}$ the long way, and let γ_4 be the conjugate of γ_2. We will bound the absolute value of the integral on each segment separately, so we need not worry about the orientations. The value of η is chosen to be less than $R - 1$, and $\xi < \pi/2$ is chosen as large as is necessary to make γ contained in the Camembert region.

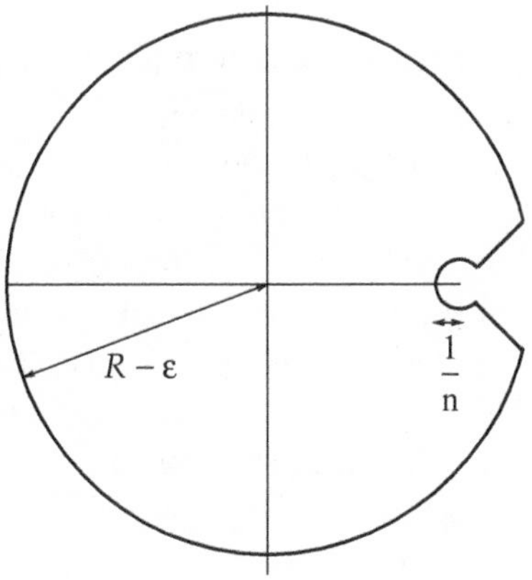

Figure 3.2 The contour γ.

On γ_1, the modulus of f is at most $Kn^{-\alpha}$, the modulus of z^{-n-1} is at most $(1 - n^{-1})^{-n-1} \le 2e$, and the integral of $|dz|$ is at most $2\pi n^{-1}$, leading to a contribution of size at most $6Kn^{-\alpha-1}$. On γ_3 the z^{-n-1} factor in the integral reduces the modulus to at most $C(\eta)(1+\eta)^{-n}$, which is of course $O(n^{-N})$ for any N. Because the order of the nth coefficient of g is $n^{-\alpha-1}$, we are in good shape so far.

By symmetry, we need now only do the computation for γ_2. Set $\omega = e^{i\xi}$ and parametrize the integral as $z = 1 + (\omega/n)t$ for $t = 1$ to En for a constant, $E = |\beta - 1|$. We have $|f(z)| \le K|z-1|^\alpha = K(t/n)^\alpha$ and

$$|z^{-n-1}| = \left|1 + \frac{\omega t}{n}\right|^{-n-1}$$

so

$$\int_{\gamma_2} |f(z)||z^{-n-1}||dz| \le \int_1^{En} K\left(\frac{t}{n}\right)^\alpha \left|1 + \frac{\omega t}{n}\right|^{-n-1} \frac{dt}{n}$$

$$\le Kn^{-\alpha-1} \int_1^\infty t^\alpha \left|1 + \frac{\omega t}{n}\right|^{-n-1} dt. \qquad (3.4.2)$$

We need to see that the integral in (3.4.2) is bounded above for sufficiently large n. The bound $|1 + \omega t/n| \ge 1 + \mathrm{Re}\{\omega t/n\} = 1 + (t/n)\cos(\xi)$ implies an upper bound of

$$J_n := \int_1^\infty t^\alpha \left(1 + \frac{t\cos(\xi)}{n}\right)^{-n} dt$$

for the integral in (3.4.2). The integrand is monotone decreasing in n, and clearly finite for $n > 1 + \alpha^+$, so the decreasing limit is

$$J := \lim_{n\to\infty} J_n = \int_1^\infty t^\alpha e^{-t\cos(\xi)}\, dt,$$

which is finite. We have now bounded all four integrals by multiples of $n^{-\alpha-1}$, so the proof of the first assertion is complete.

The proof of the second assertion is contained in here as well. When $|f| \le Kg$, then the integral over γ_1 is bounded above by $6Kn^{-\alpha-1}$, the integral over γ_3 is $o(n^{-\alpha-1})$, and the integral over γ_2 is bounded by JK. Furthermore, the contributions to each of these four integrals from parts of γ at distance greater than δ from 1 are $o(n^{-\alpha-1})$ for any fixed $\varepsilon > 0$. If $f(z) = o(g(z))$ at $z = 1$, then for any $\varepsilon > 0$, there is a δ such that $|f(z)| \le \varepsilon|g(z)|$ when $|1 - z| \le \delta$. It follows that $a_n \le (2J + 6 + o(1))\varepsilon n^{-\alpha-1}$. This is true for every $\varepsilon > 0$, whence $a_n = o(n^{-\alpha-1})$.

Finally, the third assertion is an immediate consequence of the first two assertions. □

Example 3.4.3 (Catalan numbers) Let $a_n := \frac{1}{n+1}\binom{2n}{n}$ be the nth Catalan number. The generating function for these was shown in Example 2.3.1 to be given by

$$f(z) := \sum_{n=0}^{\infty} a_n z^n = \frac{1-\sqrt{1-4z}}{2z} = \frac{1-2\sqrt{\frac{1}{4}-z}}{2z}.$$

There is an algebraic singularity at $r = 1/4$, near which the asymptotic expansion for f begins

$$f(z) = 2 - 4\sqrt{\frac{1}{4}-z} + 8\left(\frac{1}{4}-z\right) - 16\left(\frac{1}{4}-z\right)^{3/2} + O\left(\frac{1}{4}-z\right)^2.$$

Note that $f/\sqrt{1/4-z}$ is not analytic in any disk of radius $1/4+\varepsilon$, because both integral and half-integral powers appear in f, but f is analytic in a Camembert-shaped region. Theorem 3.4.2 thus gives (note that the integral powers of $[1-z]$ do not contribute):

$$\begin{aligned} a_n &\sim \left(\frac{1}{4}\right)^{1/2-n} n^{-3/2}\frac{-4}{\Gamma(-1/2)} + \left(\frac{1}{4}\right)^{3/2-n} n^{-5/2}\frac{-16}{\Gamma(-3/2)} + O(n^{-7/2}) \\ &= 4^n n^{-3/2}\frac{(-4)(\frac{1}{4})^{1/2}}{\Gamma(-1/2)} + 4^n n^{-5/2}\frac{(-16)(\frac{1}{4})^{3/2}}{\Gamma(-3/2)} + O(n^{-7/2}) \\ &= 4^n\left(\frac{n^{-3/2}}{\sqrt{\pi}} - n^{-5/2}\frac{3}{2\sqrt{\pi}} + O(n^{-7/2})\right). \end{aligned}$$

■

Example 3.4.4 (branching random walk: logarithmic singularity) For an example including a logarithmic term, recall from Example 2.2.12 the implicit equation

$$\phi(z) = [(1-p)z + p\phi(\phi(z))]^2.$$

This characterizes the probability generating function for the number, X, of particles to reach the origin in a binary branching nearest-neighbor random walk with absorption at the origin. Aldous showed (personal communication) that there is a critical value p satisfying $16p(1-p) = 1$, such that for all greater p, X is sometimes infinite, whereas for lesser p, X is never infinite. At the critical value, X is always finite, and it is of interest to know the likelihood of large values of X. See (Aldous, 2013) for more details.

To prepare for the transfer theorem, we show that

$$\phi(z) = 1 - \frac{1-z}{4p} - (c + O(1))\frac{1-z}{\log(1/(1-z))}, \tag{3.4.3}$$

where $c = \log(1/(4p))/(4p)$ and $z \in [0, 1]$. If we knew this for all z in a Camembert region, we would be able to deduce from this and (3.4.1) that

$$a_n \sim cn^{-2}(\log n)^{-2},$$

so that X has a first moment but not a "$1 + \log$" moment. At present, we have (3.4.3) only on the unit interval and can apply only weaker Tauberian theorems. It is probably true in a Camembert region, and the resulting open problem is recommended to the interested reader.

To show (3.4.3), fix $0 < z_0 < 1$ and let $z_n = \phi^{(-n)}(z_0)$ so that $z_n \uparrow 1$. The recursion for ϕ gives

$$z_n = ((1-p)z_{n+1} + pz_{n-1})^2.$$

Changing variables to $y_n = 1 - z_n$ gives

$$y_n = 1 - ((1-p)(1-y_{n+1}) + p(1-y_{n-1}))^2$$

$$= 1 - (1 - ((1-p)y_{n+1} + py_{n-1}))^2.$$

Solving for y_{n+1} gives

$$y_{n+1} = \frac{1 - \sqrt{1 - y_n} - py_{n-1}}{1-p}.$$

Setting $x_n = y_n/(4p)^n$ and using $16p(1-p) = 1$ gives

$$x_{n+1} = 2x_n - x_{n-1} + O(y_n)^2.$$

Verifying first that y_n is small, we then have $x_n \sim An + B$, whence $y_n \sim (4p)^n(An + B)$. We may write this as

$$y_{n+1} = 4py_n + (1 + o(1))\frac{y_{n+1}}{n+1} = 4py_n + (1 + o(1))\frac{y_{n+1}}{\log y_{n+1}/\log(4p)}.$$

Let $z = 1 - y_{n+1}$ so $\phi(z) = 1 - y_n$. We then have

$$1 - \phi(z) = \frac{1-z}{4p} - (1 + o(1))\frac{1-z}{4p}\frac{\log(4p)}{\log(1-z)}$$

for all real $z \uparrow 1$, proving (3.4.3). ∎

Notes

One of the earliest and most well-known uses of generating function analysis to obtain asymptotics was Hardy and Ramanujan's derivation of asymptotics for the number of partitions of an integer (Hardy and Ramanujan, 2000a). The original argument used a Tauberian theorem and the behavior of the generating function $f(s)$ as $s \uparrow 1$ through real values, although later follow-up work (see, for example, Hardy and Ramanujan, 2000b) used circle methods. It seems difficult to trace the use of branchpoint methods or smooth saddle point methods, although Hayman (1956) was perhaps the earliest influential work in this area.

The exposition in this chapter does not follow any one source, although it owes a debt to Chapter 11 of Henrici (1991) and to the beautiful publication by Flajolet and Odlyzko (1990). A good reference book for this material is the text by Flajolet and Sedgewick (2009). The book by Dobrushkin (2010) is also worth reading.

Exercises

3.1 (number of set partitions)
Use the exponential generating function $f(z) = \exp(e^z - 1)$ for number a_n of partitions of $[n]$ (Example 2.5.7) to derive the estimate

$$a_n = (\log n + O(1))^n.$$

3.2 (Exercise 2.6 continued)
Find the approximate value of the least positive zero of the generating function f from Exercise 2.6. You can do this with or without rigorous bounds. Prove that the least positive zero has the least modulus of any singularity of f (hint: use the fact that the coefficients a_n of f are positive), and use this to estimate the limsup logarithmic growth rate $\limsup_{n\to\infty} n^{-1} \log a_n$. Prove that this limsup is equal to the liminf, so the limit exists.

3.3 (tricky radius of convergence)
Sometimes, even when f is given explicitly, it is not so obvious how to compute the radius of convergence to obtain the limsup exponential coefficient behavior as in (3.1.6). The Taylor coefficients of the function

$$f(z) := \frac{\arctan \sqrt{2e^{-z} - 1}}{\sqrt{2e^{-z} - 1}}$$

were shown by Wilf (1994) to yield rational approximations to π. Ward (2010) gives an asymptotic analysis. See if you can do the first step of this: find the radius of convergence of the Taylor series for f at zero.

3.4 (extending to a Camembert region)
Open problem: Is the generating function ϕ from Example 3.4.4 analytic in a Camembert region?

PART II

Mathematical Background

4

Fourier-Laplace Integrals in One Variable

4.1 Statement of Main Result

This chapter is devoted to the proof of the standard saddle point approximation

$$\int_\gamma A(z)\exp(-\lambda\phi(z))\,dz \sim A(z_0)\sqrt{\frac{2\pi}{\phi''(z_0)\lambda}}\exp(-\lambda\phi(z_0)) \tag{4.1.1}$$

and some generalizations. The most general univariate result we will obtain is the following theorem, which is proved in Section 4.3. It concerns the asymptotic evaluation of the integral of $A(z)\exp(-\lambda\phi(z))$ as $\lambda \to \infty$. The functions A and ϕ are called the ***amplitude*** and ***phase*** functions, respectively (although in the case where $\phi = i\rho$ is purely imaginary, the term "phase" usually denotes ρ rather than $i\rho$).

Theorem 4.1.1 (univariate Fourier-Laplace asymptotics) *Let A and ϕ be analytic functions on a neighborhood $\mathcal{N} \subseteq \mathbb{C}$ of the origin. Let*

$$A(z) = \sum_{j=0}^{\infty} b_j z^j$$

$$\phi(z) = \sum_{j=0}^{\infty} c_j z^j$$

be the power series for A and ϕ and let $l \geq 0$ and $k \geq 2$ be the indices of the least nonvanishing terms in the respective series, i.e., $b_l \neq 0$, $c_k \neq 0$ and $b_j = 0$ for $j < l$ and $c_j = 0$ for $j < k$. Let $\gamma : [-\varepsilon, \varepsilon] \to \mathbb{C}$ be any smooth curve with $\gamma(0) = 0 \neq \gamma'(0)$ and assume that $\text{Re}\{\phi(\gamma(t))\} \geq 0$ with equality *only at $t = 0$. Denote*

$$\mathcal{I}_+(\lambda) := \int_{\gamma|_{[0,\varepsilon]}} A(z)\exp(-\lambda\phi(z))\,dz\,;$$

$$\mathcal{I}(\lambda) := \int_{\gamma} A(z)\exp(-\lambda\phi(z))\,dz\,;$$

$$C(k,l) := \frac{\Gamma((1+l)/k)}{k}.$$

Then there are asymptotic expansions

$$\mathcal{I}_+(\lambda) = \sum_{j=l}^{\infty} a_j C(k,j)(c_k\lambda)^{-(1+j)/k} \tag{4.1.2}$$

$$\mathcal{I}(\lambda) = \sum_{j=l}^{\infty} \alpha_j C(k,j)(c_k\lambda)^{-(1+j)/k} \tag{4.1.3}$$

with the following explicit description.

(i) *a_j is a polynomial in the values $b_l, \ldots, b_j, c_k^{-1}, c_{k+1}, \ldots, c_{k+j-l}$ explicitly constructed in the proof, the first two values of which are $a_l = b_l$ and $a_{l+1} = b_{l+1} - \frac{2+l}{k}\frac{c_{k+1}}{c_k}$.*

(ii) *The choice of kth root in the expression $(c_k\lambda)^{-(1+j)/k}$ is made by taking the principal root in $x^{-1}(c_k\lambda x^k)^{1/k}$ where $x = \gamma'(0)$.*

(iii) *The numbers α_j are related to the numbers a_j when k is even by*

$$\alpha_j = \begin{cases} 2a_j & \text{if } j \text{ is even} \\ 0 & \text{if } j \text{ is odd} \end{cases}$$

and when k is odd by

$$\alpha_j = \left(1 - \zeta^{j+1}\right) a_j$$

where

$$\zeta = -\exp\left(\frac{i\pi}{k}\,\mathrm{sgn}\,\mathrm{Im}\,\{\phi(\gamma'(0))\}\right).$$

Remarks

(i) If $\phi(0) = \nu \neq 0$ but $\mathrm{Re}\{\phi(x)\}$ is still maximized at $x = 0$, then one may apply this result, replacing ϕ by $\phi - \nu$ and multiplying the outcome by $\exp(\lambda\nu)$.

(ii) The hypothesis that the minimum of $\mathrm{Re}\{\phi\}$ at 0 is strict will be removed when we reach the multivariate setting, for example in Theorem 5.4.8. In one variable, due to analyticity, either the minimum is strict or the real part of ϕ is identically zero. The analysis of a purely imaginary phase function takes place more naturally with C^∞ methods, which are discussed in Section 4.5.

(iii) For the first conclusion, involving $\mathcal{I}_+$, it is not necessary that ϕ extend to a differentiable function with nonpositive real part to the left of zero.

To those unfamiliar with stationary phase methods, this result may seem difficult to decipher, but both the statement and proof are actually quite intuitive. When A and ϕ are real, the orders of magnitude of such integrals are evident from direct integration of the orders of magnitude. Changing variables to simplify the exponent produces a full asymptotic development of the integral. When the phase is complex, one can use integration by parts to cancel the oscillation, or one can reduce to the real case by a contour shift. The latter requires stronger hypotheses (analyticity rather than smoothness) but gives stronger results (exponentially small remainders rather than rapidly decreasing remainders). To give all of the intuition, we take a route to the derivation i.e., longer than necessary. We begin with a stripped down special case, in which direct integration suffices, then give the arguments that extend this to greater generality, giving parallel arguments in the smooth and analytic categories when appropriate.

4.2 Real Integrands

The conclusion of the univariate Fourier-Laplace Theorem for real amplitude and phase functions would be an asymptotic expansion

$$\int_0^\varepsilon A(x)\exp(-\lambda\phi(x))\,dx \sim \sum_{j=l}^\infty a_j \lambda^{-(1+j)/k}$$

valid for real analytic functions A and ϕ with power series coefficients as in the univariate Fourier-Laplace Theorem. The main result of this section, Theorem 4.2.4, yields this expansion, along with further information about a_j.

When working on the real line, complex analytic techniques are not needed, and consequently we need to assume only differentiability and not analyticity. We build the argument in three steps: first, take A and ϕ to be monomials; next, keep the restriction on ϕ but remove the restriction on A; finally, remove the restriction on ϕ as well. The first step is accomplished via an exact computation, the second is achieved via a remainder estimate, and the third is deduced from the second by a change of variables.

A and ϕ Are Monomials

On the positive half-line, we can get away with a change of variables involving a fractional power. This allows us to handle the special case of monomial phase and amplitude by an exact integral, holding for any nonnegative real powers, α and β.

Substitute $y = \lambda x^\alpha$ to get

$$\int_0^\infty x^\beta \exp(-\lambda x^\alpha)\,dx = \int_0^\infty \left(\frac{y}{\lambda}\right)^{\beta/\alpha} e^{-y} \frac{1}{\alpha} \frac{y^{1/\alpha - 1}}{\lambda^{1/\alpha}}\,dy$$

$$= \frac{1}{\alpha}\lambda^{-(1+\beta)/\alpha} \int_0^\infty y^{\frac{1+\beta}{\alpha}-1} e^{-y}\,dy\,.$$

By the definition of the Γ-function, we therefore have the exact evaluation

$$\int_0^\infty x^\beta \exp(-\lambda x^\alpha)\,dx \;=\; C(\alpha, \beta)\,\lambda^{-(1+\beta)/\alpha} \tag{4.2.1}$$

$$C(\alpha, \beta) := \frac{\Gamma(\frac{1+\beta}{\alpha})}{\alpha}\,. \tag{4.2.2}$$

All the contribution to (4.2.1) comes from a neighborhood of zero: for any $\varepsilon > 0$ the contribution from $x \in [\varepsilon, \infty)$ is exponentially small in λ, so the integral over $[0, \varepsilon]$ captures the value up to an exponentially small correction:

$$\left|\int_0^\varepsilon x^\beta \exp(-\lambda x^\alpha)\,dx - C(\alpha, \beta)\,\lambda^{-(1+\beta)/\alpha}\right| \text{ decays exponentially.}$$

When β is an integer and α is an even integer, the corresponding two-sided integrals make sense as well:

$$\int_{-\infty}^\infty x^l \exp(-\lambda x^{2k})\,dx = \begin{cases} 2C(2k, l)\,\lambda^{-(1+l)/(2k)} & \text{if } l \text{ is even;} \\ 0 & \text{if } l \text{ is odd.} \end{cases} \tag{4.2.3}$$

ϕ is a Monomial, A is Anything

The results for monomials easily imply the following estimate.

Lemma 4.2.1 (big-O lemma) *Let $k, l > 0$ with k an integer. If A and ϕ are real-valued, piecewise smooth functions, with $A(x) = O(x^l)$ at $x = 0$, and $\phi(x) \sim x^k$ at $x = 0$ and vanishing in $[0, \varepsilon]$ only at 0, then*

$$\int_0^\varepsilon A(x) \exp(-\lambda\phi(x))\,dx = O(\lambda^{(-l-1)/k})$$

as $\lambda \to \infty$.

Proof Pick K such that $|A(x)| \le K|x|^l$ on $[-\varepsilon, \varepsilon]$ and δ such that $|\exp(-\lambda\phi(x))| \le \exp((\delta - \lambda)|x|^k)$ on $[0, \varepsilon]$. Then

$$\left|\int_0^\varepsilon A(x) \exp(-\lambda\phi(x))\,dx\right| \le K \int_0^\varepsilon x^l \exp((\delta - \lambda)x^k)\,dx$$

$$= O\left((\lambda - \delta)^{-(1+l)/k}\right)$$

by (4.2.1). □

For monomial phase functions and general amplitude functions, we now have the following result.

Lemma 4.2.2 *Suppose that A is a real function with*

$$A(x) = \sum_{j=l}^{M-1} b_j x^j + O(x^M)$$

as $x \to 0$. Then

$$\int_0^{\varepsilon} A(x) \exp(-\lambda x^k)\, dx = \sum_{j=l}^{M-1} b_j C(k, j) \lambda^{-(1+j)/k} + O(\lambda^{-(1+M)/k})$$

where $C(k, j) = \Gamma((1 + j)/k)/k$ are the constants computed in (4.2.1).

Remark Note that the hypothesis on A is quite weak. In particular, A need not even be in the class C^1 (for example, take $A(x) = x^M \sin(x^{-M})$). If A is represented by an infinite asymptotic series (convergent or not), then an infinite asymptotic expansion for the integral follows by applying the lemma for each M.

Proof Multiply the estimate

$$A(x) - \sum_{j=0}^{M-1} b_j x^j = O(x^M)$$

by $\exp(-\lambda\phi(x))$ and integrate. Using (4.2.1) to evaluate the integral of each monomial and Lemma 4.2.1 to bound the integral of the right-hand side gives

$$\left| I - \sum_{j=0}^{M-1} \int_0^{\varepsilon} b_j x^j \exp(-\lambda x^k)\, dx \right| = O(\lambda^{(-M-1)/k}),$$

which is the conclusion of the lemma. □

General A and ϕ

A change of variables reduces the general case to Lemma 4.2.2. A bit of care is required to ensure we understand the asymptotic series for the functions involved in the change of variables.

Lemma 4.2.3 *Let $M \geq 2$ be an integer and let*

$$y(x) = c_1 x + \cdots + c_{M-1} x^{M-1} + O(x^M) \tag{4.2.4}$$

in a neighborhood of zero, where $c_1 \neq 0$. Then there is a neighborhood of zero on which y is invertible. The inverse function $x(y)$ has an expansion

$$x(y) = a_1 y + \cdots + a_{M-1} y^{M-1} + O(y^M)$$

with a_j polynomials in $c_1, \ldots, c_j$ and c_1^{-1}.

Proof Suppose $c_1 = 1$. From $y = x + O(x^2)$, we see that $y \sim x$ at zero; hence $x = y + O(x^2) = y + O(y^2)$. Now let $2 \le n < M$ and suppose inductively that $x = y + a_2 y^2 + \cdots + a_{n-1} y^{n-1} + O(y^n)$, where $a_2, \ldots, a_{j-1}$ are polynomials in $c_2, \ldots, c_{j-1}$. Let a be an indeterminate, and plug in the value of y in (4.2.4) to the quantity

$$x - (y + a_2 y^2 + \cdots + a_{n-1} y^{n-1} + a y^n).$$

The result is a polynomial in x, whose coefficients in degrees $1, \ldots, n-1$ vanish due to the induction hypothesis, plus a remainder of $O(x^M)$. The coefficient of the x^n term may be written as $a - P(a_2, \ldots, a_{n-1}, c_2, \ldots, c_n)$, where P is a polynomial. By induction, this is a polynomial in $c_2, \ldots, c_n$. Setting a_n equal to this polynomial, we see that

$$x - y - \sum_{j=2}^{n} a_n y^n = O(x^{n+1}).$$

This completes the induction.

When $n = M - 1$, observing that $O(x^M) = O(y^M)$ completes the proof of the lemma for $c_1 = 1$. To remove the restriction on c_1, apply the case $c_1 = 1$ to represent x as a function of y/c_1, which shows that $x = \sum_{j=1}^{M-1} a_j y^j + O(y^M)$ with $c_1^j a_j$ a polynomial in $c_2, \ldots, c_j$. □

Theorem 4.2.4 *Let $k, l \le M$ be integers. Suppose that A and ϕ are real functions with ϕ of class C^M and series*

$$A(x) = \sum_{j=l}^{M-1} b_j x^j + O(x^M)$$

$$\phi(x) = \sum_{j=k}^{M} c_j x^j + O(x^{M+1})$$

as $x \to 0$, where $b_l, c_k \neq 0$. Then as $\lambda \to \infty$, the quantity $\mathcal{I}(\lambda) := \int_0^\varepsilon A(x) \exp(-\lambda\phi(x))\, dx$ has asymptotic expansion

$$\mathcal{I}(\lambda) \sim \sum_{j=l}^{M-1} a_j C(k, j)(c_k \lambda)^{-(1+j)/k} + O(\lambda^{-(1+M)/k}) \tag{4.2.5}$$

with $C(k, l) = \Gamma(\frac{1+l}{k})/k$ as in (4.2.2) and the terms a_j given by polynomials in $b_l, \ldots, b_j$ and $c_k^{-1}, c_{k+1}, \ldots, c_{k+j-l}$. The leading two terms are given by

$$a_l = b_l\,;$$

$$a_{l+1} = b_{l+1} - \frac{2+l}{k}\frac{c_{k+1}}{c_k}\,. \tag{4.2.6}$$

Proof We employ the change of variables $y = (\phi(x))^{1/k}$. Writing

$$\phi(x) = c_k x^k \left(1 + \frac{c_{k+1}}{c_k} x + \cdots + \frac{c_M}{c_k} x^{M-k} + O(x^{M+1-k})\right),$$

we see that

$$y = c_k^{1/k} x \left(1 + \cdots + \frac{c_M}{c_k} x^{M-k} + O(x^{M+1-k})\right)^{1/k}. \tag{4.2.7}$$

Using the Taylor series for $(1+u)^{1/k}$ (i.e., the binomial expansion), we see that

$$y = c_k^{1/k} \sum_{j=1}^{M} d_j x^j + O(x^{M+1})$$

where d_j are polynomials in $c_{k+1}, \ldots, c_{k+j}$ and c_k^{-1}.

By the previous lemma, the inverse function satisfies

$$x = \sum_{j=1}^{M} e_j \left(\frac{y}{c_k^{1/k}}\right)^j + O(y^{M+1}), \tag{4.2.8}$$

where e_j is a polynomial in $c_{k+1}, \ldots, c_{k+j}$. A function of class C^M with nowhere vanishing derivative has an inverse of class C^M, which justifies term-by-term differentiation, yielding

$$x'(y) = c_k^{-1/k} \sum_{j=1}^{M} j e_j \left(\frac{y}{c_k^{1/k}}\right)^{j-1} + O(y^M).$$

The change of variables formula gives

$$\mathcal{I}(\lambda) = \int_0^{y(\varepsilon)} \tilde{A}(y) \exp(-y^k)\, dy,$$

where $\tilde{A}(y) = A(x(y))x'(y)$. Plugging in the series for x and x' into the definition of $\tilde{A}$ gives

$$\tilde{A}(y) = c_k^{-1/k} \sum_{j=l}^{M-1} \tilde{b}_j \left(\frac{y}{c_k^{1/k}}\right)^j + O(y^M)$$

where $\tilde{b}_j$ is a polynomial in $b_l, \ldots, b_j, c_k^{-1}, c_{k+1}, \ldots, c_j$, to be evaluated shortly. The existence of the expansion (4.2.5) now follows from the monomial exponent case (Lemma 4.2.2).

The leading terms (4.2.6) are computed as follows. The change of variables is

$$y = c_k^{1/k} x \left(1 + \frac{c_{k+1}}{c_k} x + O(x^2)\right)^{1/k}$$
$$= c_k^{1/k} x \left(1 + \frac{c_{k+1}}{k\, c_k} x + O(x^2)\right).$$

Inverting and differentiating,

$$x = \frac{y}{c_k^{1/k}} - \frac{1}{k}\frac{c_{k+1}}{c_k}\left(\frac{y}{c_k^{1/k}}\right)^2 + O(y^3);$$

$$x'(y) = \frac{1}{c_k^{1/k}} - \frac{2}{c_k^{2/k}}\frac{c_{k+1}}{k\,c_k}\,y + O(y^2).$$

Composing and multiplying shows the coefficients of $\tilde{A}(y) = A(x(y))x'(y)$ to be

$$\tilde{b}_l = b_l$$

$$\tilde{b}_{l+1} = b_{l+1} - \frac{l+2}{k}\frac{c_{k+1}}{c_k}$$

and evaluating $\int \tilde{A}(y)\exp(-\lambda y^k)\,dy$ via Lemma 4.2.2 yields (4.2.6). □

4.3 Complex Phase

Extending the results of the previous section to complex amplitudes is trivial – by linearity of the integral, the result holds separately for $\mathrm{Im}\{A\}$ and $\mathrm{Re}\{A\}$, and these may then be recombined to give the result for complex A. When it comes to complex phases, we are faced with a choice. If we assume A and ϕ are analytic in a neighborhood of zero, we are entitled to move the contour; this is the quickest justification for extending the conclusion to complex phases without much change in the formula and is the approach taken in this section.

PROOF OF THEOREM 4.1.1:

Step 1: Evaluation of the One-Sided Integral $\mathcal{I}_+$

Let $\gamma_+ : [0, \varepsilon] \to \mathbb{C}$ denote the restriction $\gamma_{[0,\varepsilon]}$ so that $\mathcal{I}_+ = \int_{\gamma_+} A(z)\exp(-\lambda\phi(z))\,dz$. Evaluate $\mathcal{I}_+$ as follows.

We employ the same change of variables $y = \phi(z)^{1/k}$ as in the proof of Theorem 4.2.4, only we will need to be careful in choosing a branch of the $1/k$ power. Formula (4.2.7) defines k different functions, one for each choice of the the kth root in the expressions $c_k^{1/k}$. It follows from Lemma 4.2.3 that each of these k functions and their inverses are analytic in a neighborhood of the origin. We will need a notation for the ***principal*** k^{th} ***root***. This is the analytic function from the plane minus the negative real half-line to the cone $K := \{z : -\pi/k < \arg(z) < \pi/k\}$ defined by $\mathfrak{p}(u^{1/k}) := z$ for the unique $z \in K$ such that $z^k = u$. Let $\boldsymbol{v}$ denote any positive real multiple of $\gamma'(0)$. Near the origin, $\phi(z) \sim c_k z^k$ and the requirement that $\mathrm{Re}\{\phi\} \geq 0$ forces the quantity $\boldsymbol{v}$ to be in the windmill-shaped set of pre-images under $c_k z^k$ of the right half-plane, shown in Figure 4.1. Define $f(x) := \mathfrak{p}(\phi(x)^{1/k})$. Because the path $\phi(\gamma_+(t))$ remains in the positive real half-plane for $0 < t \leq \varepsilon$, it also remains in the slit plane, and hence maps the image of γ_+ bi-analytically to the cone K. With this choice of $1/k$ power, the change of variables (4.2.7)

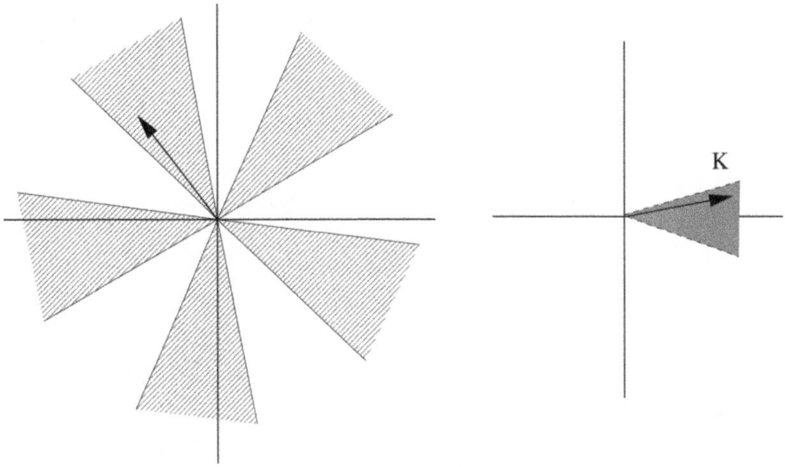

Figure 4.1 Arrows represent $\boldsymbol{v}$ and $(d/dt)|_{t=0} f(\gamma^+(t))$.

becomes

$$y = f(x) = \eta x \left(1 + \cdots + \frac{c_M}{c_k} x^{M-k} + O(x^{M-1-k})\right)^{1/k}, \tag{4.3.1}$$

where $\eta = \boldsymbol{v}^{-1}\, \mathfrak{p}(c_k \boldsymbol{v}^k)^{1/k}$ and the branch of the $1/k$ power of the series in parentheses is the one that fixes 1. Thus $\eta = f'(0)$ and

$$(d/dt)_{t=0} f(\gamma_+(t)) = \eta \boldsymbol{v}\,. \tag{4.3.2}$$

The inverse function, $x = g(y)$, is given by taking $c_k^{1/k} = \eta$ in (4.2.8). This choice fulfills property (ii) in the Theorem.

As in the proof of Theorem 4.2.4, we then have

$$\mathcal{I}_+ = \int_{\tilde{\gamma}} \tilde{A}(y) \exp(-\lambda y^k)\, dy, \tag{4.3.3}$$

where $\tilde{\gamma} = f \circ \gamma_+$ is the image of γ_+ under the change of variables.

Let $p = f(\gamma(\varepsilon))$ denote the endpoint of $\tilde{\gamma}$. Let $p' > 0$ denote the real part of p, let α be the line segment $[0, p']$, and let β denote the line segment $[p', p]$. The contour $\tilde{\gamma}$ is homotopic to $\alpha + \beta$ (see Figure 4.2), whence $\int_{\tilde{\gamma}} h(z)\, dz = \int_\alpha h(z)\, dz + \int_\beta h(z)\, dz$ for any analytic function h. On compact subsets of K, $\mathrm{Re}\{y^k\}$ is bounded from below by a positive constant. It follows that on β, there are positive C and ρ for which

$$\left|\tilde{A}(y) \exp(-\lambda y^k)\right| \le C e^{-\rho\lambda}.$$

(The reason we chose the principal value is so that β would lie inside K.) We conclude that

$$\mathcal{I}_+ = \int_\alpha \tilde{A}(y) \exp(-\lambda y^k)\, dy + R$$

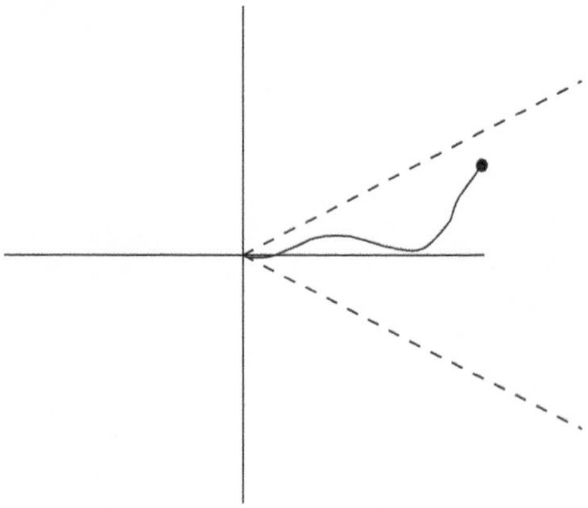

Figure 4.2 The path $\tilde{\gamma}$ in the cone K and the line segments α and β.

for a remainder R that decays exponentially. Applying Theorem 4.2.4 (with complex amplitude) to $\int_\alpha$ gives the asymptotic series

$$\mathcal{I}_+ \sim \sum_{j=l}^{\infty} a_j C(k, j) c_k^{-(1+j)/k} \lambda^{-(1+j)/k},$$

which shows that the expansion of $\mathcal{I}_+$ satisfies conclusions (i) and (ii).

Step 2: Evaluation of the Two-Sided Integral $\mathcal{I}$

To reduce this to the problem solved in step 1, define the contour $\gamma_- : [0, \varepsilon] \to \mathbb{C}$ by $\gamma_-(t) := \gamma(-t)$. The orientation of γ_- is from 0 to $-\varepsilon$ so it appears with sign reversed; in other words,

$$\mathcal{I} = \mathcal{I}_+ - \mathcal{I}_-$$

where

$$\mathcal{I}_- := \int_{\gamma_-} A(z) \exp(-\lambda \phi(z))\, dz\,.$$

The integral for $\mathcal{I}_-$ has nearly the same data as the integral for $\mathcal{I}_+$. The functions A and ϕ are identical. The only difference between the two integrals is the contour. The contour affects the integral only via the choice of η in (4.3.1). Denoting the two choices by η_+ and η_-, we know that $\eta_- = \eta_+/\zeta$ for some ζ with $\zeta^k = 1$. Denoting the respective inverse functions by g_+ and g_-, we see that $g_-(y) = g_+(\zeta y)$. The two changes of variables produce amplitudes $\tilde{A}_+$ and $\tilde{A}_-$ in (4.3.3) satisfying

$$\begin{aligned}
\tilde{A}_+(y) &= A(g_+(y)) \cdot g_+'(y)\,; \\
\tilde{A}_-(y) &= A(g_-(y)) \cdot g_-'(y) \\
&= A(g_+(\zeta y)) \cdot \zeta g_+'(\zeta y) \\
&= \zeta \tilde{A}_+(\zeta y)\,.
\end{aligned}$$

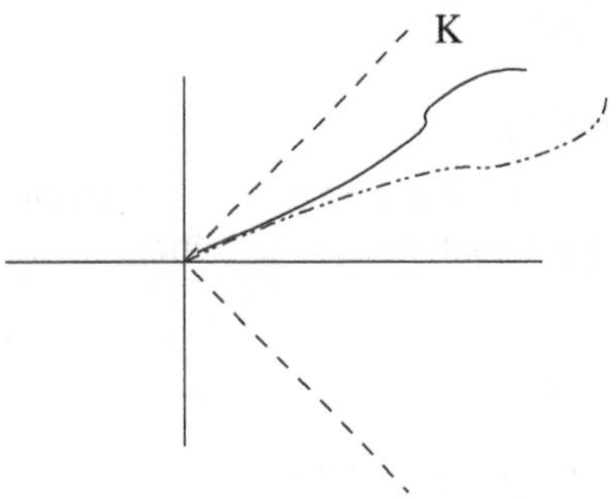

Figure 4.3 k is even: $\phi(\gamma_+)$ is shown solid, and $\phi(\gamma_-)$ is shown dotted.

The coefficients of the power series for $\tilde{A}_+$ and $\tilde{A}_-$ are therefore related by $[y^j]\tilde{A}_- = \zeta^{j+1}[y^j]\tilde{A}_+$. The asymptotic expansions of $\tilde{A}_\pm$ are integrated term by term in (4.3.3), which implies that the coefficients α_j for the two-sided integral I are related to the coefficients a_j for the one-sided integral I_+ via $\alpha_j = (1 - \zeta^{j+1})a_j$. Thus part (iii) of Theorem 4.1.1 is reduced to the correct identification of ζ. The evaluation of ζ breaks into two cases, depending on the parity of k.

Suppose first that k is even. Because $\phi(z) \sim c_k z^k$, the image of the smooth curve γ under ϕ does a U-turn at the origin, with the tangents to the images $\phi(\gamma_-(t))$ and $\phi(\gamma_+(t))$ coinciding at $t = 0$ (see Figure 4.3). Because γ_- reverses the orientation of the parametrization, we see that $\boldsymbol{v}_- := \gamma_-'(0) = -\boldsymbol{v}_+$. The powers $\boldsymbol{v}_-^k$ and $\boldsymbol{v}_+^k$ coincide, whereby

$$\eta_- = \boldsymbol{v}_-^{-1}\, \mathfrak{p}(c_k \boldsymbol{v}_-^k)^{1/k} = -\boldsymbol{v}_+^{-1}\, \mathfrak{p}(c_k \boldsymbol{v}_+^k)^{1/k} = -\eta_+ .$$

When k is even, therefore, ζ takes the value -1. This leads to $\alpha_j = 2a_j$ for even j and $\alpha_j = 0$ for odd j, completing the proof of the theorem for even k.

When k is odd, the images of γ_+ and γ_- under ϕ point in opposite directions (see Figure 4.4). Because both are in the closed right half-plane, this implies that one is in the positive imaginary direction and one is in the negative imaginary direction. Thus the argument of the tangent to $\phi(\gamma_+)$ is $\sigma\pi/2$,

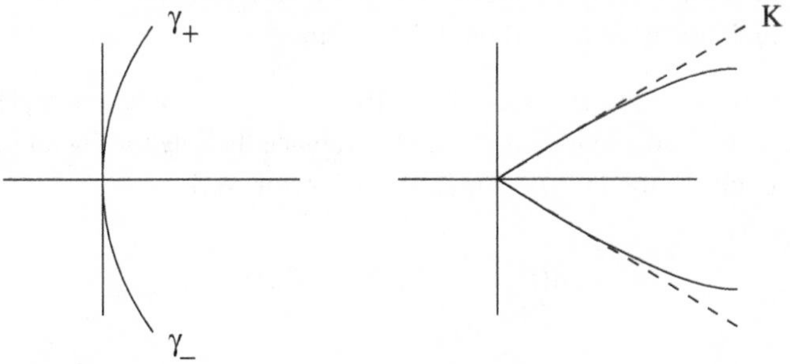

Figure 4.4 k is odd: $\phi(\gamma_+)$ and $\phi(\gamma_-)$ and their principal $1/k$ powers.

where the sign σ is given by

$$\sigma := \operatorname{sgn} \operatorname{Im} \left\{ \phi(\gamma'(0)) \right\}.$$

The argument of the tangent to $\phi(\gamma_-)$ is $-\sigma\pi/2$ and thus differs from the argument of $\phi(\gamma_+)$ by $-\sigma\pi$. Mapping by the principal k^{th} root shrinks the difference in arguments by a factor of k, thus

$$\mathfrak{p}(c_k \boldsymbol{v}_-^k)^{1/k} = e^{-i\pi\sigma/k}\, \mathfrak{p}(c_k \boldsymbol{v}_+^k)^{1/k}.$$

Again, the reversal of parametrization implies $\boldsymbol{v}_- = -\boldsymbol{v}_+$, whence

$$\eta_- = (-1) \cdot e^{-i\pi\sigma/k} \eta_+ = \frac{\eta_+}{\zeta},$$

with $\zeta = -e^{i\pi\sigma/k}$ as in the statement of the theorem.

4.4 Classical Methods: Steepest Descent (Saddle Point) and Watson's Lemma

This section contains some classical results that may be proved using the machinery of Sections 4.2 and 4.3. Lemma 4.2.2 with $k = 1$ is a special case of Watson's Lemma (smooth amplitude). The usual statement is as follows:

Proposition 4.4.1 (Watson's Lemma) *Let $A : \mathbb{R}^+ \to \mathbb{C}$ have asymptotic development*

$$A(t) \sim \sum_{m=0}^{\infty} b_m t^{\beta_m}$$

with $-1 < \operatorname{Re}\{\beta_0\} < \operatorname{Re}\{\beta_1\} < \cdots$ *and* $\operatorname{Re}\{\beta_m\} \uparrow \infty$. *Then the Laplace transform has asymptotic series*

$$L(\lambda) := \int_0^\infty A(t) e^{-\lambda t}\, dt \sim \sum_{m=0}^{\infty} b_m \Gamma(\beta_m + 1) \lambda^{-(1+\beta_m)}$$

as $\lambda \to \infty$.

Remark This result is similar in spirit to Darboux's Theorem (Theorem 3.3.2) and is a conceptual precursor to Darboux's Theorem.

Proof We reproduce the argument from Bleistein and Handelsman (1986, Section 4.1). It is by now obvious that we may replace the integral by an integral on $[0, \varepsilon]$, introducing only an exponentially small error. Writing

$$A(t) = \sum_{m=0}^{N} b_m t^{\beta_m} + R_N(t)$$

for $R_N = O(t^{\operatorname{Re}\{\beta_{m+1}\}})$ at 0, we may integrate term by term to get the first N terms of the expansion (up to an exponentially small correction for truncating the integral on $[0, \varepsilon]$), then use Lemma 4.2.1 to see that the remainder satisfies

$$\left| \int_0^\varepsilon R_n(t) e^{-\lambda t} \, dt \right| = O\left(\lambda^{-\operatorname{Re}\{\beta_m\}-1}\right),$$

proving the proposition. □

Our Morse-theoretic approach to the evaluation of integrals subsumes the classical *saddle point method* (also called the *steepest descent method*). Nevertheless, having summarized how one computes asymptotics near a point where $\phi' = 0$, it would be wasteful to leave the scene without a brief discussion of the method of steepest descent as it is elementarily understood. We present this as a method, stating no theorems but giving instructions and an example.

Consider again the integral

$$\mathcal{I}(\lambda) = \int_\gamma A(z) \exp(-\lambda \phi(z)) \, dz,$$

only now suppose that ϕ' does not vanish on γ. As we have seen before, $\lambda^{-1} \log \mathcal{I}(\lambda)$ has a limsup of at most $\nu := \sup_{z \in \gamma} \operatorname{Re}\{\phi(z)\}$. This is sharp when ϕ' vanishes somewhere on the support of A in γ, but otherwise is not expected to be sharp. The saddle point method says to deform the contour so as to pass through a point x where ϕ' vanishes. From our Morse theoretic analyses, we know that this can always be done and solves the problem of minimizing ν. The saddle point method, as commonly understood, merely says to attempt to deform the contour so as to pass through such a point. The phrase "steepest descent" comes from the fact that the real part of ϕ must have a local maximum on the contour at x, rather than a minimum or inflection point (if x is not a maximum, then it cannot be the highest point on the contour, so we know from Morse theory that there is a higher critical point the contour must pass through). Having deformed the contour to pass through x in the right direction, one then applies the univariate Fourier-Laplace Theorem, with $\phi(0) \neq 0$ as in the remark following the theorem.

Example 4.4.2 Consider the univariate power series $f(z) = (1-z)^{-1/2}$. By the binomial theorem, this generates the numbers $a_n := (-1)^n \binom{-1/2}{n} \sim \sqrt{1/(\pi n)}$. Let us instead evaluate this via a contour integral. Because we understand meromorphic integrands the best, we change variables to $z = 1 - y^2$ with $dz = -2\,y\,dy$. Then, letting C be a small circle around the origin, oriented counter-clockwise,

$$\begin{aligned} a_n &= \frac{1}{2\pi i} \int_C z^{-n-1} (1-z)^{-1/2} \, dz \\ &= \frac{1}{2\pi i} \int_E (1-y^2)^{-n-1} y^{-1} (-2y) \, dy \\ &= \frac{i}{\pi} \int_E (1-y^2)^{-n-1} \, dy \, . \end{aligned}$$

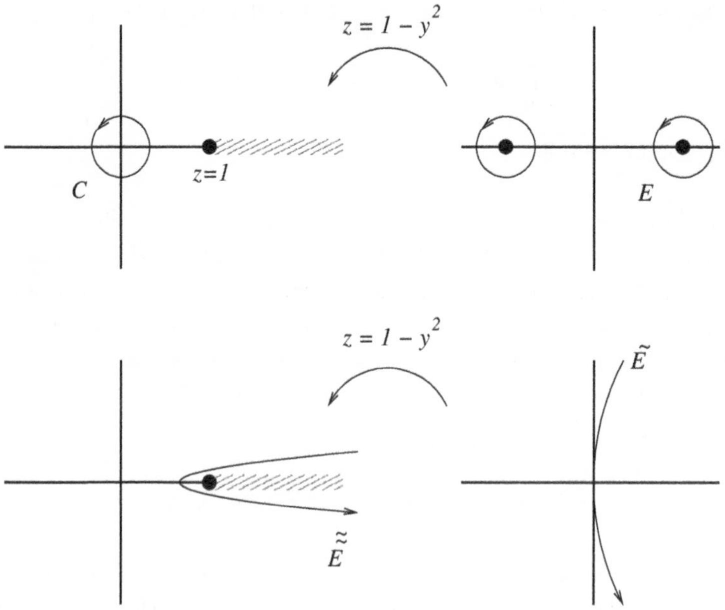

Figure 4.5 Above: the contours C and E. Below: $\tilde{E}$ and $\tilde{\tilde{E}}$.

As shown in Figure 4.5, E is a small circle in the y-plane, oriented counterclockwise, around either the point $+1$ or the point -1, because either of these contours maps to a small contour around 0 in the z-plane. Let us take a circle around $+1$. In the y-plane, there is a critical point for $\phi(y) := -\log(1-y^2)$ at the origin. The contour E may be deformed to a contour $\tilde{E}$ passing through the origin in the downward direction. It is easy to see that any contour separating 1 and -1 must intersect the segment $(-1, 1)$, so this deformation does indeed produce a minimax height contour. Changing variables to $y = -it$ gives $a_n = \frac{i}{\pi} \int (1+t^2)^{-n-1} (-i)\, dt$. Because $\phi(y) = \phi(-it) \sim -t^2$ near the origin, the integral then becomes asymptotically $\sqrt{2/(\pi n \phi''(0))} = \sqrt{1/(\pi n)}$. ■

Remark The function $(1-z)^{-1/2}$ is analytic on the slit plane $\mathbb{C} \setminus \{x \in \mathbb{R} : x \geq 1\}$, which we may view as half of the Riemann surface $\mathfrak{R}$ obtained by gluing two copies of the slit plane, with the upper half of each attaching along the slit to the lower half of the other. The change of variables $z = 1 - y^2$ is the map from $\mathfrak{R}$ to $\mathbb{C}$. The saddle point contour $\tilde{E}$, when mapped to the z-plane, comes in to $+1$ along one copy of the slit, does a U-turn, and goes back along the other copy of the slit. Perturbing slightly gives a hairpin-shaped contour $\tilde{\tilde{E}}$ that may be drawn in the slit plane. This explains the origin of the Camembert contour in Figure 3.1.

4.5 Analytic Versus C^∞ Category

The results of the previous sections hold when the amplitude and phase are assumed only to be smooth, not necessarily analytic. This indicates there should be arguments that use smooth techniques, such as partitions of unity and integration by parts, rather than contour deformation. Such an approach to evaluating stationary phase integrals has been developed and used extensively by harmonic analysts, who are chiefly interested in the case where ϕ is purely imaginary. Note that this is not covered by the assumptions of the univariate Fourier-Laplace Theorem: the contour decomposition requires that $\text{Re}\{\phi\}$ be strictly positive away from zero. Our chief reason for including this section is that any treatment of Fourier-Laplace integrals that bypasses smooth methods is pedagogically and historically incomplete. Results in this section are used only once or twice for the analysis of generating functions in this book.

When the exponent $i\phi(z)$ is imaginary, the modulus of the integrand is equal to $|A(z)|$, so it is no longer true that one may cut off the integral outside of an interval $[-\varepsilon, \varepsilon]$ and expect to introduce negligible remainders. Instead, one assumes that A has compact support, then uses smooth partitions of unity to reduce to integrals over small intervals. Note that neither partitions of unity nor compactly supported functions exist in the analytic category; however, when the contour of integration γ is a closed curve, any amplitude function has compact support on γ, so both the analytic and the smooth methods apply and may be compared.

In this section we give asymptotics for

$$\mathcal{I}(\lambda) := \int_a^b A(z) \exp(i\lambda\phi(z))\, dz,$$

where A and ϕ are smooth and A is supported on a compact sub-interval of (a, b); we use the term "phase" to denote ϕ rather than $i\phi$. The steps are similar to the steps of the proof of Theorem 4.2.4 except for the insertion of a localization step at the beginning and the introduction of a damping term in the step where both amplitude and phase are monomial:

- Localization
- Big-O estimate
- Monomials (with damping)
- Monomial phase
- Full theorem

Localization Lemma in C^∞

Following Stein (1993), we begin with a localization principle.

Lemma 4.5.1 (localization lemma) *Suppose $\phi'(x) \neq 0$ for all $x \in (a, b)$. Then $\mathcal{I}(\lambda)$ is rapidly decreasing, i.e.,*

$$\mathcal{I}(\lambda) = O(\lambda^{-N}) \qquad \text{as } \lambda \to \infty$$

for any $N \geq 0$.

Proof The smooth vanishing of A at the endpoints allows us to integrate by parts without introducing boundary terms. Integrate by parts with $dU = i\lambda\phi' e^{i\lambda\phi}\, dx$ and $V = A/(i\lambda\phi')$ to get

$$\mathcal{I}(\lambda) = -\int_a^b e^{i\lambda\phi(x)} \frac{d}{dx}\left(\frac{A}{i\lambda\phi'}\right)(x)\, dx.$$

For any $N \geq 1$, we may repeat this N times to obtain

$$\mathcal{I}(\lambda) = \int_a^b e^{i\lambda\phi(x)}(-\lambda^{-N})\mathcal{D}^N(A)(x)\, dx, \tag{4.5.1}$$

where $\mathcal{D}$ is the differential operator $f \mapsto (d/dx)(f/i\phi')$. Letting

$$K_N = (b-a) \sup_{a \leq x \leq b} |\mathcal{D}^n A(x)|, \tag{4.5.2}$$

we see that

$$|\mathcal{I}(\lambda)| \leq \lambda^{-N} K_N$$

which proves that $\mathcal{I}$ is a rapidly decreasing function of λ. □

Remarks

(i) Compare this to the argument in the analytic case. There, if ϕ' is nowhere vanishing, the contour can be "pushed down" along the gradient flow so that the maximum of Re$\{i\phi\}$ is strictly negative, resulting in an integral that decreases exponentially in λ.

(ii) Because ϕ' does not vanish, we may change variables to $y = \phi(x)$ and the conclusion is equivalent to the more familiar statement that the Fourier transform of the smooth function $\tilde{A}$ is rapidly decreasing.

(iii) Although the lemma is stated only for purely imaginary phase functions, the same argument in fact shows that $\mathcal{I}(\lambda)$ is rapidly decreasing whenever the real part of ϕ is nonnegative and ϕ' is nonvanishing.

We call Lemma 4.5.1 the localization lemma for the following reason. Suppose we allow ϕ' to vanish on some finite set of points $x_1, \ldots, x_d \in [a, b]$. Then the contribution to $\mathcal{I}(\lambda)$ from any closed region not containing some x_i is rapidly decreasing, so the asymptotics for $\mathcal{I}(\lambda)$ may be read off as the sum of contributions local to each x_i. Indeed, for each i, let $[a_i, b_i]$ be tiny intervals containing x_i, with all intervals disjoint, and let $\xi_1, \ldots, \xi_d$ be a partition of unity subordinate to $\{[a_i, b_i] : 1 \leq i \leq d\}$. Once we see how to obtain asymptotics in a neighborhood of x_i containing no other critical points, we can write $A = A_0 + \sum_{i=1}^d A\xi_i$, so that the support of A_0 contains no x_i. By the localization lemma, $\int e^{i\lambda\phi(x)} A_0(x)\, dx$ is rapidly decreasing. It follows that as long as the integrals $\mathcal{I}_i(\lambda) := \int_a^b e^{i\lambda\phi(x)} A_i(x)\, dx$ sum to something not rapidly decreasing, the asymptotic development of $\mathcal{I}(\lambda)$ is gotten by summing the developments of $\mathcal{I}_i(\lambda)$.

Our main result for one-variable purely oscillating integrals will be the asymptotic development of integrals defined as follows.

Theorem 4.5.2 *Let ϕ and A be smooth real functions with A having compact support in (a, b) whose closure contains zero. Let $k \geq 2, l \geq 0$ be integers, and suppose the power series for A and ϕ are given by $\{b_j\}$ and $\{c_j\}$ as in Theorem 4.2.4, with $c_k > 0$. Suppose that ϕ' vanishes in $[a, b]$ at 0 but nowhere else. Let $\tilde{A} := (A \circ g) \cdot g'$, where g is the inverse function to $x \mapsto (\phi/c_k)^{1/k}$. Then as $\lambda \to \infty$ there is an asymptotic development*

$$\mathcal{I}(\lambda) := \int_a^b A(x) \exp(-i\lambda\phi(x))\, dx \sim \sum_{j=l}^{\infty} \alpha_j C(k, j)(i\, c_k\, \lambda)^{-(1+j)/k}.$$

The coefficients α_j are obtained from the power series coefficients $a_0, \ldots, a_j$ for $\tilde{A}$ exactly as in part (iii) of the univariate Fourier-Laplace Theorem. The constant in the $O(\lambda^{-(N+1)/k})$ remainder term is bounded by a continuous function of the suprema of the first $N + 1$ derivatives of ϕ and A on the support of A. The $1/k$ power of $i\, c_k \lambda$ is the principal value.

The Big-O Lemma in C^∞

The following smooth counterpart to Lemma 4.2.1 is proved by showing that the main contribution comes from an interval of size $\lambda^{-1/k}$. The increase in length over the very short proof of Lemma 4.2.1 is due to the need to keep track of a partition of unity function and its derivatives.

Lemma 4.5.3 *If η is smooth and compactly supported and $l \geq 1$ and $k \geq 2$ are integers, then*

$$\left| \int_{-\infty}^{\infty} e^{i\lambda x^k} x^l \eta(x)\, dx \right| \leq C\lambda^{-(l+1)/k} \tag{4.5.3}$$

for a constant C depending on k, l and the first l derivatives of η.

Proof Let α be a nonnegative smooth function equal to its maximum of 1 on $|x| \leq 1$ and vanishing on $|x| \geq 2$. Choose an $\varepsilon > 0$ and rewrite (4.5.3) as

$$\int e^{i\lambda x^k} x^l \eta(x)\alpha(x/\varepsilon)\, dx + \int e^{i\lambda x^k} x^l \eta(x)[1 - \alpha(x/\varepsilon)]\, dx. \tag{4.5.4}$$

The absolute value of the first integrand is at most $|x|^l \cdot (\sup_{|x|\leq 2} |\eta(x)|) \cdot \mathbf{1}_{|x|\leq 2\varepsilon}$, which yields an integral of at most $C_1 \varepsilon^{l+1}$, where $C_1 = \frac{2^{l+1}}{l+1} \sup_{|x|\leq 2} |\eta(x)|$.

The second integral will be done by parts, and to prepare for this, we examine the iteration of the operator $D := (d/dx)(\cdot/x^{k-1})$ applied to the function $x^l \eta(x)(1 - \alpha(x/\varepsilon))$. The result will be a sum of monomials, each monomial being a product of a power of x, a derivative of η, a derivative of α, and a power of ε. In fact, if

(a, b, c, d) is shorthand for $x^a \eta^{(b)}(x)\alpha^{(c)}(x/\varepsilon)\varepsilon^d$, and $a \geq 0$, then

$$D(a, b, c, d) = (a - k + 1)(a - k, b, c, d) + (a - k + 1, b + 1, c, d) + (a - k + 1, b, c + 1, d - 1).$$

By induction, we see that $D^N(a, b, c, d)$ is the sum of terms $C \cdot (r, s, t, u)$ with $r + u \geq a + d - kN, s \leq b + N, t \leq c + N$, and C is bounded above by the factorial $\max\{kN, a\}!$. In particular, because $\varepsilon \leq x$, we may replace positive powers of ε by the same power of x to arrive at the upper bound:

$$\left|D^N\left[x^l \eta(x)(1 - \alpha(x/\varepsilon))\right]\right| \leq \mathbf{1}_{|x|\geq\varepsilon} C|x|^{l-kN} \tag{4.5.5}$$

where C is the product of $\sup_{j\leq N, |x|\in(1,2)} \eta^{(j)}(x)$ and $\sup_{j\leq N, |x|\in(1,2)} \alpha^{(j)}(x)$.

Now we fix an $N \geq 1$ and integrate the second integrand of (4.5.4) by parts N times, each time integrating $-ik\lambda x^{k-1} e^{i\lambda x^k}$ and differentiating the rest. The resulting integral is

$$\int e^{i\lambda x^k}(-ik\lambda)^{-N} D^N\left[x^l \eta(x)(1 - \alpha(x/\varepsilon))\right] dx.$$

By (4.5.5), the modulus of the integrand is at most $C\mathbf{1}_{|x|\geq\varepsilon}|x|^{l-kN}(k\lambda)^{-N}$, which integrates to at most $C_2\lambda^{-N}\varepsilon^{l-kN+1}$. Set $\varepsilon = \lambda^{-1/k}$ and add the bounds on the two integrals to obtain an upper bound on $\int e^{i\lambda x^k} g(x)\, dx$ of $(C_1 + C_2)\lambda^{-(l+1)/k}$. We have also shown that C_1 and C_2 depend only on k, l, the first l derivatives of η, and the first l derivatives of α. Thus, taking α to be a fixed, convenient function, the lemma is proved. □

This yields an immediate corollary.

Corollary 4.5.4 *Let $\phi(x) = x^k$. If a smooth function g vanishes in an neighborhood of 0 and decreases rapidly at infinity (or is compactly supported), then $\mathcal{I}(\lambda) = \int e^{i\lambda x^k} g(x)\, dx$ is rapidly decreasing.*

A and ϕ are Monomials and A is Damped

In this section, we prove the C^∞ version of Lemma 4.2.2. However, because a monomial amplitude function does not have compact support, we introduce a damping function, which will later need to be removed. Define

$$\mathcal{I}(\lambda, k, l, \delta) := \int_{-\infty}^{\infty} e^{i\lambda x^k} e^{-\delta|x|^k} x^l\, dx\,. \tag{4.5.6}$$

Lemma 4.5.5 *As $\lambda \to \infty$, there is an asymptotic development*

$$\mathcal{I}(\lambda, k, l, \delta) \sim \lambda^{-(l+1)/k} \sum_{j=0}^{\infty} C(j, k, l, \delta)\lambda^{-j}\,.$$

The constants in the Nth remainder term remain bounded (in fact go to 0) as $\delta \to 0$.

Proof Let $z = (\delta - i\lambda)^{1/k}x$, where we choose the principal branch of the $1/k$ power. The half-line integral, which we will denote $\mathcal{I}_+(\lambda, k, l, \delta)$, may be written as

$$\int_0^{\infty(\delta - i\lambda)^{1/k}} e^{-z^k}(\delta - i\lambda)^{-l/k} z^l \frac{dz}{(\delta - i\lambda)^{1/k}} \, .$$

Now we may rotate the contour back to the real line. Specifically, for fixed λ, as $M \to \infty$, the difference between the above integral taken from 0 to $M(\delta - i\lambda)^{1/k}$ and the integral along the positive real line segment $[0, M|\delta - i\lambda|^{1/k}]$ is the integral of an exponentially small function of M along an arc of length $O(M)$; the difference therefore goes to zero and we obtain

$$\mathcal{I}_+(\lambda, k, l, \delta) = (\delta - i\lambda)^{-(l+1)/k} \int_0^\infty e^{-x^k} x^l \, dx \, ,$$

where the $(l + 1)$ power of the principal $1/k$ power is used. The definite integral has value $C(k, l)$. Writing $(\delta - i\lambda)^{-(l+1)/k}$ as $(-i\lambda)^{-(l+1)/k}(1 + \delta i/\lambda)^{-(l+1)/k}$ and using the binomial theorem gives

$$\mathcal{I}_+(\lambda, k, l, \delta) = C(k, l) e^{i\pi(l+1)/(2k)} \lambda^{-(l+1)/k} \sum_{j=0}^\infty (i\delta)^j \binom{-(l+1)/k}{j} \lambda^{-j} .$$

Denote

$$C_+(j, k, l, \delta) = k^{-1} \Gamma\left(\frac{l+1}{k}\right) e^{i\pi(l+1+jk)/(2k)} \binom{-(l+1)/k}{j} \delta^j \qquad (4.5.7)$$

and add the analogous computation for $\mathcal{I}_-(j, k, l, \delta)$ to prove the lemma with

$$C(j, k, l, \delta) := C_+(j, k, l, \delta) + C_-(j, k, l, \delta) \, ,$$

the remainder terms evidently going to zero as $\delta \to 0$. □

ϕ is a Monomial, A is Anything

Theorem 4.5.6 *Let $\phi(x) = x^k$. Let A be smooth and compactly supported with 0 in the closed support, and let $\{b_j\}$ be the power series coefficients at 0, with l denoting the index of the first nonvanishing term b_l. Then $\mathcal{I}_+ := \int_0^\infty A(x) \exp(i\lambda\phi(x)) \, dx$ has asymptotic development*

$$\mathcal{I}_+ \sim \sum_{j=l}^\infty b_j C(k, j)(i\lambda)^{-(1+j)/k} .$$

The constant in the $O(\lambda^{-N/k})$ remainder term is bounded in terms of the suprema of the first N derivatives of A near 0. A similar result holds for the two-sided *integral, $\mathcal{I}$, with coefficients α_j obtained by plugging b_j in for a_j in the conclusion to the univariate Fourier-Laplace Theorem.*

Proof Let U be a smooth function i.e., 1 on the support of A and vanishes outside of a compact set. Fix $N \geq 1$ and $\delta > 0$ and define the polynomial $P(x) = P_{N,\delta}(x)$ to be the sum of the Taylor series for $e^{\delta x^k} A(x)$ through the x^N term. Let $b_{j,\delta}$ denote the Taylor coefficients of $P_{N,\delta}$. Define the normalized remainder term $R(x) = R_{N,\delta}(x)$ by $e^{\delta x^k} A(x) = P(x) + x^{N+1} R(x)$. Now represent I_+ as $B_1 + B_2 + B_3$, where

$$B_1 := \int_0^\infty e^{i\lambda x^k} e^{-\delta x^k} P(x)\, dx \ ;$$

$$B_2 := \int_0^\infty e^{i\lambda x^k} x^{N+1} e^{-\delta x^k} R(x) U(x)\, dx \ ;$$

$$B_3 := \int_0^\infty e^{i\lambda x^k} e^{-\delta x^k} P(x)(U(x) - 1)\, dx \ .$$

By Lemma 4.5.3 with $\eta(x) = e^{-\delta x^k} R(x) U(x)$ and $l = N + 1$, we know that the magnitude of B_2 is bounded by $K\lambda^{-(l+2)/k}$. Similarly, by Corollary 4.5.4, we see that B_3 is rapidly decreasing as $\lambda \to \infty$. Furthermore, in both cases, K may be bounded in terms of k, l, and the first l derivatives of A, the bound being uniform over δ in a neighborhood of 0. It follows that the asymptotic series for $\mathcal{I}_+$, up to the $\lambda^{-(l+1)/k}$ term, may be obtained by taking $\delta \to 0$ in B_1.

Because P is a finite sum of monomials, we may use Lemma 4.5.5 to compute B_1, which recovers the hypothesis of the theorem. As before, we may sum results for $\mathcal{I}_+$ and $\mathcal{I}_-$ to prove the result for $\mathcal{I}$. □

General A and ϕ

Because $i\phi$ always lies along the imaginary axis, we may use a diffeomorphic change of variables to change ϕ into $ic_k x^k$, under which the contour remains along the imaginary axis (thus there is no need for arguments about moving the contour).

Proof of Theorem 4.5.2 By assumption, $\phi(x) = c_k x^k (1 + \theta(x))$ where $\theta(x) = O(|x|)$. Let $y = x(1 + \theta(x))^{1/k}$. This is a diffeomorphism in a neighborhood of 0, and we write $x = g(y)$ to denote its inverse. Then $c_k y^k = \phi(x)$, and so we may change variables to write

$$\int e^{i\lambda\phi(x)} A(x)\, dx = \int e^{i\lambda c_k y^k} \tilde{A}(y)\, dy.$$

The result now follows from Lemma 4.2.2. □

Notes

Our chief sources for Sections 1 and 2 were Bleistein and Handelsman (1986) and Wong (2001); see also Henrici (1991), which was used heavily in Chapter 3. Although it certainly follows from the extensive analyses in, for example, Bleistein and Handelsman (1986, Chapter 7), the two types of integrals, Fourier and Laplace, are seldom treated together, and we have never seen the univariate Fourier-Laplace Theorem stated in exactly this form. We have also never seen a derivation by purely complex analytic methods.

Watson's Lemma may be found in many places. The version here agrees with the statements in Bleistein and Handelsman (1986, Section 4.1) and Henrici (1991, Section 11.5). The method of steepest descent is described very nicely in de Bruijn (1981). Our treatment is more akin to Henrici (1991, Section 11.8); see also Bleistein and Handelsman (1986, Chapter 7), especially for a couple of the exercises.

Section 4.5 borrows heavily from Stein (1993). We have attempted to fill in some details. For instance, our proof of Lemma 4.5.3 is summarized as "A simple computation shows..." in Stein (1993, bottom of page 335), which also omits details of how the argument for $k = 2$ extends to greater values of k (see, e.g., Stein 1993, p. VIII.1.3.3). Despite its omission of detail in elementary arguments, Stein's book is a beautifully written modern classic and is a recommended addition to anyone's bookshelf.

The use of complex analytic methods to prove Lemma 4.5.5 also follows Stein (1993) (see 1.3.3 and step 1 of 1.3.1 in Chapter 8 of Stein).

Exercises

4.1 (degenerate case)
Although Theorem 4.1.1 assumes $k \geq 2$, the results are valid in the degenerate case $k = 1$. The expansion for $\mathcal{I}_+$ given in part (i) by (4.1.2) holds when $k = 1$. What is ζ in the case $k = 1$ and what expansion for $\mathcal{I}$ holds as a consequence?

4.2 (next term in the expansion)
Let $k = 2$ and $l = 0$ in Theorem 4.1.1. The theorem then gives

$$\mathcal{I} = b_0\sqrt{\frac{\pi}{c_2}}\lambda^{-1/2} + a_2\lambda^{-3/2} + O(\lambda^{-5/2}).$$

Compute the coefficient a_2 in terms of b_0, b_1, b_2, c_2, c_3 and c_4.

4.3 (Bessel function)
The Bessel function of the first kind is defined by

$$J_m(r) = \frac{1}{2\pi}\int_0^{2\pi} \exp(ir\sin\theta - im\theta)\, d\theta,$$

where m is a fixed parameter (you may assume it is a positive integer). Use Theorem 4.5.2 to find the two leading terms of an asymptotic series for $J_m(r)$ in decreasing powers of r.

4.4 (Airy function)

The *Airy function* is defined by

$$\mathrm{Ai}(x) = \frac{1}{2\pi} \int_{-\infty}^{\infty} e^{i(xt + t^3/3)}\, dt.$$

Find an asymptotic expression for $\mathrm{Ai}(x)$ as $x \to \infty$ in $\mathbb{R}^+$. Step 1: change variables by $t = ix^{1/2}u$. Step 2: find the critical points and deform the contour to pass through one or more of them. Step 3: compute the expansion on a compactly supported interval; then argue that this converges as the limits of integration go to infinity.

5

Fourier-Laplace Integrals in More than One Variable

5.1 Overview

In this chapter, we generalize the work in the previous chapter on the asymptotic evaluation of the saddle point integral

$$\mathcal{I}(\lambda) := \int_C A(z) \exp(-\lambda\phi(z)),\, dz, \tag{5.1.1}$$

where the amplitude A and phase ϕ are analytic functions of a vector argument z and C is a d-chain in $\mathbb{C}^d$. In one variable, the comprehensive result Theorem 4.1.1 covers all degrees of degeneracy of the phase function (the parameter k) and all degrees of vanishing of the amplitude function (the parameter l). The range of possibilities for the phase function ϕ in higher dimensions is much greater. We are concerned only with the quadratic case. In one dimension, this boils down to taking $k = 2$. In higher dimensions, we assume nonsingularity of the ***Hessian matrix*** $\mathcal{H} := \left(\partial^2\phi/dz_j dz_k\right)$. The Taylor series for any sufficiently differentiable function ϕ is

$$\phi(z) = \phi(\mathbf{0}) + z^T \nabla \phi(\mathbf{0}) + \frac{1}{2} z^T \mathcal{H} z + O(|z|^3),$$

hence the Hessian matrix represents (twice) the quadratic term in the phase, and its nonsingularity is a generalization of nonvanishing of the quadratic term in the univariate case. The initial part of the development is the same as in the univariate case. Let $S(\boldsymbol{x}) := x_1^2 + \cdots + x_d^2$ denote the standard quadratic, generalizing the special phase function x^2 in the univariate case; we use $\boldsymbol{x}$ rather than z to highlight this assumption. A result when A is monomial and ϕ is the standard quadratic (Corollary 5.2.3) is coupled with a big-O result (Proposition 5.2.4), allowing us to integrate term by term and obtain asymptotics for the standard phase function:

Theorem 5.1.1 (Standard phase) *Let $A(\boldsymbol{x}) = \sum_{\boldsymbol{r}} a_{\boldsymbol{r}} \boldsymbol{x}^{\boldsymbol{r}}$ be any real analytic function defined on a neighborhood $\mathcal{N}$ of the origin in $\mathbb{R}^d$. Let*

$$\mathcal{I}(\lambda) := \int_{\mathcal{N}} A(\boldsymbol{x}) e^{-\lambda S(\boldsymbol{x})}\, d\boldsymbol{x}\,. \tag{5.1.2}$$

Then

$$\mathcal{I}(\lambda) \sim \sum_n \sum_{|\mathbf{r}|=n} a_{\mathbf{r}} \beta_{\mathbf{r}} \lambda^{-(|\mathbf{r}|+d)/2}$$

as an asymptotic series expansion in increasing $|\mathbf{r}|$, *where* $\beta_{\mathbf{r}} = 0$ *if any* r_j *is odd, and*

$$\beta_{2\mathbf{m}} = \pi^{d/2} \prod_{j=1}^{d} \frac{(2m_j)!}{m_j! 4^{m_j}}$$

otherwise.

We then extend to the following result concerning a general complex phase, with the assumption that the real part has a strict minimum at the origin. Arguing again by contour deformation, we will prove:

Theorem 5.1.2 (Re$\{\phi\}$ has a strict minimum) *Suppose that the real part of* ϕ *is strictly positive except at the origin and that its Hessian matrix* $\mathcal{H}$ *is nonsingular there. Let A be any analytic function not vanishing at the origin and define* $\mathcal{I}(\lambda)$ *by* (5.1.1). *Then*

$$\mathcal{I}(\lambda) \sim \sum_{\ell \geq 0} c_\ell \lambda^{-d/2-\ell},$$

where

$$c_0 = A(0) \frac{\pi^{d/2}}{\sqrt{\det\left(\frac{1}{2}\mathcal{H}\right)}},$$

and the choice of sign is defined by taking the product of the principal square roots of the eigenvalues of $\mathcal{H}$.

The last set of results in the multivariate case departs from the framework of the univariate case. All the results in the univariate case assumed a strict minimum of the real part of the phase function, until the very end, when we proved results for strictly imaginary phase functions. These last results were proved quite differently, using C^∞ methods, which allowed the introduction of bump functions. Having two completely different proofs for the two cases was not of great concern because the univariate case has a dichotomy: an analytic function on $\mathbb{R}$ whose real part has a minimum at zero either has a strict minimum there or has real part vanishing everywhere. In the multivariate case, this is no longer true. Furthermore, we need to integrate over regions such as rectangles and annuli, on which the real part of the phase is nonvanishing clear to the boundary. In one dimension this would lead to boundary terms large enough to appear in the asymptotics. In more than one variable, the boundary has positive dimension, and boundary contributions are avoided when the phase is not stationary along the boundary.

There is some overhead even in stating these results. First, one must define appropriate chains of integration; these will be chains supported on Whitney stratified spaces. Next, to find the necessary deformations of these chains, one must use vector field constructions relying on semicontinuity notions for stratified spaces; these constructions are staples of stratified Morse theory. In the end, we prove the following result, which is more precisely stated as Theorem 5.4.8.

Theorem (critical point decomposition for stratified spaces) *Let A and ϕ be analytic functions on a neighborhood of a stratified space $\mathcal{M} \subseteq \mathbb{C}^d$. If ϕ has finitely many critical points on $\mathcal{M}$, then*

$$\mathcal{I}(\lambda) \sim (\pi\lambda)^{d/2} \sum_{\boldsymbol{x}} A(\boldsymbol{x}) e^{\lambda\phi(\boldsymbol{x})} \det((1/2)\mathcal{H}(\boldsymbol{x}))^{-1/2},$$

where $\mathcal{H}(\boldsymbol{x})$ is the Hessian matrix for ϕ at $\boldsymbol{x}$, and the sum is over critical points $\boldsymbol{x}$, at which the real part of ϕ is minimized.

5.2 Standard Phase

As in the one-dimensional case, we begin with the simplest phase function and a monomial amplitude. We first state an explicit formula for the one-dimensional monomial integral in the case $k = 2$ and $l = 2n$.

Proposition 5.2.1

$$\int_{-\infty}^{\infty} x^{2n} e^{-x^2}\, dx = \beta_{2n} := \sqrt{\pi}\, \frac{(2n)!}{n!\, 4^n}\, .$$

Proof For $n = 0$, this is just the standard Gaussian integral. By induction, assume now the result for $n - 1$. Integrate by parts to get

$$\begin{aligned} \int x^{2n} e^{-x^2}\, dx &= \int \frac{-x^{2n-1}}{2} \left(-2x\, e^{-x^2}\, dx\right) \\ &= \frac{2n-1}{2} \int x^{2n-2}\, e^{-x^2}\, dx \\ &= \frac{2n-1}{2} \sqrt{\pi}\, \frac{(2n-2)!}{(n-1)!\, 4^{n-1}} \end{aligned}$$

by the induction hypothesis. This is equal to $\sqrt{\pi}(2n)!\, /\, (n!\, 4^n)$, completing the induction. □

Corollary 5.2.2

$$\int_{-\infty}^{\infty} x^{2n}\, e^{-\lambda x^2}\, dx = \beta_{2n} \lambda^{-1/2-n}\, .$$

Proof Changing variables by $y = \lambda^{1/2} x$ yields

$$\int_{-\infty}^{\infty} \lambda^{-n} y^{2n} e^{-y^2}\, \frac{dy}{\lambda^{1/2}}.$$ □

Let $S(\boldsymbol{x}) := \sum_{j=1}^{d} x_j^2$ denote the standard quadratic.

Corollary 5.2.3 (monomial integral) *Let $\mathbf{r}$ be any d-vector of nonnegative integers. Then*

$$\int_{\mathbb{R}^d} \boldsymbol{x}^{\boldsymbol{r}} e^{-\lambda S(\boldsymbol{x})}\, d\boldsymbol{x} = \beta_{\boldsymbol{r}} \lambda^{-(d+|\boldsymbol{r}|)/2},$$

where $\beta_{\boldsymbol{r}} = \prod_{j=1}^{d} \beta_{r_j}$ if all the components r_j are even and is zero otherwise.

Proof The integral factors into

$$\prod_{j=1}^{d} \left[\int_{-\infty}^{\infty} x_j^{r_j} e^{-\lambda x_j^2}\, dx_j \right],$$

reducing this to the result of Proposition 5.2.1. □

Proposition 5.2.4 (big-O estimate) *Let A be any smooth function satisfying $A(\boldsymbol{x}) = O(|\boldsymbol{x}|^r)$ at the origin. Then the integral of $A(\boldsymbol{x})e^{-\lambda S(\boldsymbol{x})}$ over any compact set K may be bounded from above by*

$$\int_K A(\boldsymbol{x}) e^{-\lambda S(\boldsymbol{x})}\, d\boldsymbol{x} = O(\lambda^{-(d+r)/2}).$$

The implied constant on the right goes to zero as the constant in the hypothesis $A(\boldsymbol{x}) = O(|\boldsymbol{x}|^r)$ goes to zero.

Proof Because K is compact and $A(\boldsymbol{x}) = O(|\boldsymbol{x}|^r)$ at the origin, it follows that there is some constant c for which $|A(\boldsymbol{x})| \le c|\boldsymbol{x}|^r$ on all of K. Let K_0 denote the intersection of K with the ball $|\boldsymbol{x}| \le \lambda^{-1/2}$ and for $n \ge 1$ let K_n denote the intersection of K with the shell $2^{n-1}\lambda^{-1/2} \le |\boldsymbol{x}| \le 2^n \lambda^{-1/2}$. On K_0

$$|A(\boldsymbol{x})| \le c\lambda^{-r/2}$$

while trivially

$$\int_{K_0} e^{-\lambda S(\boldsymbol{x})}\, d\boldsymbol{x} \le \int_{K_0} d\boldsymbol{x} \le c_d \lambda^{-d/2}.$$

Thus

$$\left| \int_{K_0} A(\boldsymbol{x}) e^{-\lambda S(\boldsymbol{x})}\, d\boldsymbol{x} \right| \le c' \lambda^{-(r+d)/2}.$$

For $n \ge 1$, on K_n, we have the upper bounds

$$|A(\boldsymbol{x})| \le 2^{rn} c \lambda^{-r/2}$$

$$e^{-\lambda S(\boldsymbol{x})} \le e^{-2^{2n-2}}$$

$$\int_{K_n} d\boldsymbol{x} \le 2^{dn} c_d \lambda^{-d/2}.$$

Letting $c'' := c \cdot c_d \cdot \sum_{n=1}^{\infty} 2^{(d+r)n} e^{-2^{2n-2}} < \infty$, we may sum to find that

$$\sum_{n=0}^{\infty} \left| \int_{K_n} A(\boldsymbol{x}) e^{-\lambda S(\boldsymbol{x})} \, d\boldsymbol{x} \right| \leq (c' + c'') \lambda^{-(r+d)/2},$$

proving the lemma. □

Proof of Theorem 5.1.1 Write $A(\boldsymbol{x})$ as a power series up to degree N plus a remainder term:

$$A(\boldsymbol{x}) = \left(\sum_{n=0}^{N} \sum_{|\boldsymbol{r}|=n} a_{\boldsymbol{r}} \boldsymbol{x}^{\boldsymbol{r}} \right) + R_N(\boldsymbol{x}),$$

where $R_N(\boldsymbol{x}) = O(|\boldsymbol{x}|^{N+1})$. Using Corollary 5.2.3 to integrate all the monomial terms and Proposition 5.2.4 to bound the integral of $R_N(\boldsymbol{x}) e^{-\lambda S(\boldsymbol{x})}$ shows that

$$\mathcal{I}(\lambda) = \sum_{n=0}^{N} \sum_{|\boldsymbol{r}|=n} a_{\boldsymbol{r}} \beta_{\boldsymbol{r}} \lambda^{-(n+d)/2} + O(\lambda^{-(N+1+d)/2}),$$

which proves the asymptotic expansion. □

5.3 Real Part of Phase Has a Strict Minimum

Let $\mathcal{N}$ be a neighborhood of the origin in $\mathbb{R}^d$. We say that the function $\phi : \mathcal{N} \to \mathbb{C}$ is analytic if ϕ is represented by a power series that converges on $\mathcal{N}$. Such a function may be extended to a holomorphic function on a neighborhood $\mathcal{N}_{\mathbb{C}}$ of the origin in $\mathbb{C}^d$. Suppose $\phi(\mathbf{0}) = 0$ and the real part of ϕ is nonnegative on $\mathcal{N}$. The gradient of ϕ must vanish at the origin. We say that ϕ has a (quadratically) nondegenerate critical point at the origin if the quadratic part of ϕ is nondegenerate. Recall that the quadratic part of ϕ is a quadratic form represented by one half the Hessian matrix. By nondegeneracy of a quadratic form, we mean nonsingularity of the Hessian; by determinant of a quadratic form, we mean the determinant of half of the Hessian.

We review how the Hessian behaves under changes of variables. If $\psi : \mathbb{C}^d \to \mathbb{C}^d$ is a bi-holomorphic map, and if ϕ has vanishing gradient at $\psi(\boldsymbol{x})$ and Hessian matrix $\mathcal{H}$ there, then the Hessian matrix $\tilde{\mathcal{H}}$ of $\phi \circ \psi$ at $\boldsymbol{x}$ is given by

$$\tilde{\mathcal{H}} = J_{\psi}^T \, \mathcal{H} \, J_{\psi} \tag{5.3.1}$$

where J_{ψ} is the Jacobian matrix of the map ψ at $\boldsymbol{x}$.

The first key lemma is that, under the assumption of nondegeneracy of the Hessian, we can change variables so that ϕ becomes the standard quadratic form.

Lemma 5.3.1 *Define* $Q := (1/2)\mathcal{H}$. *There is a bi-holomorphic change of variables* $\boldsymbol{x} = \psi(\boldsymbol{y})$ *such that* $\phi(\psi(\boldsymbol{y})) = S(\boldsymbol{y}) := \sum_{j=1}^{d} y_j^2$. *The differential* $J_{\psi} = d\psi(\mathbf{0})$ *will satisfy* $(\det J_{\psi})^2 = (\det Q)^{-1}$.

Remark This is known as the Morse Lemma. The proof here is adapted from the proof of the real version given in Stein (1993, VIII:2.3.2).

Proof Taking the last conclusion first, use (5.3.1) to see that the Hessian of the standard form S is equal to $J_\psi^T \mathcal{H} J_\psi$, where $\mathcal{H}$ is the Hessian matrix of ϕ. The Hessian of S is twice the identity matrix, so dividing by 2 and taking determinants gives $|J_\psi|^2|Q| = 1$.

To prove the change of variables, the first step is to write

$$\phi(\boldsymbol{x}) = \sum_{j,k=1}^{d} x_j x_k \phi_{j,k},$$

where the functions $\phi_{j,k} = \phi_{k,j}$ are analytic and satisfy $\phi_{j,k}(\boldsymbol{0}) = (1/2)\mathcal{H}_{j,k}$. It is obvious from a formal power series viewpoint that this can be done because the summand $x_j x_k \phi_{j,k}$ can be any power series with coefficients indexed by the orthant $\{\boldsymbol{r} : \boldsymbol{r} \geq \boldsymbol{e}_j + \boldsymbol{e}_k\}$; these orthants cover $\{\boldsymbol{r} : |\boldsymbol{r}| \geq 2\}$, so we may obtain any function ϕ vanishing to order 2; matching coefficients on the terms of order precisely 2 shows that $\phi_{j,k}(\boldsymbol{0}) = (1/2)\mathcal{H}_{j,k}$.

More constructively, we may give a formula for $\phi_{j,k}$. There is plenty of freedom, but a convenient choice is to take

$$x_j x_k \phi_{j,k}(\boldsymbol{x}) := \sum_{|\boldsymbol{r}|\geq 2} \frac{r_j(r_k - \delta_{j,k})}{|\boldsymbol{r}|(|\boldsymbol{r}| - 1)} a_{\boldsymbol{r}} \boldsymbol{x}^{\boldsymbol{r}} .$$

For fixed $\boldsymbol{r}$, it is easy to check that

$$\sum_{1\leq j,k\leq d} \frac{r_j(r_k - \delta_{j,k})}{|\boldsymbol{r}|(|\boldsymbol{r}| - 1)} = 1,$$

whence $\phi = \sum x_j x_k \phi_{j,k}$. Alternatively, the following analytic computation from Stein (1993) verifies that $\phi = \sum_{j,k} x_j x_k \phi_{j,k}$. Any function with f and f' vanishing at zero satisfies

$$f(t) = \int_0^1 \frac{d}{ds} f(ts)\, ds = \int_0^1 (1-s) \frac{d^2}{ds^2} f(ts)\, ds,$$

the second identity following from an integration by parts. Fix $\boldsymbol{x}$ and apply this with $f(t) = (d/dt)\phi(t\boldsymbol{x})$ to obtain

$$\phi(\boldsymbol{x}) = \int_0^1 (1-s) \frac{d^2}{ds^2} \phi(s\boldsymbol{x})\, ds.$$

The multivariate chain rule gives

$$\frac{d^2}{ds^2} \phi(s\boldsymbol{x}) = \sum_{j,k} x_j x_k \frac{\partial^2 \phi}{\partial x_j \partial x_k}(s\boldsymbol{x});$$

plug in $\phi = \sum_{\boldsymbol{r}} a_{\boldsymbol{r}} \boldsymbol{x}^{\boldsymbol{r}}$ and integrate term by term using $\int_0^1 (1-s)s^{n-2}\, ds = \frac{1}{n(n-1)}$ to see that $\phi = \sum_{j,k} x_j x_k \phi_{j,k}$.

The second step is an induction. Suppose first that $\phi_{j,j}(\mathbf{0}) \neq 0$ for all j. The function $\phi_{1,1}^{-1}$ and a branch of the function $\phi_{1,1}^{1/2}$ are analytic in a neighborhood of the origin. Set

$$y_1 := \phi_{1,1}^{1/2} \left[x_1 + \sum_{k>1} \frac{x_k \phi_{1,k}}{\phi_{1,1}} \right] .$$

Expanding, we find that the terms of y_1^2 of total degree at most 1 in the terms $x_2, \ldots, x_d$ match those of ϕ, and therefore,

$$\phi(\boldsymbol{x}) = y_1^2 + \sum_{j,k \geq 2} x_j x_k h_{j,k} \tag{5.3.2}$$

for some analytic functions $h_{j,k}$ satisfying $h_{j,k}(\mathbf{0}) = (1/2)\mathcal{H}_{j,k}$. Similarly, if

$$\phi(\boldsymbol{x}) = \sum_{j=1}^{r-1} y_j^2 + \sum_{j,k \geq r} x_j x_k h_{j,k},$$

then setting

$$y_r := \phi_{r,r}^{1/2} \left[x_r + \sum_{k>r} \frac{y_k h_{r,k}}{h_{r,r}} \right]$$

gives

$$\phi(\boldsymbol{x}) = \sum_{j=1}^{r} y_j^2 + \sum_{j,k \geq r+1} x_j x_k \tilde{h}_{j,k}$$

for some analytic functions $\tilde{h}_{j,k}$ still satisfying $h_{j,k}(\mathbf{0}) = (1/2)\mathcal{H}_{j,k}$. By induction, we arrive at $\phi(\boldsymbol{x}) = \sum_{j=1}^{d} y_j^2$, finishing the proof of the Morse Lemma in the case where each $\mathcal{H}_{j,j}$ is nonzero.

Finally, if some $\mathcal{H}_{j,j} = 0$, because $\mathcal{H}$ is nonsingular, we may always find some real orthogonal map U such that the Hessian $U^T \mathcal{H} U$ of $\phi \circ U$ has no vanishing diagonal entries. We know there is a ψ_0 such that $(\phi \circ U) \circ \psi_0 = S$, and taking $\psi = U \circ \psi_0$ finishes the proof in this case (in fact this is true almost surely for a Gaussian random orthogonal matrix). □

Proof of Theorem 5.1.2 The power series allows us to extend ϕ to a neighborhood of the origin in $\mathbb{C}^d$. Under the change of variables ψ from the previous lemma, we see that

$$\begin{aligned} \mathcal{I}(\lambda) &= \int_{\psi^{-1}C} A \circ \psi(\boldsymbol{y}) e^{-\lambda S(\boldsymbol{y})} (\det d\psi(\boldsymbol{y}))\, d\boldsymbol{y} \\ &:= \int_{\psi^{-1}C} \tilde{A}(\boldsymbol{y}) e^{-\lambda S(\boldsymbol{y})}\, d\boldsymbol{y}, \end{aligned}$$

where C is a neighborhood of the origin in $\mathbb{R}^d$ with the standard orientation. We need to check that we can move the chain $\psi^{-1}C$ of integration back to the real plane. If we can, then applying the expansion from Theorem 5.1.1 and noting that the terms with odd values of $|\mathbf{r}|$ all vanish yields the desired expansion in powers $\lambda^{-d/2-\ell}$.

Let $h(z) := \text{Re}\{S(z)\}$. The chain $C' := \psi^{-1}(C)$ lies in the region $\{z \in \mathbb{C}^d : h(z) > 0\}$ except when $z = 0$, and in particular, $h \geq \varepsilon > 0$ on $\partial C'$. Let

$$H(z, t) := \text{Re}\{z\} + (1 - t)i \ \text{Im}\{z\} .$$

In other words, H is a homotopy from the identity map to the map π projecting out the imaginary part of the vector z. For any chain σ, the homotopy H induces a chain homotopy, $H(\sigma)$ supported on the image of the support of σ under the homotopy H and satisfying

$$\partial H(\sigma) = \sigma - \pi\sigma + H(\partial\sigma).$$

With $\sigma = C'$, observing that $S(H(z, t)) \geq S(z)$, we see there is a $(d + 1)$-chain $\mathcal{D}$ with

$$\partial\mathcal{D} = C' - \pi C' + C''$$

and C'' supported on $\{h > \varepsilon\}$. Stokes' Theorem tells us that for any holomorphic d-form ω,

$$\int_{\partial\mathcal{D}} \omega = \int_{\mathcal{D}} d\omega = 0$$

and, consequently, that

$$\int_{C'} \omega = \int_{\pi C'} \omega - \int_{C''} \omega .$$

When $\omega = \tilde{A}e^{-\lambda S}\, d\mathbf{y}$, the integral over C'' is $O(e^{-\lambda\varepsilon})$, giving

$$\mathcal{I}(\lambda) = \int_{\pi C'} \tilde{A}(\mathbf{y})e^{-\lambda S(\mathbf{y})}\, d\mathbf{y} + O(e^{-\varepsilon\lambda}) .$$

Up to sign, the chain $\pi C'$ is a disk in $\mathbb{R}^d$ with the standard orientation plus something supported in $\{h > \varepsilon\}$. To see this, note that π maps any real d-manifold in $\mathbb{C}^d$ diffeomorphically to $\mathbb{R}^d$ wherever the tangent space is transverse to the imaginary subspace. The tangent space to the support of C' at the origin is transverse to the imaginary subspace because $S \geq 0$ on C', whereas the imaginary subspace is precisely the negative d-space of the index-d form S. The tangent space varies continuously, so in a neighborhood of the origin, π is a diffeomorphism. Observing that $\tilde{A}(\mathbf{0}) = A(\mathbf{0}) \det(d\psi(\mathbf{0})) = A(\mathbf{0})(\det((1/2)\mathcal{H}))^{-1/2}$ finishes the proof up to the choice of sign of the square root.

The map $d\pi \circ d\psi^{-1}(\mathbf{0})$ maps the standard basis of $\mathbb{R}^d$ to another basis for $\mathbb{R}^d$. Verifying the sign choice is equivalent to showing that this second basis is positively oriented if and only if $\det(d\psi(\mathbf{0}))$ is the product of the principal square roots of the eigenvalues of $\mathcal{H}$ (it must be either this or its negative). Thus we will be finished by applying the following lemma (with $\alpha = \psi^{-1}$).

Lemma 5.3.2 *Let $W \subseteq \mathbb{C}^d$ be the set $\{z : \mathrm{Re}\{S(z)\} > 0\}$. Pick any $\alpha \in GL_d(\mathbb{C})$ mapping $\mathbb{R}^d$ into $\overline{W}$ and let $M := \alpha^T \alpha$ be the matrix representing $S \circ \alpha$. Let $\pi : \mathbb{C}^d \to \mathbb{R}^d$ be projection onto the real part. Then $\pi \circ \alpha$ is orientation preserving on $\mathbb{R}^d$ if and only if $\det \alpha$ is the product of the principal square roots of the eigenvalues of M (rather than the negative of this).*

Proof First suppose $\alpha \in GL_d(\mathbb{R})$. Then M has positive eigenvalues, so the product of their principal square roots is positive. The map π is the identity on $\mathbb{R}^d$, so the statement boils down to saying that α preserves orientation if and only if it has positive determinant, which is true by definition. In the general case, let $\alpha_t := \pi_t \circ \alpha$, where $\pi_t(z) = \mathrm{Re}\{z\} + i(1-t)\,\mathrm{Im}\{z\}$. As we saw in the previous proof, $\pi_t(\mathbb{R}^d) \subseteq \overline{W}$ for all $0 \le t \le 1$, whence $M_t := \alpha_t^T \alpha_t$ has eigenvalues with nonnegative real parts. The product of the principal square roots of the eigenvalues is a continuous function on the set of nonsingular matrices with no negative real eigenvalues. The determinant of α_t is a continuous function of t, and we have seen it agrees with the product of principal square roots of eigenvalues of M_t when $t = 1$ (the real case), so by continuity, this is the correct sign choice for all $0 \le t \le 1$; taking $t = 0$ proves the lemma. □

5.4 Localization

Our aim is to integrate $A(\boldsymbol{x})e^{-\lambda\phi(\boldsymbol{x})}$ over a compact chain C. Consider, as an example, the chain $C \times I$, where C is the unit circle in $\mathbb{C}^1$ and I is the interval $[-1, 1]$. As an example, suppose the phase function is given by

$$\phi(e^{i\theta}, \alpha) := \theta^2 + i\alpha.$$

The real part of ϕ is nonnegative but vanishes along the entire line segments $\{0\} \times I$. However, there is only one critical point, namely $(0, 0)$, because the gradient of ϕ has nonvanishing imaginary part elsewhere. How can we see that the main contribution to $\mathcal{I}(\lambda)$ occurs near $(0, 0)$? In this example, foliating by circles, we may use Lemma 4.5.1 to see that the integral is small away from the median circle, $C \times \{0\}$.

The point of this section is to give a general argument localizing the integral for $\mathcal{I}(\lambda)$ to neighborhoods of critical points. First, we must extend the definition of a critical point to spaces more general than manifolds. We then show that, away from critical points, we may deform the chain of integration to where the real part of the phase is strictly positive. In the example, C will be deformed to a new chain where the real part of ϕ vanishes only at $(0, 0)$. However, showing that the integrals over the old and new chains agree is somewhat tricky. Recall from (A.1) that if $H : C \times [0, 1]$ is a homotopy between the original chain C and a new chain C', then

$$\partial H = C' - C + \partial C \times \sigma,$$

where σ is the standard 1-simplex. This last term may cause difficulty if C has a nontrivial boundary and the real part of the phase has minima on ∂C.

The main result of this section is Theorem 5.4.8, localizing the integral to critical points in the manner just described. To do this, we need to define a suitable class of chains and then develop some geometric properties of these. We begin with the classical notion of stratified spaces.

Whitney Stratifications

Many interesting spaces, such as algebraic varieties, are not manifolds. The next best thing is if a space is built nicely out of parts that are manifolds. The following discussion of stratification summarizes a longer discussion in Appendix C.

Let I be a finite partially ordered set and define an I-decomposition of a topological space Z to be a partition of Z into a disjoint union of sets $\{S_\alpha : \alpha \in I\}$ such that

$$S_\alpha \cap \overline{S_\beta} \neq \emptyset \iff S_\alpha \subseteq \overline{S_\beta} \iff \alpha \leq \beta .$$

Definition 5.4.1 (Whitney stratification) *Let Z be a closed subset of a smooth manifold $\mathcal{M}$. A* Whitney stratification *of Z is an I-decomposition such that*

(i) Each S_α is a manifold in $\mathbb{R}^n$.
(ii) If $\alpha < \beta$, if the sequences $\{x_i \in S_\beta\}$ and $\{y_i \in S_\alpha\}$ both converge to $y \in S_\alpha$, if the lines $l_i = \overline{x_i\, y_i}$ converge to a line l and the tangent planes $T_{x_i}(S_\beta)$ converge to a plane T of some dimension, then both l and $T_y(S_\alpha)$ are contained in T.

Associated with the definition of a stratification is the stratified notion of a critical point.

Definition 5.4.2 (smooth functions and their critical points) *Say that a function $\phi : \mathcal{M} \to \mathbb{C}$ on a stratified space $\mathcal{M}$ is smooth if it is smooth when restricted to each stratum. A point $p \in \mathcal{M}$ is said to be critical for the smooth function ϕ if and only if the restriction $d\phi_{|S}$ vanishes, where S is the stratum containing p.*

Whitney stratifications are ideal for the topological study of algebraic hypersurfaces because of the following classical result first proved by Whitney in the work that introduced Whitney stratifications (Whitney, 1965, Theorem 18.11) (in that article, they are called *regular stratifications*).

Proposition 5.4.3 *Every algebraic variety in $\mathbb{R}^d$ or $\mathbb{C}^d$ admits a Whitney stratification.*

The simplest example is a smooth manifold, $\mathcal{M}$. This is a Whitney stratified space with a single stratum, namely $\mathcal{M}$. The next simplest example is that of a space $\mathcal{V}$ for which one may find a finite subset E such that $\mathcal{V} \setminus E$ is a smooth manifold. The strata $(\mathcal{V} \setminus E, E)$ form a Whitney stratification. An algebraic variety

$\mathcal{V}$ whose singular locus is a smooth manifold $\mathcal{V}'$ may be stratified as $(\mathcal{V} \setminus \mathcal{V}', \mathcal{V}')$. However, if the singular locus itself has a finite, nonempty singular locus, E, it is not always true that $(\mathcal{V} \setminus \mathcal{V}', \mathcal{V}' \setminus E, E)$ is a Whitney stratification of $\mathcal{V}$; one might need to decompose the middle stratum further. See the appendices for more detail.

The second Whitney condition is difficult to read and impossible to remember, but basically it says that the strata fit together nicely. A well-known but difficult result is the local product structure of a stratified space: a point $\boldsymbol{p}$ in a k-dimensional stratum S of a stratified space $\mathcal{M}$ has a neighborhood in which $\mathcal{M}$ is homeomorphic to some product $S \times X$. According to Goresky and MacPherson (1988), a proof may be found in mimeographed notes of Mather from 1970; it is based on Thom's Isotopy Lemma, which takes up fifty pages of the same mimeographed notes.

Tangent Vector Fields

Our aim is to define a vector field along which to push a given embedding of a stratified space $\mathcal{M}$, so as to decrease the real part of ϕ everywhere, except at critical points, where we can do no better than to remain still. To do this, we begin with some basics about the tangent bundle.

The tangent space $T_{\boldsymbol{x}}(\mathcal{M})$ at a point $\boldsymbol{x}$ of the stratified space $\mathcal{M}$ is defined to be the tangent space $T_{\boldsymbol{x}}(S)$ where S is the stratum containing $\boldsymbol{x}$. To talk about continuity of vector fields, we need these spaces to fit together into a bundle. In the case where $\mathcal{M}$ is embedded in and inherits the analytic structure of $\mathbb{C}^d$, we may do precisely that. The local homeomorphism to a product, mentioned in the previous paragraph, is induced by the embedding. Each $T_{\boldsymbol{x}}(\mathcal{M})$ is naturally identified with a subspace of $T_{\boldsymbol{x}}(\mathbb{C}^d)$. A smooth section of the tangent bundle of $\mathcal{M}$ is simply a smooth vector field $f : \mathcal{M} \to \mathbb{C}^d$ such that $f(\boldsymbol{x}) \in T_{\boldsymbol{x}}(S)$ when $\boldsymbol{x}$ is in the stratum S. The product structure also gives us locally constant vector fields (though not in any natural way). The next two lemmas take advantage of this.

Lemma 5.4.4 *Let f be a smooth section of the tangent bundle to S, i.e., $f(s) \in T_s(S)$ for $s \in S$. Then each $s \in S$ has a neighborhood in $\mathcal{M}$ on which f may be extended to a smooth section of the tangent bundle.*

Proof In a local parametrization of $\mathcal{M}$ by $S \times X$, given $s \in S$, one may transport any vector $\boldsymbol{v} \in T_s(S)$ to any tangent space $T_{(s,x)}(\mathcal{M})$. Extend f by $f(s, x) := f(s)$. □

Let $\mathcal{M}$ be a real stratified space embedded in $\mathbb{C}^d$. This means that each stratum S is a subset of $\mathbb{C}^d$, and each of the chart maps ψ from a neighborhood in $\mathbb{R}^k$ to some k-dimensional stratum $S \subseteq \mathbb{C}^d$ is analytic (the coordinate functions are convergent power series) with a nonsingular differential. It follows that ψ may be extended to a holomorphic map on a neighborhood of the origin in $\mathbb{C}^k$, whose range we denote by $S \otimes \mathbb{C}$; choosing a small enough neighborhood, we may arrange for $S \otimes \mathbb{C}$ to be a complex k-manifold embedded in $\mathbb{C}^d$.

Lemma 5.4.5 (vector field near a noncritical point) *Let $\boldsymbol{x}$ be a point of the stratum S of the stratified space, $\mathcal{M}$ and suppose $\boldsymbol{x}$ is not critical for the function ϕ. Then there is a vector $\boldsymbol{v} \in T_{\boldsymbol{x}}(S \otimes \mathcal{M})$ such that* $\text{Re}\{d\phi(\boldsymbol{v})\} > 0$ *at $\boldsymbol{x}$. Furthermore, there is a continuous section f of the tangent bundle in a neighborhood $\mathcal{N}$ of $\boldsymbol{x}$ such that* $\text{Re}\{d\phi(f(\boldsymbol{y}))\} > 0$ *at every $\boldsymbol{y} \in \mathcal{N}$.*

Proof By noncriticality of $\boldsymbol{x}$, there is a $\boldsymbol{w} \in T_{\boldsymbol{x}}(S)$ with $d\phi(\boldsymbol{w}) = u \neq 0$ at $\boldsymbol{x}$. Multiply $\boldsymbol{w}$ component-wise by $\overline{u}$ to obtain $\boldsymbol{v}$ with $\text{Re}\{d\phi(\boldsymbol{v})\} > 0$ at $\boldsymbol{x}$. Use any chart map for $S \otimes \mathbb{C}$ near $\boldsymbol{x}$ to give a locally trivial coordinatization for the tangent bundle and define a section f to be the constant vector $\boldsymbol{v}$; then $\text{Re}\{d\phi(f(\boldsymbol{y}))\} > 0$ on some sufficiently small neighborhood of $\boldsymbol{x}$ in S. Finally, extend to a neighborhood of $\boldsymbol{x}$ in $\mathcal{M}$ by Lemma 5.4.4. □

Although we are working in the analytic category, the chains of integration are topological objects, for which we may use C^∞ methods (in what follows, even C^1 methods will do). In particular, a partition of unity argument enhances the local result above to a global result.

Lemma 5.4.6 (global vector field, in the absence of critical points) *Let $\mathcal{M}$ be a compact stratified space and ϕ a smooth function on $\mathcal{M}$ with no critical points. Then there is a global section f of the tangent bundle of $\mathcal{M}$ such that the real part of $d\phi(f)$ is everywhere positive.*

Proof For each point $\boldsymbol{x} \in \mathcal{M}$, let $f_{\boldsymbol{x}}$ be a section as in the conclusion of Lemma 5.4.5, on a neighborhood $U_{\boldsymbol{x}}$. Cover the compact space $\mathcal{M}$ by finitely many sets $\{U_{\boldsymbol{x}} : \boldsymbol{x} \in F\}$, and let $\{\psi_{\boldsymbol{x}} : \boldsymbol{x} \in F\}$ be a smooth partition of unity subordinate to this finite cover. Define

$$f(\boldsymbol{y}) = \sum_{\boldsymbol{x} \in F} \psi_{\boldsymbol{x}}(\boldsymbol{y}) f_{\boldsymbol{x}}(\boldsymbol{y}) .$$

Then f is smooth; it is a section of the tangent bundle because each tangent space is linearly closed; the real part of $d\phi(f(\boldsymbol{y}))$ is positive because we took a convex combination in which each contribution was nonnegative and at least one was positive. □

Another partition argument gives the final version – the one we will actually use – of this result.

Lemma 5.4.7 (global vector field, vanishing only at critical points) *Let $\mathcal{M}$ be a compact stratified space and ϕ a smooth function on $\mathcal{M}$ with finitely many critical points. Then there is a global section f of the tangent bundle of $\mathcal{M}$ such that the real part of $d\phi(f)$ is nonnegative and vanishes only when $\boldsymbol{y}$ is a critical point.*

Proof Let $\mathcal{M}_\varepsilon$ be the compact stratified space resulting in the removal of an ε-ball around each critical point of ϕ. Let f_ε be a vector field as in the conclusion of Lemma 5.4.7 with $\mathcal{M}$ replaced by $\mathcal{M}_\varepsilon$. Let c_n be a positive real number, small enough so that the magnitudes of all partial derivatives of $c_n f_{1/n}$ of order up to n are at most 2^{-n}. In the topology of uniform convergence of derivatives of bounded order, the series $\sum_n c_n f_n$ converges to a vector field f with the required properties. □

Saddle Point Theorem, Final Version

Let $\mathcal{M}$ be a compact stratified space of dimension d embedded in $\mathbb{C}^d$, and let $\phi : \mathcal{M} \to \mathbb{C}$ be analytic. Let $\boldsymbol{x}$ be an isolated critical point in a stratum S of dimension d. We have seen that ϕ extends holomorphically to a neighborhood of $\boldsymbol{x}$ in $\mathbb{C}^d$. Let $\mathcal{H}(\boldsymbol{x})$ denote the Hessian matrix for the function ϕ at $\boldsymbol{x}$. We expect the integral $\mathcal{I}(\lambda)$ of $e^{-\lambda\phi(x)}$ over $\mathcal{M}$ to have a contribution of $(2\pi\lambda)^{-d/2}/\sqrt{\det \mathcal{H}}$ near the point $\boldsymbol{x}$. Summing over $\boldsymbol{x}$ leads to the following result.

Theorem 5.4.8 (critical point decomposition for stratified spaces) *Let $\mathcal{M}$ be a compact stratified space of dimension d embedded in $\mathbb{C}^d$, and let A and ϕ be analytic functions on a neighborhood of $\mathcal{M}$. Suppose that ϕ has finitely many critical points on $\mathcal{M}$, all in strata of dimension d and all quadratically nondegenerate. Let G be the subset of these at which the real part of ϕ is minimized and assume without loss of generality that this minimal value is zero. Let C be a chain representing $\mathcal{M}$. Then the integral*

$$\mathcal{I}(\lambda) := \int_C A(\boldsymbol{z}) e^{-\lambda\phi(\boldsymbol{z})}\, d\boldsymbol{z}$$

has an asymptotic expansion

$$\mathcal{I}(\lambda) \sim \sum_{\ell=0}^{\infty} c_\ell \lambda^{-d/2-\ell}.$$

If A is nonzero at some point of G, then the leading term is given by

$$c_0 = \pi^{d/2} \sum_{\boldsymbol{x}\in G} A(\boldsymbol{x}) e^{\lambda\phi(\boldsymbol{x})} (\det((1/2)\mathcal{H})(\boldsymbol{x}))^{-1/2}. \tag{5.4.1}$$

Proof Let f be a tangent vector field as given by Lemma 5.4.7. Such a field gives rise to a differential flow, which, informally, is the solution to $d\boldsymbol{p}/dt = f(\boldsymbol{p})$. To be more formal, let $\boldsymbol{x}$ be a point in a stratum S of $\mathcal{M}$. Via a chart map in a neighborhood of $\boldsymbol{x}$, we solve the ODE $d\Phi(t)/dt = f(\Phi(t))$ with initial condition $\Phi(0) = \boldsymbol{x}$, obtaining a trajectory Φ on some interval $[0, \varepsilon_x]$ i.e., supported on S. Doing this simultaneously for all $\boldsymbol{x} \in \mathcal{M}$ results in a map

$$\Phi : \mathcal{M} \times [0, \varepsilon] \to \mathbb{C}^d,$$

with $\Phi(\boldsymbol{x}, t)$ remaining in $S \otimes \mathbb{C}$ when $\boldsymbol{x}$ is in the stratum S. The map Φ satisfies $\Phi(\boldsymbol{x}, 0) = \boldsymbol{x}$ and $(d/dt)\Phi(\boldsymbol{x}, t) = f(\Phi(\boldsymbol{x}, t))$. The fact that this may be defined up to time ε for some $\varepsilon > 0$ is a consequence of the fact that the vector field f is bounded and that a small neighborhood of $\mathcal{M}$ in $\mathcal{M} \otimes \mathbb{C}$ is embedded in $\mathbb{C}^d$. Because f is smooth and bounded, $\boldsymbol{x} \mapsto \Phi(\boldsymbol{x}, \varepsilon)$ is a local diffeomorphism for sufficiently small $\varepsilon > 0$.

The flow reduces the real part of ϕ everywhere except the critical points, which are rest points. Consequently, it defines a homotopy $H(\boldsymbol{x}, t) := \Phi(\boldsymbol{x}, t/\varepsilon)$ between C and a chain C' on which the minima of the real part of ϕ occur precisely on the set G. Recall that H induces a chain homotopy C_H with $\partial C_H = C' - C + \partial C \times \sigma$, where σ is a standard 1-simplex. Let ω denote the holomorphic d-form $A(\boldsymbol{z}) \exp(-\lambda \phi(\boldsymbol{z}))\, d\boldsymbol{z}$. Because ω is a holomorphic d-form in $\mathbb{C}^d$, we have $d\omega = 0$. Now, by Stokes' Theorem,

$$\begin{aligned} 0 &= \int_{C_H} d\omega \\ &= \int_{\partial C_H} \omega \\ &= \int_{C'} \omega - \int_C \omega - \int_{\partial C \times \sigma} \omega. \end{aligned}$$

The chain $\partial C \times \sigma$ is supported on a finite union of spaces $S \otimes C$, where S is a stratum of dimension at most $d - 1$. Recall (see Exercise A.5) that the integral of ω vanishes over such a chain. Therefore, the last term on the right drops out, resulting in

$$\int_C \omega = \int_{C'} \omega.$$

Outside of a neighborhood of G, the magnitude of the integrand is exponentially small, so we have shown that there are d-chains $C_{\boldsymbol{x}}$ supported on arbitrarily small neighborhoods $\mathcal{N}(\boldsymbol{x})$ of each $\boldsymbol{x} \in G$ such that

$$\mathcal{I}(\lambda) - \sum_{\boldsymbol{x} \in G} \int_{C_{\boldsymbol{x}}} \omega$$

is exponentially small. To finish that proof, we need only show that each $\int_{C_{\boldsymbol{x}}} \omega$ has an asymptotic series in decreasing powers of λ whose leading term, when $A(\boldsymbol{x}) \neq 0$, is given by

$$c_0(\boldsymbol{x}) = \pi^{d/2} A(\boldsymbol{x}) e^{\lambda \phi(\boldsymbol{x})} (\det((1/2)\mathcal{H})(\boldsymbol{x}))^{-1/2}. \tag{5.4.2}$$

The d-chain $C_{\boldsymbol{x}}$ may by parametrized by a map $\psi_{\boldsymbol{x}} : B \to \mathcal{N}(\boldsymbol{x})$, mapping the origin to $\boldsymbol{x}$, where B is the open unit ball in $\mathbb{R}^d$. By the chain rule,

$$\int_{C_{\boldsymbol{x}}} \omega = \int_B [A \circ \psi](\boldsymbol{x}) \exp(-\lambda[\phi \circ \psi(\boldsymbol{x})]) \det d\psi(\boldsymbol{x})\, d\boldsymbol{x}.$$

The real part of the analytic phase function $\phi \circ \psi$ has a strict minimum at the origin, so we may apply Theorem 5.1.2. We obtain an asymptotic expansion whose first term is

$$(\pi \lambda)^{d/2} [A \circ \psi](\mathbf{0}) (\det ((1/2) M_{\boldsymbol{x}}))^{-1/2}, \tag{5.4.3}$$

where $M_{\boldsymbol{x}}$ is the Hessian matrix of the function $\phi \circ \psi$. The term $[A \circ \psi](\mathbf{0})$ is equal to $A(\boldsymbol{x})$. The Hessian matrix of $\phi \circ \psi$ at the origin is given by $M_{\boldsymbol{x}} = d\psi(\mathbf{0})\, \mathcal{H}(\boldsymbol{x})\, d\psi(\mathbf{0})$. Thus

$$\det M_{\boldsymbol{x}} = (\det d\psi(\mathbf{0}))^2 \det \mathcal{H}(\boldsymbol{x}),$$

and plugging into (5.4.3) yields (5.4.2), up to the choice of sign for each $\boldsymbol{x} \in G$. □

Remark In one dimension, let $\phi(z) = -z^2$ and let $\mathcal{M}$ be an interval about zero on the imaginary axis. Then $I(\lambda) = \int_{\mathcal{M}} e^{-\lambda \phi(z)}\, dz = \pm i/\sqrt{2\pi\lambda}$ according to whether $\mathcal{M}$ is oriented up or down the imaginary axis. There does not seem to be a canonical way to relate the sign choice on the square root to the eigenvalues of $\mathcal{H}$ and the orientation of $\mathcal{M}$. Nevertheless, it is easy to give a prescription for choosing the sign that involves choosing an arbitrary map. Let ψ parametrize $\mathcal{M}$ by a patch of $\mathbb{R}^d$ with the standard orientation; then we take

$$\det(\mathcal{H})^{-1/2} := (\det \mathcal{H}(\phi \circ \psi))^{-1/2} \det J_{\psi},$$

where the square root on the right is the product of principal square roots of the eigenvalues.

5.5 Examples and Extensions

The following example occurs in Chapter 10 in connection with the asymptotic evaluation of coefficients $a_{r,s}$ of the generating function $F(x, y) := \dfrac{1}{P(x, y) Q(x, y)}$ in the direction $s/r \sim \mu$.

Example 5.5.1 Let $\mathcal{M} = S \times I$, where I is the interval $[-1, 1]$ and S is the circle $\mathbb{R}/(2\pi\mathbb{Z})$. Suppose the phase function ϕ has positive real part vanishing precisely on $\{0\} \times I$ with a unique critical point at $(0, p_0)$ and quadratic approximation

$$\phi(x, p) = Kx^2 + iLx(p - p_0) + O(|x|^3 + |p - p_0|^3) \tag{5.5.1}$$

near $(0, p_0)$, where $K > 0$ and L are real numbers. Note that the strip $\{0\} \times I$ on which the phase function vanishes extends out to the bounding circles of the cylinder $\mathcal{M}$, so we are not in a case where the magnitude of the integrand is small away from the critical point.

The Hessian matrix at $(0, p_0)$ is $\begin{bmatrix} 2K & iL \\ iL & 0 \end{bmatrix}$. The determinant of half the Hessian is equal to $L^2/4$, and from Theorem 5.4.8 we conclude that

$$\mathcal{I}(\lambda) = \int_{\mathcal{N}\times I} e^{-\lambda\phi(x)}\, dx$$

$$\sim \frac{1}{\pi\lambda\,|L|}$$

(where the choice of sign $\sqrt{L^2} = |L|$ is somewhat arbitrary and depends on properly orienting $\mathcal{N} \times I$). ■

Critical Points on the Boundary

In Theorem 5.4.8, the assumption that all stationary points be interior (in d-dimensional strata) is crucial. The contribution of a critical point in lower dimensional strata will in general be different. The details are different in every case. One case that has been used in generating function analysis is worth citing, namely the halfspace case. Suppose that $\mathcal{M}$ is locally diffeomorphic to a d-dimensional halfspace and that a critical point p is interior to a $(d-1)$-dimensional stratum, lying therefore on the border of the halfspace. If the one-sided derivatives of ϕ vanish in the normal directions at p, then a result similar to Theorem 5.4.8 holds for the leading term, but with the coefficient multiplied by precisely $1/2$. Such a result is stated as Pemantle and Wilson (2010, Corollary 2.4). This result is applied to generating function analysis in Pemantle and Wilson (2004, Lemma 4.7).

Nonisolated Critical Points

It is possible for a phase function and its gradient to vanish on an entire submanifold. Indeed this occurs in natural examples in algebraic statistics; see Lin and Pemantle (2013) for a case study. It is often not too difficult to work out via explicit changes of variable what happens in these cases. Taxonomy of all possible cases is complicated, and we did not find it worthwhile to go into anything further here, despite the existence of applications.

Notes

In the case of purely real or imaginary phase, these results are fairly standard; see Bleistein and Handelsman (1986) and Wong (2001) for real phase or Stein (1993) for imaginary phase. We have not seen the complex phase result Theorem 5.1.2 stated before. The remaining results in this section, although not entirely unexpected, are new. In particular, the existence of a deformation to localize to critical points even when the real part of the phase is not strictly minimized

(Theorem 5.4.8) seems new. Such localization in the C^∞ category is certainly not new (the same method as in the proof of Stein [1993, Theorem VIII:2.2] may be applied, for example), but the C^∞ results are weaker, giving rapid decay rather than exponential decay and not allowing for further contour deformation after the localization. A number of the results in this section appeared in Pemantle and Wilson (2010).

Exercises

5.1 (halfspace asymptotics)
Prove the "halfspace result" proved in Pemantle and Wilson (2010, Corollary 2.4): for critical points on the boundary of $\mathcal{M}$, when $\mathcal{M}$ is locally diffeomorphic to a halfspace and ϕ has nonvanishing one-sided normal derivative, the conclusion of Theorem 5.4.8 holds, but with the coefficient multiplied by $1/2$.

5.2 (non-isolated critical points)
Consider the integral

$$\int_{-\varepsilon}^{\varepsilon} \int_0^1 e^{-\lambda\phi(\theta,t)} dt d\theta,$$

where $\phi(\theta, t) = (1-t)g_1(\theta) + tg_1(\theta)$ and each g_i is analytic and vanishes to order 2 at $\theta = 0$, where its second derivative is positive.
Calculate the first-order asymptotic in λ in terms of derivatives of g_1, g_2 at 0.

6

Techniques of Symbolic Computation via Gröbner Bases

6.1 Solving Systems of Polynomial Equations

The critical point equations, (8.3.1) and (8.3.2), are algebraic equations whose solution is one step in the process of producing asymptotics for $a_{\boldsymbol{r}}$. Even when F is rational, these algebraic points are specified only as common solutions to sets of polynomial equations, so what does it mean to "find" them? One could at this point include a discussion of numerical methods. Neither of us is an expert in these, and besides, there is a different point to be made here. The point $z(\boldsymbol{r})$ determines the exponential growth rate, but computations of the exact leading term require further computations for which $z(\boldsymbol{r})$ is an input. It is good practice to maintain analytic forms for the inputs through as much of the computation as possible to take advantage of algebraic simplifications. Even if one is content to remain at the level of exponential growth rates, it would be desirable to maintain analytic expressions such as $z(\boldsymbol{r})$ to do calculus on them.

As algebraic geometers have long known, the best way to keep track of algebraic numbers is via the ideals of polynomials that annihilate them. In the last twenty years, the field of computational algebra has burgeoned, providing algorithms for manipulating these ideals and settling questions such as ideal membership and equality of ideals. During the latter part of this period, these results have been implemented, so that packages for manipulating polynomial ideals are now available in many different computing platforms. The present section is devoted to explaining how to use these in the context of computing multivariate asymptotics for the coefficients of rational generating functions. The computations are only truly effective if the coefficients are finitely specifiable. Thus the remainder of this section will work over $\mathbb{Q}[z]$ instead of $\mathbb{C}[z]$, although most of the theory is equally valid over any field of characteristic zero.

We have concentrated on the platform with which we are most familiar, namely Maple (version 14). The Gröbner basis package must be loaded with the command `with(Groebner)`. The Mathematica package is similar in many respect to Maple, and most of what we do in Maple can be replicated in Mathematica with only syntactic adaptation. Serious computational algebraists often use more powerful packages such as Singular (part of Sage) and Macaulay, but Maple has a more friendly user interface and is more versatile and widespread. Those who get in so deeply that they need greater power can consult up-to-date references. The remainder of this section explains term orders, Gröbner bases, and their use in computations over zero-dimensional ideals. The exposition somewhat follows Cox, Little, and O'Shea (2005, Chapter 1).

Term Orders

The univariate polynomial division algorithm for $p(z)/q(z)$ produces a quotient and a remainder: $p = aq + r$. The remainder, r, always has degree less than the degree of q. This works because one can divide the leading term of q into the leading term of p to find a multiple of q whose subtraction will cancel the leading term, and this may be continued until the leading term of the remainder is so small that it is not divisible by the leading term of q.

To duplicate this feat in several variables, one needs to extend the natural partial order on monomials to a total order. The extension must be compatible with multiplication, and there must be no infinite descents. Thus we define

Definition 6.1.1 *A* ***monomial order*** *on* $\mathbb{Q}[z]$ *is any relation* $>$ *on the set of monomials* $z^{\mathbf{r}}$ *satisfying:*

(i) $>$ *is a total ordering*
(ii) $>$ *is a well ordering*
(iii) if $\alpha, \beta, \gamma \in (\mathbb{Z}^+)^d$ *and* $z^\alpha > z^\beta$ *then* $z^{\alpha+\gamma} > z^{\beta+\gamma}$.

One common term order is the ***lexicographic term order***, where $z^\alpha > z^\beta$ if and only if for some $j \le d$, $\alpha_j > \beta_j$, whereas $\alpha_i = \beta_i$ for all $i < j$. Another is the total degree order, in which $\alpha > \beta$ if and only if either the degree of α is greater than the degree of β or the degrees are equal and $\alpha > \beta$ in the lexicographic order.

Definition 6.1.2 *Let* $>$ *be any monomial order. For* $f \in \mathbb{Q}[z]$*, let* $\mathrm{LT}(f)$ *denote the leading term of* f *with respect to the order* $>$.

Monomial orders do what they were intended to do: given a polynomial p, a set of polynomials $[q_1, \ldots, q_k]$ and monomial order $>$, there is an algorithm to produce a representation $p = \sum a_i q_i + r$ with $\mathrm{LT}(r)$ not divisible by any $\mathrm{LT}(q_i)$. One such algorithm is implemented in Maple as `normalf(p,` $[q_1, \ldots, q_k]$`, order)`, where `order` is an order such as `plex[x,y]` or `tdeg[y,z,x]` from a list of implemented monomial orders.

Gröbner Bases

Let I be an ideal in $\mathbb{Q}[z]$ and let $>$ be a monomial order.

Definition 6.1.3 *A* ***Gröbner basis*** *for the ideal I with respect to the monomial order $>$ is a basis $\{g_1, \ldots, g_k\}$ for I with the property that for any nonzero $f \in I$, $\mathrm{LT}(f)$ is divisible by $\mathrm{LT}(g_i)$ for some i. The basis is called* ***reduced*** *if for all $j \neq i$, no monomial of g_i is divisible by $\mathrm{LT}(g_j)$.*

It turns out that reduced Gröbner bases are unique (Cox, Little, and O'Shea, 2007, Proposition 6 of 2.7), they are algorithmically computable, and they have been implemented in Maple via the command

```
Basis([p1,...,pk], order).
```

The choice of monomial order has effects on computation time that are not fully understood. It also has important effects on the composition of the resulting Gröbner basis. The following proposition gives an example of this.

Proposition 6.1.4 *Let I be an ideal in $\mathbb{Q}[\mathbf{z}]$. The following conditions are equivalent.*

(i) The set $V(I)$ of common solutions to all polynomials in I is a finite subset of $\mathbb{C}^d$.
(ii) $\mathbb{C}[\mathbf{z}]/I$ is a finite dimensional vector space over $\mathbb{C}$.
(iii) Given a monomial order, there are finitely many monomials not divisible by a leading term of the Gröbner basis for I.

Furthermore, if these conditions are met, then there is a univariate polynomial in I whose roots are precisely the values of z_d of the last coordinates of the roots $\mathbf{z}$ of I.

Proof Assume (i). Let S be the set of last coordinates of points in $V(I)$ and let $f = \prod_{a \in S}(z_d - a)$ be the univariate polynomial vanishing precisely at points of S. Then f vanishes on I, so by Hilbert's Nullstellensatz, some power of f is in the ideal generated over $\mathbb{C}$ by I. Using the lexicographic Gröbner basis $\mathcal{B}$ over $\mathbb{C}$, this means some power of z_d is divisible by the leading term of some element of $\mathcal{B}$, hence $\mathcal{B}$ contains a polynomial in z_d alone. However, the question of whether g is in the span of all products of elements of I up to degree N is a question of linear algebra in the coefficients, so if the answer for some N is "yes" over $\mathbb{C}$, then it is "yes" over $\mathbb{Q}$. Therefore, some power of f is in $\mathbb{Q}[z_d]$, and taking the radical derives the final conclusion of the proposition from (i).

For each z_j, $1 \le j \le d$, the same argument shows that some power of z_j is a leading term of an element of $\mathcal{B}$, although for $j < d$ it does not follow that $\mathcal{B}$ contains a polynomial in z_j alone. This is, however, good enough to imply (iii), which implies (ii): the dimension of the vector space $\mathbb{C}[\mathbf{z}]/I$ is equal to the number of such monomials, and in fact these are a basis for $\mathbb{C}[\mathbf{z}]/I$ over $\mathbb{C}$.

Finally, to show that (ii) implies (i), consider the set T_j of monomials $\{z_j^k : k = 0, 1, 2, \ldots\}$. By (ii), these are linearly dependent in $\mathbb{C}[\mathbf{z}]/I$; hence some finite linear combination vanishes in $\mathbb{C}[\mathbf{z}]/I$, or equivalently, there is a polynomial $g_j(z_j) \in I$. Then g_j annihilates the jth coordinate of every $\mathbf{z} \in V(I)$, and hence the number of possible values for the jth coordinate of a point of $V(I)$ is at most $\deg(g_j)$ for each j, and there are at most $\prod_j \deg(g_j)$ points in $V(I)$. □

The lexicographic basis, although not in practice very computationally efficient, has the property that it contains a (univariate) polynomial $f \in \mathbb{Q}[z_d]$ whenever I is zero-dimensional . We call f the ***elimination polynomial*** for z_d.

Computing Modulo a Zero-Dimensional Ideal: Elimination Method

A computation we need to do again and again is to compute an algebraic function of a quantity $\boldsymbol{x}$, which is itself algebraic. To see what is involved, let us consider a simple univariate example.

Example 6.1.5 (algebraic function of an algebraic number) Suppose x is a root of the polynomial

$$P(x) := x^3 - x^2 + 11x - 2 = 0$$

and we need to compute $g(x) := x^5/(867x^4 - 1)$. Because x is the root of a cubic, we could solve for radicals. Not only is this messy, but when plugging into g, the resulting expression would be simplified by Maple only to N/D, where

$$N = -\left(\left(172 + 36\sqrt{1641}\right)^{2/3} - 128 - 2\sqrt[3]{172 + 36\sqrt{1641}}\right)^5$$

and

$$D = 15552\sqrt[3]{172 + 36\sqrt{1641}}\left(-1778217\left(172 + 36\sqrt{1641}\right)^{2/3}\right.$$
$$+ 40749\left(172 + 36\sqrt{1641}\right)^{2/3}\sqrt{1641} - 284577144$$
$$- 6707112\sqrt{1641} + 5144692\sqrt[3]{172 + 36\sqrt{1641}}$$
$$\left.+ 1076796\sqrt[3]{172 + 36\sqrt{1641}}\sqrt{1641}\right).$$

This is far from the simplest expression for this quantity. Also it evaluates in Maple to $0.1935445\ldots$ which is off in the sixth place.

We do much better if we realize that $y := g(x)$ must itself be algebraic. In fact, the pair (x, y) solves the system $\{P(x) = 0, (867x^4 - 1)y - x^5 = 0\}$. The command

```
Basis ([P , y*(867 *x^4 - 1) - x^5], plex(x,y));
```

produces a basis whose first element is the elimination polynomial

$$\theta(y) := 11454803y^3 - 2227774y^2 + 2251y - 32. \qquad (6.1.1)$$

This expresses y as the root of a cubic. Solving this in floating point will now be accurate to more than six places. It can also be expressed as the simpler radical:

$$\frac{1}{393637535306427}\sqrt[3]{A + B\sqrt{C}} + \frac{4885622710417}{3}\frac{1}{\sqrt[3]{A + B\sqrt{C}}} + \frac{2227774}{34364409}$$

where A, B, and C are integers of many digits each. ■

More generally, now, let us suppose that the vector $\boldsymbol{x}$ is the solution to $\{p_1(\boldsymbol{x}, z) = \cdots = p_d(\boldsymbol{x}, z) = 0\}$, where z is a vector of parameters and the ideal $J := \langle p_1, \ldots, p_d \rangle$ is zero-dimensional over the ring $\mathbb{C}(z)[\boldsymbol{x}]$ of polynomials in $\boldsymbol{x}$ whose coefficients are rational functions of the parameters z. We wish to compute a general algebraic function $A(\boldsymbol{x})$. In the simple example, we had a rational function $y(\boldsymbol{x}) = Q(x)/R(x)$, resulting in the polynomial equation $R(x)y - Q(x) = 0$; in the general case, we will simply have an implicit polynomial relation $Q(A, \boldsymbol{x}) = 0$ for some polynomial $Q \in \mathbb{C}[s, x_1, \ldots, x_d]$. Because $\boldsymbol{x}$ may be a multivalued function of z and A may be a multivalued function of $\boldsymbol{x}$, the best we can hope for algebraically is to find the minimal polynomial for $A(\boldsymbol{x})$ in terms of the parameters z. The solution to this will be a collection of algebraic conjugates, from among which one must choose based on specified choices of branches for $\boldsymbol{x}$ and A.

At this level, the computation is very short: the ideal $J \cup \{A\}$ has solutions $\{(x_1, \ldots, x_d, A(\boldsymbol{x}))\}$ as $\boldsymbol{x}$ varies over solutions to J.

Example 6.1.6 (multivariate) Suppose that (x, y) solves the equations $rx + r^2xy + sy^2 - rsx = 0 = sx - ry$. Let $A(x, y)$ solve $A = xA^2 + y$. Then the Maple code

```
p1 := x*r + x*y*r^2 + y^2*s - x*r*s;
p2 := x*s - y*r;
Q := x*A^2 - A + y;
Basis([p1, p2, Q], plex(x,y,A));
```

produces a basis whose first element is the elimination polynomial

$$-A^3r^3 + A^3r^3s - A^2s^3 - A^2r^3s - r^2sA + s^2r^2A.$$

We do not care about the solution $A = 0$, as this forces $x = y = 0$; we therefore remove one factor of A from the elimination polynomial. The resulting polynomial expresses A in the minimal way as in algebraic function of r and s. ■

Although this is straightforward, we have sometimes had trouble getting the computation to halt. The in-principle complexity of a Gröbner basis computation is doubly exponential, and although in practice it is usually much faster, the run times can be unpredictable. The following alternative method in the case where A is a rational function is guaranteed to take only polynomial time once a Gröbner basis for J has been computed.

Matrix Method

Let J be a zero-dimensional ideal and Q be a polynomial. We return to the problem of computing $P(z)/Q(z)$, where $z \in V(J)$ is a solution to J and P and Q are polynomials. Because z is algebraic, so is $P(z)/Q(z)$; therefore, there are polynomials in $\mathbb{Q}[z]$ that annihilate $P(z)/Q(z)$, and we take the computation of such a polynomial to be the goal. Note that this will not distinguish for which z the quantity $P(z)/Q(z)$ has been computed – for irreducible varieties, these are all algebraically conjugate and satisfy the same polynomials.

Pick a Gröbner basis $\mathcal{B}$ and enumerate the monomials not divisible by a leading term of any member of the basis. This results in a list $A := \{z^r : r \in \mathcal{A}\}$ for some set $\mathcal{A}$ whose cardinality is the complex vector space dimension of $\mathbb{C}[z]/J$. If $r, s \in \mathcal{A}$, then either $r + s \in \mathcal{A}$ or else z^{r+s} may be reduced, via the Maple command `normalf(`z^{r+s}`, `$\mathcal{B}$`, order)` to a linear combination of elements of A. In other words, the vector space W spanned by A over $\mathbb{C}$ has an algebra structure, and we know how to determine coefficients $\{c_{n,m} : n \in (\mathbb{Z}^d)\cdot m \in A\}$ with $c_{n,m} = \delta_{n,m}$ when $n \in A$ and such that

$$z^r \cdot z^s = \sum_{m \in A} c_{r+s,m} z^m .$$

A matrix representation for this algebra (in terms of multiplication on the right) is obtained by mapping each z^r to the operator of multiplication by z^r. Thus z^r maps to $M(r)$ where $M(r)$ is a square matrix indexed by $\mathcal{A}$ and

$$M(r)_{s,m} := c_{r+s,m}$$

is the coefficient of z^m in z^{r+s}.

Once we compute $M(r)$ for each $r \in A$, we may add, subtract, and multiply these matrices to obtain a matrix for multiplication by any polynomial $P(z)$. Furthermore, we may invert a matrix $M(Q(z))$ to obtain a matrix representing division by $Q(z)$. Thus the matrix $M := M(P(z))[M(Q(z))]^{-1}$ represents multiplication by $P(z)/Q(z)$ where $z \in V(J)$. Let L be any univariate polynomial. Then $L(P(z)/Q(z)) = 0$ for all $z \in V(J)$ if and only if $L(M)$ is the zero matrix. The minimal polynomial satisfied by $P(z)/Q(z)$ for all $z \in V(J)$ is the minimal polynomial for M, which may be computed by Maple's `minpoly` command. If $V(J)$ is irreducible, this is the minimal polynomial for each $P(z)/Q(z)$. If $V(J)$ is not irreducible, then of course to get the minimal polynomial for a particular $P(z)/Q(z)$, one must specify a component of $V(J)$.

Example 6.1.5 continued: The monomials $\{1, x, x^2\}$ form a basis for $\mathbb{C}[x]/P$. Multiplication by x is represented by a matrix already in rational canonical form:

$$M(x) = \begin{bmatrix} 0 & 1 & 0 \\ 0 & 0 & 1 \\ 2 & -11 & 1 \end{bmatrix} .$$

The matrix representing $y = g(x)$ is given by $T := M^5/(867M^4 - I)$. The Maple code

```
MinimalPolynomial(T,y);
```

then returns the polynomial $\theta(y)$ from (6.1.1).

6.2 Examples of Gröbner Basis Computation

It will be more interesting to do examples later once we have more complicated formulae, such as Theorem 9.5.7, which estimates a_r up to a factor of $(1 + o(1))$ in the "smooth point" case, where the geometry of $\{H = 0\}$ is the simplest (here, H is the denominator of a rational generating function of interest). For now, however, there is plenty we can learn about computing the locations of the critical points themselves. Let $\mathcal{V}$ denote the set $\{H = 0\}$. Peeking ahead to Section 8.3, we find that smooth critical points (those where ∇H does not vanish) are given by the equations $H = 0$ along with $\nabla H \parallel \boldsymbol{r}$. Here, the positive real vector parameter $\boldsymbol{r}$ matters only up to scalar multiples and represents the direction of indices in which asymptotics are desired. The geometric statement $\nabla H \parallel \boldsymbol{r}$ is shorthand for $d - 1$ independent equations $r_1 \partial H/\partial x_j = r_j \partial H/\partial x_1$ for $2 \le j \le d$.

When $H = \prod_{j=1}^{k} H_j$ is a product of square free factors whose varieties intersect transversely at $\boldsymbol{x}$, we call $\boldsymbol{x}$ a *transverse multiple point.* If all points of $\mathcal{V}$ are transverse multiple points, then the critical point equations are $H(\boldsymbol{x}) = 0$, along with the requirement that $\boldsymbol{r}$ be in the span of $\{\nabla H_j(\boldsymbol{x})\}$, where j runs over only those values for which $H_j(\boldsymbol{x}) = 0$. (Check that this reduces to the stated equations in the case $k = 1$ of smooth points!) A quick dimension check shows that on each stratum of $\mathcal{V}$ (a stratum being determined by the subset of functions H_j that vanish), we expect a zero-dimensional set of solutions.

Let us consider an illuminating special case. Suppose that $H = H_1 H_2$. When looking for smooth critical points, should we look separately on the components H_j, or should we forge ahead with the critical point equations (8.3.1)–(8.3.2)? Observe that where H_1 vanishes, the gradient of H

$$\nabla H = H_1 \nabla H_2 + H_2 \nabla H_1 = H_2 \nabla H_1$$

is parallel to the gradient of H_1. Theoretically, therefore, it does not make a difference. Computationally, however, removing the extraneous factor of H_2 can only speed up the computation of smooth critical points on H_1. Thus it is better, though not necessary, to recognize when H factors.

Example 6.2.1 (simplifying quadratics) In Chapter 12, we discuss the generating function for the number a_{nk} of distinct subsequences of length k of the string of length n that cyclically repeats the letters $1, \dots, d$. The generating function is given in Flaxman, Harrow, and Sorkin (2004, Equation 7) as

$$F(x,y) = \sum a_{nk} x^k y^n = \frac{1}{1 - y - xy(1 - y^d)}.$$

We will compute the case $d = 3$, so $F = 1/(1 - y - xy(1 - y^3))$. Fix $1 < \lambda < (d+1)/2$. Let us compute coefficients in the direction $n/k = \lambda$; i.e., $\hat{\boldsymbol{r}} = \frac{1}{1+\lambda}(1, \lambda)$.

Suppose we forget to check whether H factors. First, if we check for singularities, we will find one at $(1, 1)$. Second, we solve the critical point equations for a smooth point $z(\lambda) = (x(\lambda), y(\lambda))$ given by (8.3.3):

$$1 - y - xy(1 - y^3) = 0\,;$$
$$g := ky(-1 - x(1 - y^3) + 3xy^3) + nxy(1 - y^3) = 0\,.$$

The Maple command `Basis([H, g],plex(x,y))` returns a basis $\{g_1^*, g_2^*\}$, where

$$g_1^*(y) := (1 - y)^2[(n - 3k)y^2 + (n - 2k)y + (n - k)]$$

and g_2^* has leading term $x^1 y^0$. The factorization of g_1^* is another tip that H may factor. Going back and checking, we find that H does indeed factor into $(1 - y)$ and $(1 - xy(1 - y^d)/(1 - y)$. It is easy to see that there are no smooth critical points on the component $y = 1$, so we compute on the other component. Redoing the computation for $1 - xy(1 - y^d)/(1 - y)$ yields the basis $\mathcal{B} := \{g_1, g_2\}$ where $g_1 = [(n - 3k)y^2 + (n - 2k)y + (n - k)]$ g_2 still has a pure x term. Before continuing, observe several points:

(i) In Chapter 9, we see how to arrive at this more transparently.
(ii) In this case, y is quadratic over the rationals, and one could use the quadratic formula to solve by radicals. When $d \geq 6$, however, and in practice when $d \geq 4$, one cannot do this.
(iii) Even in the present case, $d = 3$, solving by radicals and plugging into the polynomial for x as a function of y will yield an expression i.e., correct but difficult for Maple to simplify.[1]

Accordingly, we continue without solving for y. The leading terms of $\mathcal{B}$ are y^2 and x. There are exactly two monomials not divisible by one of these, namely 1 and y. In the basis $\{1, y\}$ for $\mathbb{C}[x, y]/\langle\mathcal{B}\rangle$, multiplication by y is particularly simple: 1 goes to y and y goes to

$$y^2 = (\lambda - 1)/(3 - \lambda) + y(\lambda - 2)/(3 - \lambda).$$

[1] By hand, one can force Maple to repeatedly multiply parts of the expression by their algebraic conjugates.

Thus

$$M(y) = \begin{bmatrix} 0 & 1 \\ \frac{\lambda-1}{3-\lambda} & \frac{\lambda-2}{3-\lambda} \end{bmatrix}.$$

Using `minpoly`, we may verify that the minimal polynomial for this is g_1. What about x? From the equation $H = 0$ we know that

$$x = \frac{1}{y(1+y+y^2)} = \frac{(3-\lambda)^2}{(4-\lambda)y + (\lambda - 1)}.$$

Computing $(3-\lambda)^2[(4-\lambda) * M(y) + (\lambda - 1)]^{-1}$ gives

$$M(x) = \begin{bmatrix} 1/3\,\frac{-10\,\lambda+11+2\,\lambda^2}{\lambda-1} & 1/3\,\frac{(-4+\lambda)(-3+\lambda)}{\lambda-1} \\ 4/3 - 1/3\,\lambda & -1 + 1/3\,\lambda \end{bmatrix}$$

and computing the minimal polynomial p_x for x in an indeterminate, t, gives

$$p_x(t) = 3t^2 - \frac{14 - 14\lambda + 3\lambda^2}{\lambda - 1}t + \frac{(3-\lambda)^3}{\lambda - 1}.$$

Now it is permitted to solve for radicals to express the exponential order of a_{nk} as a function of n and k:

$$x(\lambda) = \frac{3\,\lambda^2 - 14\,\lambda + 14 + \sqrt{-3\,\lambda^4 + 36\,\lambda^3 - 152\,\lambda^2 + 256\,\lambda - 128}}{6\,\lambda - 6}$$

$$y(\lambda) = \frac{\lambda - 2 + \sqrt{-8 + 12\,\lambda - 3\,\lambda^2}}{6 - 2\lambda}$$

$$a_{nk} \approx x\left(\frac{n}{k}\right)^{-k} y\left(\frac{n}{k}\right)^{-n}.$$

■

Example 6.2.2 (an arrangement point) Suppose that the denominator H of the generating function F factors as $H = H_1 H_2 H_3$, where

$$H_1 = 1 - x - y - xy$$
$$H_2 = 1 - 2x - xy - y^2$$
$$H_3 = 1 - \frac{1}{4}x - \frac{3}{2}y - \frac{1}{4}y^2.$$

Let us see what we can determine about $\mathcal{V}$. First, we check whether the divisors H_j are all smooth. We may check the smoothness of H_j by computing

```
Basis([H[j] , diff(H[j],x) , diff(H[j],y)] , tdeg(x,y));
```

here, we are free to use the term order `tdeg` i.e., fastest for computation, because all we want to know is whether the ideal in each case is equal to `[1]` (i.e., whether the intersection of the three equations is empty). We see that it is. Next, we check for a common intersection point. The command

```
gb := Basis([H[1], H[2], H[3]],plex(y,x))
```

returns `gb := [1 - 4x - x^2, 1 - 2y - x]`. Here we used the term order `plex(y,x)` to get an elimination basis, which we stored in `gb` for later use. We see from this that there are two common intersection points, whose x-values are the two roots $-2 \pm \sqrt{5}$ of the quadratic $1 - 4x - x^2$ and whose y-coordinate is $(1 + x)/2$, i.e., respectively, $(-1 \pm \sqrt{5})/2$. We let $\boldsymbol{p}$ denote the solution in the positive quadrant and $\boldsymbol{q}$ the solution in the negative quadrant.

The points $\boldsymbol{p}$ and $\boldsymbol{q}$ cannot be transverse multiple points because they are intersections of three curves in $\mathbb{C}^2$. However, if at least these curves are pairwise transverse, then the points satisfy the criteria for what is termed in Chapter 10 *arrangement points*. We therefore check whether the gradients are pairwise linear independent at $\boldsymbol{p}$. Execute the commands

```
uj := [diff(H[j],x),diff(H[j],y)];
```

for $j = 1, 2, 3$, followed by

```
Dij := ui [1] * uj [2] - uj [1] * ui [2];
```

for $(i, j) = (1, 2), (1, 3), (2, 3)$. To check linear dependence of the first two curves at $\boldsymbol{p}$, we need to evaluate D_{12} at $\boldsymbol{p}$. In this case, we have an explicit expression for $\boldsymbol{p}$, but in general we may not; however, we may always evaluate by reducing D_{12} modulo the ideal of $\boldsymbol{p}$:

```
NormalForm(D12,gb,plex(y,x));
```

If $D_{12}(\boldsymbol{p}) = 0$, then reducing modulo a Gröbner basis for the ideal defined by $\boldsymbol{p}$, with respect to the same term order, must return zero. The value returned is $-(1 + 3x)/2$, so we see that the first two curves intersect transversely at $\boldsymbol{p}$. Incidentally, because $\boldsymbol{q}$ is the algebraic conjugate of $\boldsymbol{p}$, the computation at $\boldsymbol{q}$ is identical, and transversality at one implies transversality at the other.

The results of Chapter 10 require that we find a linear relation for H_1, H_2, and H_3 over the analytic functions in a neighborhood of $\boldsymbol{p}$. The Maple command `Normal-Form` may be used to compute the remainder, R, of f modulo $g_1, \ldots, g_n$. When $\{g_1, \ldots, g_n\}$ is a Gröbner basis, the optional fourth argument of `NormalForm` may be used to store the quotients p_j such that $f = R + \sum_j p_j g_j$. Unfortunately, when $\{g_1, \ldots, g_n\}$ is not a Gröbner basis, the quotients returned in the fourth argument of `NormalForm` do not satisfy this equation. Until this is corrected, we must write our own subroutine to do this. Let us assume that we have written such a routine; call it `ReduceWitness` (see Exercise 6.1). If H_1 were in $\langle H_2, H_3 \rangle$, then computing `ReduceWitness(H1 , [H2, H3], plex(x, y));` would find g_2 and g_3 for which $H_1 = g_2 H_2 - g_3 H_3$.

In the local ring at $\boldsymbol{p}$, it is indeed true that $H_1 \in \mathcal{I} := \langle H_2, H_3 \rangle$, but this can and does fail in the polynomial ring because $\mathcal{I}$ is not prime there. In fact, the curves

$\mathcal{V}_2$ and $\mathcal{V}_3$ intersect in four points, only two of which are in $\mathcal{V}_1$. To see this, load the PolynomialIdeals package and compute

```
gb23 := Basis([H[1],H[2]] , plex(y,x));
extra23 := Quotient(gb23,PolynomialIdeal(gb));
elimpoly := extra23 [2];
```

This returns the ideal of the "extra" points:

```
extra23 := ⟨3x − 8 − 2y, x^2 − 8x + 20⟩;
```

The generators for this ideal are nonvanishing on $\boldsymbol{p}$, and multiplying H_1 by any of these, say the elimination polynomial $x^2 - 8x + 20$, produces a polynomial that vanishes wherever H_2 and H_3 vanish, hence is in the ideal gb23. Then, computing

```
ReduceWitness(elimpoly * H[1] , gb , plex(y,x))
```

produces g_2 and g_3 such that elimpoly $(H_1 - g_2 H_2 - g_3 H_3) = 0$ as desired. ■

6.3 D-Modules: Computing with D-Finite Functions

In Chapter 2 we saw that several classes of generating functions are closed under addition and multiplication. One such class is the algebraic functions. If $F, G \in \mathbb{C}[[z_1, \ldots, z_d]]$ are algebraic over $\mathbb{C}[z_1, \ldots, z_d]$, then the fact that $F + G$ and FG are algebraic is in fact an effective fact in the following sense. We must be given algebraic functions F and G in some canonical form. Given that these are supposed to be algebraic, it makes sense to take as inputs definitions that witness the algebraicity of F and G. Specifically, let us take as inputs a polynomial $P \in \mathbb{C}[z_1, \ldots, z_d][x]$ for which $P(F) = 0$ and a polynomial $Q \in \mathbb{C}[z_1, \ldots, z_d][x]$ for which $Q(G) = 0$. Then we may use Gröbner basis computations to find a polynomial in $\mathbb{C}[z_1, \ldots, z_d][x]$ annihilating $F + G$ and another one annihilating FG (see Exercises 6.2 and 6.3).

When it comes to D-finite functions, one may similarly ask whether operations known to preserve D-finiteness may be carried out effectively. The answer is yes, although the implementation is still evolving, so we do not describe this in depth. To make sense of this, let us specify the problem as follows. By analogy with the case of algebraic functions, we should take the inputs to be D-finite functions, specified in a form that bears witness to their being D-finite. Unfortunately, the definition of D-finiteness for multivariate functions is that their derivatives generate a finite dimensional vector space. We would like a more effective definition, such as the univariate definition, which gives a linear differential equation (2.4.1). In the multivariate case, a D-finite function F does satisfy such a differential equation, but one typically needs F to satisfy more than one such equation to guarantee that

F is D-finite and to compute differential equations satisfied by expressions involving F. We are led to consider an algebraic structure on all possible differential equations.

Consider $2d$ operations on $\mathbb{C}[[z_1, \ldots, z_d]]$: the first d of these are multiplication by $z_1, \ldots, z_d$, respectively; the second d are differentiation $(\partial/\partial z_1), \ldots, (\partial/\partial z_d)$, respectively. Denote the first d by $x_1, \ldots, x_d$ and the last d by $\partial_1, \ldots, \partial_d$. Every commutation relation among these $2d$ operators is trivial except for the commutation relation

$$\partial_j x_j = x_j \partial_j + 1 .$$

This motivates the following definition, which may be found in Coutinho (1995, Chapter 1).

Definition 6.3.1 (Weyl algebra) *Let U_d denote the free algebra over $\mathbb{C}$ generated by the symbols $x_1, \ldots, x_d, \partial_1, \ldots, \partial_d$, and let A_d denote the quotient of this by the two-sided ideal generated by the set*

$$\{[x_i, x_j], [\partial_i, \partial_j], [x_i, \partial_j], [x_i, \partial_i] - 1 : i \neq j\}$$

where $[u, v]$ denotes the commutator $uv - vu$. The algebra A_d is called the Weyl algebra and is isomorphic to the ring of differential operators on $\mathbb{C}[z_1, \ldots, z_d]$ that are linear over $\mathbb{C}(z_1, \ldots, z_d)$ (which we then extend to view as linear operators on $\mathbb{C}[[z_1, \ldots, z_d]]$).

Suppose that $P, Q \in A_d$ annihilate $F \in \mathbb{C}[[z_1, \ldots, z_d]]$. Then $\alpha P + \beta Q$ annihilate F for $\alpha, \beta \in \mathbb{C}$ and RP annihilates F for any $R \in A_d$. Consequently, the annihilator, $\mathcal{I}$, of F in A_d is a left ideal of A_d. There is a condition on the annihilator of F, which we do not define here, which implies D-finiteness of F. An ideal is said to be ***holonomic*** if a certain quotient that may be constructed forms a finite dimensional vector space; see Saito, Sturmfels, and Takayama (2000, Definition 1.4.8) for the full definition. The annihilator of F is holonomic if and only if F is D-finite. Furthermore, holonomicity is algorithmically checkable (Algorithm 1.4.17 of Saito, Sturmfels, and Takayama, 2000).

It turns out that the theory of Gröbner bases may be adapted almost without alteration for certain cases where non-commutativity is limited, and these include the Weyl algebra. This is laid out in Sections 1.1 and 1.2 of Saito, Sturmfels, and Takayama (2000) and apparently originated with Castro (1984) and Galligo (1985). Another treatment, in the slightly more general context of ***Ore algebras***, appears in Chyzak and Salvy (1998) and is attributed to Kandri-Rody and Weispfenning (1990) and Kredel (1993).

The implementation of Gröbner basis techniques for non-commutative algebras is more complete in systems other than Maple, such as Singular and CoCoA, but there is not time to get into those systems here! Maple 14 does have a package called `Ore_algebra`, which can apparently do the computations necessary, for example, to find a basis for the annihilator of $F + G$ given bases for the annihilators of F and G. An example of how to do this is given in Chyzak and Salvy (1998, Section 2.2).

Notes

The theory of Gröbner bases has been around for decades. Their popularization only now has two apparent reasons. First, computing power has caught up to theory, so it is now possible to do large computations with a reasonable expectation that they will halt. Second, the renaissance of the theory coincided with the publication of several early texts which are gems of exposition. Two splendid volumes, Cox, Little, and O'Shea (2005) and Cox, Little, and O'Shea (2007), arose from use in undergraduate research programs dating back to the 1990s, and the monograph by Sturmfels (2002) arose from a CBMS lecture series circa 2000. There has been a recent proliferation of texts on computational algebra, each serving a somewhat different purpose, but these initial, delightful volumes did wonders for the subject.

Elimination theory can also be based on the method of *resultants*. This older method is sometimes useful when Gröbner basis algorithms fail to terminate quickly and has been generalized to higher dimensions. See Sturmfels (2002) and Cox, Little, and O'Shea (2005) for concrete details and Gel'fand, Kapranov, and Zelevinsky (1994) for a far-reaching generalization of the theory.

The fact that the annihilator in the Weyl algebra of a function is holonomic if and only if the function is D-finite seems to have been proved first by Kashiwara (1978); we quoted it from Chyzak and Salvy (1998, Section 2). For ideals that are not the annihilator of a function, the notions of holonomy and D-finiteness do not exactly correspond.

Exercises

6.1 (witnessing $f = R \mod I$)
Adapt the algorithm described in Cox, Little, and O'Shea (2007, Chapter 2, Section 3) for producing the quotients $p_1, \ldots, p_n$ as well as the remainder R such that $f = R + \sum_{j=1}^{n} p_j g_j$, where the polynomials f and $g_1, \ldots, g_n$ are given, in the case where $\{g_1, \ldots, g_n\}$ are not a Gröbner basis for the ideal I they generate. It is up to you whether to allow the term order to be input or simply to work with respect to a single-term order such as `plex`. Some helpful pseudocode is given on page 64 of Cox, Little, and O'Shea (2007).

6.2 (effective addition of algebraic functions)
The `Basis` command in Maple's `Groebner` package, if given a term ordering `plex`(a, b) and inputs in variables $a, b, c, d \ldots$, will treat this as a computation over $\mathbb{C}(c, d, \ldots)[a, b]$, i.e. polynomials in a and b with coefficients in a rational function field. Use the substitution $F = (s + d)/2$ and $G = (s - d)/2$ and the `Basis` command to find a polynomial annihilating $F + G$.

6.3 (effective multiplication of algebraic functions)
Can you repeat Exercise 6.2 but for multiplication rather than addition?

6.4 (small algebraic sets are nice)

Suppose that a set of algebraic equations has a single solution $\mathbf{z} = (z_1, \ldots, z_d)$. Prove that $\mathbf{z}$ is a rational point by showing that each coordinate z_j is the solution to a univariate algebraic equation having only one solution. Similarly, if a zero-dimensional variety consists of two points $\{\mathbf{z}, \mathbf{w}\}$, then either it is reducible and the two points are rational or it is irreducible and both are quadratic and algebraically conjugate.

7
Cones, Laurent Series, and Amoebas

We introduce some notation that will be useful throughout this book. Without ambiguity, we may extend the logarithm and exponential functions coordinate-wise to vectors:

$$\log \boldsymbol{z} := (\log z_1, \ldots, \log z_d)$$
$$\exp(\boldsymbol{x}) := (e^{x_1}, \ldots, e^{x_d})$$

Another useful notation is the coordinate-wise log-modulus:

$$\operatorname{Re}\log \boldsymbol{z} := \operatorname{Re}\{\log \boldsymbol{z}\} := (\log |z_1|, \ldots, \log |z_d|).$$

Our ultimate goal is to evaluate the multivariate Cauchy integral (1.3.1). First, however, we give a word of explanation about this chapter, which comprises a few more digressions, into cones and their duals, properties of Laurent series, domains of convergence, and polynomial amoebas. The relevance of Laurent series as generating functions is self-evident. Section 2.1 gave a brief treatment of formal power series that barely touched on analytic properties. The reasons for the comparatively long treatment here of Laurent series are that (1) analytic properties are necessary to establish properties of the formal power series, (2) rigorous developments of these properties for Laurent series, although well known, appear in print rarely or never, and (3) these subsume results for ordinary power series, thus paying off several debts of rigor to preceding chapters. Amoebas are less central, but they provide a context in which the early results on minimal points of multivariate generating functions can be understood and generalized.

Regarding cones and duality, these notions arise throughout Fourier theory. In our setting, denoting the logarithms of the radii of the torus of integration in Cauchy's integral formula by $b_1, \ldots, b_d$, the Cauchy integral becomes

$$T = T_{\boldsymbol{b}} := \{\boldsymbol{z} : |z_j| = e^{b_j} \text{ for all } 1 \leq j \leq d\}.$$

The change of variables $z = \exp(\boldsymbol{b} + i\boldsymbol{x})$, $d\boldsymbol{z} = i^d \boldsymbol{z}\, d\boldsymbol{x}$, turns this into an integral over the flat torus $\mathbf{T}_{\text{flat}} := \mathbb{R}^d/(2\pi\mathbb{Z})^d$:

$$a_r = \left(\frac{1}{2\pi}\right)^d e^{-\boldsymbol{r}\cdot\boldsymbol{b}} \int_{\mathbf{T}_{\text{flat}}} \exp(-i\boldsymbol{r}\cdot\boldsymbol{x})\tilde{F}(\boldsymbol{x})\, d\boldsymbol{x}.$$

Here, $\tilde{F}(\boldsymbol{x}) := F(e^{b+ix})$ and we recognize the integral as a Fourier transform. The presence of the quantity $\boldsymbol{r}\cdot\boldsymbol{x}$ shows that the index vector $\boldsymbol{r} \in \mathbb{Z}^d$ plays a role dual to that of $\boldsymbol{x} \in \mathbb{R}^d$ (or $\mathbb{C}^d$). It will be helpful to keep this duality in mind and to build notation that reflects this. Accordingly, let $(\mathbb{R}^d)^*$ denote a copy of $\mathbb{R}^d$ with a basis dual to the standard basis of $\mathbb{R}^d$, and for $\boldsymbol{x} \in \mathbb{R}^d$ and $\boldsymbol{r} \in (\mathbb{R}^d)^*$, use the interchangeable notations $\langle \boldsymbol{r}, \boldsymbol{x}\rangle$ or $\boldsymbol{r}\cdot\boldsymbol{x}$ to denote the pairing. We denote vectors in $\mathbb{R}^d$ as column vectors and vectors in $(\mathbb{R}^d)^*$ as row vectors, so a third possibility for the inner product is the notation $\boldsymbol{r}\boldsymbol{x}$.

7.1 Cones and Dual Cones

We review some basic facts about convexity that are useful throughout the book. A subset C of $\mathbb{R}^d$ is ***convex*** if it is closed under taking convex combinations: if $x, y \in X$ then $(1-t)x + ty \in C$ for all t with $0 \le t \le 1$. In other words, the line segment joining x and y lies in C. The intersection of all convex supersets of a set A is its ***convex hull*** $\mathtt{hull}(A)$. The convex hull of a set equals the convex hull of the subset of *extreme points*: those points x that do not lie in the interior of any line segment contained in C. A ***supporting hyperplane*** V to C is a hyperplane that intersects C and such that C lies entirely on one side of V — the last condition is equivalent to saying that $v \cdot x \ge 0$ for all $x \in C$ or $v \cdot x \le 0$ for all $x \in C$, where v is a normal to V. A ***convex cone*** is a subset closed under addition and under multiplication by positive scalars. Every open (closed) convex cone is the intersection of all open (closed) half-spaces that contain it. Let K be an open convex cone in $\mathbb{R}^d$. The (closed) convex ***dual cone*** $\mathrm{K}^* \subseteq (\mathbb{R}^d)^*$ is defined to be the set of vectors $\boldsymbol{v} \in (\mathbb{R}^d)^*$ such that $\boldsymbol{v}\cdot\boldsymbol{x} \ge 0$ for all $\boldsymbol{x} \in \mathrm{K}$. Familiar properties of the dual cone are:

$$K \subseteq L \Rightarrow K^* \supseteq L^*; \tag{7.1.1}$$

$$(K \cap L)^* = \mathtt{hull}(K^* \cup L^*). \tag{7.1.2}$$

The tangent cone is a generalization of the tangent space of a manifold to spaces with singularities. Suppose that $\boldsymbol{x}$ is a point on the boundary of a convex set C. Then the intersection of all halfspaces that contain C and have $\boldsymbol{x}$ on their boundary is a closed affine cone with vertex $\boldsymbol{x}$ (a translation by $\boldsymbol{x}$ of a closed cone in $\mathbb{R}^d$) that contains C. Translating by $-\boldsymbol{x}$ and taking the interior gives the (open) solid ***tangent cone*** to C at $\boldsymbol{x}$, denoted by $\mathtt{tan}_{\boldsymbol{x}}(C)$. An alternative definition is:

$$\mathtt{tan}_{\boldsymbol{x}}(C) = \{\boldsymbol{v} : \boldsymbol{x} + \varepsilon\boldsymbol{v} \in C \text{ for all sufficiently small } \varepsilon > 0\}.$$

The (closed) ***normal cone*** to C at $\boldsymbol{x}$, denoted $\texttt{normal}_{\boldsymbol{x}}(C)$, is the convex dual cone to the negative of the tangent cone:

$$\texttt{normal}_{\boldsymbol{x}}(C) = (-\,\texttt{tan}_{\boldsymbol{x}}(C))^* .$$

Equivalently, it corresponds to the set of linear functionals on C that are maximized at $\boldsymbol{x}$, or to the set of outward normals to support hyperplanes to C at $\boldsymbol{x}$. The following notation for the degree of vanishing of a function and the leading homogeneous part of a function at a point will be useful.

Definition 7.1.1 (degree of vanishing, homogeneous part) *For any locally analytic function $f : \mathbb{C}^d \to \mathbb{C}$ and any point $z \in \mathbb{C}^d$, we let* $\deg(f, z)$ *denote the* ***degree of vanishing*** *of f at z:*

$$\deg(f, z) := \sup\{n : f(z + \boldsymbol{w}) = O(|\boldsymbol{w}|^n)\} .$$

We let $\texttt{hom}(f, z)$ *denote the sum of all monomials of minimal degree in the power series for $f(z + \cdot)$, and we call this the* ***homogeneous part*** *of f at z. Thus*

$$f(z + \boldsymbol{w}) = \texttt{hom}(f, z)(\boldsymbol{w}) + O\left(|\boldsymbol{w}|^{\deg(f,z)+1}\right) .$$

When $z = \mathbf{0}$, we may omit z from the notation: thus $\texttt{hom}(f) := \texttt{hom}(f, \mathbf{0})$.

Remark The degree of vanishing is zero if $f(z) \neq 0$ and in general is the least degree of any term in the power series expansion of $f(z + \cdot)$.

The term *tangent cone* has a different meaning in algebraic contexts, which we shall also require. (The term *normal cone* has an algebraic meaning as well, which we will not need.) To avoid confusion, we define the ***algebraic tangent cone*** of f at $\boldsymbol{x}$ to be $\mathcal{V}_{\texttt{hom}(f,z)}$ and denote this by $\texttt{algtan}_{\boldsymbol{x}}(f)$. Note that the algebraic tangent cone is not a solid cone but is in general a hypersurface (it often contains the *boundary* of the solid tangent cone). An equivalent but more geometric definition is that the algebraic tangent cone is the union of lines through $\boldsymbol{x}$ that are the limits of secant lines through $\boldsymbol{x}$; thus for a unit vector $\boldsymbol{u}$, the line $\boldsymbol{x} + t\boldsymbol{u}$ is in the algebraic tangent cone if there are $\boldsymbol{x}_n \in \mathcal{V}_f$ distinct from but converging to $\boldsymbol{x}$ for which $(\boldsymbol{x}_n - \boldsymbol{x})/||\boldsymbol{x}_n - \boldsymbol{x}|| \to \pm\boldsymbol{u}$. This equivalence and more is contained in the following results.

We let S_1 denote the unit sphere $\{(z_1, \ldots, z_d) : |z_1|^2 + \cdots + |z_d|^2 = 1\}$ and let $S_r := r S_1$ denote the sphere of radius r.

We recall a standard notion of distance between sets. Given a metric space (X, d), we can define a metric space whose points are the compact subsets of X and where the ***Hausdorff metric*** ρ is defined stepwise by

$$d(x, Y) = \inf_{y \in Y} d(x, y)$$

$$d(Z, Y) = \sup_{z \in Z} d(z, Y)$$

$$\rho(Z, Y) = \max\{d(Z, Y), d(Y, Z)\}.$$

Lemma 7.1.2 (algebraic tangent cone is the limiting secant cone) *Let Q be a polynomial vanishing to degree $m \geq 1$ at the origin and let $A = \mathtt{hom}(Q)$ be its homogeneous part; in particular,*

$$Q(z) = A(z) + R(z)$$

where A is a nonzero homogeneous polynomial of degree m and $R(z) = O(|z|^{m+1})$. Let Q_ε denote the polynomial

$$Q_\varepsilon(z) := \varepsilon^{-m} Q(\varepsilon z) = A(z) + R_\varepsilon(z),$$

where $R_\varepsilon(z) = \varepsilon^{-m} R(\varepsilon z) \to 0$ as $\varepsilon \to 0$. Let $\mathcal{V}_\varepsilon := \mathcal{V}_{Q_\varepsilon} \cap S_1$ denote the intersection of $\{Q_\varepsilon = 0\}$ with the unit sphere, and let $\mathcal{V}_0$ denote the intersection of S_1 with the zero set of A. Then $\mathcal{V}_\varepsilon \to \mathcal{V}_0$ in the Hausdorff metric as $\varepsilon \to 0$.

Proof On any compact set, in particular S_1, $R_\varepsilon \to 0$ uniformly. If $z^{(n)} \to z$ and $z^{(n)} \in \mathcal{V}_{1/n}$ then for each n,

$$|A(z^{(n)})| = |Q_{1/n}(z^{(n)}) - R_{1/n}(z^{(n)})| = |R_{1/n}(z^{(n)})| \to 0\,.$$

Hence $A(z) = 0$ by continuity of A, and we see that any limit point of $\mathcal{V}_\varepsilon$ as $\varepsilon \to 0$ is in $\mathcal{V}_0$. Conversely, fix a unit vector $z \in \mathcal{V}_0$. The homogenous polynomial A is not identically zero; therefore, there is a projective line along which A has a zero of finite order at z. Let $\gamma : \mathbb{C} \to S_1$ denote any analytic curve through z along which A has a zero of finite order at z. The univariate holomorphic function $\gamma \circ Q_\varepsilon$ converges uniformly to $\gamma \circ A$; therefore, by Hurwitz's Theorem, for ε sufficiently small, there are k zeros of Q_ε converging to z as $\varepsilon \to 0$. In particular, z is a limit point of $\mathcal{V}_\varepsilon$ as $\varepsilon \to 0$. □

Let A be any homogeneous polynomial. Corresponding to each smooth point z of $\mathcal{V}_A$ is an affine hyperplane $z^* := \{z' : (z' - z) \cdot \nabla A(z) = 0\}$ tangent to $\mathcal{V}_A$ at z. The closure of the set of all such points $z^* \in (\mathbb{C}^d)^*$ is an algebraic variety, and its annihilating polynomial A^* is called the ***algebraic dual*** to A. For example, if A is a quadratic form given by $A(\boldsymbol{x}) = \boldsymbol{x}^T M \boldsymbol{x}$, then A^* is a quadratic form whose matrix relative to the dual basis is M^{-1}.

7.2 Laurent Series

The ring of Laurent polynomials in d variables is the ring $\mathbb{C}[z_1, z_1^{-1}, \ldots, z_d, z_d^{-1}]$. In what follows, it will be convenient to extend our scope to consider generating functions that are rational over the ring of Laurent polynomials. Not only does this allow us to deal more naturally with generating functions such as the Aztec Diamond generating function $\dfrac{z/2}{1 - (x + x^{-1} + y + y^{-1})z/2 + z^2}$, but it is the natural level of generality in which to discuss amoebas, which in turn are the best level of generality to discuss power series and their domains of convergence.

Let $\mathcal{L}(z)$ denote the complex vector space of formal linear combinations of monomials z^r as r ranges over all of $\mathbb{Z}^d$. We call these ***formal Laurent series***. The space $\mathcal{L}(z)$ is a module over the ring of Laurent polynomials: if f is a Laurent polynomial and G is a formal Laurent series, then the coefficient of z^r in fG involves only finitely many terms of G and is therefore well defined. Note, however, that some elements of $\mathcal{L}(z)$ have nontrivial annihilators; for example, in one variable, if $G = \sum_{n\in\mathbb{Z}} z^n$ then $(1-z)G = 0$. Also, because the set of pairs (α, β) summing to $\gamma \in \mathbb{Z}^n$ is infinite, there is no natural product structure on $\mathcal{L}(z)$. We will see that convergent Laurent series are much better behaved, but first let's look at the canonical example of why rational functions may have more than one Laurent series representation.

Example 7.2.1 (Laurent series for $z/(1-z)$) Let G_1 be the Laurent series $\sum_{n\geq 1} z^n$; this is convergent on $\mathcal{D}_1 := \{z : |z| < 1\}$. Let $G_2 := \sum_{n\leq 0} -z^n$. Then G_2 is convergent on $\mathcal{D}_2 := \{z : |z| > 1\}$. We have $(1-z)G_1 = (1-z)G_2 = z$. Each series converges to $z/(1-z)$ uniformly on its domain. The intersection of the two domains is, of course, empty. ■

Turning now to the study of convergent Laurent series, let $\mathcal{D}$ be an open simply connected domain, and let $\mathcal{L}(z)(\mathcal{D})$ denote the subspace of $\mathcal{L}(z)$ consisting of series that are absolutely convergent, uniformly on compact subsets of the domain $\mathcal{D}$. When discussing convergence of Laurent series, we always mean uniform convergence on compact sets. The following general facts about domains of convergence of Laurent series are stated (without proof) as Proposition 1.5 of Gel'fand, Kapranov, and Zelevinsky (1994, Chapter 6). We have provided proofs of most of these because of the difficulty of finding such proofs in the literature; probably this is not essential to one's understanding and may be skipped on first reading.

Theorem 7.2.2 (domains of convergence of Laurent series)

(i) Let $G(z) = \sum_{r\in\mathbb{Z}^d} a_r z^r$ be a Laurent series. Then the open domain of convergence of G has the form $\mathcal{D} = \mathrm{Re}\log^{-1}(B)$ for some convex open set $B \subseteq \mathbb{R}^d$.

(ii) The function g defined by the series G is holomorphic in $\mathcal{D}$.

(iii) Conversely, if g is a holomorphic function on $\mathcal{D} := \mathrm{Re}\log^{-1}(B)$, with B convex and open in $\mathbb{R}^d$, then there is a unique Laurent series $G \in \mathcal{L}(z)(\mathcal{D})$ converging to ϕ. The coefficients of G are given by Cauchy's integral formula:

$$a_r := \left(\frac{1}{2\pi i}\right)^d \int_{T(x)} z^{-r-1} g(z)\, dz \tag{7.2.1}$$

where $T(x)$ is the torus $\mathrm{Re}\log^{-1}(x)$ for any $x \in B$.

With these facts established, we may define multiplication in $\mathcal{L}(z)(\mathcal{D})$ as follows. Let ι denote the identification map from $\mathcal{L}(z)(\mathcal{D})$ to the space of holomorphic functions on $\mathcal{D}$. Then ι is invertible, and holomorphicity of the product of holomorphic functions allows us to define $G \cdot H := \iota^{-1}(\iota(G) \cdot \iota(H))$. Similarly, if f is everywhere nonvanishing on $\mathcal{D}$, then there is a unique Laurent series identified with the holomorphic function $1/f$. We may therefore specify a formal Laurent series (e.g., the Aztec Diamond generating function) as a quotient of Laurent polynomials, provided that we specify a domain on which the denominator is nonvanishing.

The proof of Theorem 7.2.2 requires the development of a few well-known facts about series of holomorphic functions.

Proposition 7.2.3 (uniqueness) *Let $\sum_{\mathbf{r}} a_{\mathbf{r}} \mathbf{z}^{\mathbf{r}}$ be a Laurent series converging uniformly to zero on the torus $T_{\mathbf{x}} := \{\exp \mathbf{x} + i\mathbf{y} : \mathbf{y} \in \mathbb{R}^d\}$. Then $a_{\mathbf{r}} = 0$ for all $\mathbf{r}$.*

Proof Assume without loss of generality that $\mathbf{x} = \mathbf{0}$. Then $\sum_{\mathbf{r}} a_{\mathbf{r}} \exp(i\mathbf{r} \cdot \mathbf{y}) \to 0$ uniformly on $(\mathbb{R}/(2\pi\mathbb{Z}))^d$. Thus $\sum a_{\mathbf{r}} e^{i\mathbf{r}\cdot\mathbf{y}}$ is a Fourier series for the zero function. By uniqueness of Fourier series expansions, $a_{\mathbf{r}} = 0$ for all $\mathbf{r}$. □

Proposition 7.2.4 (identity theorem) *If analytic functions f and g on a connected domain $D \subseteq \mathbb{C}^n$ agree on an open subset, then they agree on all of D.*

Proof The set K where they agree is a closed subset of D because it is the inverse image of $\{0\}$ under the continuous function $f - g$. Let the closure of the interior of K be denoted $K' \subseteq K$. Choose any $z_0 \in K'$. The functions f and g agree at z_0, and by definition of analyticity, partial derivatives of all orders exist for each function at z_0, and each function is equal to the limit of its Taylor expansion in a neighborhood of z_0. The partial derivatives, hence the Taylor expansions, are determined by values in any open set with z_0 on the boundary, hence by values in the interior of K, and hence are the same for the two functions. It follows that the two functions agree in a neighborhood of z_0. Thus z_0 is in the interior of K'. Because $z_0 \in K'$ was arbitrary, K' is open. It is closed as well, so by connectedness of D, we see that $K' = D$. □

Proposition 7.2.5 *The uniform limit of analytic functions on a domain $D \subseteq \mathbb{C}^n$ is analytic.*

Sketch of proof Stokes' Theorem implies that the integral over ∂C of a holomorphic d-form must vanish. In fact, the converse is true: if $\int_C f\, dz$ vanishes whenever C is the boundary of a $(d+1)$-simplex, then f is analytic. The integral of the uniform limit of functions is the limit of the integrals, which is zero, proving that the limit is holomorphic. □

Proposition 7.2.6 (logarithmic convexity of domains of convergence) *Let $F := \sum_{\mathbf{r}} a_{\mathbf{r}} \mathbf{z}^{\mathbf{r}}$ be a formal Laurent series, and let $\mathcal{D}$ be its open domain of convergence, i.e., the interior of the set of $\mathbf{z}$ for which $\sum_{\mathbf{r}} |a_{\mathbf{r}} \mathbf{z}^{\mathbf{r}}| < \infty$. Then $\mathcal{D} = \operatorname{Re} \log^{-1}(B)$ for some open convex set $B \subseteq \mathbb{R}^d$.*

Proof Convergence depends on z only through the moduli of the components, and hence the domain of convergence is invariant under $z_j \mapsto e^{i\theta} z_j$, hence is the union of tori $T(\boldsymbol{x}) = \{e^{\boldsymbol{x}+i\mathbb{R}^d}\}$, and hence equal to $\operatorname{Re}\log^{-1}(B)$ for some B. Clearly, if $\boldsymbol{x} \in B$ then

$$\sup_{\boldsymbol{r}} |a_{\boldsymbol{r}}| e^{\boldsymbol{r}\cdot\boldsymbol{x}} < \infty, \tag{7.2.2}$$

because a series with unbounded terms cannot converge uniformly. However, if

$$\sup_{\boldsymbol{r}} |a_{\boldsymbol{r}}| e^{\boldsymbol{r}\cdot\boldsymbol{x}'} < C \tag{7.2.3}$$

for all $\boldsymbol{x}'$ in some neighborhood of $\boldsymbol{x}$, then for some $\varepsilon > 0$, this holds whenever $\boldsymbol{x}' = \boldsymbol{x} \pm \varepsilon \boldsymbol{e}_j$ and $1 \le j \le d$. Let $|\boldsymbol{r}|$ denote $\max_j |r_j|$. When the maximum value is $|r_j|$ and $r_j > 0$, let $\boldsymbol{x}' = \boldsymbol{x} + \varepsilon \boldsymbol{e}_j$, and when the maximum is $|r_j|$ and $r_j < 0$, let $\boldsymbol{x}' = \boldsymbol{x} - \varepsilon \boldsymbol{e}_j$. In either case,

$$\sup_{|\boldsymbol{r}|\ge k} |a_{\boldsymbol{r}}| e^{\boldsymbol{r}\cdot\boldsymbol{x}} = \sup_{|\boldsymbol{r}|\ge k} |a_{\boldsymbol{r}}| e^{\boldsymbol{r}\cdot(\boldsymbol{x}-\boldsymbol{x}')} e^{\boldsymbol{r}\cdot\boldsymbol{x}'} \le e^{-\varepsilon k} \sup_{|\boldsymbol{r}|\ge k} |a_{\boldsymbol{r}}| e^{\boldsymbol{r}\cdot\boldsymbol{x}'} \le e^{-\varepsilon k} C\,.$$

This is summable; therefore, the interior of the set of $\boldsymbol{x}$ satisfying (7.2.2) is contained in the domain of convergence. We conclude that the open domain of convergence is the union over C of the set of $\boldsymbol{x}'$ satisfying (7.2.3). Fixing C, the set $B(C)$ of $\boldsymbol{x}'$ satisfying (7.2.3) is an open set. Being the intersection of open halfspaces, it is convex. For $C' > C$, the set $B(C')$ contains $B(C)$, and hence $B = \cup_C B(C)$. Being the union of open convex sets, B is open, and because the union is increasing, B is convex. □

Proof of Theorem 7.2.2 We have just proved (i) in Proposition 7.2.6. Statement (ii) is Proposition 7.2.5 because $\iota(G)$ is the uniform limit on compact subsets of $\mathcal{D}$ of the series of partial sums. Uniqueness in (iii) is Proposition 7.2.3. It remains to show that (7.2.1) defines a Laurent series $G := \sum_{\boldsymbol{r}} a_{\boldsymbol{r}} \boldsymbol{z}^{\boldsymbol{r}}$ converging to g.

By holomorphicity, the integral (7.2.1) defining $a_{\boldsymbol{r}}$ is independent of the choice $T(\boldsymbol{x})$ of chain of integration. Fix $\boldsymbol{x} \in B$ and choose $\varepsilon > 0$ small enough so that $\boldsymbol{x} \pm \varepsilon \boldsymbol{e}_j \in B$ for all $1 \le j \le d$. The modulus of $g(\boldsymbol{z})$ is bounded on the finite union of tori $T(\boldsymbol{x} \pm \varepsilon \boldsymbol{e}_j)$. By (7.2.1), and the same argument as in the previous proof, we see that $|a_{\boldsymbol{r}} \boldsymbol{z}^{\boldsymbol{r}}| \le K \exp(-\varepsilon |\boldsymbol{r}|)$ for all $\boldsymbol{z} \in T(\boldsymbol{x})$. For a slightly smaller value of ε, this holds for all $\boldsymbol{z} \in T(\boldsymbol{x}')$, for $\boldsymbol{x}'$ in some neighborhood $\mathcal{N}(\boldsymbol{x})$ of $\boldsymbol{x}$. If $K \subseteq B$ is any compact set, covering with finitely many neighborhoods $\mathcal{N}(\boldsymbol{x})$ shows that such a bound holds for all $\boldsymbol{z} \in \operatorname{Re}\log^{-1}(K)$. In particular, the series G converges uniformly on compact subsets of $\mathcal{D}$. Once we show that $\iota(G) = g$ on some subset of $\mathcal{D}$ with nonempty interior, the theorem follows from the identity theorem.

Let B' be a closed rectangle $\prod_{j=1}^d [u_j, v_j]$ contained in the open domain of convergence. For $d = 1$, the proof that $\sum_{n=-\infty}^{\infty} a_n z^n$ converges uniformly to f when f is holomorphic on any annulus containing $\{z : e^a \le |z| \le e^b\}$ and $a_n := (2\pi i)^{-1} \int_\gamma z^{-n-1} f(z)\, dz$ may be found in most complex variable texts. For example, 1.11 of Conway (1978, Chapter V) uses the Cauchy kernel representation

$$f(z) = \frac{1}{2\pi i}\int_{\gamma_2}\frac{f(w)}{w-z}\,dw - \frac{1}{2\pi i}\int_{\gamma_1}\frac{f(w)}{w-z}\,dw,$$
$$:= f_2(z) - f_1(z)$$

where γ_1 and γ_2 are, respectively the inner and outer boundaries of the annulus. The function f_2 is holomorphic on the disk of radius e^b, and the function $f_1(1/z)$ is holomorphic on the disk of radius e^{-a}, so the usual power series expansions show that $f_2(z) = \sum_{n\geq 0} a_n z^n$ while $f_1(z) = \sum_{n\leq -1} a_n z^n$.

In d variables, when $B = \prod_{j=1}^d [u_j, v_j]$, we require a representation for $f(z)$ analogous to the Cauchy kernel representation, expressing f as the sum of 2^d integrals over the d-dimensional faces $\text{Re}\log^{-1}(B)$, which is a product of d annuli. The details are omitted. □

Remark 7.2.7 Let $K \subseteq \mathbb{R}^d$ be a cone containing the origin, and let K_Z denote the intersection of K with $\mathbb{Z}^d$. Suppose that $\{\boldsymbol{x} + \boldsymbol{y} : \boldsymbol{x}, \boldsymbol{y} \in K_Z \setminus \{\boldsymbol{0}\}\}$ is a proper subset of K_Z. Then each $\boldsymbol{x} \in K$ is in only finitely many of the sets $\{\boldsymbol{0}\}, K_Z \setminus \{\boldsymbol{0}\}, K_Z \setminus \{\boldsymbol{0}\} + K_Z \setminus \{\boldsymbol{0}\}, \ldots$. It follows that the space of formal Laurent series whose coefficients vanish outside K_Z forms a ring under the formal product operation. All facts about formal power series transfer to this case. For instance, if the coefficients $\{a_r\}$ of the series G are supported on K and $a_0 = 0$, then $1/(1-G) := 1 + G + G^2 + \cdots$ is a well-defined formal Laurent series inverting $(1-G)$, even if the coefficients of G grow sufficiently rapidly that G is not convergent on any domain. As an example, consider the function $1/(1-(x + x^{-1} + y + y^{-1})z/2 + z^2)$ from Example 11.3.5. The coefficients are supported on the cone $K := (r, s, t) : |r| + |s| \leq t$. Coefficients of the formal convolution will be finite sums, whence the formal convolution is well defined without reference to any domain of convergence.

7.3 Amoebas

If f is any Laurent polynomial, we define the ***amoeba*** of f by

$$\texttt{amoeba}(f) := \{\text{Re}\log z : f(z) = 0\}$$

to be the set of log-moduli of zeros of f. The simplest example is the amoeba of a linear function, such as $f = 2 - x - y$, shown in Figure 7.1(a). The amoeba of a product is the union of amoebas, as shown in Figure 7.1(b). The following result is stated in Gel'fand, Kapranov, and Zelevinsky (1994, Chapter 6), Theobald (2002, Theorem 2), and Forsberg, Passare, and Tsikh (2000, Theorem 1.1).

Proposition 7.3.1 *The connected components of* $\mathbb{R}^d \setminus \texttt{amoeba}(f)$ *are convex. The components are in one-to-one correspondence with Laurent series expansions of* $1/f$.

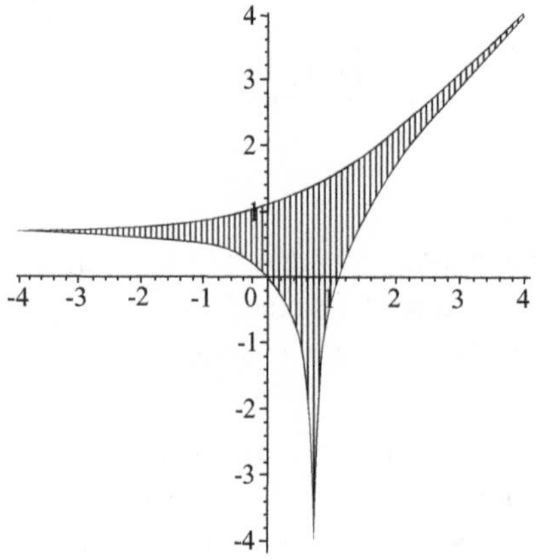

(a) amoeba(2 – x – y)

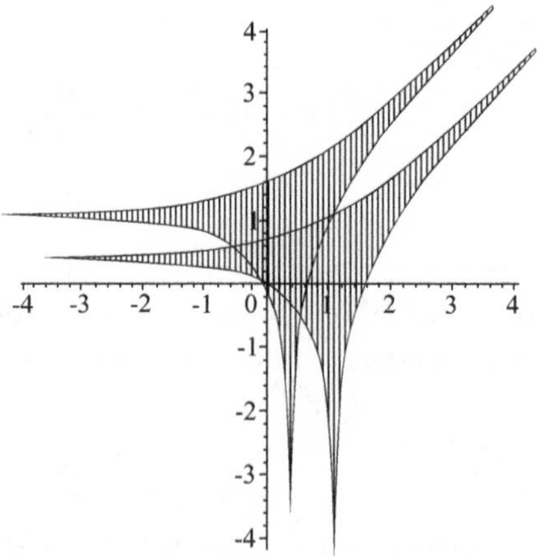

(b) amoeba(3 – 2x – y)(3 – x – 2y)

Figure 7.1 Two amoebae.

Proof Let $\{B_\alpha : \alpha \in I\}$ be the components of $\mathbb{R}^d \setminus \texttt{amoeba}(f)$. Let G_α denote the Laurent series (7.2.1), which converges to the holomorphic function $1/f$ on B_α. Any $\boldsymbol{x} \in \partial B_\alpha$ is in $\texttt{amoeba}(f)$, meaning that f vanishes somewhere on $T(\boldsymbol{x})$, so no Laurent expansion of $1/f$ can converge on $T(\boldsymbol{x})$. It follows that B_α is the domain of convergence of G_α, whence by part (i) of Theorem 7.2.2, B_α is convex. Applying parts (i), (ii), and (iii) of Theorem 7.2.2 in order to any Laurent expansion of $1/f$ shows that it is equal to some G_α. Finally, $G_\alpha \neq G_\beta$ when $\alpha \neq \beta$ because domains of convergence are convex and the convex hull of $B_\alpha \cup B_\beta$ intersects $\texttt{amoeba}(f)$. □

A ***polytope*** P is the convex hull $\texttt{hull}(E)$ of a finite collection of points $E \subseteq \mathbb{R}^d$. The extreme points of $\texttt{hull}(E)$, which form a subset of E, are called ***vertices of the polytope*** P. If $x \in \partial P$, then x is a vertex if and only if the normal cone $\texttt{normal}_x(P)$ has non-empty interior. It is easy to see that the interiors of $\texttt{normal}_x(P)$ are disjoint as x varies over the vertices of P and that the union of the closures is all of $\mathbb{R}^d$. [Note that we used $\mathbb{R}^d$ for the ambient space of the normal cone, rather than $(\mathbb{R}^d)^*$; i.e. because we will consider polytopes in $(\mathbb{R}^d)^*$.]

The set of exponents of monomials of a Laurent polynomial, f, is a finite subset $E \subseteq \mathbb{Z}^d \subseteq (\mathbb{R}^d)^*$. The ***Newton polytope*** is defined to be the convex hull in $(\mathbb{R}^d)^*$ of all exponents of monomials of f:

$$\mathrm{P}(f) := \texttt{hull}\{\boldsymbol{r} : a_{\boldsymbol{r}} \boldsymbol{z}^{\boldsymbol{r}} \text{ is a nonzero monomial of } f\}.$$

Proposition 7.3.2 (Gel'fand, Kapranov, and Zelevinsky, 1994, Ch. 6, Prop. 1.7 and Cor. 1.8) *The vertices (extreme points) of* $\mathrm{P}(f)$ *are in bijective* *correspondence with connected components of* $\mathbb{R}^d \setminus \texttt{amoeba}(f)$ *containing an affine convex cone with non-empty interior.*

Proof Let $\boldsymbol{p} \in E$ be a vertex of $\mathrm{P}(f)$. We may write

$$f(z) = a_p z^p \cdot \left(1 + \sum_{p \neq r \in E} \frac{a_r}{a_p} z^{r-p}\right) := a_p z^p(1 + g(z)).$$

The Laurent series

$$G = a_p^{-1} z^{-p} \left(1 - g + g^2 - \cdots\right)$$

is formally well defined (see Remark 7.2.7).

Claim: There is a translation $\boldsymbol{b} + \texttt{normal}_{\boldsymbol{p}}(\mathrm{P}(f))$ of the normal cone to the Newton polytope at $\boldsymbol{p}$, such that G converges to the holomorphic function $1/f$ for all z with $\mathrm{Re}\log z \in \boldsymbol{b} + \texttt{normal}_{\boldsymbol{p}}(\mathrm{P}(f))$.

Proof: By its explicit description, the series G converges to $1/f$ wherever $|g(x)| < 1$. Fix any $\boldsymbol{u}$ in the interior of $\texttt{normal}_{\boldsymbol{p}}(\mathrm{P}(f))$, so by definition, $\boldsymbol{u} \cdot (\boldsymbol{r} - \boldsymbol{p}) < 0$ for any $\boldsymbol{r} \neq \boldsymbol{p}$ in the Newton polytope. Applying this to the finitely many $\boldsymbol{r} \neq \boldsymbol{p}$ in E, we see that we may choose $\lambda > 0$ large enough so that

$$\log \sum_{p \neq r \in E} |a_r| + \lambda \sum_{p \neq r \in E} \boldsymbol{u} \cdot (\boldsymbol{r} - \boldsymbol{p}) < 0.$$

This implies that $|g(z)| < 1$ whenever $\mathrm{Re}\log|z| = \lambda\boldsymbol{u}$. In fact, if $\mathrm{Re}\log z = \lambda\boldsymbol{u} + \boldsymbol{v}$ for $\boldsymbol{v} \in \texttt{normal}_{\boldsymbol{p}}(\mathrm{P}(f))$ then $\boldsymbol{v} \cdot (\boldsymbol{r} - \boldsymbol{p}) < 0$ implies that $|g(z)| < 1$ as well. We conclude that $|g(z)| < 1$ when $\mathrm{Re}\log z \in \lambda\boldsymbol{u} + \texttt{normal}_{\boldsymbol{p}}(\mathrm{P}(f))$.

Now that we have a Laurent expansion of $1/f$ convergent on a translate of $\texttt{normal}_{\boldsymbol{p}}(\mathrm{P}(f))$, we let $B(\boldsymbol{p})$ denote the component of the complement of $\texttt{amoeba}(f)$ containing this affine cone. If two cones have intersecting interior, then their affine cones intersect as well. The closures of the cones $\texttt{normal}_{\boldsymbol{p}}(\mathrm{P}(f))$ cover $\mathbb{R}^d$ as $\boldsymbol{p}$ varies over vertices of $\mathrm{P}(f)$, and hence any projective cone with non-empty interior intersects some $\texttt{normal}_{\boldsymbol{p}}(\mathrm{P}(f))$. Translates of projective cones with intersecting interiors intersect, and hence any component of the complement of $\texttt{amoeba}(f)$ containing an affine cone with non-empty interior intersects some $B(\boldsymbol{p})$, and hence is equal to some $B(\boldsymbol{p})$. To finish the argument, we refer to the short argument in Gel'fand, Kapranov, and Zelevinsky (1994, page 196) showing that the components $B(\boldsymbol{p})$ are distinct for distinct vertices $\boldsymbol{p}$. □

Let B be any convex set containing a translate of the projective cone K. If $\boldsymbol{x} \in \partial B$, then $\tan_{\boldsymbol{x}}(B) \supseteq K$. [To see this, note that there is a $\boldsymbol{b}$ such that $\lambda\boldsymbol{u} + \boldsymbol{b} \in B$ for any $\lambda > 0$, $\boldsymbol{u} \in K$; any half-space containing B is of the form $\{\phi < c\}$ for some $c \geq 0$ and ϕ linear, and we see that $\phi(u) \leq 0$, hence $\boldsymbol{u} \in B$.] From this, we obtain:

Proposition 7.3.3 *If the component B of the complement of* $\texttt{amoeba}(f)$ *corresponds to a vertex of the Newton polytope, i.e., if $B = B(\boldsymbol{p})$, then $\boldsymbol{x} \in \partial B(\boldsymbol{p})$ implies* $\texttt{tan}_{\boldsymbol{x}}(B) \subseteq \texttt{normal}_{\boldsymbol{p}}(\mathrm{P}(f))$.

A case arising frequently with generating functions is when the Laurent polynomial f is an ordinary polynomial with nonzero constant term. In this case, $\mathbf{0}$ is a vertex of $\mathrm{P}(f)$. All other points of $\mathrm{P}(f)$ are in the nonnegative orthant, so $N_{\mathbf{0}}(\mathrm{P}(f))$ contains the negative orthant. Another common case is when f has a nonzero constant term and there are weights $w_1, \ldots, w_d \in \mathbb{R}^+$ such that for any nonconstant monomial $a_{\boldsymbol{m}}\boldsymbol{x}^{\boldsymbol{m}}$ appearing in f, $\sum w_i m_i > 0$. This case is very similar to the previous case. Again, $\mathbf{0}$ is a vertex of $\mathrm{P}(f)$. In both cases, assuming without loss of generality that the constant term of f is 1, one may write $1/f = 1/(1+g) = 1 - g + g^2 - \cdots$, in which each monomial appears only finitely often. In fact, an invertible affine change of coordinates $\boldsymbol{m} \mapsto \boldsymbol{Lm}$ maps $\mathrm{P}(f)$ into the nonnegative orthant. For example, if $f = 1 - (x + x^{-1} + y + y^{-1})z/2 + z^2$ as in the Aztec diamond generating function, then $(i, j, k) \mapsto (i + k, j + k, k)$ maps $\mathrm{P}(f)$ into the nonnegative orthant. This corresponds to the change of variables $z = xyz'$, which maps f to the ordinary polynomial $1 - (x^2y + xy^2 + x + y)z'/2 + x^2y^2(z')^2$. In either of these cases, the domain of convergence of $1/f$ is $|g| \le 1$. The component of the complement of the amoeba that corresponds to the vertex $\mathbf{0}$ of the Newton polytope is the one containing the (affine image of the) negative orthant.

Legendre Transform

The ***Legendre transform*** of a convex function $f : \mathbb{R}^d \to \mathbb{R}$ is the function $f^* : (\mathbb{R}^d)^* \to \mathbb{R}$ defined by

$$f^*(\boldsymbol{v}) = \sup_{\boldsymbol{x}} \langle \boldsymbol{v}, \boldsymbol{x} \rangle - f(\boldsymbol{x}) .$$

It is also often called the ***convex dual*** of f and satisfies the duality relation $f^{**} = f$; see, for example, Rockefellar (1966). Legendre transforms are intimately connected with exponential rates of growth and decay. For example, in probability theory, the *rate function* $f(\lambda)$, defined as the Legendre transform of the logarithm of the moment generating function, gives the rate of exponential decay of the probability of the mean n IID variables to exceed λ.

Legendre transforms typically arise in logarithmic coordinates. The application of primary concern to us is the Legendre transform of an amoeba boundary. Let B be a component of the boundary of an amoeba, hence convex. For $\boldsymbol{r} \in (\mathbb{R}^d)^*$, define

$$\beta^*(\boldsymbol{r}) := \inf\{-\boldsymbol{r} \cdot \boldsymbol{x} : \boldsymbol{x} \in B\} . \tag{7.3.1}$$

The infimum, when achieved, occurs on the boundary of B. Therefore, this is the negative of the Legendre transform of the convex function i.e., 1 on B and ∞ on B^c. In the next chapter, we see how this gives a natural and often sharp bound on the exponential rate of growth or decay of the coefficients of a rational function.

Notes

The study of amoebas and the origin of the term "amoeba" are generally credited to Gelfand, Kapranov, and Zelevinsky (1994). This seminal text on discriminants devotes much of Chapter 6 to amoebas and Newton polytopes. Their development of basic results on amoebas (Gel'fand, Kapranov, and Zelevinsky, 1994, Section 6.1) begins by quoting without proof some basic facts about Laurent series akin to Theorem 7.2.2 (their Proposition 1.5 of Chapter 6). The reference they give, namely Krantz (2001), proves these only for ordinary power series, and the resulting wild-goose chase led us to write down more complete developments of these basic facts. The first author learned many of these when sitting in on a graduate course at Penn given by L. Matusevich in Fall 2004. Other helpful sources include Mikhalkin (2000), Theobald (2002), and Mikhalkin (2004).

A good part of the theory of amoebas of algebraic hypersurfaces goes through for analytic hypersurfaces. This and its applications to statistical physics, for example, make up the content of Passare, Pochekutov, and Tsikh (2011). Because the theory of amoebas of analytic hypersurfaces is still being formed, we mostly avoid its use.

Exercises

7.1 (finding Laurent series)
The complement of the amoeba of the polynomial $2 - x - y$ has three components, as shown in Figure 7.1. Find the three corresponding Laurent series.

7.2 (strict convexity of amoeba)
Which bivariate polynomials $f(x, y)$ have flat spots in their amoebas? More precisely, give a simple necessary condition on f for $\texttt{amoeba}(f)$ to fail to be strictly convex, in other words, for $\texttt{amoeba}(f)$ to contain a line segment in its boundary.

7.3 (strict convexity of amoeba)
Consider the trivariate polynomial

$$f(x, y, z) := z^2 - (1 - xy)z - 1 .$$

Are the components of the complement of its amoeba strictly convex?

PART III

Multivariate Enumeration

8

Overview of Analytic Methods for Multivariate Generating Functions

Part III returns to the problem at the heart of this book, namely how to make judicious choices of contours of integration to asymptotically evaluate the Cauchy integral (1.3.1). This may be summarized in one sentence as: push the chain of integration down to a *critical point*, then interpret the integral locally as an instance of a type i.e. classically understood. Depending on the geometry near the critical point, this step may take one of several forms. Each of the three chapters following this one details the process from start to finish in one of three cases, which together cover most known examples of rational generating functions. Extensions beyond this class are discussed briefly in Chapter 13.

These three chapters are self-contained, developing the theory in full rigor for the respective three cases. The present chapter is a complement, in which motivation and intuition are given for all the development in subsequent chapters. Its mission is to expand the six steps given at the end of Section 1.3 into a comprehensible program. In some cases we give heuristics without proof or give proofs that rely on Morse theory (included in Appendixes B and C). In one case, the heuristic relies on a conjecture not yet solved. Nevertheless, these heuristics provide firm guidance regarding how to accomplish the ensuing analyses. When these methods are fully developed in later chapters, only vestiges of the Morse-theoretic infrastructure remain. The proofs there rely on this as a guide but use explicit deformations that appear somewhat mysterious when the motivation is removed.

Throughout Part III of this book, the following notational conventions are in force. The d-variable function $F(z)$ is the quotient of analytic functions $G(z)/H(z)$, usually polynomials, with the denominator H vanishing on a variety we denote by $\mathcal{V} = \mathcal{V}_H$. We consider a component B of the complement of $\mathtt{amoeba}(H)$ and the Laurent series expansion $F(z) = \sum_r a_r z^r$ that converges on B. For those who prefer to simplify and deal with ordinary power series, one may always take B to be the component containing a ray $(-\infty, b] \cdot (1, \dots, 1)$ and $\boldsymbol{r}$ to run over $\mathbb{N}^d$.

When H is a general analytic function rather than a polynomial, we require that the conclusions of the theorems in Section 7.3 hold for H (see the notes at the end of Chapter 7).

Given a vector index $\mathbf{r} \in \mathbb{N}^d$, the d-form $\omega := \mathbf{z}^{-\mathbf{r}-\mathbf{1}} F(\mathbf{z}) d\mathbf{z}$ is the integrand of the Cauchy integral; its domain of analyticity is $\mathcal{M} := (\mathbb{C}^*)^d \setminus \mathcal{V}$. The unitized vector $\mathbf{r}/|\mathbf{r}|$ representing the direction of $\mathbf{r}$ is denoted $\hat{\mathbf{r}}$; unless we say otherwise, $|\mathbf{r}|$ is taken to be the ℓ^1-norm $\sum_{j=1}^d |r_j|$. We will seek to compute asymptotics for the Maclaurin coefficients $a_{\mathbf{r}}$ of F as $\hat{\mathbf{r}}$ varies over a neighborhood of an arbitrary fixed direction $\hat{\mathbf{r}}_*$. For $\mathbf{x} \in \mathbb{R}^n$, denote by $\mathbf{T}(\mathbf{x})$ the torus $\exp(\mathbf{x} + i\mathbb{R}^d)$. Thus $\mathbf{T}(\operatorname{Re}\log \mathbf{z}_*)$ is the torus through $\mathbf{z}_*$.

8.1 Exponential Rate

The crudest level at which nontrivial estimation of $a_{\mathbf{r}}$ normally occurs is the exponential level, namely statements of the form $\log |a_{\mathbf{r}}| \sim g(\mathbf{r})$ as $\mathbf{r} \to \infty$ in some specified way. If there is an oscillatory term, such an asymptotic statement might be violated when the modulus of the oscillatory factor in $a_{\mathbf{r}}$ is exceptionally small. These estimates are most useful when they are uniform, as $\hat{\mathbf{r}}$ varies over some neighborhood of some direction $\hat{\mathbf{r}}_*$. It therefore makes sense to smooth the exponential rate somewhat, replacing the rate function g by the limsup neighborhood rate function $\overline{\beta}$.

Definition 8.1.1

$$\overline{\beta}(\hat{\mathbf{r}}_*) = \inf_{\mathcal{N}} \limsup_{\mathbf{r}\to\infty, \hat{\mathbf{r}}\in\mathcal{N}} |\mathbf{r}|^{-1} \log |a_{\mathbf{r}}|, \tag{8.1.1}$$

where $\mathcal{N}$ varies over a system of open neighborhoods of $\hat{\mathbf{r}}_*$ whose intersection is the singleton $\{\hat{\mathbf{r}}_*\}$.

Example 8.1.2 Suppose that $a_{rs} = \binom{r+s-1}{s} - \binom{r+s-1}{r}$, corresponding to the bivariate generating function $\sum_{i,j} a_{ij} x^i y^j = (x-y)/(1-x-y)$. Then $a_{rr} = 0$, so the naive definition of exponential rate would yield $-\infty$ in that direction, whereas our definition yields $\log 2$. ■

The following upper bound on $\overline{\beta}$ is trivial but important. For any $\mathbf{x} \in B$, convergence of the series $\sum_{\mathbf{r}} a_{\mathbf{r}} \mathbf{z}^{\mathbf{r}}$ implies the magnitude of the terms goes to zero. Because $\mathbf{z} = \exp(\mathbf{x} + i\mathbf{y})$, $a_{\mathbf{r}} = o(\exp(-\mathbf{r}\cdot\mathbf{x}))$ as $\mathbf{r} \to \infty$, meaning that the set of $\mathbf{r}$ such that $a_{\mathbf{r}} \geq \varepsilon \exp(-\mathbf{r}\cdot\mathbf{x})$ is finite for any $\varepsilon > 0$. Recall the amoeba Legendre transform defined by

$$\beta^*(\mathbf{r}) := \inf\{-\mathbf{r}\cdot\mathbf{x} : \mathbf{x} \in B\};$$

this function depends on the polynomial H and the component B of $\texttt{amoeba}(H)^c$, but we suppress this in the notation. Taking the infimum over $\boldsymbol{x} \in B$ leads immediately to

$$\overline{\beta}(\boldsymbol{r}) \leq \beta^*(\boldsymbol{r}). \tag{8.1.2}$$

The quantity β^* is *semialgebraic* and hence effectively computable. It is therefore of great interest to know when $\overline{\beta} = \beta^*$, which will enable us to compute the exponential rate. This is related to the question of whether a *dominating point* can be found. We begin by investigating the infimum in (7.3.1) and whether the infimum is achieved on $\overline{B}$. First, the infimum could equal $-\infty$. This case is trivial.

Proposition 8.1.3 *If H is a Laurent polynomial and the infimum of $-\boldsymbol{r} \cdot \boldsymbol{x}$ on a component B of* $\texttt{amoeba}(H)^c$ *is $-\infty$, then $a_{\boldsymbol{r}} = 0$.*

Proof Let $B^{(\varepsilon)}$ denote the set of points whose ε-neighborhood is contained in B. If $-\boldsymbol{r} \cdot \boldsymbol{x}$ is unbounded from below, then choose $\boldsymbol{x}_n \in B^{(\varepsilon)}$ with $-\boldsymbol{r} \cdot \boldsymbol{x}_n \leq -n$. Because the set $B^{(\varepsilon)}$ and the function $-\boldsymbol{r} \cdot \boldsymbol{x}$ are semialgebraic, we may choose $\{\boldsymbol{x}_n\}$ so that there is a polynomial lower bound $|H(\boldsymbol{x}_n)| \geq n^{-\alpha}$ for some α and hence

$$|\boldsymbol{z}^{-1} F(\boldsymbol{z})| \leq P(n) \text{ on } \mathbf{T}(\boldsymbol{x}_n)$$

for some polynomial P. Estimating $a_{\boldsymbol{r}}$ via Cauchy's integral formula on $\mathbf{T}(\boldsymbol{x}_n)$ and using $|\boldsymbol{z}^{-\boldsymbol{r}}| \leq e^{-n}$ on $\mathbf{T}(\boldsymbol{x}_n)$ then gives

$$\begin{aligned} |a_{\boldsymbol{r}}| &= \left| \left(\frac{1}{2\pi i}\right)^d \int_{\mathbf{T}(\boldsymbol{x}_n)} \boldsymbol{z}^{-\boldsymbol{r}-\boldsymbol{1}} F(\boldsymbol{z})\, d\boldsymbol{z} \right| \\ &\leq \frac{1}{(2\pi)^d} |\mathbf{T}(\boldsymbol{x}_n)|\, e^{-n} P(n). \end{aligned}$$

The $(2\pi)^{-d}$ term cancels the volume of the torus, and we finish by observing that $e^{-n} P(n) \to 0$. □

Remark Leaving the realm of quotients of Laurent polynomials, we might consider a more general meromorphic function G/H with H analytic and with a power series converging on a domain B. When $-\boldsymbol{r} \cdot \boldsymbol{x}$ is unbounded from below on B, the coefficients $a_{\boldsymbol{r}}$ decay super-exponentially. We have relatively little to say in this generality, as our methods are developed chiefly for estimates in the exponential regime. For cases such as this, with super-exponentially decaying coefficients, we suggest to try a saddle point method directly: some multivariate version of Hayman's method (Hayman, 1956), the univariate version of which was explained in Section 3.2.

The set of $\boldsymbol{r}$ for which $\beta^*(\boldsymbol{r}) > -\infty$ needs a name; we call this cone Ξ in accordance with references such as Pemantle and Wilson (2008). When the infimum in (7.3.1) is finite, it will be achieved unless B has an asymptote in the direction normal to $\boldsymbol{r}$ (in this case, one might say it is achieved at a projective point of B). Assuming no asymptote, the infimum is achieved uniquely unless B fails to be strictly convex and its boundary contains a line segment. Thus the most common case will be the *nonflat* case, defined as follows.

Definition 8.1.4 (nonflat directions) *The direction $\hat{\boldsymbol{r}}_* \in \Xi$ is* ***nonflat*** *if the infimum in* (7.3.1) *is attained at a unique point, $\boldsymbol{x}_{\min}(\hat{\boldsymbol{r}}_*)$, of $\overline{B}$, which we call the* ***minimizing point*** *for $\hat{\boldsymbol{r}}_*$.*

Remark We often write simply $\boldsymbol{x}_{\min}$ when $\hat{\boldsymbol{r}}_*$ is understood. The point $\boldsymbol{x}_{\min}$ must lie on the boundary of B because extrema of linear functions are never attained in the interior of any set.

Consider a nonflat direction $\hat{\boldsymbol{r}}_* \in \Xi$ with minimizing point $\boldsymbol{x}_{\min}$. We deduced (8.1.2) directly from the definition of the domain of convergence. The chain of integration for the Cauchy integral may be taken as $\mathbf{T}(\boldsymbol{x})$ for any $\boldsymbol{x} \in B$. Sending $\boldsymbol{x} \to \boldsymbol{x}_{\min}$ gives a second proof that $\overline{\beta}(\boldsymbol{r}) \leq \beta^*(\boldsymbol{r})$. Using the multivariate Cauchy formula is slightly less elementary than the observation that the terms of a convergent power series must go to zero, but it leads to a more general conclusion: if we can deform the chain of integration *beyond* $\mathbf{T}(\boldsymbol{x}_{\min})$, to a chain on which $-\boldsymbol{r} \cdot \boldsymbol{x}$ is bounded above by some $c < \beta^*(\boldsymbol{r})$, we can deduce that $\overline{\beta} \leq c < \beta^*$. Conversely, if we cannot, then we have strong evidence that $\overline{\beta}(\boldsymbol{r}) = \beta^*(\boldsymbol{r})$. It is tempting to conjecture that $\overline{\beta}(\boldsymbol{r}) = \beta^*(\boldsymbol{r})$ if and only if the chain $\mathbf{T}(\boldsymbol{x})$ for $\boldsymbol{x} \in B$ cannot be deformed into a chain supported on $\{\boldsymbol{x} : -\boldsymbol{r} \cdot \boldsymbol{x} < \beta^*(\boldsymbol{r})\}$. This is shown to be false as stated in Pemantle (2010, Section 5). However, a modified form of this may be possible (Pemantle, 2010, Conjecture 5.1). Investigating such deformations requires we leave the realm of easy-to-visualize chains such as tori and more toward a view that is more topological than geometric. This is the subject of the remainder of this chapter.

8.2 Morse Theory Redux

The domain of holomorphy of a rational function F is the complement of the zero set of the denominator. It is an open subset of $\mathbb{C}^d$, namely the manifold $\mathcal{M} := \mathbb{C}^d \setminus \{z : (z_1 \cdots z_d)H(z) = 0\}$ obtained by removing the coordinate hyperplanes and the singular variety $\mathcal{V}$. We recall a consequence of Stokes' Theorem: if ω is a d-form, holomorphic on a domain $\mathcal{D}$ in $\mathbb{C}^d$, then $\int_C \omega$ depends only on the homology class of C in $H_d(\mathcal{D})$; see Theorem A.3.8 of Appendix A. Letting $\omega = z^{-r-1} F(z)\, dz$ be the Cauchy integrand and $\mathcal{D} = \mathcal{M}$, we see that $\int_C z^{-r-1} F(z)\, dz$ depends only on the homology class of C in $H_d(\mathcal{M})$. This is step (ii) of the procedure outlined at the end of Section 1.3.

Fix $\hat{\boldsymbol{r}}_*$ and define the ***height function***

$$h(\boldsymbol{x}) := h_{\hat{\boldsymbol{r}}_*}(\boldsymbol{x}) := -\hat{\boldsymbol{r}}_* \cdot \operatorname{Re} \log \boldsymbol{x} \,.$$

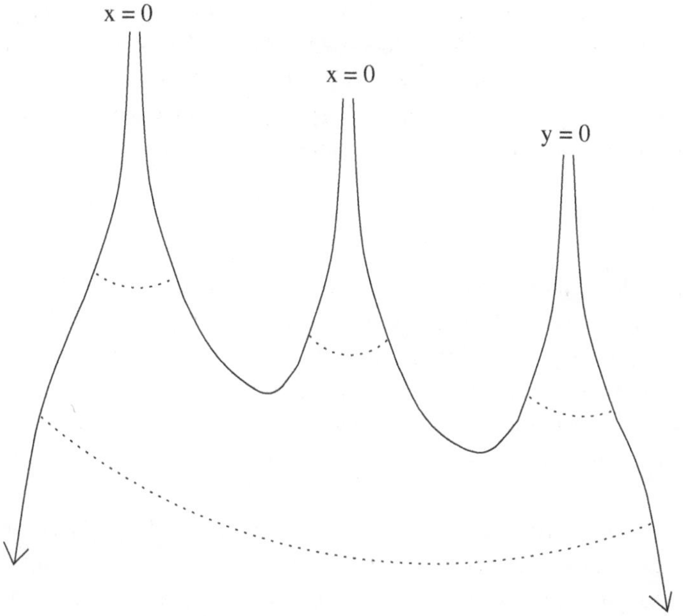

Figure 8.1 Simplified drawing of $\mathcal{V}$, with the vertical axis representing h.

The height function $|\boldsymbol{r}|\, h$ is a good surrogate for the log magnitude of the integrand $z^{-r-1} F(z)$ because it captures the part that goes to infinity with $\boldsymbol{r}$, leaving only the factor $z^{-1} F(z)$, which is bounded on compact subsets of $\mathcal{M}$. We may rephrase the question of deformation of the chain of integration as one of finding the minimax height chain: what chain C of integration, homologous in $H_d(\mathcal{M})$ to $\mathbf{T}(\boldsymbol{x})$ for $\boldsymbol{x} \in B$, achieves the least value of $\max_{\boldsymbol{x} \in C} h(\boldsymbol{x})$? Of particular interest is whether we can we make this maximum less than β^*.

Duality relates the topology of $\mathcal{M}$ to the topology of $\mathcal{V}$, which is the more classical object of Morse theory. It is helpful to envision a two-dimensional example in which the singular variety $\mathcal{V}$ is drawn in such a way that the up direction corresponds to height. We cannot draw all of the ambient $\mathbb{C}^2$, but the complex algebraic curve $\mathcal{V}$ will be a surface of two real dimensions, which is depicted as sitting in a three-dimensional ambient space as a surrogate for $\mathbb{C}^2$. The complex curve $\mathcal{V}$ will intersect the coordinate axes in a finite number of points, at which h will be infinite and will have finitely many points at infinity, at which the height will tend to $-\infty$. Figure 8.1 depicts an example.

The algebraic topology of a space endowed with a smooth height function h and the question of minimax height of a cycle representing a given homology class in that space is known as Morse theory. We now summarize the essentials of this theory as it pertains to our problem. A more complete development is given in Appendixes B and C. To reduce confusion among those who have some knowledge of topology, we point out that chains that are homotopic in a space $\mathcal{M}$ are also homologous, but not necessarily vice versa. Evaluation of the Cauchy integral requires only that chains be homologous, and Morse theory guarantees only that chains be homologous, whence we work with homology rather than homotopy. Most of the time, however, we prove homology by demonstrating a homotopy, and it is safe to read "homologous" as "homotopic" except when noted.

Smooth Morse Theory

The basic Morse theory, named after Marston Morse and more fully developed in the seminal work of Milnor (1968), concerns a compact manifold $\mathcal{V}$ endowed with a smooth height function $h : \mathcal{V} \to \mathbb{R}$. Let $\mathcal{V}^c$ denote the subset $\{x \in \mathcal{V} : h(x) \leq c\}$ of points at height at most c. The traditional purpose of Morse theory is to tell how $\mathcal{V}^c$ changes as c increases from its minimum to it maximum value. The fundamental Morse Lemma is that the topology does not change between critical values. The second main result is a description of how the topology changes at critical values. These two results are respectively Lemma B.1.2 and Theorem B.1.3 of Appendix B. The second result is usually stated under the further assumption that h be a *Morse function*, meaning that the critical points should be quadratically nondegenerate (see Definition B.1.1). We sometimes remove this assumption, leading to a slightly modified conclusion. For now we concentrate on the implications of the Morse Lemma, returning to the description of the attachment in Section 8.5.

For us what is important is the implications for the minimax height of a cycle representing a given homology class. Suppose $\mathcal{V}$ has k critical points with distinct critical values $c_1 > \cdots > c_k$. The Morse Lemma is proved by showing that for any interval $[a, b]$ containing no critical values of h, the set $\mathcal{V}^b$ retracts homotopically into $\mathcal{V}^a$. Any cycle C supported on $\mathcal{V}^b$ is carried by this retraction to a cycle in $\mathcal{V}^a$. Given a d-cycle C, let $h_*(C)$ denote the infimum of c such that C is homologous in $H_d(\mathcal{V})$ to a cycle supported on $\mathcal{V}^c$. It follows that $h_*(C)$ is always a critical value c_j of h.

Let C be a cycle in $\mathcal{V}$. Either $h_*(C) = c_1$ or C is homologous to a cycle supported on $\mathcal{V}^{c_1-\varepsilon}$. A necessary and sufficient condition for the latter is that the homology class $[C]$ vanishes when projected to the relative homology group $H_d(\mathcal{V}, \mathcal{V}^{c_1-\varepsilon})$ for some $\varepsilon > 0$. When this homology class in nonzero, we say there is a topological obstruction at height c_1. Inductively, $h_*(C) = c_j$, where j is the least index for which a topological obstruction occurs. The obstruction is local to the critical point $\boldsymbol{p}$ of height c_j in the sense that there is a retraction of $\mathcal{V}^{c_j+\varepsilon}$, all trajectories of which reach height $c_j - \varepsilon$ except for those that enter a neighborhood of $\boldsymbol{p}$. This argument is formalized in Lemma B.2.1. The cycle representative C_* achieving the minimax height h_* is unique when considered as a homology class in the pair $(X^{h_*}, X^{h_*-\varepsilon})$. Letting $\boldsymbol{p}$ denote the unique critical point at height h_*, this pair is homotopy equivalent to the pair $(X^{h_*-\varepsilon} \cup N(\boldsymbol{p}), X^{h_*-\varepsilon})$, where $N(\boldsymbol{p})$ denotes an arbitrarily small neighborhood of $\boldsymbol{p}$. This pair is denoted $X^{\boldsymbol{p},\mathrm{loc}}$ (see Definition B.1.6). Summing up, we have the following lemma.

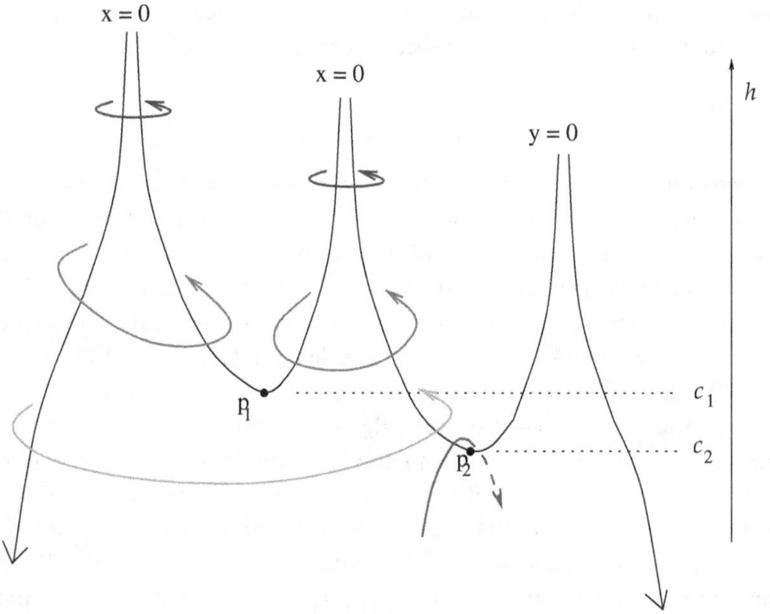

Figure 8.2 A homology class traveling down a surface that has two saddles, p_1 and p_2, at respective heights $c_1 > c_2$.

Lemma 8.2.1 (quasi-local cycle) *Let $\mathcal{V}$ be a compact manifold of homology dimension d, and let $h : \mathcal{V} \to \mathbb{R}$ be smooth. Suppose that h has finitely many critical points with distinct critical values $c_1 > \cdots > c_k$. Let C be any cycle in $H_d(\mathcal{V})$. Then:*

(i) The minimax height $h_*(C)$ of the class $[C]$ is equal to c_j for some j.
*(ii) Let **p** be the unique critical point of h at height h_*. Then C is homologous to a cycle C_* supported on the union of $\mathcal{V}^{c_j-\varepsilon}$ and an arbitrarily small neighborhood of **p**. The value of j and the homology class of C_* in the space $X^{\mathbf{p},\mathrm{loc}}$ is uniquely determined by C.*
(iii) j may be characterized as the least index i for which the image of C in $H_d(\mathcal{V}, \mathcal{V}^{c_i})$ vanishes.

Remark If the quadratic part of h near the critical point $\mathbf{z}$ at height c_j is nondegenerate (in particular, if h is a Morse function), then the relative homology group $H_d(\mathcal{V}^{c_j+\varepsilon}, \mathcal{V}^{c_j-\varepsilon})$ is cyclic and C_* is a generator.

To illustrate, suppose that C is the upper cycle with two components, pictured in Figure 8.2. Two critical values $c_1 > c_2$ are shown. For any $c > c_1$, the retraction of $\mathcal{V}$ to $\mathcal{V}^c$ carries C to a cycle homotopic to C, such as the middle cycles in the figure. In this case, because the two components are in opposite directions near the saddle p_1 at height c_1, there is no topological obstruction to continuing below p_1. The two components merge, forming the lower cycle (notice that this homology is not a homotopy). At the point p_2, there is a topological obstruction, and the maximum height of C cannot be pushed below height c_2.

Stratified Morse Theory

The second main result of Morse theory, describing the attachment, will help us to understand the nature of the quasi-local cycle C_*. Before embarking on this, however, we need to remove some of the assumptions. The basic Morse theory was extended in the 1980s to stratified spaces (Goresky and MacPherson, 1988). A second important extension, to noncompact spaces such as the complements of manifolds and stratified spaces, is also detailed in Goresky and MacPherson (1988). We discuss these extensions in turn, as we will require them both.

Recall from Section 5.4 that every algebraic variety is a Whitney stratified space. Such a space $\mathcal{V}$ is the disjoint union of manifolds of various dimensions. The function $h : \mathcal{V} \to \mathbb{R}$ is said to be smooth if it is smooth when restricted to each stratum. The point $\boldsymbol{p} \in \mathcal{V}$ is said to be critical if it is a critical point of the function $h|_S$, where S is the stratum of $\mathcal{V}$ containing $\boldsymbol{p}$. In general $\mathcal{V}$ will not be compact; we will require that h be a ***proper map***, meaning that the inverse image of any compact set is compact; this assumption is automatically satisfied for any continuous h when $\mathcal{V}$ is compact.

Lemma 8.2.2 (Fundamental Lemma for stratified spaces) *Let $\mathcal{V}$ be a stratified space and let $h : \mathcal{V} \to \mathbb{R}$ be a smooth, proper function with finitely many distinct critical values. If h has no critical values in $[a, b]$, then X^a is a strong deformation retract of X^b. In particular, the homotopy types of X^t are all naturally identified for $a \le t \le b$, and any cycle in $\mathcal{V}^b$ is homotopic to a cycle in $\mathcal{V}^a$.*

Again, it follows that any cycle C may be pushed downward until it reaches a topological obstruction at a critical point $\boldsymbol{p}$, at which point it becomes a quasi-local cycle having height at most $h_*(C) - \varepsilon$, except in a neighborhood of $\boldsymbol{p}$.

Nonproper Morse Theory

The actual space $\mathcal{M}$ through which we wish to deform chains of integration is the complement in $\mathbb{C}^d$ of an algebraic variety $\mathcal{V}$. The height function $h : \mathcal{M} \to \mathbb{R}$ is never proper. Still, one might expect from topological duality that the critical values of h on $\mathcal{V}$ are the only values of c at which the topology of $\mathcal{M}^{c+\varepsilon}$ and $\mathcal{M}^{c-\varepsilon}$ can differ. Indeed this is the case. The following is proved in Goresky and MacPherson (1988) (see the discussion in Section C.3 of the appendixes).

Lemma 8.2.3 (Fundamental Lemma for complements of stratified spaces) *Let $\mathcal{M}$ denote the complement in $(\mathbb{C}^d)^*$ of a stratified space $\mathcal{V}$. If the smooth, proper function $h : \mathcal{V} \to \mathbb{R}$ has no critical values in $[a, b]$ on $\mathcal{V}$, then X^a is a strong* *deformation retract of* X^b. *In particular, any cycle in* $\mathcal{M}^b$ *is homotopic to a cycle in* $\mathcal{M}^a$ *and the three conclusions of Lemma 8.2.1 hold.*

We would like to use this version of the Morse Lemma to execute step (iii) of the program in Section 1.3. Unfortunately, the function h is usually not proper, even on $\mathcal{V}$. The difficulty is that h can sometimes approach a finite limit as $\boldsymbol{x} \to \infty$ on $\mathcal{V}$, and in fact once $d \geq 3$, this occurs generically. If there is a compactification of $\mathcal{V}$ such that h extends continuously as a function to the extended reals $[-\infty, \infty]$, then Lemma 8.2.3 still holds, with the provision that the strata at infinity may contain new critical points. The existence of such a compactification is conjectured in Pemantle (2010, Conjecture 2.11), but it is still unknown whether this holds.

If the conjecture fails, it is possible that the deformation of $\mathcal{M}^b$ to $\mathcal{M}^a$ cannot take place without some trajectories shooting off to infinity. The conjecture says, essentially, that this does not happen unless there are projective critical points: points on the projective variety where the gradient of h vanishes. Such cases do arise (see DeVries, 2011). In the absence of these, the conjecture would guarantee that the only obstructions to lowering C occur at affine critical points. Because Morse theory is serving only as a guide, there is nothing to stop us from assuming this conjecture and proceeding with our intended analysis, verifying our deformations for each new class of problems. Any explicit chains of integration that we find will work even if is not proved that we must find them.

Non-Morse Morse Theory

Before turning to the analysis of critical points, we clean up a detail or two about our assumptions on h. It is traditional to assume that h is a *Morse function*, meaning that the critical points are nondegenerate and the critical values are distinct. These assumptions enter when computing the nature of the attachments. The Morse Lemma, in all its forms, remains true when assuming only that h has isolated critical points. Dropping the assumption of distinct critical points, the pair $(X^c, X^{c-\varepsilon})$ will in general be homotopy equivalent to a direct sum of local pairs, denoted by $X^{\boldsymbol{p},\mathrm{loc}}$ and defined in the following Lemma; see also (B.1).

Lemma 8.2.4 (quasi-local cycles when critical values are not distinct) *Let* $\mathcal{M} = (\mathbb{C}^d)^* \setminus \mathcal{V}$ *be the complement of a stratified space and let* $h : (\mathbb{C}^d)^* \to \mathbb{R}$ *be smooth and proper. Let* $\boldsymbol{p}_{i,1}, \ldots, \boldsymbol{p}_{i,n_i}$ *denote the set of critical points with critical value* c_i. *Then*

(i) The minimax height $h_*(C)$ *of the class* $[C]$ *is equal to* c_j *for some* j.
(ii) The cycle C *is homologous to a cycle* C_* *supported on the union of* $\mathcal{M}^{c_j-\varepsilon}$ *and an arbitrarily small neighborhood of* $\{\boldsymbol{p}_{j,1}, \ldots, \boldsymbol{p}_{j,n_j}\}$.
(iii) j *may be characterized as the least index* i *for which the image of* C *in* $H_d(\mathcal{M}, \mathcal{M}^{c_i})$ *vanishes.*

(iv) *The homology group* $H_d(\mathcal{M}^{c_j+\varepsilon}, \mathcal{M}^{c_j-\varepsilon})$ *is naturally the direct sum, induced by inclusion, of the homology groups* $H_d(X^{p_{j,i},\mathrm{loc}})$ *for* $1 \le i \le n_j$*, where* $X^{p,\mathrm{loc}}$ *denotes the pair* $(X^{c_j-\varepsilon} \cup N(\boldsymbol{p}), X^{c_j-\varepsilon})$ *and* $N(\boldsymbol{p})$ *is an arbitrarily small neighborhood of* $\boldsymbol{p}$.

(v) *The cycle* C_* *may be written as a sum* $\sum_{z\in\texttt{contrib}} C_*(z)$*, where* `contrib` *is the set of* $z = \boldsymbol{p}_{j,i}$ *for which the projection of* C *to* $H_d(\mathcal{M}^{z,\mathrm{loc}})$ *is nonzero.*

Remark 8.2.5 The Cauchy integral is unchanged if we vary the chain of integration over a homology class in $H_d(\mathcal{M})$. Relative homology classes are a coarser partition of chains of integration and do not leave integrals invariant. It is true, however, that chains in the same relative homology class in $H_d(\mathcal{M}, \mathcal{M}^{c_j-\varepsilon})$ have integrals differing by terms that are $O(\exp[(c-\varepsilon)|\boldsymbol{r}|])$ and hence are exponentially smaller than $\exp[(c_j + o(1))|\boldsymbol{r}|]$ (see Proposition A.3.10 in Appendix A). Once we establish that the main term of the integral has limsup logarithmic magnitude $(c_j + o(1))\,|\boldsymbol{r}|$, it follows that the error term introduced within a relative homology class is negligible. The useful consequence for us is that the topological invariant $C_* = \sum_{z\in\texttt{contrib}} C_*(z)$ determines the asymptotics of $a_{\boldsymbol{r}}$ up to an exponentially smaller remainder.

8.3 Critical Points

We return to the analysis of a Laurent series for $F = G/H$, convergent on a component B of the complement of the amoeba of H. Let $T = \mathbf{T}(\boldsymbol{x})$ for some $\boldsymbol{x} \in B$. Fixing $\hat{\boldsymbol{r}}_*$, the function $h = h_{\hat{\boldsymbol{r}}_*}$ is constant on T, with common value $b := -\hat{\boldsymbol{r}}_* \cdot \boldsymbol{x}$. List the critical values of h that are at most b in descending order $b > c_1 > c_2 > \cdots$. Assuming Conjecture 2.11 of Pemantle (2010), the cycle C is homologous to the sum of one or more quasi-local cycles at critical points at height $h_*(C)$. The critical points are an algebraic set, of dimension zero except in degenerate cases, and are effectively computable from H. To see this, we first observe that there is an effectively computable Whitney stratification for $\mathcal{V}$ (see the end of Section C.1 in Appendix C for a description of the procedure in Kaloshin [2005]). The strata of $\mathcal{V}$ are smooth manifolds, each being of some dimension $k \le d-1$. A stratum S of dimension k is a k-dimensional algebraic variety $\overline{S}$ minus possibly some varieties of smaller dimensions. Any irreducible k-dimensional complex algebraic variety has a representation as the intersection of $d-k$ algebraic hypersurfaces. At generic points, these intersect transversely. Representing $\overline{S}$ in this way, we may assume without loss of generality that all points at which the intersection is not transverse are in $\overline{S} \setminus S$; i.e., they have been placed in lower dimensional strata. Thus $\overline{S}$ has a representation as the intersection of $d-k$ algebraic hypersurfaces[1] $\mathcal{V}_{f_1}, \ldots, \mathcal{V}_{f_{d-k}}$, intersecting transversely at every

[1] Ravi Vakil offers the following justification for this fact: affine d-space is nonsingular, so the maximal ideal of the local ring at the generic point of X is generated by $d-k$ elements; these $d-k$ rational

point of S, with the polynomials f_j being effectively computable and having nonvanishing gradient at every point of S.

Vanishing of $dh|_S$ at x is equivalent to the vector $\hat{\boldsymbol{r}}_*$ being in the span of the $d-k$ vectors $\{\nabla_{\log} f_i(x) \, : \, 1 \le i \le d-k\}$, where

$$\nabla_{\log} f(\boldsymbol{x}) := (x_1 \partial f/\partial x_1, \ldots, x_d \partial f/\partial x_d)$$

is the gradient of f with respect to $\log \boldsymbol{x}$. Let M be the $(d-k+1) \times d$ matrix whose rows are these $d-k$ gradients together with $\hat{\boldsymbol{r}}_*$. At all points of S, the submatrix of M consisting of the first $d-k$ rows has rank $d-k$. The condition that the span of the $d-k$ gradients contain $\hat{\boldsymbol{r}}_*$ is equivalent to the vanishing of the k determinants M_{d-k+i}, $1 \le i \le k$, where M_{d-k+i} contains the first $d-k$ columns of M together with the $(d-k+i)^{th}$ column. This gives the d ***critical point equations***:

$$f_i = 0\,, \quad i = 1, \ldots, d-k\,; \tag{8.3.1}$$

$$\det(M_{d-k+i}) = 0\,, \quad i = 1, \ldots, k\,. \tag{8.3.2}$$

Smooth critical points are the most common special case. The smooth points of $\mathcal{V}$ are in a stratum S of dimension $k := d-1$ with $\overline{S} = \mathcal{V}$. The defining equation for $\overline{S}$ is $f_1 := H = 0$. Being in the span of $\{\nabla_{\log} f_1\}$ means being parallel to $\nabla_{\log} f_1$, leading to $d-1$ equations for vanishing 2×2 subdeterminants:

$$\begin{aligned} H &= 0 \\ r_1 x_2 \frac{\partial H}{\partial x_2} &= r_2 x_1 \frac{\partial H}{\partial x_1} \\ &\vdots \\ &\vdots \\ r_1 x_d \frac{\partial H}{\partial x_d} &= r_d x_1 \frac{\partial H}{\partial x_1} \end{aligned} \tag{8.3.3}$$

Alternatively, one may write this compactly as follows.

$$\begin{gathered} H = 0 \\ \nabla_{\log} H \parallel \hat{\boldsymbol{r}}_* \end{gathered}$$

The map between directions $\hat{\boldsymbol{r}}$ or $\boldsymbol{r}$ and critical points is fundamental. Some useful notation is as follows. Given a vector $\boldsymbol{r}$, let `critical`$(\boldsymbol{r})$ denote the set of all critical points $\boldsymbol{z} \in \mathcal{V}$ for $h_{\hat{\boldsymbol{r}}}$. Observe that `critical`$(\boldsymbol{r})$ depends only on the

functions are regular in a Zariski-neighborhood of the generic point of X and therefore intersect transversely along X in a smaller Zariski-neighborhood.

direction $\hat{\boldsymbol{r}}$. In the other direction, given $z \in \mathcal{V}$, we denote by $\mathbf{L}(z)$ the linear space normal in logarithmic coordinates to the stratum S of $\mathcal{V}$ that contains z. Is is clear that

$$z \in \texttt{critical}(\boldsymbol{r}) \Longleftrightarrow \boldsymbol{r} \in \mathbf{L}(z)$$

and that this defines a binary relation on the pair $(\mathcal{V}, (\mathbb{R}^d)^*)$. The following examples illustrate the definition and computation of this relation. Low-dimensional notational conventions are in use: (x, y, z) instead of (x_1, x_2, x_3) and (r, s, t) instead of (r_1, r_2, r_3). Thus, for example, $\hat{\boldsymbol{r}}$ denotes $(\hat{r}, \hat{s}, \hat{t}) := (r/|r+s+t|, s/|r+s+t|, t/|r+s+t|)$.

Example 8.3.1 (binomial coefficients continued) Recall from Example 2.2.2 that the binomial coefficients $\binom{r+s}{r,s}$ have generating function $\dfrac{1}{1-x-y}$. With $H = 1-x-y$, we find $\nabla H = (-1,-1)$, which never vanishes, so the variety $\mathcal{V}$ is smooth. The equations (8.3.3) are

$$1 - x - y = 0$$

$$-sx = -ry.$$

The solution is $x = \dfrac{r}{r+s}$, $y = \dfrac{s}{r+s}$. Thus, in this example, $\texttt{critical}(\boldsymbol{r}) = \hat{\boldsymbol{r}}$. ■

Example 8.3.2 (Delannoy numbers continued) Continuing from Example 2.2.6, we recall that the denominator for the Delannoy generating function is given by $H = 1 - x - y - xy$. The gradient is $(-1-y, -1-x)$. To check that $\mathcal{V}$ is smooth, we check that $-1-y$, $-1-x$, and H never simultaneously vanish. This is verified by the Gröbner basis computation

```
Groebner[Basis]([-1-x,-1-y,1-x-y-x*y],plex(x,y));,
```

which returns the basis `[1]`. See Sections 6.1 and 6.2 for more on the use of the Maple package `Groebner`. The critical point equations are the two following equations:

$$1 - x - y - xy = 0$$

$$sx(1+y) = ry(1+x)$$

Solving this with

```
Groebner[Basis]([s*x*(1+y)-r*y*(1+x),1-x-y-x*y], plex(x,y));
```

yields $[sy^2 - s + 2ry, s - sy - r + rx]$. The first of the two polynomials is the elimination polynomial for y. Dividing through by $(r+s)$, we see these polynomials are homogeneous in (r, s). Solving the elimination polynomial for y gives

$$y = \frac{-r \pm \sqrt{r^2 + s^2}}{s}.$$

Setting the second basis polynomial equal to zero gives x as a function of y: $x = (sy + r - s)/r$. One may avoid messing around with quadratics by computing an elimination polynomial for x directly:

$$x = \frac{-s \pm \sqrt{r^2 + s^2}}{r}.$$

This gives four possible (x, y) pairs, of which two solve the second critical point equation: the two positive roots go together and the two negative roots go together. Thus `critical`(r, s) consists of the two points

$$\left(\frac{\sqrt{r^2+s^2} - s}{r}, \frac{\sqrt{r^2+s^2} - r}{s}\right) \text{ and } \left(\frac{-\sqrt{r^2+s^2} - s}{r}, \frac{-\sqrt{r^2+s^2} - r}{s}\right). \tag{8.3.4}$$

We see later that the first point is the one that determines the asymptotics of a_{rs}. ■

To encounter a nontrivial example that is not a smooth point, we require one more variable ($d = 3$) and one greater codimension ($d - k = 2$).

Example 8.3.3 (two intersecting planes) Let $H = H_1 H_2$ with $H_1 := 4 - 2x - y - z$ and $H_2 := 4 - x - 2y - z$. These two planes intersect in the line ℓ containing the points $(0, 0, 4)$ and $(\frac{4}{3}, \frac{4}{3}, 0)$. Thus the strata are $S_1 := \mathcal{V}_1 \setminus \ell$, $S_2 := \mathcal{V}_2 \setminus \ell$ and $S_3 := \ell$. Critical points in the stratum S_2 are obtained by solving the equations (8.3.3) for (x, y, z) in terms of (r, s, t). Solving

$$\begin{aligned} 4 - x - 2y - z &= 0 \\ tx &= rz \\ 2ty &= sz \end{aligned}$$

gives $x = 4\hat{r}$, $y = 2\hat{s}$, $z = 4\hat{t}$. Finding the critical point on S_1 is analogous and gives $x = 2\hat{r}$, $y = 4\hat{s}$, $z = 4\hat{t}$. Critical points on S_3 are obtained by solving (8.3.1) and (8.3.2), as follows.

$$\begin{aligned} 4 - x - 2y - z &= 0 \\ 4 - 2x - y - z &= 0 \\ ryz + sxz - 3txy &= 0 \end{aligned}$$

Solving these yields $x = y = (4/3)(\hat{r} + \hat{s})$, $z = 4\hat{t}$, which is the unique point on l at which (r, s, t) lies in the plane spanned by the logarithmic tangents $(x, 2y, z)$ and $(2x, y, z)$ to the planes defined by H_1 and H_2.

To summarize, `critical`(r, s, t) consists of the three points

$$\left\{(4\hat{r}, 2\hat{s}, 4\hat{t}), (2\hat{r}, 4\hat{s}, 4\hat{t}), \left(\frac{4}{3}(\hat{r} + \hat{s}), \frac{4}{3}(\hat{r} + \hat{s}), 4\hat{t}\right)\right\}.$$

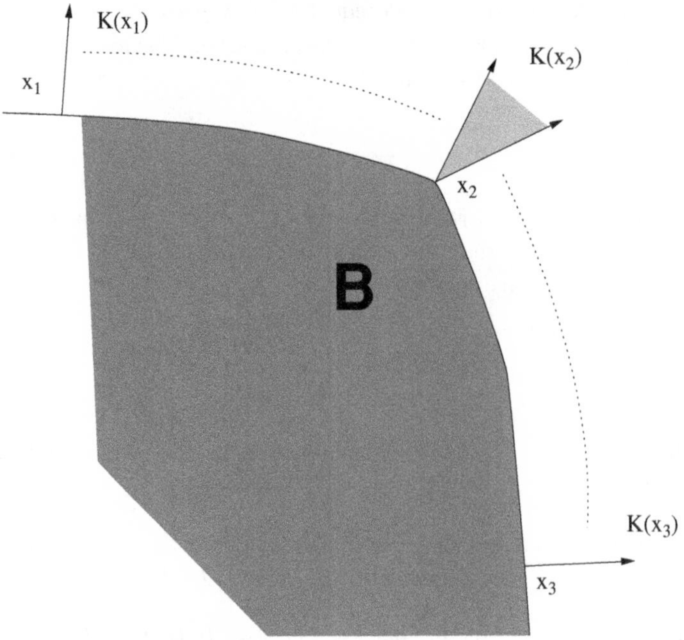

Figure 8.3 One chamber is a cone with interior; the rest are rays.

For $z \in \ell$, the space $\mathbf{L}(z)$ is the span of the normal vectors to $\log H_1$ and $\log H_2$ at $\log z$; for $z \in S_j$ with $j = 1$ or 2, the space $\mathbf{L}(z)$ is $\langle \boldsymbol{r} \rangle$, where $\boldsymbol{r}$ is normal to $\log S_j$ at $\log z$. ■

8.4 Minimal Points

Let $F = G/H, \mathcal{V}, \mathcal{M}$, and the component B of the complement of $\texttt{amoeba}(H)$ be given. Recall that the open domain of convergence $\mathcal{D}$ of the Laurent series for F on B is the union of the tori $\mathbf{T}(\boldsymbol{x})$ for $\boldsymbol{x} \in B$. Let C denote the cycle $\mathbf{T}(\boldsymbol{x})$ for some $\boldsymbol{x} \in B$. Pick a nonflat direction $\hat{\boldsymbol{r}}_* \in \Xi$, let $h := h_{\hat{\boldsymbol{r}}_*}$, let $\boldsymbol{x}_{\min}$ denote the minimizing point for h on $\overline{B}$, and let $h_* := h(\boldsymbol{x}_{\min}) = \beta^*(\hat{\boldsymbol{r}}_*)$. Critical points $z \in \partial\mathcal{D}$ are called ***minimal points***. The set of minimal points in $\texttt{critical}(\hat{\boldsymbol{r}}_*)$ is denoted by $\texttt{minimal}(\hat{\boldsymbol{r}}_*)$. Minimal points play an important role in determining whether $\overline{\beta}(\hat{\boldsymbol{r}}_*)$ is equal to $\beta^*(\hat{\boldsymbol{r}}_*)$ and in computing the asymptotics of $a_{\boldsymbol{r}}$ in this case. Early work using explicit contour deformations by Pemantle and Wilson (2002) and Pemantle and Wilson (2004) relied heavily on minimal points.

Recall from Section 8.1 the set Ξ of $\boldsymbol{r}$ such that $-\boldsymbol{r} \cdot \boldsymbol{x}$ is bounded from below on B. This is a convex dual to the set B. There is a natural chamber decomposition into the outward normal cone to B at $\boldsymbol{x}$ (see Section 7.1). The decomposition $\Xi = \bigcup_{\boldsymbol{x} \in \partial B} \texttt{normal}_{\boldsymbol{x}}(B)$ is illustrated in Figure 8.3.

Locally Oriented Minimal Points

An important step in determining the set $\texttt{contrib}$ is to complexify the decomposition in Figure 8.3. This partition of Ξ associates points $\boldsymbol{x} \in \partial B$ with directions $\boldsymbol{r}$ of asymptotics potentially "governed" by the real point $\boldsymbol{x}$. The quotation marks signify that the Cauchy integral takes place in $\mathbb{C}^d$, not $\mathbb{R}^d$, localized to some critical point $z = \exp(\boldsymbol{x} + i\boldsymbol{y})$. The association of z with $\boldsymbol{r}$ will be more informative than the association of $\boldsymbol{x}$ with $\boldsymbol{r}$. It is not so obvious how to do this. In fact, it requires Gårding's theory of hyperbolic functions (Gårding, 1950). With each minimal point $z = \exp(\boldsymbol{x} + i\boldsymbol{y}) \in \texttt{minimal}(\boldsymbol{r})$, we will associate a cone $\mathbf{K}(z) \supseteq \texttt{tan}_{\boldsymbol{x}}(B)$. Its dual, $\mathbf{N}(z) := \mathbf{K}(z)^*$, will be a subset of $\texttt{normal}_{\boldsymbol{x}}(B)$. This defines a finer relation: say $z \in \texttt{local}(\boldsymbol{r})$ if and only if $\boldsymbol{r} \in \mathbf{N}(z)$. We call these points ***locally oriented***. As with $\texttt{critical}$ and $\texttt{minimal}$, the set $\texttt{local}(\boldsymbol{r})$ depends on $\boldsymbol{r}$ only through $\hat{\boldsymbol{r}}$. The somewhat lengthy construction is carried out in Chapter 11. Until then, consider $\mathbf{K}(z)$ and $\texttt{normal}(z)$ to have been defined by some black box so that the set $\texttt{local}(\hat{\boldsymbol{r}})$ is defined as well and satisfies

$$\texttt{local}(\hat{\boldsymbol{r}}) \subseteq \texttt{minimal}(\hat{\boldsymbol{r}}) \subseteq \texttt{critical}(\hat{\boldsymbol{r}}). \tag{8.4.1}$$

Among the points in $\texttt{minimal}(\boldsymbol{r})$, only those in $\texttt{local}(\boldsymbol{r})$ have the potential to contribute to asymptotics in direction $\boldsymbol{r}$. In many cases, we do not need to worry about computing $\texttt{local}(\boldsymbol{r})$. For most applications of smooth point asymptotics (which means most applications to date), if we compute all the minimal points, we can figure out by hand which one(s) contribute to the asymptotics. Readers unlikely to venture beyond the case of smooth points can safely skip the rest of this section, in which we give intuition and examples as to why $\mathbf{N}(z)$ might be a proper subset of $\texttt{normal}_{\boldsymbol{x}}(B)$.

We have already seen that any topological obstruction to replacing $\mathbf{T}(\boldsymbol{x})$ by a cycle supported on $\mathcal{M}^{h_*-\varepsilon}$ occurs at a minimal point. Let us see now intuitively why some minimal points do not pose an obstruction. Near each point $z \in \mathcal{V} \cap \mathbf{T}(\boldsymbol{x}_{\min})$, the image of variety $\mathcal{V}$ under the $\operatorname{Re}\log$ map avoids B. This map may or may not cover a neighborhood of $\boldsymbol{x}_{\min}$ in ∂B. If it does so, we call $\boldsymbol{x}_{\min}$ ***well covered***, and we call z a ***covering point***. The intuition is that the local geometry of B is captured by a covering point, so one expects that the geometric quantity β^* might agree with the topological quantity h_* if $\boldsymbol{x}_{\min}$ is well covered. We see in the proof of Proposition 8.4.3 that if $\boldsymbol{x}_{\min}$ is well covered, then $\mathbf{N}(z) = \texttt{normal}_{\boldsymbol{x}}(B)$. If $\boldsymbol{x}_{\min}$ is not well covered, then each minimal point has a geometry that "comes off the boundary" in some places, and it is possible therefore that none of these points obstructs the movement of $\mathbf{T}(\boldsymbol{x}_{\min})$ below height h_*. The following example is taken from Baryshnikov and Pemantle (2011, Example 2.19).

Example 8.4.1 (two complex lines with ghost intersection) Let $H = L_1 L_2 = (3 - x - 2y)(3 + 2x + y)$. The variety $\mathcal{V}_H$ is shown in Figure 8.4(a). The amoeba of H is the union of $\texttt{amoeba}(3 - x - 2y)$ and $\texttt{amoeba}(3 + 2x + y)$, the latter of which is identical to $\texttt{amoeba}(3 - 2x - y)$ because the amoeba of $F(-x, -y)$ is the

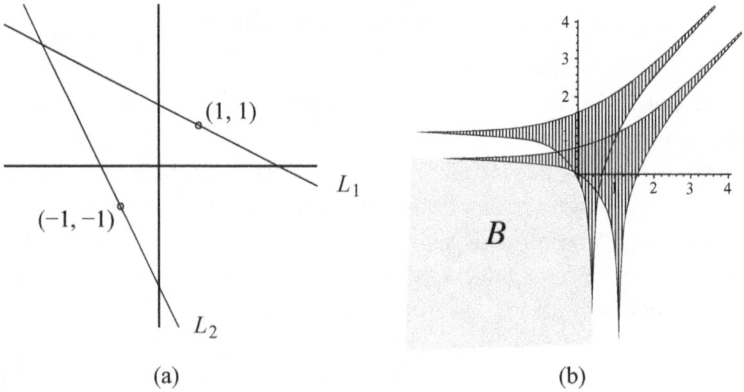

Figure 8.4 The zero set of $(3 - x - 2y)(3 + 2x + y)$ from Example 8.4.1 and the OPS component.

same as the amoeba of $F(x, y)$. The component B of $\mathbb{R}^d \setminus \texttt{amoeba}(H)$ containing the negative quadrant corresponds to the ordinary power series. An enlargement of this component is shown in Figure 8.4(b). When $\boldsymbol{x} = (0, 0)$, the linearization of H at $\boldsymbol{x}$ is just $\ell_1\ell_2 := (x + 2y)(2x + y)$, the zero set of which contains the two rays forming the boundary of

$$\texttt{tan}_{\boldsymbol{x}}(B) = \{(u, v) \in \mathbb{R}^2 : 2u + v < 0 \text{ and } u + 2v < 0\}.$$

There are two points $z \in \mathcal{V}_H$ in $\operatorname{Re}\log^{-1}(0, 0)$, namely, $(1, 1)$ and $(-1, -1)$. The first is in $\mathcal{V}_{L_1}$ and the second is in $\mathcal{V}_{L_2}$. Locally, if H were equal to just L_1, then $(1, 1)$ would be a critical point, the cone $\texttt{tan}_{\boldsymbol{x}_{\min}}(B)$ there would be the halfspace $\{(u, v) \in \mathbb{R}^2 : u + 2v < 0\}$ and $\texttt{normal}_{\boldsymbol{x}_{\min}}(B)$ would be the ray $\{(t, 2t) : t > 0\}$. This means that $\mathbf{N}(1, 1)$ should be this ray. Similarly, if H were to equal L_2, then we should have $\mathbf{N}(-1, -1) = \{(2t, t) : t > 0\}$. These two degenerate cones are strictly contained in the cone $\texttt{normal}_{\boldsymbol{x}_{\min}}(B)$, which is the cone bounded by the two rays. The term "ghost intersection" refers to the fact that the two curves $\operatorname{Re}\log \mathcal{V}_{L_1}$ and $\operatorname{Re}\log \mathcal{V}_{L_2}$ intersect at $(0, 0)$, but the lines $\mathcal{V}_{L_1}$ and $\mathcal{V}_{L_2}$ have different imaginary parts and have no intersection on the unit torus, although they do intersect at $(-3, 3)$. Consequently, the point $(0, 0)$ is not a well-covered point of $\mathbb{R}^2$.

Near each of the two minimal points $(1, 1)$ and $(-1, -1)$, the variety $\mathcal{V}$ looks like a complex line, but these two have different slopes. Under the log-modulus map, one coincides with ∂B in the second quadrant but comes off of ∂B in the fourth quadrant, and the other does the opposite.

In terms of obstruction, there is an obstruction at $(1, 1)$ for the height function $h_{\hat{\boldsymbol{r}}}$ when $\hat{\boldsymbol{r}} = (1/3, 2/3)$ and an obstruction at $(-1, -1)$ for the height function $h_{\hat{\boldsymbol{r}}}$ when $\hat{\boldsymbol{r}} = (2/3, 1/3)$, but no minimal critical point of $\mathcal{V}$ is an obstruction for $h_{\hat{\boldsymbol{r}}}$ when $\hat{\boldsymbol{r}}$ lies between these two directions. Contrast this to what happens when the signs in the second factor are flipped to $(3 - 2x - y)$, as in Figure 7.7.1(b) and

Example 10.1.3 (later). The amoeba is the same, but now the lines intersect at $(1, 1)$, which is minimal. Now $\mathbf{N}(1, 1)$ is the whole cone bounded by the rays $(t, 2t)$ and $(2t, t)$, and the crossing lines are an obstruction for any direction in this cone. ■

Generalizing Example 8.4.1 is tricky because we are working with complex structures projected down to real structures. The defining property of the cone $\mathtt{tan}_{x_{\min}}(B)$ is in terms of the projected set (the amoeba), whereas we will replace this by a collection of cones $\{\mathbf{K}(z) : z \in \mathtt{minimal}(\hat{r}_*)\}$, whose definition involves looking at more than the real projection of $\mathcal{V}$ near z. The precise definition of these cones requires the theory of hyperbolic functions and is deferred to Chapter 11 (see Definition 11.1.9). Intuitively, the cone $\mathbf{K}(z)$ is a real cone, which would be "on one side of $\mathcal{V}$ near z" if we were working with a real algebraic variety instead of a complex one. The cone $\mathbf{K}(z)$ always contains $\mathtt{tan}_{x_{\min}}(B)$, whereby its dual $\mathbf{N}(z)$ is always contained in $\mathtt{normal}_{x_{\min}}(B)$. The following result, summing up the properties of the families $\{\mathbf{K}(z)\}$ and $\{\mathbf{N}(z)\}$, is proved in Chapter 11.

Theorem 8.4.2 *Let $x_{\min}$ be the minimizing point for direction $\hat{r}_*$ on the component B. There is a family of cones $\{\mathbf{K}(z) : z \in \mathbf{T}(x_{\min})\}$ varying semi-continuously with z, such that the following hold.*

(i) Let $\mathbf{N}(z)$ be the dual to $\mathbf{K}(z)$. If $\hat{r}_* \in \mathbf{N}(z)$, then z is a critical point for $h = h_{\hat{r}_*}$.
(ii) With $\mathtt{local}(\hat{r})$ defined as in (8.4.1), if $\mathtt{local}(\hat{r})$ is empty, then the cycle C can be deformed to a cycle C_* below height h_*, hence $\overline{\beta}(\hat{r}_*) < \beta^*(\hat{r}_*)$.
(iii) If $\mathtt{local}$ is nonempty, then usually $\overline{\beta} = \beta^*$.

Conclusion (iii) is meaningless of course, but because this chapter is informal, we have inserted this intuition. This proposition tells us, for instance, that in Example 8.4.1, for any direction strictly between the rays of slopes 1/2 and 2, the set $\mathtt{local}$ is empty and the proposition tells us that $\overline{\beta} < \beta^*$. Concerning the likelihood of finding $\mathtt{local}$ to be non-empty, one has the following result, which is a version of Pringsheim's Theorem.

Proposition 8.4.3 *If the Laurent coefficients a_r of $F = G/H$ on B are all nonnegative, then for any nonflat $\hat{r}_* \in \Xi$, the positive point $z_* = \exp(x_{\min})$ is a critical point in $\mathbf{T}(x_{\min})$ with $\hat{r}_* \in \mathbf{N}(z)$.*

Remark Because $\mathtt{normal}_{x_{\min}}(B)$ is the set of r for which $x_{\min}$ is the minimizing point, nonnegativity of the coefficients implies the reverse containment $\mathtt{normal}_{x_{\min}(r)}(B) \subseteq \mathbf{N}(z_*(r))$ (whence the two are equal).

Proof To see this, note first that there must be some singularity on the torus $\mathbf{T}(x_{\min})$. Otherwise, the domain of holomorphy of F would contain a neighborhood of $\mathbf{T}(x_{\min})$, and the Cauchy integral could be expanded beyond $\mathbf{T}(x_{\min})$, resulting in convergence of the Laurent series beyond $\mathbf{T}(x_{\min})$ and contradicting

$\boldsymbol{x}_{\min} \in \partial B$. The function F is rational, hence meromorphic, and thus this singularity must be a pole. In particular, $\sum a_{\boldsymbol{r}} \boldsymbol{z}^{\boldsymbol{r}}$ is not absolutely convergent on $\mathbf{T}(\boldsymbol{x}_{\min})$. Nonnegativity of the coefficients $a_{\boldsymbol{r}}$ together with this lack of convergence implies that $\sum a_{\boldsymbol{r}} \boldsymbol{z}_*^{\boldsymbol{r}} = +\infty$. Meromorphicity of F now implies that $\boldsymbol{z}_*$ is a pole of F. Moreover, the same argument shows that $\exp(\boldsymbol{x}')$ is a pole of F for any $\boldsymbol{x}' \in \partial B$. In particular, the image of any neighborhood of $\boldsymbol{z}_*$ in $\mathcal{V}$ under Re log covers a neighborhood of $\boldsymbol{x}_{\min}$ in ∂B. In other words, the point $\boldsymbol{x}_{\min}$ is well covered by $\boldsymbol{z}_*$. The cone $\mathbf{K}(\boldsymbol{z}_*)$ is a component of the complement of $\operatorname{Re}\log[U]$, where U is a neighborhood of $\boldsymbol{z}_*$ in $\mathcal{V}$. Because $\operatorname{Re}\log[U]$ contains a neighborhood of $\boldsymbol{x}_{\min}$ in ∂B, we see that $\mathbf{K}(\boldsymbol{z}_*) = \mathbf{K}(\boldsymbol{x}_{\min})$. Hence $\mathbf{N}(\boldsymbol{z}_*)$ is the dual to $\mathbf{K}(\boldsymbol{x}_{\min})$, and in particular it contains $\hat{\boldsymbol{r}}_*$. □

8.5 Describing the Quasi-Local Cycles

Theorem 8.4.2 gives sufficient conditions to conclude $\overline{\beta}(\hat{\boldsymbol{r}}_*) < \beta^*(\hat{\boldsymbol{r}}_*)$. The only way we know to go the other way and prove that these two quantities are equal is to compute the asymptotics of $a_{\boldsymbol{r}}$ for $\boldsymbol{r}/|\boldsymbol{r}|$ in a neighborhood of $\hat{\boldsymbol{r}}_*$. There are two ways to go about this: one can use piecemeal contour shifting and residue arguments in special cases, or one can attempt to harness stratified Morse theory to obtain general results. Our approach is a compromise. We use the Morse theory to understand what is going on and to motivate constructions of explicit deformations.

Recall from part (iv) of Lemma 8.2.4 that when one or more critical points $\boldsymbol{z}_j$ obstruct the passage of C below some level c (the common value of $h(\boldsymbol{z}_j)$), the projection of C to $H_d(\mathcal{M}, \mathcal{M}^{c-\varepsilon})$ may be written as the sum of cycles $C_*(\boldsymbol{z}_j)$, where each $C_*(\boldsymbol{z}_j)$ is quasi-local to $\boldsymbol{z}_j$, meaning it is supported on the union of $\mathcal{M}^{c-\varepsilon}$ with a small neighborhood of $\boldsymbol{z}_j$. We have put off until now discussing the second main result of Morse theory, which describes the quasi-local cycles. The result is usually framed in terms of a description of the attachment that transforms $\mathcal{M}^{c-\varepsilon}$ into a space whose homotopy type is that of $\mathcal{M}^{c+\varepsilon}$. This result characterizes a basis for the homology groups of the pair $(\mathcal{M}^{c+\varepsilon}, \mathcal{M}^{c-\varepsilon})$ local to each $\boldsymbol{z}_j$; the class $[C]$ must then be a sum of these basis elements, although if there is more than one generator, it is not always clear which sum. A full development of the underlying theory is given in the appendixes. Our brief treatment here is less formal.

Let S be a stratum of $\mathcal{V}$. Near any point $x \in S$ there is a local product structure: a neighborhood $\mathcal{N}$ of x in $\mathcal{V}$ is diffeomorphic to a product $N \times B_k$, where k is the dimension of the stratum S as a real manifold, B_k is a k-ball, and N is the *normal slice*, i.e., the intersection of a neighborhood of x in $\mathcal{V}$ with a plane normal to S. Define the *normal link* $\tilde{N}$ to be the intersection of a neighborhood U of x in $\mathcal{M}$ with a plane normal to S (recall that $\mathcal{M}$ is the complement of $\mathcal{V}$). These definitions do not require complex structure and are illustrated by the following example in $\mathbb{R}^3$.

Example 8.5.1 Let $\mathcal{V}$ be three planes in $\mathbb{R}^3$, all intersecting in a line, S, and let $\mathcal{M} := \mathcal{V}^c$. Then $k = 1$ and N has the shape of an asterisk. The normal link $\tilde{N}$ is

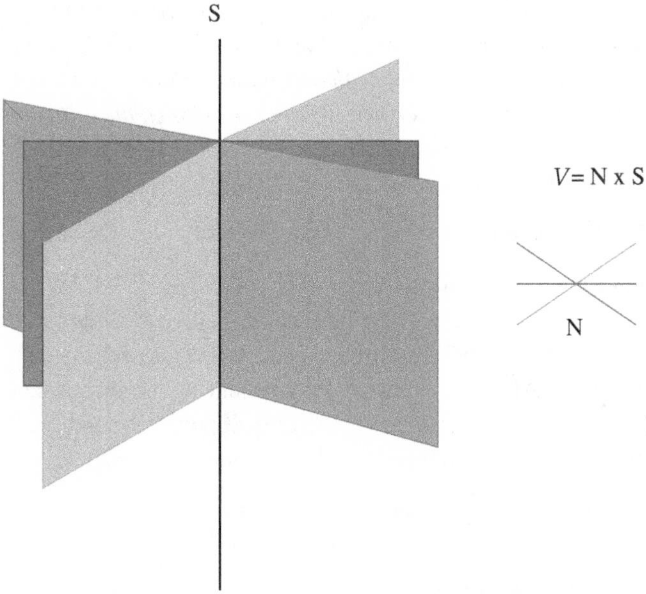

Figure 8.5 A configuration of three collinear planes and its normal slice.

a disk with an asterisk removed, which has the homotopy type of six points. See Figure 8.5. ■

Definition 8.5.2 (normal Morse data) *Let x be a critical point of S at height c. The* normal Morse data *at the point x is the topological pair* N-data $:= (\tilde{N}, \tilde{N} \cap \mathcal{M}^{c-\varepsilon})$*, the homotopy type of which is constant when ε is sufficiently small and positive.*

There is also a definition of the tangential Morse data, denoted T-data, which in general will depend on the index of the critical point $x \in S$. In our context, h is always a harmonic function, which implies that the index of any critical point is half the dimension of the manifold. The dimension of a complex k-manifold is $2k$ whence the index of every critical point is k. In this special case, the definition boils down to

$$\texttt{T-data} := (B_k, \partial B_k).$$

This is realized as the ***downward subspace*** at x: there are orthonormal coordinates in which the quadratic approximation to h near x is diagonal with k positive and k negative directions, and we may take B_k to be a patch tangent to the k-dimensional subspace of negative directions. The category of topological pairs has a product, namely $(A, B) \times (C, D) := (A \times B, A \times D \cup B \times C)$. The attachment is a product, in this sense, of the normal and tangential Morse data.

Theorem 8.5.3 (fundamental decomposition of stratified Morse theory) *Let $\mathcal{M}$ be the complement of a stratified space $\mathcal{V}$ with smooth, harmonic height function h. Let S be a stratum of dimension k with an isolated critical point $x \in S$. Assume the function h is* **Morse**: *i.e., the critical values are distinct, the critical points are nondegenerate, and the tangency assumption is satisfied in Definition C.2.1. Then for sufficiently small $\varepsilon > 0$,*

$$\left(\mathcal{M}^{c+\varepsilon}, \mathcal{M}^{c-\varepsilon}\right) \simeq \texttt{N-data} \times \texttt{T-data} = \left(\tilde{N}, \tilde{N} \cap \mathcal{M}^{c-\varepsilon}\right) \times \left(B_k, \partial B_k\right) . \quad (8.5.1)$$

The second component in this product is a subspace of $\mathcal{V}$. When evaluating the Cauchy integral, we will be integrating first over `N-data`, leaving an integral over `T-data`. As before, if the critical values are not distinct, then the pair $(\mathcal{M}^{c-\varepsilon} \cup B(z), \mathcal{M}^{c-\varepsilon})$ is homotopy equivalent to the direct sum of pairs $(\mathcal{M}^{c-\varepsilon} \cup B(z), \mathcal{M}^{c-\varepsilon})$, as z varies over the critical points at height c. The right-hand side of (8.5.1) must then be replaced by the more general

$$\left(\mathcal{M}^{c+\varepsilon}, \mathcal{M}^{c-\varepsilon}\right) \simeq \bigoplus_{z} \texttt{N-data}(z) \times \texttt{T-data}(z) \quad (8.5.2)$$

$$= \bigoplus_{z} \left[\left(\tilde{N}(z), \tilde{N}(z) \cap \mathcal{M}^{c-\varepsilon}\right) \times \left(B_k(z), \partial B_k(z)\right)\right] .$$

Definition 8.5.4 (quasi-local cycle in $\mathcal{V}$) *Let z be a critical point for h on $\mathcal{V}$ in a stratum S of dimension k at which h is locally Morse (quadratically nondegenerate). Let $C(z)$ denote the relative cycle in $H_k(\mathcal{V}, \mathcal{V}^{c-\varepsilon})$, which is the downward k-patch in S through the critical point z and the generator for the cyclic group $H_k(S, S^{h(z)-\varepsilon})$.*

Putting this together with the general Morse theoretic decomposition of Lemma 8.2.4 gives a representation of C that we use extensively in what follows.

Theorem 8.5.5 (Morse decomposition of the relative cycle for the Cauchy integral) *In the notation of Lemma 8.2.4,*

$$C_* = \sum_{z \in \texttt{contrib}} C_*(z) = \sum_{z \in \texttt{contrib}} \texttt{N-data}(z) \times \texttt{T-data}(z) .$$

If h is Morse at the point z, then $C(z)$ is a relative cycle representing `T-data`. *If h is Morse at every point $z \in$* `contrib`, *then*

$$C_* = \sum_{z \in \texttt{contrib}} \texttt{N-data}(z) \times C(z).$$

The subject of the next three chapters is the development of the Cauchy integral on these cycles in the three respective cases: when $\boldsymbol{p}$ is a smooth point, when $\boldsymbol{p}$ is a multiple point, and when $\boldsymbol{p}$ is a cone point. Here follow brief descriptions of the Morse data in the case of a smooth point and a transverse multiple point; the remaining cases require more intricate descriptions and analyses.

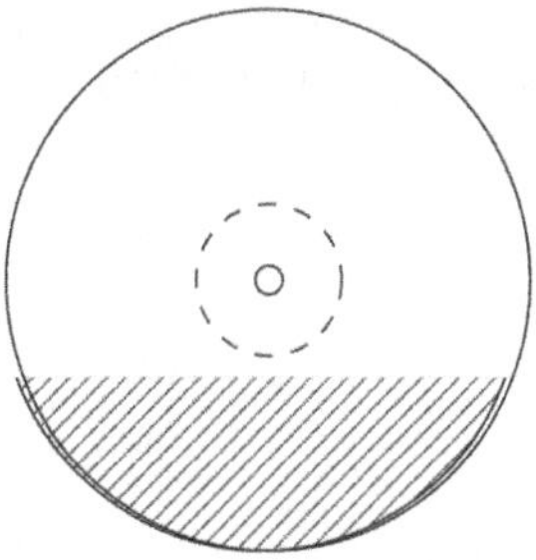

Figure 8.6 The normal Morse data at a smooth point is homotopy equivalent to a circle (dotted line).

Example 8.5.6 (quasi-local cycle at a smooth point) Suppose $\mathbf{p}$ is a smooth point of $\mathcal{V}$. Then $k = d - 1$ and $\nabla H(\mathbf{p})$ is nonvanishing (assuming H is squarefree). Near $\mathbf{p}$, the stratum S coincides with $\mathcal{V}$ and has real codimension 2. The normal slice N is a point, and the normal link is a punctured disk. The normal Morse data is a punctured disk modulo a patch near the bottom, which is the homotopy type of a circle (see Figure 8.6). The tangential Morse data is a k-patch in the downward direction, whose boundary lies entirely in $\mathcal{M}^{c-\varepsilon}$. When $d = 2$ and $k = 1$, the set $\mathcal{V}$ is a complex algebraic curve of real dimension 2 and $\mathbf{p}$ is a classical saddle point with one steepest ascent axis and one steepest descent axis. The pair `T-data` is a rainbow with $\mathbf{p}$ at its apex and both boundary points below height $c - \varepsilon$. The product `N-data` $\times$ `T-data` is the product of a circle with this rainbow. One may visualize this as a piece of macaroni, with the rainbow as the spine in the hollow center of the macaroni; see Figure 8.7. The pair $(\mathcal{M}^{c+\varepsilon}, \mathcal{M}^{c-\varepsilon})$ has the homotopy type of this macaroni modulo its end-circles. There is a single homology generator

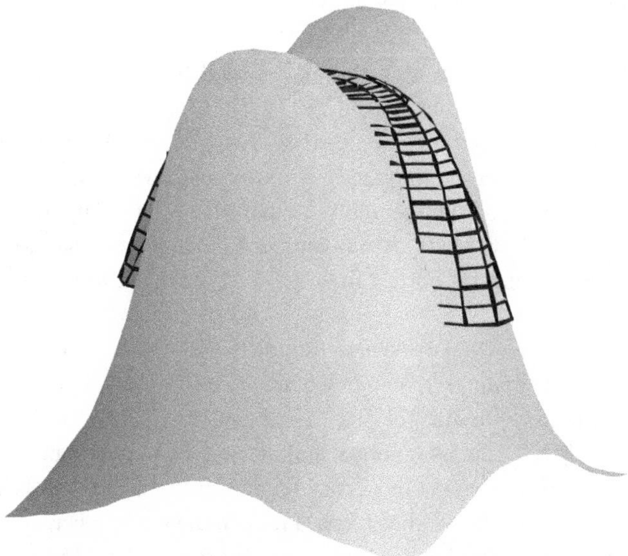

Figure 8.7 The quasi-local cycle at a smooth point.

Figure 8.8 Transverse multiple point when $d = k = 2$: two planes in fourspace.

in dimension 2, namely the pair itself. This means that either there is no obstruction to passing C below height c or C is homologous to this macaroni. ■

The normal space at a point $\boldsymbol{p}$ in a stratum S of complex codimension k is a complex k-space. We say that $\boldsymbol{p}$ is a ***transverse multiple point*** if N is diffeomorphic to k transversely intersecting complex hyperplanes in this complex k-space. Technically this includes smooth points ($k = 1$), although we will usually assume $k \geq 2$.

Figure 8.8 illustrates this when $d = k = 2$ and the two maximal strata are complex lines (real planes) in $\mathbb{C}^2$ ($\mathbb{R}^4$). The two planes intersect at a single point. The normal slice is locally the union of these two planes. The normal link is the complement of two intersecting planes in four space. In coordinates where one plane is the first two coordinates and the other plane is the last two, this is the product of two copies of $\mathbb{C} \setminus \{0\}$. As in Figure 8.6, each of these may be contracted radially to a circle. Thus the normal link is homotopic to a 2-torus.

Example 8.5.7 (quasi-local cycle at a multiple point) Suppose $\boldsymbol{p}$ is a transverse multiple point in a stratum of complex codimension k. Locally, the pair $(\mathbb{C}^d, \mathcal{V})$ is diffeomorphic to the product of S with k copies of $(\mathbb{C}, \{0\})$. Thus $\tilde{N}$ is diffeomorphic to $(\mathbb{C}, \{0\})^{d-k}$, and the normal link has the homotopy type of a k-torus. The picture would look something like Figure 8.7: the tangential data (smooth manifold drawn with solid shading) would remain the same, whereas the normal data (product of a circle and an arc, drawn in wire-frame) would be replaced by a product of a torus with the arc. ■

8.6 Evaluating the Integral on a Quasi-Local Cycle

Going back to the program at the end of Section 1.3, Sections 8.5, 8.3, and 8.4 complete, respectively, step (iii), step (iv), and as much as we are ready to do of step (v). The last installment of this overview chapter is a brief discussion of step (vi). The general approach starts with the Cauchy integral, written as $\int_C \omega$, where $\omega = z^{-r-1} F(z)\, dz$ is a d-form holomorphic on $\mathcal{M}$ and C is the class in $H_d(\mathcal{M})$ of a small torus linking the coordinate axes.

Morse Theory (Lemma 8.2.4) allows us to push the chain of integration in the class C down to a minimax height relative cycle $C_* \in H_d(\mathcal{M}, \mathcal{M}^{c_*-\varepsilon})$. The fundamental decomposition (Theorem 8.5.3) allows us to write $C_*(z)$ as a product $\texttt{T-data}(z) \times \texttt{N-data}(z)$. When h is locally Morse at every $z \in \texttt{contrib}$, Theorem 8.5.5 yields a relative cycle representative for C of the form

$$\sum_{z \in \texttt{contrib}} C_*(z) = \sum_{z \in \texttt{contrib}} \texttt{N-data}(z) \times C(z).$$

Recall from Remark 8.2.5 that integrals may be evaluated over relative cycles, provided one accepts an error term of an exponentially smaller magnitude. Thus

$$\begin{aligned} (2\pi i)^d a_r &\sim \int_{C_*} w^{-r-1} F(w)\, dw \\ &\sim \sum_{z \in \texttt{contrib}} \int_{C_*(z)} w^{-r-1} F(w)\, dw \\ &\sim \sum_{z \in \texttt{contrib}} \int_{C(z)} \left[\int_{\texttt{N-data}} w^{-r-1} F(w)\, dw \right]. \end{aligned} \tag{8.6.1}$$

The asymptotic equalities must be interpreted as holding up to an additive term of $O\left(e^{(c_*-\varepsilon)|r|}\right)$. We continue with a computation that will be spelled out in the next chapter. In the cases we are able to evaluate, the inner integral is some kind of residue. The factor w^{-r} may be pulled through the residue, so that the residue at z of $w^{-r-1} F(w)\, dw$ is equal to $y^{-r} \eta\, dy$ for some $\eta(y)$ holomorphic on the smooth points of $\mathcal{V}$. Thus

$$(2\pi i)^d a_r \sim \sum_{z \in \texttt{contrib}} \int_{C(z)} y^{-r} \eta\, dy. \tag{8.6.2}$$

The magnitude of the outer integral over $C(z)$ will be largest roughly where $|y^{-r}|$ is largest, which occurs at $y = z$. If Res is well behaved, the outer integral will be a saddle-point integral of a type that can be automatically evaluated, leading to

$$a_r \sim \sum_{z \in \texttt{contrib}} \Phi_z(r), \tag{8.6.3}$$

where Φ_z is a reasonably simple formula and may be computed without determining whether $z \in W$. In the cases of smooth points and multiple points, Res is in fact well enough behaved; cone points will be seen to take a lot more work. These three cases make up the next three chapters; brief outlines are as follows.

Smooth Points

Let $\boldsymbol{p}$ be a smooth critical point for h on $\mathcal{V}$. The function F, hence the integrand $\omega := F(\boldsymbol{z})\boldsymbol{z}^{-\boldsymbol{r}-\boldsymbol{1}}\, d\boldsymbol{z}$, has a simple pole on $\mathcal{V}$. Each normal slice intersects $\mathcal{V}$ in a single point, hence ω has a single simple pole in each slice. The inner integral is therefore a simple residue. We are more precise about this in Chapter 9. The quasi-local cycle $C(\boldsymbol{z})$ is diffeomorphic to a $(d-1)$-ball at the center of which $\boldsymbol{z}^{-\boldsymbol{r}}$ has a stationary phase point and maximal magnitude. The iterated integral is therefore a standard saddle-point integral and obeys a well-known asymptotic estimate when $\boldsymbol{r} \to \infty$ with $\hat{\boldsymbol{r}} \to \hat{\boldsymbol{r}}_*$. In the case where h is Morse, one obtains

$$a_{\boldsymbol{r}} \sim C(\hat{\boldsymbol{r}})\, |\boldsymbol{r}|^{(1-d)/2} \boldsymbol{z}^{-\boldsymbol{r}} . \tag{8.6.4}$$

Other behaviors of h lead to more complicated expressions, but always ones whose exponential order is $\exp[(c + o(1))|\boldsymbol{r}|]$.

Multiple Points

Let $\boldsymbol{p}$ be a multiple point for h on a stratum S of complex codimension k. The function F, hence the integrand $\omega := F(\boldsymbol{z})\boldsymbol{z}^{-\boldsymbol{r}-\boldsymbol{1}}\, d\boldsymbol{z}$, has an order k pole in the form of a *normal crossing*; in the language of resolution of singularities, the singularity comes already resolved. The form ω has an order k singularity of normal form in each slice. The inner integral in (8.6.1) is therefore a k-fold iterated residue. This is not hard to define, but we will not do so now. Those who have not seen this before will see it developed in Chapter 10. The upshot again is an integral of the form $\int \text{Res}(\omega) = \int \boldsymbol{z}^{-\boldsymbol{r}} \eta$ over a $(d-k)$- dimensional patch on which h is maximized at $\boldsymbol{p}$. The outer integral is again a saddle-point integral, this time in $d-k$ dimensions, leading to estimates of the form

$$a_{\boldsymbol{r}} = C(\hat{\boldsymbol{r}})|\boldsymbol{r}|^{-(d-k)/2} \boldsymbol{z}^{-\boldsymbol{r}}, \tag{8.6.5}$$

as $\boldsymbol{r} \to \infty$ with $\hat{\boldsymbol{r}}$ in some neighborhood of $\hat{\boldsymbol{r}}_*$. Setting $k = 1$ gives (8.6.4) as a special case.

Both (8.6.4) and (8.6.5) have error terms whose magnitude is of order one less power of $\boldsymbol{r}$. Both also have full asymptotic developments in negative powers of $|\boldsymbol{r}|$. Setting $k = d$ is also interesting. This is a complete intersection, in which case the remainder term is exponentially small – we give details in Chapter 10.

Cone Points

When $\mathcal{V}$ has an isolated singularity with nontrivial monodromy, it is more difficult to determine the quasi-local cycle and integrate over it. In the few such cases we can handle, the local geometry of $\mathcal{V}$ at the singularity $\boldsymbol{p}$ is that of a cone. An example of this is the so-called ***cube grove*** creation function

$$F(x, y, z) = \frac{1}{1 + xyz - (1/3)(x + y + z + xy + xz + yz)}. \tag{8.6.6}$$

The variety $\mathcal{V}$ is smooth except at the single point $(1, 1, 1)$. There, after an orthogonal affine change of variables and a translation of the origin to $(1, 1, 1)$, the denominator of F looks asymptotically like $2xy + 2xz + 2yz$, hence the word ***cone point***. This isolated singularity is a stratum unto itself; thus there is no outer integral. The inner integral, however, is a beast. Chapter 11 is devoted to unraveling such integrals.

Notes

Asymptotic formulae in the presence of smooth strictly minimal points first appeared in Pemantle and Wilson (2002), followed by formulae for strictly minimal multiple points in Pemantle and Wilson (2004). These results excluded the case where $\mathcal{V}$ intersects $\mathbf{T}(\boldsymbol{x}_{\min})$ in an infinite set, the so-called *toral* case. Irrelevance of these noncritical nuisance points was shown in Baryshnikov, Brady, Bressler, and Pemantle (2010) and then in greater generality in Baryshnikov and Pemantle (2011). An expanded treatment of the methods outlined in this section is given in Pemantle (2010).

Multivariate analogs of Hayman's formula have been pursued by several authors; see e.g., Bender and Richmond (1996), Drmota, Gittenberger, and Klausner (2005), Gittenberger and Mandlburger (2006).

Exercises

8.1 (exponential rate definition)
Carry out the computation in Example 8.1.2 to show that $\overline{\beta}(\hat{\boldsymbol{r}}_*) = \log 2$, where $\hat{\boldsymbol{r}}_*$ represents the diagonal direction.

8.2 (enough minimal points in the combinatorial case)
Prove the following result. Suppose that $F = G/H$ and that $H = 1 - P$, where P has nonnegative coefficients and is ***aperiodic*** (there is no proper additive subsemigroup of $\mathbb{N}^d$ outside of which the coefficients of P vanish identically). Then each direction for which there exists a contributing point is nonflat, and the corresponding contributing point is strictly minimal and lies in the positive orthant.

9

Smooth Point Asymptotics

In this chapter we give asymptotics for a_r when the dominant critical point or points are smooth points. By far the most common case, and the one occupying the majority of the chapter, is when the pole at z is simple and the critical point z for h on $\mathcal{V}$ is quadratically nondegenerate. We prove a number of formulae of the form (8.6.3)

$$a_r \sim \sum_{z \in W} \Phi_z^{(2)}(r) .$$

The superscript (2) indicates quadratic nondegeneracy. The quantity $\Phi_z^{(2)}$ is an invariant of the quadruple $(\hat{r}, G, H, z)$, where z is a quadratically nondegenerate critical point of $h_{\hat{r}}$ on $\mathcal{V}_H$. In particular, $\Phi_z^{(2)}$ may be evaluated at a critical point z without determining whether $z \in \texttt{contrib}$ or whether there is even any decomposition $C_* = \sum_{z \in \texttt{contrib}} C_*(z)$. Several quite different expressions for $\Phi_z^{(2)}$ will be derived.

(9.2.11) $\Phi_z^{(2)}(r) = (2\pi)^{(1-d)/2} (\det \mathcal{H}(\hat{r}))^{-1/2} \dfrac{G(z)}{z_k \partial H / \partial z_k(z)} r_k^{(1-d)/2} z^{-r}$

(9.5.10) $\Phi_z^{(2)}(r) = \dfrac{G}{\sqrt{2\pi}} x^{-r} y^{-s} \sqrt{\dfrac{-y H_y}{s Q}}$

(9.5.3) $\Phi_z^{(2)}(r) = (2\pi)^{(1-d)/2} \mathcal{K}^{-1/2} \dfrac{G(z)}{|\nabla_{\log} H(z)|} |r|^{(1-d)/2} z^{-r}$

(9.5.16) $\Phi_z^{\text{non}}(r) = (2\pi)^{(1-d)/2} \dbinom{-r_k}{\ell - 1} (\det \mathcal{H}(\hat{r}))^{-1/2} \dfrac{G(z)}{z_k^{\ell} (\partial H / \partial z_k)^{\ell}(z)} r_k^{(1-d)/2} z^{-r}$

The first is valid for any dimension. The second is specific to dimension $d = 2$ and gives an explicit expression

$$Q(x, y) := -y^2 H_y^2 x H_x - y H_y x^2 H_x - x^2 y^2 (H_y^2 H_{xx} + H_x^2 H_{yy} - 2 H_x H_y H_{xy})$$

in place of the Hessian determinant $\det \mathcal{H}(\hat{r})$ arising in the first expression. The third expression is valid in any dimension and is given in terms of geometric invariants, namely the gradient and the curvature, $\mathcal{K}$, of $\mathcal{V}$ in logarithmic coordinates.

The fourth is a generalization of the first to the case where H is an ℓth power for some integer $\ell \geq 1$.

The change of geometry at a quadratically degenerate smooth point leads to substantially different asymptotic behavior. When $d = 2$, we are able to give the somewhat universal formula

$$\Phi^{(k)}_{(x,y)}(\boldsymbol{r}) = \frac{\Gamma(1/k)}{k\pi} \frac{G(x,y)}{y\partial H(x,y)/\partial y} c^{-1/k} s^{-1/k} x^{-r} y^{-s}, \tag{9.5.14}$$

where the superscript (k) denotes that the degree h vanishes to order k. As d increases, the possible geometries multiply rapidly and go beyond the analyses worked out in Chapter 5. If the need arises, analytic results from a source such as Varchenko (1977) can be coupled with Theorem 9.2.1 and Proposition 9.2.5 to derive asymptotics when $d \geq 3$ and the dominant critical point is a smooth and quadratically degenerate.

9.1 Smooth Points

Let us take a moment to pin down what it means to stipulate that "the dominant critical points are smooth points." Recall from Section 8.3 that the chain of integration in the Cauchy integral is homologous to the sum of one or more quasi-local cycles defined at critical points. We will asymptotically evaluate the quasi-local integral at any smooth critical point, yielding the various formulae for $\Phi^{(2)}$. If all the quasi-local cycles of maximum height appearing in the integral are centered at smooth points, this will be sufficient to produce an asymptotic series for $a_{\boldsymbol{r}}$.

Although we have used the term "smooth point" already in an informal manner, we clarify the notion as follows: we say that a point $\boldsymbol{p} \in \mathcal{V}$ is a smooth point if $\mathcal{V}$ is a manifold in a neighborhood of $\boldsymbol{p}$. If $F = G/H$ with H squarefree, then this is equivalent to $\nabla H \neq 0$. Our definition allows for a power of H to vanish. However, for most of the chapter, we assume that in fact $\boldsymbol{p}$ is a simple pole or, in other words, that $\nabla H \neq \mathbf{0}$. This is because we do not have examples of useful generating functions with square factors in the denominator. The theory we develop for simple smooth points goes through almost unchanged for higher powers; accordingly, we will score some cheap generality by including modifications to the formulae for higher powers.

Phenomenologically, quadratically nondegenerate smooth point asymptotics seem somewhat limited at first. The invariant $\Phi^{(2)}_{\boldsymbol{z}}$ is of the form $C_{\hat{\boldsymbol{r}}}|\boldsymbol{r}|^{(1-d)/2}\boldsymbol{z}^{-\boldsymbol{r}}$. The coefficient C and the basepoint $\boldsymbol{z}$ depend only on the unitized vector $\hat{\boldsymbol{r}}$ and vary continuously except at certain degenerate points. Asymptotics of this form are known in the physics literature as ***Ornstein-Zernike*** behavior. For the direction $\boldsymbol{r}$ that maximizes the magnitude of $\boldsymbol{z}(\hat{\boldsymbol{r}})^{-\boldsymbol{r}}$ over some cross-section of vectors (e.g., unit vectors or vectors with last component equal to 1), the behavior is Gaussian over neighborhoods $\{\boldsymbol{r} + \Delta\boldsymbol{r} : |\Delta\boldsymbol{r}| = o(|\boldsymbol{r}|^{2/3})\}$. Gaussian limit theorems are discussed in Pemantle and Wilson (2010, Section 3.5) and are surveyed in Section 9.6 that follows. For now, the point is that this very specific type of asymptotic limits a priori the scope of the method, as only sums of multivariate sequences with Gaussian behavior can possibly be governed by nondegenerate smooth points.

Nevertheless, the wealth of applications for smooth points appears quite large. Of the twenty worked applications in Pemantle and Wilson (2010), for example, all but a few require only smooth point analysis. This may be due to the fact that smoothness of $\mathcal{V}$ is a generic property of H, although applications seem to have an uncanny knack for avoiding generic behavior. Most of the analyses of multivariate generating functions prior to 2000, such as the results of Bender *et al.* throughout the 1980s and 1990s, concern only Ornstein-Zernike behavior (many have the phrase "central limit" in the title).

There are two methods by which smooth point integrals have been evaluated: surgery and residue forms. Historically, the surgical approach came first. The Cauchy integral over a torus just inside $\mathbf{z}$ is chopped into a piece near $\mathbf{z}$ and a piece staying away from $\mathbf{z}$. The former integral is compared with the integral of a similar piece pushed beyond $\mathbf{z}$. The result is an integral over $d-1$ of the coordinates of a residue in the remaining coordinate. This approach has the drawback of requiring a non-canonical separation of the coordinates into one coordinate in which a residue will be taken and $d-1$ coordinates over which one integrates. The formulae do not, therefore, appear natural. Also, this point of view makes it less obvious how one is carrying out the program described in the previous chapter. On the positive side, the proofs are elementary, requiring only undergraduate complex analysis, and lead directly to explicit expressions for asymptotic estimates of $a_{\mathbf{r}}$. Another drawback of surgery is that it requires the critical points to be minimal. We see later that this is unnecessarily restrictive.

The approach by residue forms, in contrast, is coordinate-free, leading to formulae that are more intrinsic to the problem and make intuitive sense. No minimality hypothesis is required, whence the method is more general. For computation, one often has to pass to a coordinate system, but this is not necessarily a drawback because waiting until the end to coordinatize will usually lead to simplification and greater understanding. The main drawback to analyzing the Cauchy integral via residue forms is that the method is not elementary and requires a first course in differential forms (Appendix A). For this reason, we have structured this chapter to include both methods.

Chapter Outline

Sections 9.2 and 9.3 give parallel derivations of the basic result for smooth points. The main work is in identifying `contrib`, although one also has to massage the integral into a usable form. In the case where there are minimal smooth points, the surgical approach is able to produce a complete solution, i.e., an identification of `contrib`. If there is a minimal smooth point $z \in \mathbf{T}(\boldsymbol{x}_{\min}(\hat{\boldsymbol{r}}))$, then the set `contrib` consists precisely of those critical points on the minimal torus $\mathbf{T}(\boldsymbol{x}_{\min})$. Theorem 9.2.1 chops up the Cauchy integral into a piece that will become Φ_z and another piece i.e., provably small. Proposition 9.2.5 shows the first piece to be a Fourier-Laplace integral. The remaining results in Section 9.2 prove (9.2.11) for various configurations of minimal points.

Section 9.3 takes the residue approach, which is to identify the inner integral as the residue form $\mathrm{Res}(\omega)$, an explicit formula for which is given in Proposition 9.3.3. Theorem 9.3.2 is the analogue of Theorem 9.2.1, showing that integration over the intersection class σ approximates a_r to within a difference i.e., exponentially smaller. Lemma 9.3.6 shows σ to be none other than $\sum_{z \in \texttt{contrib}} C(z)$. Finally, Theorem 9.3.7 incorporates an analogue of Proposition 9.2.5, writing the residue integral as a Fourier-Laplace integral, then gives an asymptotic formula that can be manipulated into (9.2.11).

Section 9.4 is devoted to the general problem of determining `contrib`. In the case where there are minimal points, we already know how to do this; Theorem 9.4.2 re-derives this result in a more general context. The remainder of Section 9.4 concerns the determination of `contrib` in the bivariate setting. When $d = 2$ and $\mathcal{V}$ is smooth, an algorithm is given, Algorithm 9.4.7, that always succeeds in computing `contrib`.

Section 9.5 gives four more formulae. The first is the alternative formula for $\Phi^{(2)}$ in terms of curvature (9.5.3). The second is the explicit formula (9.5.10) in terms of partial derivatives of H. The third, (9.5.14), addresses the case where $d = 2$ and h is quadratically degenerate. The fourth, (9.5.16), addresses the case when H is not squarefree.

The final section of this chapter applies the foregoing results to prove probabilistic limit laws for asymptotics governed by smooth points. Such limit laws have been around for decades, appearing in the titles of early publications on multivariate generating function asymptotics such as Bender and Richmond (1983). Weak laws of large numbers are given in Theorems 9.6.1 and 9.6.3. A local central limit theorem is given in Theorem 9.6.6.

9.2 The Smooth Point Integral via Surgery

In this section, our standing assumption on F is that $F = G/H = \sum_r a_r z^r$ is a Laurent series convergent on a component B of the complement of $\texttt{amoeba}(H)$; here H may be any analytic function (see the notes for Chapter 7). We recall that the minimal points are those critical points $z = \exp(\boldsymbol{x} + i\boldsymbol{y})$ of $\mathcal{V}$ whose log-moduli $\boldsymbol{x}$ lie on ∂B. The surgical approach was first carried out in Pemantle and Wilson (2002), where minimal points were classified as ***strictly minimal***, ***finitely minimal***, or ***torally minimal*** according to whether the cardinality of the intersection $\mathcal{V}_1$ of $\mathcal{V}$ with $\mathbf{T}(\boldsymbol{x})$ is equal to one, a finite number greater than one, or infinity. In the infinite case, all examples that have been studied satisfy the following ***torality hypothesis*** given in Baryshnikov, Brady, Bressler, and Pemantle (2010, Proposition 2.1):

> **Torality:** The analytic function H satisfies the torality hypothesis on $\mathbf{T}(\boldsymbol{x})$ if $H(z) = 0$ and $\operatorname{Re}\log z_j = x_j$ for $1 \le j \le d-1$ imply $\operatorname{Re}\log z_d = x_d$. (9.2.1)

One example of a function satisfying the torality hypothesis is the bilinear function $1 - \beta(x+y) + xy$; another example is the denominator $\det(I - z_{d+1}MU)$ of the generating function for any quantum random walk; here M is a diagonal matrix of monomials in $z_1, \ldots, z_d$ and U is unitary; see Example 9.5.6. For ease of exposition, we carry out most of the computations for strictly minimal points. This is easily extended to finitely minimal points and to minimal points satisfying the torality hypothesis. Until further notice, we therefore assume strict minimality and adopt the following notation. Fix $\boldsymbol{z}_* = \exp(\boldsymbol{x}_* + i\,\boldsymbol{y}_*)$ and assume that

- $\mathcal{V}_1 := \mathcal{V} \cap \mathbf{T}(\boldsymbol{x}_*) = \{\boldsymbol{z}_*\}$.
- H is squarefree.
- $\boldsymbol{z}_*$ is a smooth point, i.e., $\nabla H(\boldsymbol{z}_*) \neq 0$.

Let

$$\hat{\boldsymbol{r}}_* := \frac{\nabla_{\log} H(\boldsymbol{z}_*)}{|\nabla_{\log} H(\boldsymbol{z}_*)|}$$

be the normalized logarithmic gradient of H at $\boldsymbol{z}_*$. We may assume without loss of generality that $\hat{\boldsymbol{r}}_*$ has no zero components, because coefficients of Laurent expansions of F with some component equal to zero are components of specializations of F to having some zero arguments. It follows that $z_j \neq 0$ for all j. The following construction, quoted from Pemantle and Wilson (2002, Lemma 4.1), represents $a_{\boldsymbol{r}}$ as a saddle-point integral local to $\boldsymbol{z}_*$, as $|\boldsymbol{r}| \to \infty$ and $\hat{\boldsymbol{r}}$ varies in some neighborhood of $\hat{\boldsymbol{r}}_*$.

Because $\boldsymbol{z}_*$ is a smooth point, we know $\nabla H(\boldsymbol{z}_*) \neq \mathbf{0}$, and we may pick a coordinate k such that $\partial H/\partial z_k(\boldsymbol{z}_*) \neq 0$. Let $\rho := |(\boldsymbol{z}_*)_k|$ denote the modulus of the kth coordinate of $\boldsymbol{z}_*$. This approach parametrizes $\mathcal{V}$ locally by the $(d-1)$ coordinates other than the kth one. It will be convenient to let a superscript circle denote the projection of a d-vector onto these $d-1$ coordinates and to write $\boldsymbol{z} = (\boldsymbol{z}^\circ, z_k)$, even though the second coordinate is not appended but inserted into position k. In keeping with this notation, we write T° for the $(d-1)$-dimensional torus $T(\boldsymbol{x}_*^\circ)$ through $\boldsymbol{z}_*^\circ$.

Let g parametrize $\mathcal{V}$ by $\boldsymbol{z}^\circ$ in a neighborhood of $\boldsymbol{z}_*$. Let us be more explicit about this. By the implicit function theorem, there are a real number $\delta \in (0, \rho)$, a neighborhood $\mathcal{N}$ of $\boldsymbol{z}_*^\circ$ in T° and an analytic function $g : \mathcal{N} \to \mathbb{C}$ such that for $\boldsymbol{z}^\circ \in \mathcal{N}$,

(i) $H(\boldsymbol{z}^\circ, g(\boldsymbol{z}^\circ)) = 0$
(ii) $\rho \le |g(\boldsymbol{z}^\circ)| < \rho + \delta$ with equality only if $\boldsymbol{z}^\circ = \boldsymbol{z}_*^\circ$
(iii) $H(\boldsymbol{z}^\circ, w) \neq 0$ if $w \neq g(\boldsymbol{z}^\circ)$ and $|w| < \rho + \delta$

Let C_1 denote the circle of radius $\rho - \delta$ centered at the origin of the complex plane, and let C_2 denote the circle of radius $\rho + \delta$. Write the Cauchy integral as an iterated integral:

$$a_r = \left(\frac{1}{2\pi i}\right)^d \int_{T^\circ} (z^\circ)^{-r^\circ} \left[\int_{C_1} w^{-r_k} F(z^\circ, w) \frac{dw}{w}\right] \frac{dz^\circ}{z^\circ}. \tag{9.2.2}$$

The key observation is that the inner integral is small away from z_*°. Indeed, for each fixed $z^\circ \neq z_*^\circ$, the function $F(z^\circ, \cdot)$ has radius of convergence greater than ρ. Hence the inner integral is $O(|w_*| + \varepsilon)^{-r_k}$ for some $\varepsilon > 0$. By continuity of the radius of convergence, a single $\varepsilon > 0$ may be chosen for all compact sets K not containing z_*°. Going back to the Cauchy integral, we see that

$$\left|z_*^r (a_r - I)\right| \to 0 \tag{9.2.3}$$

exponentially, where I is the integral in (9.2.2) with T° replaced by any neighborhood of z_*° in T°. Once we have replaced T° by a small neighborhood of z_*°, we may compare the inner integral over C_1 to the integral of the same integrand over C_2. Note that we cannot do this without first "cutting out" a small neighborhood.

We have not made use of g yet, but we do so now, choosing the neighborhood that defines I to be the neighborhood $\mathcal{N}$ on which properties (i)–(iii) hold for g. We will compare I to another integral I' in which the contour C_1 for the inner integral is replaced by C_2:

$$I := \left(\frac{1}{2\pi i}\right)^d \int_{\mathcal{N}} (z^\circ)^{-r^\circ} \left[\int_{C_1} w^{-r_k} F(z^\circ, w) \frac{dw}{w}\right] \frac{dz^\circ}{z^\circ}$$

$$I' := \left(\frac{1}{2\pi i}\right)^d \int_{\mathcal{N}} (z^\circ)^{-r^\circ} \left[\int_{C_2} w^{-r_k} F(z^\circ, w) \frac{dw}{w}\right] \frac{dz^\circ}{z^\circ}.$$

The inner integrand has a unique pole in the annulus $\rho - \delta \leq |w| \leq \rho + \delta$, occurring at $w = g(z^\circ)$. Letting

$$\Psi(z^\circ) := \operatorname{Res}\left(\frac{F(z^\circ, w)}{w}; w = g(z^\circ)\right), \tag{9.2.4}$$

we see that the difference between the inner integrals is $(2\pi i)R(z^\circ)$, where the residue $R(z^\circ)$ is defined by

$$R(z^\circ) := \operatorname{Res}\left(g(z^\circ)^{-r_k} \frac{F(z^\circ, w)}{w}; w = g(z^\circ)\right) \tag{9.2.5}$$

$$= g(z^\circ)^{-r_k} \operatorname{Res}\left(\frac{F(z^\circ, w)}{w}; w = g(z^\circ)\right)$$

$$= g(z^\circ)^{-r_k} \Psi(z^\circ).$$

From the magnitude of the integrand defining I', we see that

$$\left|z_*^{\boldsymbol{r}} I'\right| \to 0 \tag{9.2.6}$$

exponentially in $|\boldsymbol{r}|$ for $\hat{\boldsymbol{r}}$ in some neighborhood of $\hat{\boldsymbol{r}}_*$. Putting together (9.2.3) and (9.2.6) proves the following estimate, which may be thought of as the computational analogue of the fact that one can integrate in relative homology at the expense of an exponentially small error (see Remark 8.2.5).

Theorem 9.2.1 (reduction to residue integral) *Define*

$$\chi := I - I' = \left(\frac{1}{2\pi i}\right)^{d-1} \int_{\mathcal{N}} (\boldsymbol{z}^\circ)^{-\boldsymbol{r}^\circ} g(\boldsymbol{z}^\circ)^{-r_k} \Psi(\boldsymbol{z}^\circ) \frac{d\boldsymbol{z}^\circ}{\boldsymbol{z}^\circ}, \tag{9.2.7}$$

with Ψ *given by* (9.2.4). *Under the condition that* $\boldsymbol{z}_*$ *is a smooth, strictly minimal, critical point for* $\hat{\boldsymbol{r}}_*$, *the quantity* $a_{\boldsymbol{r}}$ *is well estimated by* χ *in the sense that there is a neighborhood of* $\boldsymbol{r}_*$ *such that*

$$\left|z_*^{\boldsymbol{r}}\,(a_{\boldsymbol{r}} - \chi)\right| \to 0$$

exponentially in $|\boldsymbol{r}|$ *as* $\hat{\boldsymbol{r}}$ *varies over this neighborhood.*

Remark 9.2.2 This result requires only meromorphicity in a neighborhood of $\bigcup_{\boldsymbol{x}\in B} \mathbf{T}(\boldsymbol{x})$. In fact, one could reduce this to a neighborhood of $\bigcup_{\boldsymbol{x}\in\mathcal{N}} \mathbf{T}(\mathrm{Re}\log \boldsymbol{z})$.

Modification for Finitely Minimal Points

It is easy to modify this to apply in the case of finitely minimal points. If $\mathcal{V}$ intersects $\mathbf{T}(\boldsymbol{x}_*)$ in a finite set E, we let $\{\mathcal{N}_{\boldsymbol{p}} : \boldsymbol{p} \in E\}$ denote neighborhoods defined as was the neighborhood $\mathcal{N}$, near each point $\boldsymbol{p}$. The following corollary has exactly the same proof as Theorem 9.2.1.

Corollary 9.2.3 (reduction for finitely minimal points) *We have* $|\boldsymbol{z}^{\boldsymbol{r}}(a_{\boldsymbol{r}} - \chi)| \to 0$ *exponentially in* $|\boldsymbol{r}|$ *as* $\hat{\boldsymbol{r}}$ *varies over a neighborhood of* $\hat{\boldsymbol{r}}_*$, *where* Ψ *is given by* (9.2.4) *and*

$$\chi := I - I' = \left(\frac{1}{2\pi i}\right)^{d-1} \sum_{\boldsymbol{p}\in E} \int_{\mathcal{N}_{\boldsymbol{p}}} (\boldsymbol{z}^\circ)^{-\boldsymbol{r}^\circ} g(\boldsymbol{z}^\circ)^{-r_k} \Psi(\boldsymbol{z}^\circ) \frac{d\boldsymbol{z}^\circ}{\boldsymbol{z}^\circ}. \qquad \square$$

Modification under Torality Hypothesis

If instead of finite minimality we assume the torality hypothesis (9.2.1), then we can reduce to an integral on the torus with no surgery required. In the following proposition, g is the multivalued function solving for z_k as a function of $\boldsymbol{z}^\circ$; the number of values, counted with multiplicities, is the degree m of z_k in H. Except on a lower dimensional set where two values coincide, these vary analytically with $\boldsymbol{z}^\circ$. The multivalued integrand should be interpreted as a sum over all m values.

Corollary 9.2.4 (reduction under torality) *Let H satisfy the torality hypothesis on $\mathbf{T}(\boldsymbol{x})$, and suppose that all poles of H on $\mathbf{T}(\boldsymbol{x})$ are simple. Then*

$$a_{\boldsymbol{r}} = \chi := \left(\frac{1}{2\pi i}\right)^{d-1} \int_{T^\circ} (\boldsymbol{z}^\circ)^{-\boldsymbol{r}^\circ} g(\boldsymbol{z}^\circ)^{-r_k} \Psi(\boldsymbol{z}^\circ) \frac{d\boldsymbol{z}^\circ}{\boldsymbol{z}^\circ} .$$

The function Ψ is given by (9.2.4).

Remark In this case, dimension is reduced by one without localizing. The localization occurs when we apply the multivariate saddle-point results of Chapter 5, which show that the $(d-1)$-dimensional integral is determined by the behavior of g and Ψ near critical points $\boldsymbol{p} \in T^\circ$.

Proof This time we may take C_1 to be the circle of radius of $\rho - \delta$ and C_2 to be the circle of radius $\rho + \delta$ for any $\delta \in (0, \rho)$. The inner integral will be the sum of simple residues at points $g(\boldsymbol{z}^\circ)$ for any $\boldsymbol{z}^\circ$, and the proof is completed the same way as Theorem 9.2.1. □

Fourier-Laplace Integral Formulae

Theorem 9.2.1 and its modifications for finite points and toral points (Corollaries 9.2.3 and 9.2.4) give us an estimate χ for $a_{\boldsymbol{r}}$ in a variety of cases. To evaluate this using saddle-point machinery, we would like it in a standard form. Change variables to $z_j = (z_*)_j e^{i\theta_j}, 1 \leq j \leq d,\ j \neq k$ and note that $dz_j = iz_j d\theta_j$. Let $\mathcal{N}'$ denote the diffeomorphic image of $\mathcal{N}$ under this change of variables, in other words, an arbitrarily small neighborhood of the origin in $\mathbb{R}^{d-1}$, and let ϕ and A denote the functions $\log(g/g_*)$ and Ψ respectively after the change of variables, centered by $(i/r_k)\boldsymbol{r}^\circ \cdot \boldsymbol{\theta}$:

$$\phi(\boldsymbol{\theta}) := \log \frac{g(\boldsymbol{z}_*^\circ \exp(i\boldsymbol{\theta}))}{g(\boldsymbol{z}_*^\circ)} + \frac{i}{r_k}\boldsymbol{r}^\circ \cdot \boldsymbol{\theta} \tag{9.2.8}$$

$$A(\boldsymbol{\theta}) := \Psi(\boldsymbol{z}_*^\circ \exp i\boldsymbol{\theta}) . \tag{9.2.9}$$

The following result completes the job of expressing $a_{\boldsymbol{r}}$ as a standard saddle-point integral.

Proposition 9.2.5 (reduction to Fourier-Laplace integral) *Another expression for χ in Theorem 9.2.1 is*

$$\chi = \left(\frac{1}{2\pi}\right)^{d-1} \boldsymbol{z}_*^{-\boldsymbol{r}} \int_{\mathcal{N}'} e^{-r_k \phi(\boldsymbol{\theta})} A(\boldsymbol{\theta})\, d\boldsymbol{\theta} ,$$

with ϕ and A given by (9.2.8)–(9.2.9). *When $\hat{\boldsymbol{r}} = \hat{\boldsymbol{r}}_*$, the function ϕ vanishes to order at least 2 at the origin.*

Proof Computing $(2\pi)^{d-1}\chi$ after the change of variables, and recalling $g(z_*^\circ) = (z_*)_k$, we find that

$$(2\pi)^{d-1}\,\chi = \int_{N'} \left(z_*^\circ\, e^{i\theta}\right)^{r^\circ} \left[g(z_*^\circ)\frac{g(z^\circ)}{g(z_*^\circ)}\right]^{r_k} \Psi\left(z_*^\circ e^{i\theta}\right)\, d\theta$$

$$= z_*^{-r} \int_{N'} \exp\left[-r_k\left(i r^\circ \cdot \theta + \log \frac{g(z^\circ)}{g(z_*^\circ)}\right)\right] A(\theta)\, d\theta$$

$$= z_*^{-r} \int_{N'} e^{-r_k \phi(\theta)}\, A(\theta)\, d\theta\,,$$

proving the first statement. It is obvious that $\phi(\mathbf{0}) = 0$. To see that $\nabla\,\phi(\mathbf{0})$ vanishes as well, we use the definition to compute

$$\frac{\partial \phi}{\partial \theta_j}(\mathbf{0}) = i\frac{r_j}{r_k} + \frac{i(z_*)_j\, \frac{\partial g}{\partial z_j}(z_*^\circ)}{g(z_*^\circ)}\,. \tag{9.2.10}$$

Implicitly differentiating $H(z^\circ, g(z^\circ)) = 0$, we see that (in subscript partial derivative notation) $g_j = -H_j/H_k$; substituting this into (9.2.10) and using $g(z_*^\circ) = (z_*)_k$ yields

$$\frac{\partial \phi}{\partial \theta_j}(\mathbf{0}) = i\frac{r_j}{r_k} - i\,\frac{(z_*)_j}{(z_*)_k}\frac{H_j}{H_k}.$$

The verification is finished by recalling from the critical point equations that $\nabla_{\log} H(z_*) = \lambda \boldsymbol{r}_*$ for some $\lambda \neq 0$ and plugging in $\lambda (r_j)_*$ for $(z_*)_j H_j$. □

Example 9.2.6 (binomial coefficients continued further) Recall from Example 8.3.1 that $F = 1/(1 - x - y)$ and that

$$z_* := (x_*, y_*) = \hat{\boldsymbol{r}}_* = \left(\frac{r}{r+s}, \frac{s}{r+s}\right).$$

Neither partial derivative of $H = 1 - x - y$ vanishes. Picking coordinate $k = 2$, we parametrize $\mathcal{V}$ by $(x, g(x))$, where $g(x) = 1 - x$. We compute

$$\Psi(x) = \operatorname{Res}\left(\frac{1}{y(1-x-y)}; y = 1 - x\right)$$

$$= \frac{1}{1-x}\operatorname{Res}\left(\frac{1}{1-x-y}; y = 1 - x\right) = \frac{-1}{1-x}.$$

Hence

$$A(\theta) = \frac{-1}{1 - x_* e^{i\theta}}$$

and

$$\phi(\theta) = i\theta\,\frac{r}{s} + \log\frac{1 - x_* e^{i\theta}}{1 - x_*}\,.$$

Notice that it is not obvious from the preceding formula for ϕ that $d\phi/d\theta = 0$ at $\theta = 0$, as guaranteed by Proposition 9.2.5:

$$\frac{d}{d\theta}\left(i\theta\frac{r}{s} + \log\frac{1-\frac{r}{r+s}e^{i\theta}}{1-\frac{r}{r+s}}\right) = 0\,.$$

■

When the function ϕ in Equation (9.2.8) is quadratically nondegenerate, the formula in Proposition 9.2.5 leads directly to an asymptotic estimate. The following result was proved in Pemantle and Wilson (2002, Theorem 3.5).

Theorem 9.2.7 (one quadratically nondegenerate smooth point) *Let $F = G/H = \sum_r a_r z^r$ be a Laurent series and fix $\hat{\mathbf{r}}_* \in (\mathbb{R}^d)^*$. Let $\mathcal{N}$ be a neighborhood of $\hat{\mathbf{r}}_*$ such that as $\hat{\mathbf{r}}$ varies over $\mathcal{N}$, there is a smoothly varying quadratically nondegenerate critical point $\mathbf{z}(\hat{\mathbf{r}})$ for $h_{\hat{\mathbf{r}}}$. Suppose F is meromorphic in a neighborhood of $\bigcup_{\mathbf{r}\in\overline{\mathcal{N}}} \mathbf{T}(\mathbf{z}(\hat{\mathbf{r}}))$. Define ϕ by (9.2.8) with $\hat{\mathbf{r}}$ and $\mathbf{z}(\hat{\mathbf{r}})$ in place of $\mathbf{r}$ and $\mathbf{z}_*$. Let $\mathcal{H}(\hat{\mathbf{r}})$ denote the Hessian matrix for $\phi(\hat{\mathbf{r}})$ around its critical point $\mathbf{z}(\hat{\mathbf{r}})$. Define the quantity*

$$\Phi_{\mathbf{z}}^{(2)}(\mathbf{r}) := (2\pi)^{(1-d)/2}\mathbf{z}^{-\mathbf{r}}(\det\mathcal{H}(\hat{\mathbf{r}}))^{-1/2}\frac{G(\mathbf{z})}{z_k H_k(\mathbf{z})} r_k^{(1-d)/2}\,. \tag{9.2.11}$$

Then:

(i) There is an asymptotic series

$$a_{\mathbf{r}} \sim (2\pi)^{(1-d)/2}\mathbf{z}(\hat{\mathbf{r}})^{-\mathbf{r}}\sum_{\ell=0}^{\infty} C_\ell(\hat{\mathbf{r}}) r_k^{(1-d)/2-\ell}\,. \tag{9.2.12}$$

(ii) The leading term of (9.2.12) *is $\Phi_{\mathbf{z}}^{(2)}(\mathbf{r})$.*

(iii) The expansion is uniform in a neighborhood of $\hat{\mathbf{r}}$, meaning that for each $N \geq 1$, the remainder term satisfies

$$\left|a_{\mathbf{r}} - \mathbf{z}(\hat{\mathbf{r}})^{-\mathbf{r}}\sum_{\ell=0}^{N-1} C_\ell(\hat{\mathbf{r}}) r_k^{(1-d)/2-\ell}\right| = O\left(\mathbf{z}(\hat{\mathbf{r}})^{-\mathbf{r}} r_k^{(1-d)/2-N}\right)$$

uniformly as $\mathbf{r} \to \infty$ with $\hat{\mathbf{r}} \in \mathcal{N}$.

Proof By assumption, the critical point equations at $\hat{\mathbf{r}}_*$ define a finite set of values $\mathbf{z}(\hat{\mathbf{r}}_*)$. Any zero-dimensional variety varies smoothly with the parameters as long as there is no double solution. The set of $\hat{\mathbf{r}}$ for which the corresponding variety of $\mathbf{z}$ has dimension greater than zero is closed and hence avoids a neighborhood of $\hat{\mathbf{r}}_*$. The sets on which strict minimality and vanishing of $\det\mathcal{H}$ fail are also closed. The Hessian matrix is the Jacobian of the logarithmic gradient, whence nonvanishing of $\det\mathcal{H}$ implies nonsingularity of the logarithmic gradient map. This is the map taking $\mathbf{z} \in \mathcal{V}$ to $\mathbf{r} \in \mathbb{RP}^{d-1}$, whence by the inverse function theorem it has a smooth inverse, implying that $\mathbf{z}(\hat{\mathbf{r}})$ varies smoothly in a neighborhood and finishing the proof of part (i).

Strict minimality of $z(\hat{r})$ implies that the real part of ϕ has a strict minimum at the origin. We may apply the basic multivariate result from Theorem 5.1.2 with $\lambda = r_k$ and the d there equal to $d-1$ in the present notation. This gives the correct power of (2π) in (9.2.11) and the factor of $\det \mathcal{H}(\hat{r})^{-1/2}$. To evaluate the remaining factor $A(\mathbf{0}) = \psi(\mathbf{0}) = \Psi(z_*^\circ)$, we compute the residue of $F(z_*^\circ, w)/w$ at $w = g(z_*^\circ) = (z_*)_k$. Because $F/w = G/(Hw)$ has a simple pole at z_*, and because H_k is nonvanishing there,

$$\operatorname{Res}\left(\frac{G}{wH}(z_*)\,;\,w\right) = \frac{G(z_*)}{(z_k)_*\,\partial H/\partial z_k(z_*)}\,.$$

The Taylor series for A varies continuously with $\hat{r}$, whence the remainder term coming from Proposition 5.2.4 is uniform in a neighborhood of $\hat{r}_*$, completing the proof. □

Extending this to minimal points that are not strictly minimal is a matter of using Theorem 5.4.8 in place of Theorem 5.1.2. Extending to finitely minimal points is an application of Corollary 9.2.3, and extending to the case of torality is an application of Corollary 9.2.4. We therefore have the following more general results. The reader interested only in ordinary power series may ignore the amoeba and take B to be the logarithmic domain of convergence of convergence.

Corollary 9.2.8 (finitely many quadratically nondegenerate smooth points) *Let $F = G/H$, let B be a component of* $\texttt{amoeba}(H)^c$, *and let $\hat{r}_*$ be a nonflat direction with minimizing point $x_{\min} \in \partial B$. Suppose that the set E_* of critical points of $h_{\hat{r}_*}$ on $\mathcal{V}_1 := \mathcal{V} \cap T(x_{\min})$ is finite and nonempty and that $\mathcal{V}$ is smooth in a neighborhood of $T(x_{\min})$. For each $z_* \in E_*$, let ϕ_* denote the parametrizing function defined in* (9.2.8) *and suppose the corresponding Hessian matrix $\mathcal{H}^*$ is nonsingular for each of these. Then these hypotheses hold for $\hat{r}$ in some compact neighborhood $\mathcal{N}$ of $\hat{r}_*$ over which $E(\hat{r})$ varies analytically. Denoting each $z \in E$ by* $\exp(x + iy)$, *on this neighborhood there is a uniform asymptotic expansion*

$$a_r \sim \exp(-r\cdot x(\hat{r}))\sum_{\ell=0}^{\infty}\left[\sum_{z\in E(\hat{r})}\exp(-ir\cdot y)\,C_{z,\ell}(\hat{r})\right] r_{k(\ell)}^{(1-d)/2-\ell}\,. \qquad (9.2.13)$$

The leading term is equal to

$$\sum_{z\in E(r)}\Phi_z^{(2)}(r)\,.$$

Remark This is an asymptotic expansion in the sense that the remainder term for the partial sum over $0 \le \ell \le N-1$ is $O(\exp(-r\cdot x)|r|^{(1-d)/2-N})$; in particular, it is not asserted that any given term is nonzero, only that some term will be nonzero.

Proof Compute the integral around each point $z_* \in E_*$ by residues, as before; summing these contributions and substituting $z_*^{-r} = \exp(-r\cdot x_{\min})\exp(-ir_*\cdot y)$ as $\exp(x_{\min} + iy)$ varies over points of E_* then recovers (9.2.13) for $\hat{r} = \hat{r}_*$. The failure of each hypothesis occurs on a closed set of $\hat{\boldsymbol{r}}$, and the formula (9.2.13) varies smoothly in $\hat{\boldsymbol{r}}$ with remainder term uniformly bounded in a neighborhood of $\hat{\boldsymbol{r}}_*$, proving the corollary. □

Corollary 9.2.9 *Let $F = G/H$ with H satisfying the torality hypothesis* (9.2.1) *on the torus $\mathbf{T}(\boldsymbol{x})$. Let $F = \sum_{\boldsymbol{r}} a_{\boldsymbol{r}} \boldsymbol{z}^{\boldsymbol{r}}$ be the Laurent series for F whose coefficients are given by the Cauchy integral on $\mathbf{T}(\boldsymbol{x} - \delta \mathbf{e}_k)$. Let $E(\hat{\boldsymbol{r}})$ denote the intersection of* $\mathtt{critical}(\hat{\boldsymbol{r}})$ *with $\mathbf{T}(\boldsymbol{x})$, and let $\hat{\boldsymbol{r}}_*$ be any point for which $E(\hat{\boldsymbol{r}}_*)$ is nonempty, the poles at points $\boldsymbol{z} \in E(\hat{\boldsymbol{r}}_*)$ are all simple, and the Hessians $\mathcal{H}(\boldsymbol{z})$ nonsingular. Then, uniformly as $\hat{\boldsymbol{r}}$ varies in a neighborhood of $\hat{\boldsymbol{r}}_*$,*

$$a_{\boldsymbol{r}} \sim (2\pi)^{(1-d)/2} \sum_{\boldsymbol{z} \in E(\hat{\boldsymbol{r}})} \boldsymbol{z}^{-\boldsymbol{r}} C_\ell(\hat{\boldsymbol{r}}) r_k^{(1-d)/2-\ell} .$$

Again, the expansion should be interpreted so as to allow some terms to vanish, and again the leading term is given by

$$a_{\boldsymbol{r}} \sim \sum_{\boldsymbol{z} \in E(\hat{\boldsymbol{r}})} \Phi_{\boldsymbol{z}}^{(2)}(\boldsymbol{r}).$$

We summarize the results of this section as follows.

Theorem 9.2.10 *Suppose $F = G/H$ and fix $\hat{\boldsymbol{r}}_* \in \Xi$. Let B be a component of* $\mathtt{amoeba}(H)^c$, *and let $\boldsymbol{x}_{\min}$ be the minimizing point for $h = h_{\hat{\boldsymbol{r}}_*}$ on ∂B. Suppose that the set E of critical points of h on $\mathcal{V} \cap \mathbf{T}(\boldsymbol{x}_{\min})$ is finite and that either $\mathcal{V} \cap \mathbf{T}(\boldsymbol{x}_{\min})$ is finite or the torality hypothesis is satisfied. Then, uniformly as $\hat{\boldsymbol{r}}$ varies over a compact set on which h is locally Morse,*

$$a_{\boldsymbol{r}} = \sum_{\boldsymbol{z} \in E} \Phi_{\boldsymbol{z}}^{(2)}(\boldsymbol{r}) + O\left(|\boldsymbol{r}|^{(-1-d)/2} \left|\boldsymbol{z}^{-\boldsymbol{r}}\right|\right).$$

9.3 The Smooth Point Integral via Residue Forms

Our second derivation of the saddle-residue integral requires the notion of a residue form. This generalizes the univariate residue $\mathrm{Res}(f; a) := \lim_{z \to a}(z - a) f$ at a simple pole of a univariate analytic function.

Definition 9.3.1 *Let ξ be a meromorphic d-form on a domain in $\mathbb{C}^d$, and assume ξ has a simple pole on a neighborhood $U \subseteq \mathcal{V}$. Let ι denote the inclusion of $\mathcal{V}$ into $\mathbb{C}^d$. Define*

$$\mathrm{Res}(\xi; U) := \iota^* \theta,$$

where $\xi = G/H$ represents ξ as a quotient of analytic functions on U and θ is any $(d-1)$-form satisfying

$$dH \wedge \theta = G \, d\boldsymbol{z} .$$

It is proved in Appendix A.5 that θ always exists and that $\iota^* \theta$ is well defined, i.e., independent of the representation of ξ as a quotient of analytic functions and independent of the particular solution θ to $dH \wedge \theta = G\,dz$. This construction is easily seen to be functorial. Thus Res is a functor that takes a meromorphic d-form with a simple pole on a neighborhood of a set $U \subseteq \mathcal{V}$ to a holomorphic $(d-1)$-form on U.

The purpose of introducing the residue form is to state the following representation of the coefficients of a rational function, which is an analogue of Theorem 9.2.1 and of the relative homology fact in Remark 8.2.5. The following result is an immediate consequence of Theorem A.5.3.

Theorem 9.3.2 *Let $F = G/H$ with a simple pole on $\mathcal{V}$. Let c be any real number, and suppose that $\mathcal{V}$ is smooth above height $c - \varepsilon$. Let T be a torus on the boundary of a polydisk small enough to avoid $\mathcal{V}$. Let Y denote $\mathcal{M}^{c-\varepsilon}$ for some $\varepsilon > 0$. Then for any $\varepsilon' < \varepsilon$,*

$$\left| a_r - \frac{1}{(2\pi i)^{d-1}} \int_{\mathtt{INT}[T;\mathcal{V}]_Y} \mathrm{Res}(\omega) \right| = O\left(e^{(c-\varepsilon')|r|} \right).$$

Here, $\mathtt{INT}[T;\mathcal{V}]_Y$ *is the intersection class defined in Section A.4.*

Proof Cauchy's integral theorem represents a_r as $(2\pi i)^{-d} \int z^{-r-1} F(z) dz$ over T. Let $\omega := z^{-r-1} F(z)\,dz$, which is holomorphic on $\mathcal{M}$. Observe that $T = 0$ in $H_d((\mathbb{C}^*)^d, Y)$. We may therefore apply Theorem A.5.3 to see that

$$\begin{aligned} a_r &= (2\pi i)^{-d} \int_T \omega \\ &= (2\pi i)^{1-d} \int_{\mathtt{INT}[T;\mathcal{V}]_Y} \mathrm{Res}(\omega) + \int_{C'} \omega, \end{aligned}$$

where C' is supported on the interior of Y. The last part of Theorem A.5.3 gives the desired estimate. □

For computation, one must introduce coordinates. Although the residue form is natural, and its integral (corresponding to the outer integral in (9.2.2)) is a scalar and also natural, splitting a form into components on orthogonal spaces is not. Before continuing with the theory, let us have a look at how the residue form might best be understood. The space of holomorphic $(d-1)$-forms on the $(d-1)$-manifold $U \subseteq \mathcal{V}$ is one-dimensional over the space of holomorphic functions. This is the space where $\iota^*\theta$ lives, so if we choose a generator ξ, the form $\mathrm{Res}(\omega; \mathbf{p})$ is a varying multiple $g(z)\xi$. The space of holomorphic $(d-1)$-forms on $\mathbb{C}^d$ is, on the other hand, d-dimensional, whence the kernel of ι^* is $(d-1)$-dimensional. This makes it difficult, without further structure, to pick θ in a natural way. Geometrically, one might take advantage of the orthogonality of $\mathcal{V}$ to the vector ∇H: pick θ to annihilate any $(d-1)$-tuple of tangent vectors containing one perpendicular to D. This defines θ up to a scalar multiple, after which (A.1) completes the specification of θ.

Conversely, G and H are defined in terms of the specific variables $z_1, \ldots, z_d$, and it may make sense to sacrifice some naturality to remain in these coordinates. Accordingly, we have the following coordinate representation of both θ and $\iota^*\theta$, in which, conveniently, dz_j denotes both a 1-form on $\mathbb{C}^d$ and its pullback to $\mathcal{V}$. Recall that $\omega = z^{-r-1} G\, dz/H$.

Proposition 9.3.3 *On a domain in $\mathcal{V}$ where $\partial H/\partial z_1$ does not vanish,*

$$\mathrm{Res}(\omega) = \frac{z^{-r} G}{\partial H/\partial z_1} dz_2 \wedge \cdots \wedge dz_d \,. \tag{9.3.1}$$

An analogous result holds with any z_k in place of z_1. Let $dz_{\hat{k}}$ denote $dz_1 \wedge \cdots \wedge \widehat{dz_k} \wedge \cdots \wedge dz_d := dz_1 \wedge \cdots \wedge dz_{k-1} \wedge dz_{k+1} \wedge \cdots \wedge dz_d$. Then

$$\mathrm{Res}(\omega) = (-1)^{k-1} \frac{z^{-r} G}{\partial H/\partial z_k} dz_{\hat{k}} \,.$$

Proof Taking θ to be the right side of (9.3.1), we compute

$$dH \wedge \theta = \left(\sum_{j=1}^{d} \frac{\partial H}{\partial z_j} dz_j \right) \wedge \left(\frac{z^{-r} G}{\partial H/\partial z_1} dz_2 \wedge \cdots \wedge dz_d \right).$$

Expanding, we find that all terms except one cancel, leaving $z^{-r} G\, dz$ as desired. The proof for z_j is the same, with the sign factor $(-1)^{k-1}$ coming from the position of dz_k in the wedge product. □

As usual for small dimensions such as $d = 2$, we drop the subscripts and see what the formula looks like. When $d = 2$, the Cauchy integrand is $\omega := x^{-r-1} y^{-s-1} (G(x,y)/H(x,y))\, dx dy$. Wherever $H_y \neq 0$, we may take $k = 2$ to obtain the description

$$\mathrm{Res}(\omega) = -x^{-r-1} y^{-s-1} \frac{G}{H_x} dy \,.$$

When $H_x \neq 0$, we could alternatively have obtained

$$\mathrm{Res}(\omega) = x^{-r-1} y^{-s-1} \frac{G}{H_y} dx \,.$$

To check that these agree, note that their difference is a multiple of $dx/H_y + dy/H_x$ which vanishes on $\mathcal{V}_H$, where the tangent vector field is the space spanned by the tangent vector $H_y(\partial/\partial x) - H_x(\partial/\partial y)$. In general, the sign factor of $(-1)^{k-1}$, introduced when the kth coordinate is selected, will drop out once the residue is integrated and will therefore not appear in formulae such as 9.2.11, which must be invariant under permuting coordinates.

Example 9.3.4 (binomial coefficients continued even further) For the binomial coefficients, $G = 1$ and $H = 1 - x - y$, which gives us $\theta = -x^{-r-1} y^{-s-1}\, dy$. Mapping by ι^* does not change the formula. Thus

$$\mathrm{Res}(\omega) = -x^{-r-1} y^{-s-1}\, dy \,.$$

■

Example 9.3.5 (Delannoy numbers continued) Here, $F = \dfrac{1}{1-x-y-xy}$. We have $G = 1$, $H = 1 - x - y - xy$ and hence

$$\text{Res}(\omega) = -x^{-r-1}y^{-s-1}\frac{dy}{1+y}.$$

Because the residue lives on $\mathcal{V}$, we may rewrite this any way we like using the relation $1 - x - y - xy = 0$. For example,

$$\text{Res}(\omega) = x^{-r-1}y^{-s-1}\frac{x\,dy}{y-1}.$$

■

The final work for this section is to put the previous result into a form of a Fourier-Laplace integral. Recall from (8.6.2) that the factor of z^{-r} may be pulled through the residue, leading to

$$\text{Res}(\omega) = z^{-r}\,\text{Res}(\eta), \tag{9.3.2}$$

where $\eta := z^{-1}F(z)\,dz$.

Lemma 9.3.6 *If the class C has a representation $C = \sum_{z\in W} C_*(z)$ in $H_d(\mathcal{M}, \mathcal{M}^{c_*-\varepsilon})$ as in Lemma 8.2.4, then*

$$\sigma = \left[\sum_{z\in W} C(z)\right]$$

in $H_d(\mathcal{M}, \mathcal{M}^{c-\varepsilon})$, where

- *$\sigma := \mathtt{INT}[T;\mathcal{V}]_Y$ is the intersection class as in Theorem 9.3.2.*
- *W is the set of critical points z for h such that $C_*(z)$ is nonvanishing defined in Lemma 8.2.4.*
- *$C(z)$ is the relative cycle in Definition 8.5.4 and is supported on $\mathcal{M}(z)$, the union of $\mathcal{M}^{c-\varepsilon}$ with a small neighborhood of z.*

Proof By the construction of the intersection cycle, the isomorphism in Proposition A.4.1 maps σ to the class $[T]$ (all this being relative to $\mathcal{M}^{c-\varepsilon}$). The isomorphism is a product with a small circle. Smoothness of $\mathcal{V}$ implies `N-data` is always a small circle, whence `N-data` $\times\sigma = [T]$. By construction of $C(z)$, we know that $\sum_z$ `N-data` $\times C(z) = [T]$. Because the product with `N-data` is an isomorphism, we conclude that $\sigma = \sum_{z\in W} C(z)$. □

Theorem 9.3.7 *Let $F = G/H$ with a simple pole on $\mathcal{V}$. Fix $\hat{\mathbf{r}}_*$ and suppose that $\mathcal{V}$ is smooth above height $h_{\hat{\mathbf{r}}_*} = c - \varepsilon$. Suppose there is a subset W of critical points of $\mathcal{V}$ at height c such that $C = \sum_{z\in W} C_*(z)$ in $H_d(\mathcal{M}, \mathcal{M}^{c-\varepsilon})$, and suppose all critical points in W are nondegenerate. Then there is a compact neighborhood $\mathcal{N}$ of $\hat{\mathbf{r}}_*$ such that as $\hat{\mathbf{r}}$ varies over $\mathcal{N}$, $W(\hat{\mathbf{r}})$ varies smoothly and contains only quadratically nondegenerate critical points. Define $\Phi_z^{(2)}$ as in (9.2.11). Then as*

$r \to \infty$ with $\hat{r}$ varying over $\mathcal{N}$, there is an asymptotic series for a_r whose first term is

$$a_r \sim \sum_{z \in W} \Phi_z^{(2)}(r)\,.$$

Proof Plugging (9.3.2) and Lemma 9.3.6 into Theorem 9.3.2 and letting $=_o$ denote equality up to a term of order $O(e^{(c-\varepsilon)|r|})$,

$$a_r =_o (2\pi i)^{-(d-1)} \int_\sigma \operatorname{Res}(\omega)$$

$$=_o (2\pi i)^{-(d-1)} \sum_{z \in W} \int_{C(z)} \operatorname{Res}(\omega)$$

$$=_o (2\pi i)^{-(d-1)} \sum_{z \in W} \int_{C(z)} \exp(-|r|\phi) \operatorname{Res} \eta,$$

where $\phi(y) = -\hat{r} \cdot \log y$ for some branch of the logarithm in a suitable neighborhood of each of the points $z \in W$. By Proposition 9.3.3, we have the explicit formula

$$\operatorname{Res}(\eta) = (-1)^{k-1} \frac{z^{-1} G}{\partial H / \partial z_k} dz_1 \wedge \cdots \widehat{dz_k} \cdots \wedge dz_d\,.$$

Denoting the right-hand side of this formula by $A(z)$, we then have

$$a_r =_o (2\pi i)^{-(d-1)} \sum_{z \in W} \int_{C(z)} e^{-|r|\phi} A(z),$$

where the real part of ϕ is maximized at the point z interior to the chain $C(z)$. Because ϕ is quadratically nondegenerate on $\mathcal{V}$ at z, we conclude from the basic multivariate saddle-point Theorem 5.1.2 that

$$a_r =_o (2\pi i)^{-(d-1)/2} \sum_{z \in W} |r|^{-(d-1)/2} A(z)\, M^{-1/2},$$

where M is the determinant of the Hessian matrix of ϕ at z. Parametrizing $\mathcal{V}$ locally by the coordinates other than z_k shows that

$$|r|^{(1-d)/2} A(z) M^{-1/2} = r_k^{(1-d)/2} \frac{G(z)}{z_k \partial H / \partial z_k} \det(\mathcal{H}(z))^{-1/2},$$

which shows that the summand corresponding to z is the quantity $\Phi(z)$ from (9.2.11). □

9.4 Effectively Computing the Intersection Class in Two Cases

According to Lemma 9.3.6, the intersection class $\mathtt{INT}[\mathcal{T}; \mathcal{V}]$ of Theorem 9.3.2 should be a sum over critical points z of quasi-local cycles $C(z)$ in $\mathcal{V}$. Assuming that each such z is a smooth point of $\mathcal{V}$, `N-data` will always be a circle, and lifting to $\mathcal{M}$ gives the identity in relative homology:

$$[T] = \sum_z C_*(z) = \sum_z \gamma_z \times C(z),$$

where γ_z is a circle in the normal slice at z (see also Example 8.5.6). In general, the Morse theoretic description of the topology of $\mathcal{M}$ requires assumptions we cannot verify. In the remainder of this section, we consider two cases in which an explicit construction computes the intersection class in terms of quasi-local cycles. The first case is when z_* is a minimal point. In this case, we recover the results proved in Section 9.2 via surgery.

The second case is when $d = 2$. In the two-dimensional case, we have an algorithm to compute the topology of $\mathcal{M}$ when the first obstruction lies beyond the domain of convergence. The relative homology group $H_d(\mathcal{M}, \mathcal{M}^{c-\varepsilon})$ is computed where $c = h_{\hat{r}}(z)$ is the height of the first obstruction, z. This is enough to estimate a_r up to terms of a lower exponential order. We examine each of these two cases in turn.

Minimal Points

Suppose $z_* = \exp(\boldsymbol{x}_{\min} + i\,\boldsymbol{y}_*)$ is a (not necessarily strictly) minimal, smooth critical point for h in the nonflat direction $\hat{\boldsymbol{r}}_*$. Let $\boldsymbol{x}'$ be a point inside B and choose $\boldsymbol{x}''$ outside of B with $\boldsymbol{x}_{\min}$ on the line segment joining $\boldsymbol{x}'$ and $\boldsymbol{x}''$. Let $\mathbf{H} : \mathbf{T}_{\text{flat}} \times [0, 1] \to \mathbb{C}^d$ denote the homotopy

$$\mathbf{H}(\boldsymbol{y}, t) = \exp(\boldsymbol{x}'' + t(\boldsymbol{x}'' - \boldsymbol{x}') + i\,\boldsymbol{y}),$$

taking $T(\boldsymbol{x}')$ to $T(\boldsymbol{x}'')$. Let $c := -\hat{\boldsymbol{r}}_* \cdot \boldsymbol{x}_{\min}$.

Lemma 9.4.1 *If $\boldsymbol{x}', \boldsymbol{x}''$ are sufficiently close to $\boldsymbol{x}_{\min}$, then the image of $\mathbf{H}$ intersects $\mathcal{V}$ transversely.*

Proof Let $z \in \mathcal{V}_1 := \mathcal{V} \cap T(\boldsymbol{x}_{\min})$. Then $T_z(\mathcal{V})$ is the complex orthogonal complement to the vector $\vec{\alpha} := \nabla H(z_*)$. Let W denote the purely imaginary vector space (viewed as a d-dimensional real vector space); the tangent space $T_z(\mathbf{H})$ is the direct sum of W with the one-dimensional subspace spanned by the pointwise product $\vec{\beta}$ of $\exp(\boldsymbol{x}_{\min})$ with $(\boldsymbol{x}'' - \boldsymbol{x}')$. If $\vec{\alpha}$ is not a complex scalar multiple of a real vector, then $T_z(\mathcal{V}) \cup W$ already spans the whole space. If $\vec{\alpha}$ is a scalar multiple of a real vector $\boldsymbol{v}$, then $T_z(\mathcal{V}) \cup T_z(\mathbf{H})$ spans the whole space unless $\boldsymbol{v}$ is parallel to β. This would violate minimality because $\boldsymbol{x}'$ lies in B. We conclude that $\mathcal{V}$ and $\mathbf{H}$ intersect transversely at all points of $T(\boldsymbol{x})$. The set at which $\mathcal{V}$ and $\mathbf{H}$ fail to intersect transversely is closed; therefore, choosing $\boldsymbol{x}''$ close enough to $\boldsymbol{x}_{\min}$, the entire intersection is transverse. Note that the choice of $\boldsymbol{x}'$ does not matter because $\mathcal{V}$ does not intersect $\operatorname{Re}\log^{-1}(B)$ at all. □

Theorem 9.4.2 *If $\boldsymbol{x}_{\min}$ is the unique minimizer in ∂B for $h = h_{\hat{r}}$ and the set of critical points $E \subseteq \mathbf{T}(\boldsymbol{x}_{\min})$ is a finite nonempty set of quadratically nondegenerate smooth points, then the intersection cycle is the sum of quasi-local cycles $C(\boldsymbol{z})$ for $\boldsymbol{z} \in E$, where $C(\boldsymbol{z})$ is a homology generator of $(\mathcal{V}^{h(\boldsymbol{x}_{\min})+\varepsilon}, \mathcal{V}^{h(\boldsymbol{x}_{\min})-\varepsilon})$, for example the descending submanifold.*

Proof There are several ways to see this. We give the most self-contained proof, then remark on two others. Assume first that E is a singleton, $\{\boldsymbol{z}_*\}$. The intersection cycle is obtained by intersecting $\mathcal{V}$ with a homotopy between the original chain of integration and a chain at height ε below $\boldsymbol{x}_{\min}$. We may take these chains to be $\mathbf{T}(\boldsymbol{x}')$ and $\mathbf{T}(\boldsymbol{x}'')$, respectively. By Lemma 9.4.1, these intersect $\mathcal{V}$ transversely in a chain σ, which is by definition a relative cycle representing the intersection class. We may assume without loss of generality that $h_{\hat{r}}$ has a strict maximum on σ at $\boldsymbol{z}_*$: if the maximum is not strict, deforming σ by the downward gradient flow for a small time will replace σ by a chain σ' homotopic to σ on which h is strictly maximized at $\boldsymbol{z}_*$. By Exercise B.3, the relative cycle σ is a homology generator. If $|E| > 1$, then applying a local version of the same homotopy at each $\boldsymbol{z} \in E$, together with the direct sum decomposition (8.5.2), proves the theorem. □

Remark The theorem follows from Proposition 10.3.6 because a smooth point is a special case of an arrangement point. The theorem also follows from the surgical approach. It is essentially the fact that the Cauchy integral is the sum of nonzero contributions near each $\boldsymbol{z} \in E$. More specifically, Corollary 9.2.3 writes the Cauchy integral for $a_{\boldsymbol{r}}$ as an explicit sum of integrals $\int_{\mathcal{N}(\boldsymbol{z})} \omega$, where $\mathcal{N}(\boldsymbol{z})$ are neighborhoods of each $\boldsymbol{z} \in E$ and ω is a residue. This means the integrand is an integral over some circle γ. In the terms of Theorem 8.5.3, the circle γ is `T-data` and $\mathcal{N}(\boldsymbol{z})$ is `N-data`. Once we see that $\mathbf{T} = \sum_{\boldsymbol{z} \in E} \gamma(\boldsymbol{z}) \times C(\boldsymbol{z})$, it follows from Lemma 9.3.6 that the intersection cycle is equal to $\sum_{\boldsymbol{z} \in E} C(\boldsymbol{z})$.

The following example illustrates the application of Theorem 9.4.2 to asymptotic estimation of $a_{\boldsymbol{r}}$.

Example 9.4.3 (Delannoy numbers continued) To apply the theorem, we first need to identify $C(\boldsymbol{z})$. For each $\hat{\boldsymbol{r}}$, by definition $C(\boldsymbol{z})$ is a local 1-manifold in $\mathcal{V}$ passing through one or more points $(x, y) \in \texttt{critical}_{\hat{r}}$ where the function $h_{\hat{r}}$ is critical. We know h will have index 1 at such a point (see the discussion following Proposition B.3.1), and thus $C(\boldsymbol{z})$ is a path element passing through such a point (x, y) in a direction in which h has a nondegenerate maximum. Let us now specialize to the positive real critical point

$$(x_0, y_0) := \left(\frac{\sqrt{r^2+s^2}-s}{r}, \frac{\sqrt{r^2+s^2}-r}{s} \right).$$

We claim that $C(\boldsymbol{z})$ passes through (x_0, y_0) in a purely imaginary direction. To see this, parametrize $\mathcal{V}$ by the y-coordinate and note that the function h is the real part of the complex analytic function $\tilde{h} := -r \log x - s \log y$, which is real for real arguments. Along the curve $\mathcal{V}$, the second derivative $\tilde{h}''$ is positive real. Thus the direction of (quadratic) greatest increase of h is the real direction, and the direction of greatest decrease is the imaginary direction. Letting $c := h(x_0, y_0) = -r \log x_0 - s \log y_0$, we conclude that

$$\begin{aligned}\int_C \omega &\sim_c 2\pi i \int_{C(z)} \text{Res}(\omega) \\ &\sim_c 2\pi i \int_{y_0-\varepsilon i}^{y_0+\varepsilon i} -x^{-1-r} y^{-1-s} \frac{dy}{1+y} \\ &= 2\pi i \int_{-\varepsilon}^{\varepsilon} \left(\frac{1-y_0-it}{1+y_0+it}\right)^{-1-r} (y_0+it)^{-1-s} \frac{i\,dt}{1+y_0+it}.\end{aligned}$$

To show that $\int_C \omega$ is an actual one-variable integral, we have written it as explicitly as possible. Again, once coordinates are introduced, it is not obvious that $t = 0$ is a stationary phase point. Another way to express the same integral leaves the contour of integration real:

$$\int_{C(z)} \text{Res}(\omega) \sim_c \int_{y_0-\varepsilon i}^{y_0+\varepsilon i} \exp\left[(r+s)h_{\hat{r}}(x(y), y)\right] \frac{1}{x(y)\,y} \frac{dy}{1+y}. \tag{9.4.1}$$

■

Dimension 2

In any number of variables, a potential program to evaluate the residue integral is as follows.

1. Explicitly compute some cycle representing the intersection class.
2. For each critical point starting at the highest, try to push the cycle below the critical point.
3. When it is not possible to do so, describe the local cycle i.e., "snagged" on the critical point.
4. Check whether this is a quasi-local cycle of the form we have already described, and if so, read off the estimate from saddle point asymptotics.

This program is not effective because the step of "pushing the cycle down" is not algorithmic. An exception is when $d = 2$, where the cycle C has codimension 1 in $\mathcal{V}$, whence, up to a time change, there is only one way for it to flow downward. Our discussion in this case will follow the recent work of DeVries, van der Hoeven, and Pemantle (2012).

Suppose the height $h(x, y) = -\hat{r} \log|x| - \hat{s} \log|y|$ of a point (x, y) is at least c. Then either x or y must have modulus at most e^{-c}. There is some $\varepsilon > 0$ for which $\mathcal{V}$ does not intersect the set $\{(x, y) \colon |x| \le \varepsilon |y| \le \varepsilon\}$, and taking any $c \ge \log(1/\varepsilon)$, we see that no connected component of $\mathcal{M}^{\ge c}$ contains both points with $|x| \le \varepsilon$ and points with $|y| \le \varepsilon$. Conversely, every connected component of $\mathcal{M}^{\ge c}$ contains points with arbitrarily large height and hence points with either $|x| < \varepsilon$ or $|y| < \varepsilon$.

Accordingly, for sufficiently large c, we may decompose $\mathcal{M}^{\geq c}$ into the disjoint union $X^{\geq c} \cup Y^{\geq c}$, where $X^{\geq c}$ is the union of connected components containing points with arbitrarily small x-coordinates, and $Y^{\geq c}$ is the union of connected components containing points with arbitrarily small y-coordinates. The set of c for which the components may be separated in this manner is clearly an upper interval (c_*, ∞), where c_* is the maximum over all paths in $\mathcal{V}$ connecting the x-axis to the y-axis of the minimum height along the path. The topology of $\mathcal{V}^{\geq c}$ cannot change with c except at critical values, whence c_* is a critical value of h. For any $c > c_*$, the set $X^{\geq c}$ inherits a natural orientation as a Riemann surface, hence defines a 2-chain. We will prove the following results:

Theorem 9.4.4

(i) Let T be a torus bounding a polydisk disjoint from $\mathcal{V}$. The cycle $\partial X^{\geq c}$ is in the homology class α if and only if $c \geq c_*$.

(ii) $h_*(\alpha) = c_*$ where $h_*(\alpha)$ is the minimum over cycles in class α of the maximum height along the cycle.

(iii) Let E be the set of critical points at height c_* which are both limits of points in $X^{\geq c_* + \varepsilon}$ as $\varepsilon \downarrow 0$ and of points in $Y^{\geq c_* + \varepsilon}$ as $\varepsilon \downarrow 0$. The class α is represented by a sum $\sum_{z \in E} C(z)$, where each $C(z)$ reaches its strict height maximum at z.

Before proving this, we record a lemma transferring between components of $\mathcal{V} \cap \{|x| \leq \varepsilon\}$ and components of $\mathcal{V}^{\geq c}$. Fix ε sufficiently small, let T be the torus with radii ε, and use the homotopy that expands the radius of the y-coordinate. Let m be the y-degree of H. The surface $\mathcal{V}$ is the graph of an m-valued algebraic function $y = y_1(x), \ldots, y_m(x)$, except at finitely many points where two of these coalesce or one becomes infinite. For sufficiently small $\varepsilon > 0$, this graph is a smooth m-fold cover of a neighborhood of the circle $\{|x| = \varepsilon\}$. The cycle representing α defined by the chosen homotopy is the graph over $\{|x| = \varepsilon\}$ and is therefore the sum of topological circles comprising this m-fold cover, with their inherited orientations as boundaries of regions in a Riemann surface. In other words, $\partial(\mathcal{V} \cap \{|x| \leq \varepsilon\})$ represents α. It is easier to work with $\partial X^{\geq c}$ as c varies than with $\mathcal{V} \cap \{|x| \leq \varepsilon\}$ as ε varies, which motivates this lemma.

Lemma 9.4.5 *For c sufficiently large and ε sufficiently small, the components of $X^{\geq c}$ are in one-to-one correspondence with the components of $\mathcal{V} \cap \{|x| \leq \varepsilon\}$ and corresponding components are homotopy equivalent.*

Proof This is part (iv) of Theorem 3.2 of DeVries, van der Hoeven, and Pemantle (2012). Components on which the magnitude of y is bounded correspond to patches around points $(0, y) \in \mathcal{V}$, in which case there is little to prove. Components with unbounded y correspond to Puiseux expansions with leading term $y = cx^{-a/b}$. The argument in this case is longer but not deeper. □

Proof of Theorem 9.4.4: Let $c_1 > c_2 > \cdots c_p$ be the critical values of h. Let W be the set of real numbers c for which the cycle $\partial X^{\geq c}$ is in the homology class α. We establish four properties of W:

(a) The set W contains some $c > c_1$.
(b) $[c_1, \infty) \subseteq W$.
(c) If $c_i \in W$ and $c_i > c_*$, then $c_i - \varepsilon \in W$ for all sufficiently small $\varepsilon > 0$.
(d) If $c \in W$ and $c_i > c > c_{i+1}$, then the entire closed interval $[c_{i+1}, c_i]$ is a subset of W.

Together these imply that $W = [c_*, \infty)$ and establish part (i) of the theorem as well as showing $h_*(\alpha) \leq c_*$. Property (a) is implied by Lemma 9.4.5. Properties (b) and (d) follow from the fact that the downward gradient flow carries $\partial X^{\geq b}$ to $\partial X^{\geq a}$ whenever $(a, b]$ contains no critical values of h. To prove property (c) we argue as follows.

Assume first that there is a single critical point (x, y) with $h(x, y) = c_i$. Let $k \geq 2$ be the order to which the derivatives of h vanish at (x, y). We know from DeVries, van der Hoeven, and Pemantle (2012) that there is a neighborhood U and an analytic parametrization $\psi : U \to \mathcal{V}$ with $h(\psi(z)) = c_i + \mathrm{Re}\{z^k\}$. Locally, the image of U is divided into $2k$ sectors with $h > c_i$ and $h < c_i$ in alternating sectors.

Figure 9.1 shows $\mathcal{V}^{\geq c}$ (shaded) for three values of c in the case $k = 2$. A circle is drawn to indicate a region of parametrization for which $h = c_i + \mathrm{Re}\{z^k\}$. In the top diagram $c > c_i$, in the middle diagram $c = c_i$ and in the bottom, $c < c_i$. The arrows show the orientation of $\partial\mathcal{V}^{\geq c}$ inherited from the complex structure of $\mathcal{V}$. The pictures for $k > 2$ are similar but with more alternations. Consider the first picture, where $c > c_i$. Because $c > c_{xy}$, each of the k-shaded regions is in $X^{\geq c}$ or $Y^{\geq c}$ but not both. Let us term these regions "x-regions" or "y-regions" accordingly. Because $c_i > c_{xy}$, this persists in the limit as $c \downarrow c_i$, which means that either all k regions are in $Y^{\geq c}$ or all k regions are in $X^{\geq c}$. In the former case, $\partial X^{\geq c}$ does not contain (x, y) for c in an interval around c_i, and the first Morse Lemma shows that $\partial X^{\geq c_i + \varepsilon}$ is homotopic to $\partial X^{\geq c_i - \varepsilon}$. In the latter case, we consider the cycle $\partial X^{\geq c_i + \varepsilon} + \partial B$ where B is the polygon shown in Figure 9.2.

Because we added a boundary, this is homologous to $\partial X^{c_i + \varepsilon}$. However, it is also homotopic to $\partial X^{c_i + \varepsilon}$: within the parametrized neighborhood, the lines may be shifted so as to coincide with $\partial X^{c_i + \varepsilon}$, whereas outside this neighborhood, the downward gradient flow provides a homotopy.

This completes the proof in the case of a single critical point at height c_i. In the case where there is more than one critical point at height c_i, we may add the boundary of a polygon separately near each critical x-point, i.e., each point i.e., a limit as $\varepsilon \downarrow 0$ of points in $X^{c_i + \varepsilon}$. □

Remark At a critical point of height $c > c_*$ where use is made of the cancellation of k local paths, the equivalence of $\partial X^{\geq c+\varepsilon}$ and $\partial X^{\geq c-\varepsilon}$ is at the level of homology,

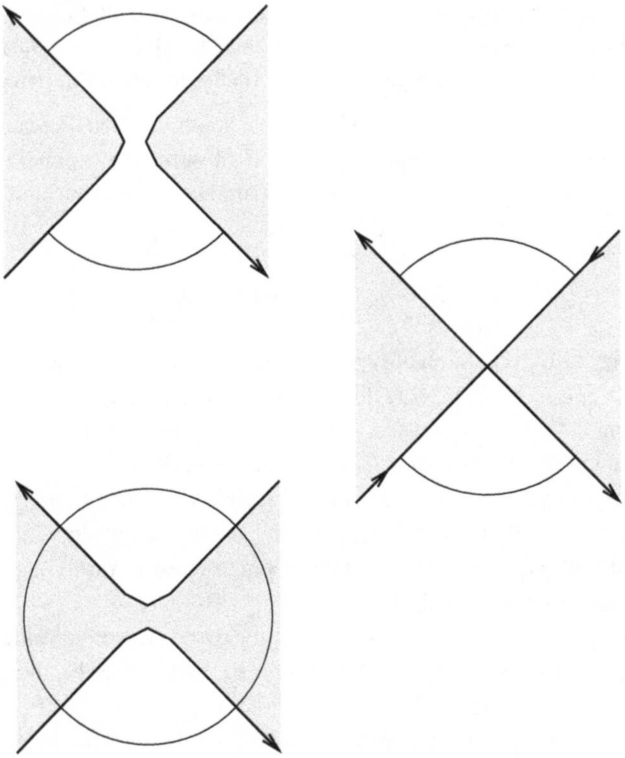

Figure 9.1 $\mathcal{V}^{\geq c}$ and its boundary for three values of c.

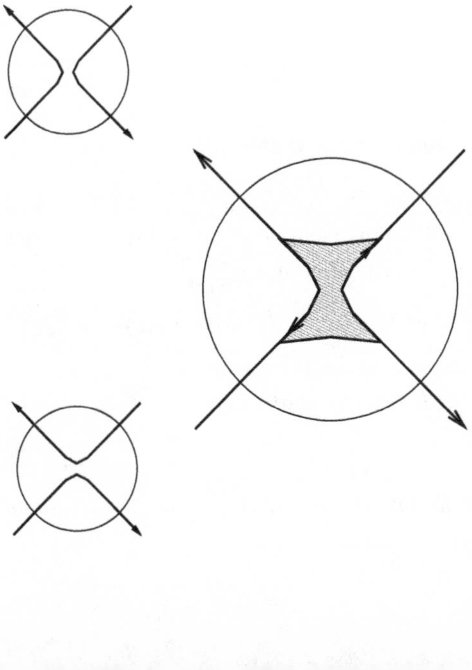

Figure 9.2 $\mathcal{V}^{\geq c_i+\varepsilon}$ and $\mathcal{V}^{\geq c_i-\varepsilon}$ differ locally by a boundary.

not homotopy. Indeed, as we lower the height past the value c, the number of components of our representing cycle will in general change. Thus we make essential use of homology rather than relying on deformation (homotopy) alone.

Suppose (x, y) is a critical point of order k at height c. Then there are precisely k components of $\mathcal{M}^{>c}$ having (x, y) in their closure, corresponding to the k components near 0 of $\text{Re}\{z^k\} > 0$. Using the parametrization ψ from the proof of Theorem 9.4.4, these regions are $\psi[U_1], \ldots, \psi[U_k]$, where

$$U_j := \left\{z : 2\pi\frac{j}{k} - \frac{\pi}{2k} < \text{Arg}(z) < 2\pi\frac{j}{k} + \frac{\pi}{2k}\right\}.$$

The boundary of U_j is a union of two segments of constant argument: the first has argument $2\pi j/k - \pi/(2k)$ and is directed toward the origin, whereas the second has argument $2\pi j/k + \pi/(2k)$ and is directed away from the origin. Rotating the second segment by an angle of $\pi/(2k)$, the image under ψ remains in $\mathcal{M}^{\leq c}$ and becomes the steepest-descent segment characterized by $\text{Arg}(z) = \pi(2j+1)/k$. Similarly, the inwardly pointing segment may be rotated by $-\pi/(2k)$ to become the segment with argument $\pi(2j-1)/k$. Denote these inwardly and outwardly oriented segments respectively by γ_j^- and γ_j^+. Because $\gamma_j^- = -\gamma_{j-1}^+$, there is cancellation modulo k when we add $\partial\psi[U_j]$ for two consecutive values of j. For $(x, y) \in E$, we may use the preceding observations to describe the quasi-local chain $C(x, y)$ of integration. The inwardly directed segment $\psi[\partial_1(U_j)]$ is present if and only if the U_j is an x-component and U_{j-1} is a y-component, whereas the outwardly directed segment $\psi[\partial_1(U_j)]$ is present if and only if U_j is an x-component and U_{j+1} is a y-component. In particular, some segment is present if and only if both types of component are present, corroborating this as the condition for $(x, y) \in E$. We summarize this in a corollary.

Corollary 9.4.6 *For $(x, y) \in E$, the chain $C(x, y)$ is given by*

$$C(x, y) = \sum_j \left[\chi(U_j) - \chi(U_{j-1})\right] \gamma_j^-,$$

where $\chi(U_j)$ is one if U_j is an x-component and zero otherwise.

We may use the characterization of c_* in Theorem 9.4.4 to give an algorithm for computing c_* and a representation of the intersection cycle as a sum of cycles $C(x, y)$ over an explicitly computed set of critical points.

Algorithm 9.4.7 (Determination of W in the smooth, bivariate case)

1. *List the critical value in order of decreasing height.*
2. *Set the provisional value of c_* to the highest critical value.*
3. *For each critical point at height c_*:*
   (a) *Compute the order k of the critical point.*
   (b) *Follow each of the k ascent paths until it is clear, whether the x-coordinate or the y-coordinate goes to zero.*

(c) *Add the point to the set* E *if and only if at least one of the* k *paths has* x*-coordinate going to zero and at least one of the* k *paths has* y*-coordinate going to zero.*

4. *If* E *is nonempty, then terminate and output* c_* *and* E.
5. *Else, if* c_* *is not the least critical value, then replace* c_* *by the next lower critical value and go to step 3.*
6. *Else, if no critical values remain, then* $c_* = -\infty$, E *is empty, and the asymptotics decay super-exponentially.*

It is fairly evident that all these steps are effective. The first step, namely ordering a finite set of algebraic numbers, is obviously effective but tricky to implement well. This is discussed in Gourdon and Salvy (1996). The approach in DeVries, van der Hoeven, and Pemantle (2012) is based on ball arithmetic (van der Hoeven, 2009). The other step i.e., tricky to implement well is computing ascent paths and ascertaining whether they go toward $\{x = 0\}$ or $\{y = 0\}$. The doctoral dissertation by DeVries (2011) handles such problems as the convergence of step size to zero.

We conclude this section with an example of the evaluation of the intersection class for a particular smooth bivariate generating function whose analysis first appeared in DeVries (2010).

Example 9.4.8 (bicolored supertrees) A combinatorial class known as ***bicolored supertrees*** is counted by the diagonal of the following bivariate generating function in DeVries (2010).

$$F(x, y) := \frac{G(x, y)}{H(x, y)} := \frac{2x^2y(2x^5y^2 - 3x^3y + x + 2x^2y - 1)}{x^5y^2 + 2x^2y - 2x^3y + 4y + x - 2}.$$

A bicolored supertree is a Catalan tree with each node replaced by a bicolor planted Catalan tree. This class was reverse-engineered in Raichev and Wilson (2012a) for its analytic properties. The analysis of its diagonal asymptotics in DeVries (2010) is far from automatic, as the asymptotics are not determined by a minimal point. Following the program outlined previously, we automate and simplify the analysis.

The first step is to list the critical points in the diagonal direction. These are points on $\mathcal{V}$ with the two coordinates of the logarithmic gradient equal. Thus we compute in Maple:

```
Basis([H , x*diff(H,x) - y*diff(H,y)] , plex(y,x));
```

and discover the following zero-dimensional ideal.

$$\langle x^5 - 8x^4 + 20x^3 - 8x^2 - 32x + 32 \; , \; x^4 - 6x^3 + 8x^2 + 16x - 48 + 128y \rangle$$

The elimination polynomial for x factors as $(x - 2)^3(x^2 - 2x - 4)$, the latter term having roots $1 \pm \sqrt{5}$. As the second polynomial has degree 1 in y, we see that each x-value corresponds to exactly one y-value. In particular, the critical points are

$$\left(1+\sqrt{5}, \frac{3-\sqrt{5}}{16}\right), \left(2, \frac{1}{8}\right), \left(1-\sqrt{5}, \frac{3+\sqrt{5}}{16}\right),$$

with the second having multiplicity 3. These are listed in order of decreasing height under the height function $h_{1/2,1/2}$. Height is easy to test in this case because it is equivalent to the absolute value of the product of the coordinates.

Setting (x, y) equal to the highest critical point, we follow the two ascent paths, finding that both approach the x-axis. A rigorous algorithm for doing this has been implemented in Mathemagix. We may bypass it in this case with the observation that the highest critical point cannot contribute to the asymptotics because the coordinates are real and of opposite sign. The factor $x^{-n}y^{-n}$ in the asymptotic formula for $a_{n,n}$ would then force the signs to alternate on the diagonal, whereas we know the diagonal terms to be positive.

Continuing to the next-highest point, we set $(x, y) = (2, 1/8)$. This point has multiplicity 3, hence order 4. Among the four ascent paths, three climb to the x-axis and one climbs to the y-axis. This identifies $(2, 1/8)$ as giving the dominant contribution to the diagonal asymptotics. According to Corollary 9.4.6, the local cycle $C(x, y)$ consists of $\gamma_{j+1}^- \cup \gamma_j^+$, where j is the index of the region whose ascent path goes to the y-axis. Among the four descent regions, such a path inhabits two consecutive ones, making a right-angle turn as it passes through the saddle.

Finally, we evaluate the univariate integral at an order-4 saddle over a path making a right-angle turn. This requires some computation. We follow DeVries (2010). To compute the residue form, we parametrize a neighborhood of $(2, 1/8)$ in $\mathcal{V}$ by the x-coordinate and use Proposition 9.3.3 with $j = 2$ to arrive at the following expression for the residue form.

$$\omega := \frac{-G}{xy(x)\partial H/\partial y} x^{-n} y(x)^{-n}\, dx$$

Moving the origin to $x = 2$, the integral may be rewritten as

$$4^n \int_\gamma A(x) e^{-n\phi(x)}\, dx,$$

where the series expansions for A and ϕ are given by

$$A(x) := -\frac{x^3}{8} - \frac{x^4}{16} + O(x^5)$$

$$\phi(x) := -\frac{x^4}{16} + O(x^6).$$

Applying Theorem 4.1.1 to evaluate the integral on the segment γ_{j+1}^- parametrized by $(i-1)x$ on $0 \le x \le \varepsilon$ gives a series for $2\pi i \int \omega$ that begins

$$4^n \left(\frac{-i}{4\pi} n^{-1} + \frac{(1+i)\sqrt{2}\Gamma(5/4)}{8\pi} n^{-5/4} + O(n^{-3/2}) \right) .$$

Similarly, on the piece γ_j^+, we parametrize by $(-i-1)x$ and obtain the complex conjugate of the previous expansion:

$$4^n \left(\frac{i}{4\pi} n^{-1} + \frac{(1-i)\sqrt{2}\Gamma(5/4)}{8\pi} n^{-5/4} + O(n^{-3/2}) \right) .$$

When the two contributions are summed, the first terms cancel, and we are left with

$$a_{n,n} \sim \frac{4^n \sqrt{2}\Gamma(5/4)}{4\pi} n^{-5/4} .$$

■

9.5 Explicit Formulae for the Leading Term

9.5.1 A Coordinate-Free Formula for Nondegenerate Points in Any Dimension

Already in two variables, the explicit expression for Q is somewhat messy. Writing down this type of expression for more variables or for higher order terms in the expansion seems pointless. However, one can write down another version of the formula for C_0 in Theorem 9.2.7 in terms of the curvature of $\mathcal{V}$. This has the advantage of being coordinate-free; it also helps with conceptual understanding in cases such as Example 9.5.6. To state this version, we need to review the definition of the Gaussian curvature of a smooth hypersurface and extend it to certain complex algebraic hypersurfaces.

Gaussian Curvature of Real Hypersurfaces. For a smooth orientable hypersurface $\mathcal{V} \subset \mathbb{R}^{d+1}$, the ***Gauss map*** $\mathcal{G}$ sends each point $p \in \mathcal{V}$ to a consistent choice of normal vector. We may identify $\mathcal{G}(p)$ with an element of the d-dimensional unit sphere S^d. For a given patch $P \subset \mathcal{V}$ containing p, let $\mathcal{G}[P] := \bigcup_{q \in P} \mathcal{G}(q)$, and denote the area of a patch P in either $\mathcal{V}$ or S^d as $A[P]$. Then the ***Gauss-Kronecker curvature*** of $\mathcal{V}$ at p is defined as

$$\mathcal{K} := \lim_{P \to p} \frac{A(\mathcal{G}[P])}{A[P]} . \tag{9.5.1}$$

When d is odd, the antipodal map on S^d has determinant -1, whence the particular choice of unit normal will influence the sign $\mathcal{K}$, which is therefore only well defined up to sign. When d is even, we take the numerator to be negative if the map $\mathcal{G}$ is orientation reversing and we have a well-defined signed quantity. Clearly, $\mathcal{K}$ is equal to the Jacobian determinant of the Gauss map at the point p.

For computational purposes, it is convenient to have a formula for the curvature of the graph of a function from $\mathbb{R}^d$ to $\mathbb{R}$.

A number of formulae are available in the literature. If Q is a homogeneous quadratic form, we let $||Q||$ denote the determinant of the Hessian matrix of Q; to avoid confusion, we point out that the diagonal elements a_{ii} of this Hessian are twice the coefficient of x_i^2 in Q. Note that $||Q||$ may be computed with respect to any orthonormal basis. For our purposes, the following formulae are the most useful, the first of which is proved in Baryshnikov, Brady, Bressler, and Pemantle (2010).

Proposition 9.5.1 ((Baryshnikov, Brady, et al., 2010, Corollary 2.4)) *Let $\mathcal{P}$ be the tangent plane to $\mathcal{V}$ at p, and let $\mathbf{v}$ be a unit normal. Suppose that $\mathcal{V}$ is the graph of a smooth function h over $\mathcal{P}$, i.e.,*

$$\mathcal{V} = \{p + \mathbf{u} + h(\mathbf{u})\mathbf{v} : \mathbf{u} \in U \subseteq \mathcal{P}\}.$$

Let Q be the quadratic part of h, i.e., $h(\mathbf{u}) = Q(\mathbf{u}) + O(|\mathbf{u}|^3)$. Then the curvature of $\mathcal{V}$ at p is given by

$$\mathcal{K} = ||Q||. \qquad \square$$

Corollary 9.5.2 (curvature of the zero set of a polynomial) *Suppose $\mathcal{V}$ is the set $\{\mathbf{x} : H(\mathbf{x}) = 0\}$ and suppose that p is a smooth point of $\mathcal{V}$, i.e., $\nabla H(p) \neq \mathbf{0}$. Let ∇ and Q denote respectively the gradient and quadratic part of H at p. Let $Q_\perp$ denote the restriction of Q to the hyperplane $\nabla_\perp$ orthogonal to ∇. Then the curvature of $\mathcal{V}$ at p is given by*

$$\mathcal{K} = \frac{||Q_\perp||}{(\sum_{j=1}^d H_j(\mathbf{x})^2)^{d/2}}. \tag{9.5.2}$$

Proof Replacing H by $|\nabla|^{-1}H$ leaves $\mathcal{V}$ unchanged and reduces to the case $|\nabla H(p)| = 1$; we therefore assume without loss of generality that $|\nabla| = 1$. Letting $\mathbf{u}_\perp + \lambda(\mathbf{u})\nabla$ denote the decomposition of a generic vector $\mathbf{u}$ into components in $\langle\nabla\rangle$ and $\nabla_\perp$, the Taylor expansion of H near p is

$$H(p + \mathbf{u}) = \nabla \cdot \mathbf{u} + Q_\perp(\mathbf{u}) + R,$$

where $R = O(|\mathbf{u}_\perp|^3 + |\lambda(\mathbf{u})||\mathbf{u}_\perp|)$. Near the origin, we solve for λ to obtain a parametrization of $\mathcal{V}$ by $\nabla_\perp$:

$$\lambda(\mathbf{u}) = Q_\perp(\mathbf{u}) + O(|\mathbf{u}|^3).$$

The result now follows from Proposition 9.5.1. $\square$

Gaussian Curvature at Minimal Points of Complex Hypersurfaces. Suppose now that H is a real polynomial in $d+1$ variables and that p is a minimal smooth point of the corresponding complex algebraic hypersurface. We are interested in the curvature at $\log p$ of the logarithmic image $\log \mathcal{V} := \{x : H \circ \exp(x) = 0\}$ of $\mathcal{V}$. The formula (9.5.2) from Corollary 9.5.2 is well defined up to a factor of ± 1, unless the denominator vanishes, which certainly may happen in general. However, when p is minimal, i.e., $p \in \partial B$ for a component B of $\texttt{amoeba}(H)^c$, then the gradient of $H \circ \exp$ is a complex scalar multiple of a real vector. This prevents the denominator from vanishing. When $F = G/H$, we may multiply top and bottom by a unit complex number so that $\nabla_{\log} H$ is real; we then define the complex curvature by the same equation, (9.5.2), with the positive square root chosen. When d is odd, the sign of the curvature corresponds to the unit normal in the direction $\hat{r}_*(p) := \nabla_{\log} H / |\nabla_{\log} H|$.

It is useful to observe that the curvature $\mathcal{K}$ is a reparametrization of the Hessian determinant in Theorem 9.2.7 and its corollaries, in the sense that they vanish together.

Proposition 9.5.3 *The quantity $\mathcal{K}$ defined by* (9.5.2) *vanishes if and only if the quantity $|\mathcal{H}|$ from Theorem 9.2.7 vanishes.*

Proof The Hessian matrix $\mathcal{H}$ in Theorem 9.2.7 and its corollaries is the matrix of second partial derivatives of the function g expressing $\log \mathcal{V}$ as a graph over the first $(d-1)$ coordinates at a point where $\partial(H \circ \exp)/\partial x_d$ does not vanish. At such a point, the tangent plane to $\log \mathcal{V}$ is not perpendicular to the x_d-plane, and reparametrizing the graph to be over the tangent plane does not change whether the Hessian is singular. The new Hessian matrix represents the quadratic form Q in Proposition 9.5.1. It follows that singularity of the Hessian matrix from Theorem 9.2.7 is equivalent to singularity of Q in Proposition 9.5.1. The general definition of $\mathcal{K}$ is defined by the formula (9.5.2) with nonvanishing denominator, which therefore also vanishes together with $|\mathcal{H}|$. □

Theorem 9.5.4 (curvature version) *Suppose for each unit vector $\hat{r}$ in a compact subset* K *of the positive unit sphere, there is an isolated, strictly minimal, smooth, critical point $z_*(\hat{r})$ for the function $h_{\hat{r}}$ on $\mathcal{V}$. Let $\mathcal{K}(\hat{r})$ denote the curvature of* $\log \mathcal{V}$ *at* $\log z_*(\hat{r})$*, and suppose this does not vanish for $\hat{r} \in$ K. Then*

$$a_r = \left(\frac{1}{2\pi |r|} \right)^{d/2} z_*^{-r} \, \mathcal{K}^{-1/2} \, \frac{G(z_*)}{|\nabla_{\log} H(z_*)|} + O\left(|r|^{(-d+1)/2} \right) \tag{9.5.3}$$

uniformly as $|r| \to \infty$ with $\hat{r} \in$ K. Here the $-1/2$ power is taken to be the reciprocal of the product of the principal square roots of the eigenvalues of Q in the negative $\hat{r}$ *direction.*

Proof By Cauchy's integral formula, and Theorem 9.4.2 with $\omega = z^{-r} F(z) dz/z$,

$$a_r = \left(\frac{1}{2\pi i} \right)^d \int_\sigma \mathrm{Res}(\omega),$$

where σ is the intersection class of T on $\mathcal{V}$. Let $\mathbf{z} = \exp(\zeta)$ so $d\mathbf{z} = \mathbf{z}d\zeta$, so, pulling a factor of $\mathbf{z}^{-\mathbf{r}} = \exp(-\mathbf{r} \cdot \zeta)$ through the residue, we get

$$a_{\mathbf{r}} = \left(\frac{1}{2\pi i}\right)^d \int_{\tilde{\sigma}} \exp(-\mathbf{r} \cdot \zeta) \operatorname{Res}(\tilde{\mathbf{F}}(\zeta)\, d\zeta), \tag{9.5.4}$$

where $\tilde{\mathbf{F}} = F \circ \exp$ and $\tilde{\sigma} = \log \sigma$.

Let $\mathcal{P} := T_{\zeta_*} \log \mathcal{V} := T_{\log \mathbf{z}_*} \log[\mathcal{V}]$ be the logarithmic tangent space. Recall that the (complex) orthogonal complement to $\mathcal{P}$ is $\mathbb{C} \cdot \hat{\mathbf{r}}$. Near ζ_* we may parametrize $\log \mathcal{V}$ by $\mathcal{P}$. Locally,

$$\log \mathcal{V} = \{\zeta_* + \zeta_\| + h(\zeta_\|)\hat{\mathbf{r}} : \zeta_\| \in \mathcal{P}\}.$$

Given ζ, we may project back to the tangent plane by $\zeta_\| := \mathbf{x} - [\hat{\mathbf{r}} \cdot (\zeta - \zeta^*)\hat{\mathbf{r}}]$. Pick an orthonormal basis $\mathbf{v}^{(2)}, \ldots, \mathbf{v}^{(d+1)}$ for $\mathcal{P}$. We use these for local coordinates, writing the general point $\zeta \in \mathbb{C}^{d+1}$ in a neighborhood of ζ_* as

$$\zeta = \zeta_* + u_1 \hat{\mathbf{r}} + \sum_{j=2}^{d+1} u_j \mathbf{v}^{(j)}.$$

Computing the residue of $F(\zeta)\, d\zeta$ in these coordinates, using Proposition 9.3.3, gives

$$\operatorname{Res}(F(\zeta)\, d\zeta) = \frac{G \circ \exp}{\partial H \circ \exp / \partial u_1}\, du_2 \wedge \cdots \wedge du_{d+1}.$$

The partial derivative in the direction of the gradient is the magnitude of the gradient. Therefore, evaluating at ζ_*,

$$\operatorname{Res}(F(\zeta)\, d\zeta)(\zeta_*) = \frac{G(\mathbf{z}_*)}{|\nabla_{\log} H(\mathbf{z}_*)|}\, dA \tag{9.5.5}$$

where $dA := d\mathbf{u}_\| = du_2 \wedge \cdots \wedge du_{d+1}$ is equal to the oriented holomorphic d-area form for $\log \mathcal{V}$ as it is immersed in $\mathbb{C}^{d+1}$.

To understand the other factor in the integrand of (9.5.4), denoting $\lambda := |\mathbf{r}|$ and $\phi(\zeta) := \hat{\mathbf{r}} \cdot \zeta$, equation (9.5.4) becomes

$$a_{\mathbf{r}} = \left(\frac{1}{2\pi i}\right)^d \int_{\tilde{\sigma}} \exp(-\lambda\phi(\zeta)) \operatorname{Res}(\tilde{\mathbf{F}}(\zeta)\, d\zeta). \tag{9.5.6}$$

Let Q denote the quadratic part of h. By Proposition 9.5.1 (or Corollary 9.5.2) and the subsequent discussion, we see that the curvature $\mathcal{K}$ of $\log \mathcal{V}$ at the point ζ_* with respect to the unit normal $\hat{\mathbf{r}}$ is given by $||Q||$. As we will see, the eigenvalues of Q do not have positive real parts, which corresponds to the fact that we will not be integrating in the real direction, so we will need one more change of coordinates to apply results on multivariate, quadratically nondegenerate, stationary phase integrals.

Let us compute $\tilde{\sigma}$. There is a lot of freedom when choosing $\mathbf{x}'$ and $\mathbf{x}''$ in the construction of σ in Lemma 9.4.1. A convenient choice is to make the segment $\overline{\mathbf{x}'\mathbf{x}''}$ parallel to $\hat{\mathbf{r}}$. The real tangent space to $\log \mathbf{H}$ is then the sum of the imaginary

$(d+1)$-space and the real 1-space in direction $\hat{\boldsymbol{r}}$. The tangent space to $\log \mathcal{V}$ is the sum of the real d-space orthogonal to $\hat{\boldsymbol{r}}$ and the imaginary d-space orthogonal to $\hat{\boldsymbol{r}}$. The tangent space to $\tilde{\sigma}$ is the intersection of these, which is the imaginary $(d-1)$-space orthogonal to $\hat{\boldsymbol{r}}$, which is just $\operatorname{Im} \mathcal{P}$. Represent $\tilde{\sigma}$ as a graph over $\operatorname{Im} \mathcal{P}$. Because $\tilde{\sigma}$ is contained in the linear space $\operatorname{Im} \mathcal{P} + \mathbb{C} \cdot \hat{\boldsymbol{r}}$, we see that locally there is a unique analytic function $\alpha : \operatorname{Im} \mathcal{P} \to \mathbb{C} \cdot \hat{\boldsymbol{r}}$ such that $\zeta + \alpha(\zeta) \in \tilde{\sigma}$. Comparing to the previous parameterization, we see that $\alpha = h$. The quadratic part of α is therefore equal to Q. Our multivariate integral formulae are in terms of real parametrizations. We therefore reparametrize $\operatorname{Im} \mathcal{P}$ by $\zeta = i\boldsymbol{y}$ and $d\zeta = i^d \, d\boldsymbol{y}$. In these coordinates, locally

$$\tilde{\sigma} = \{ i\boldsymbol{y} + h(i\boldsymbol{y}) : \boldsymbol{y} \in \operatorname{Re} \mathcal{P} \} . \tag{9.5.7}$$

Using $\hat{\boldsymbol{r}} \cdot \boldsymbol{y}_{\|} = 0$ and $\hat{\boldsymbol{r}} \cdot \hat{\boldsymbol{r}} = 1$, we obtain

$$\begin{aligned} \phi(i\boldsymbol{y} + h(i\boldsymbol{y})) &= \phi(\zeta_*) + h(i\boldsymbol{y}) \\ &= \phi(\zeta_*) + Q(i\boldsymbol{y}) + O(|\boldsymbol{y}|^3) \\ &= \phi(\zeta_*) - Q(\boldsymbol{y}) + O(|\boldsymbol{y}|^3) . \end{aligned}$$

We know, by minimality, that ϕ is a smooth phase function whose real part has a minimum on $\tilde{\sigma}$ at ζ_*, which is $\boldsymbol{y} = 0$ in the parametrization (9.5.7). Applying Theorem 5.4.8 to (9.5.6) using the evaluation (9.5.5) then gives

$$a_{\boldsymbol{r}} = \left(\frac{1}{2\pi} \right)^{d/2} \boldsymbol{z}_*^{-\boldsymbol{r}} \left(\det(-Q) \right)^{-1/2} \frac{G(\boldsymbol{z}_*)}{|\nabla_{\log} H(\boldsymbol{z}_*)|} + O\left(|\boldsymbol{r}|^{(-d+1)/2} \right) ,$$

where the square root is taken to be the reciprocal of the product of the principal square roots of the eigenvalues of $-Q$ in the positive $\hat{\boldsymbol{r}}$-direction, all of which have nonnegative real parts. The eigenvalues of $-Q$ in direction $\hat{\boldsymbol{r}}$ are the same as the eigenvalues of Q in direction $-\hat{\boldsymbol{r}}$, which finishes the proof of the theorem. □

Again, we may expand without difficulty to include the case where there are finitely many critical points on a minimizing torus.

Corollary 9.5.5 *Let $F = G/H$ be a quotient of Laurent polynomials. Suppose that for each $\hat{\boldsymbol{r}}$ in a compact subset* K *of the positive unit sphere, the function $\hat{\boldsymbol{r}} \cdot \boldsymbol{x}$ is uniquely maximized at $\boldsymbol{x}_{\min} \in \overline{B}$, a component of the complement of* amoeba(H)*, and that the set* $\mathbf{W}$ *of critical points for $\hat{\boldsymbol{r}} \cdot \boldsymbol{x}$ in $\mathcal{V}_1 := \mathcal{V} \cap T(\boldsymbol{x}_{\min})$ is finite and non-empty for all $\hat{\boldsymbol{r}} \in$ K. Let $\mathcal{K}(\boldsymbol{z})$ denote the curvature at $\log \boldsymbol{z}$ of $\log \mathcal{V}$, and suppose this is nonvanishing for all $\boldsymbol{r} \in$ K. For each $\boldsymbol{z} \in \mathbf{W}(\hat{\boldsymbol{r}})$, we denote $\boldsymbol{z} = \exp(\boldsymbol{x}_{\min} + i\boldsymbol{y})$. Then*

$$a_{\boldsymbol{r}} = \left(\frac{1}{2\pi |\boldsymbol{r}|} \right)^{d/2} e^{-\boldsymbol{r} \cdot \boldsymbol{x}} \left[\sum_{\boldsymbol{z} \in \mathbf{W}(\hat{\boldsymbol{r}})} e^{-i\boldsymbol{r} \cdot \boldsymbol{y}} \frac{G(\boldsymbol{z})}{\nabla_{\log} H(\boldsymbol{z})} \mathcal{K}(\boldsymbol{z})^{-1/2} + O(|\boldsymbol{r}|^{-1/2}) \right]$$

uniformly as $|\boldsymbol{r}| \to \infty$ with $\hat{\boldsymbol{r}} \in$ K.

Example 9.5.6 (quantum random walk) A quantum random walk (QRW) is a model for a particle moving in $\mathbb{Z}^d$ under a quantum evolution in which the randomness is provided by a unitary evolution operator on a hidden variable taking k states. States and position are simultaneously measurable, but one must not measure either until the final time n or the quantum interference is destroyed. Examples of specific computations are given in Chapter 12. Here we consider the general form of the answer and what the results of Theorem 9.5.4 and Corollary 9.5.5 tell us qualitatively about the probability profile of the particle at time n for large n.

A QRW is defined by a $k \times k$ unitary matrix U along with k vectors $\boldsymbol{v}^{(1)}, \ldots, \boldsymbol{v}^{(k)}$ in $\mathbb{Z}^d$ representing possible steps of the walk. At each time step, the particle chooses a new state $j \in [k]$ and then moves by a jump of $\boldsymbol{v}^{(j)}$. The amplitude of a transition from state i to j is $U(i, j)$. The amplitude of a path of n steps, starting in state i_0 and ending in states i_n, is $\prod_{t=0}^{n-1} U(i_t, i_{t+1})$. Suppose the particle is known to start at the origin at time zero, in state i. The amplitude of moving from $\mathbf{0}$ to $\boldsymbol{r}$ in n time steps and ending in state j is obtained by summing the amplitudes of all paths of n steps having total displacement $\boldsymbol{r}$ and ending in state j. This gives us all we need to compute; for more on the interpretation of quantum walks, see Ambainis et al. (2001) and Baryshnikov, Brady, et al. (2010).

The multiplicative nature of the amplitudes makes QRW a perfect candidate for the transfer matrix method, the univariate version of which was discussed in Section 2.2 and whose multivariate version is discussed at length in Section 12.4. Here, we just quote the relevant result. Let M denote the $k \times k$ diagonal matrix whose (j, j)-entry is the monomial $\boldsymbol{z}^{\boldsymbol{v}^{(j)}}$. Let $P(\boldsymbol{r}, n)$ be the matrix whose (i, j)-entry is the amplitude to go from the origin in state i at time zero to $\boldsymbol{r}$ in state j at time n. Define the spacetime matrix generating function by

$$F(\boldsymbol{z}) := \sum_{\boldsymbol{r},n} P(\boldsymbol{r}, n)(\boldsymbol{z}^\circ)^{\boldsymbol{r}} z_{d+1}^n \,. \tag{9.5.8}$$

Here $\boldsymbol{z}^\circ = z_1, \ldots, z_d$ are d space variables, and z_{d+1} is the time variable. The transfer matrix method easily gives

$$F(\boldsymbol{z}) = (I - z_{d+1} M U)^{-1},$$

as is given in Baryshnikov, Brady, et al. (2010, (2.2)). The entries F_{ij} are rational functions with common denominator

$$H := \det(I - z_{d+1} M U),$$

which is known (Baryshnikov, Brady, et al., 2010, Proposition 2.1) to satisfy the torality hypothesis (9.2.1) on the unit torus, $\mathbf{T}(0, \ldots, 0)$.

Whenever $\mathbf{0} \in \partial B$, we may deduce that $\beta^* \leq 0$ with equality only for $\boldsymbol{r} \in \mathbf{N}(\mathbf{0})$. In fact, from Theorem 8.4.2, we know that β^* is strictly less than zero unless $\boldsymbol{r} \in \mathbf{N}$ and $\texttt{local}(\boldsymbol{r})$ is non-empty. When $\texttt{local}(\boldsymbol{r})$ is non-empty, then we have asymptotics governed by Corollary 9.5.5, which decay polynomially. Define the feasible velocity region $R \subseteq \mathbb{R}^d$ to be the set of all $(r_1, \ldots, r_d)$ such that

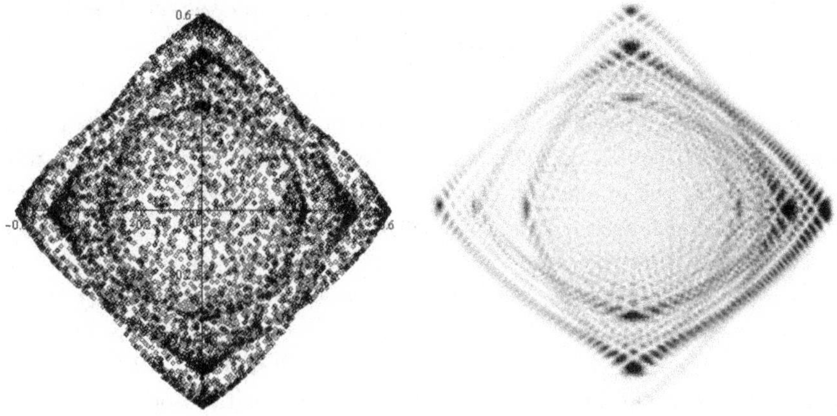

(a) Application of Gauss map to an embedded grid (b) Probability amplitudes

Figure 9.3 The $S(1/8)$ QRW.

$\overline{\beta}(r_1, \ldots, r_d, 1) = 0$; in other words, it is the set of directions in which the chance of finding the particle roughly at that rescaled point after a long time decays slower than exponentially. We then see that R is the set of all $\boldsymbol{r} \in \mathbf{N}(\mathbf{0})$ such that $\texttt{local}(\boldsymbol{r}) \neq \emptyset$.

To compute R, one may start by computing $\mathcal{V}_0 := \mathcal{V} \cap \mathbf{T}(\mathbf{0})$. For many QRWs, one finds this to be a smooth manifold diffeomorphic to one or more d-tori. At any smooth point $z \in \mathcal{V}_0$, the space $\mathbf{L}(z)$ is the line in the direction of $\nabla_{\log} H(z)$. Thus $\boldsymbol{r} \in R$ if and only if $\boldsymbol{r}$ is in the closure of the image when the logarithmic Gauss map $\nabla_{\log}$ is applied to $\mathcal{V}_0$:

$$R = \overline{\nabla_{\log}[\mathcal{V}_0]}\,.$$

This allows us to plot the feasible region by parametrizing the torus $\mathcal{V}_0$ by an embedded grid of some kind and applying $\nabla_{\log}$ to each point of the embedded grid. The result of this is shown in Figure 9.3 (a). Figure 9.3 (b) shows an intensity plot of the magnitude of the probability amplitude for the particle at time 200 for a QRW known as $S(1/8)$. The agreement of the empirical amplitudes on the right with the theoretical prediction based on the Gauss map is apparent.

What is also apparent is that not only do the regions agree, but their fine structure of darker bands and light areas agree as well. A moment's thought explains this. The image of the torus $\mathcal{V}_0$ under $\nabla_{\log}$ will be more intense in places where the Jacobian determinant of $\nabla_{\log}$ is small because the density of the image of an embedded grid is proportional to the inverse of the Jacobian determinant. The Jacobian determinant of the logarithmic Gauss map is precisely the curvature, as is clear from (9.5.1). In Theorem 9.5.4, although the $G/|\nabla_{\log} H|$ term varies a little, the dominant factor is the curvature term $\mathcal{K}^{-1/2}$. This explains why the density of the Gauss-mapped grid is a good surrogate for the probability amplitudes. ■

9.5.2 An Explicit Formula for Quadratically Nondegenerate Points When $d = 2$

The appearance of the term $\det(\mathcal{H}(\hat{\mathbf{r}}))^{-1/2}$, although conceptually powerful, prevents easy application of the formulae in Section 9.2. If one is given a rational function $F = G/H$, it would be useful to have an estimate for $a_{\mathbf{r}}$ that did not rely on the computation of a determinant of a matrix of second partial derivatives of a function defined in terms of the parametrization of $\mathcal{V}$ and a logarithmic change of coordinates. We turn now to some subcases in which explicit formulae may be written down in terms of the elementary data, namely G and H and their partial derivatives.

The first nontrivial case is a single strictly minimal smooth point in two variables. The following formula, first stated in Pemantle and Wilson (2002, Theorem 3.1), gives the leading asymptotic in terms of the partial derivatives of the numerator and denominator of the given rational function. We replicate those computations here.

Theorem 9.5.7 (smooth, $d = 2$, nondegenerate) *Let $F = G/H$ be meromorphic, and suppose that as $\hat{\mathbf{r}}$ varies in a neighborhood $\mathcal{N}$ of $\hat{\mathbf{r}}_*$, there is a smoothly varying, strictly minimal, smooth critical point $\mathbf{z}(\hat{\mathbf{r}}_*)$ in direction $\hat{\mathbf{r}}_*$. Denoting $\mathbf{z} = (x, y)$ and $\mathbf{r} = (r, s)$, we suppose also that $G(\mathbf{z}(\hat{\mathbf{r}}))$ and the expression*

$$Q(\mathbf{z}(\hat{\mathbf{r}})) := -y^2 H_y^2 x H_x - y H_y x^2 H_x - x^2 y^2 (H_y^2 H_{xx} + H_x^2 H_{yy} - 2 H_x H_y H_{xy}) \tag{9.5.9}$$

are nonzero for each $\hat{\mathbf{r}} \in \mathcal{N}$. Then as $|\mathbf{r}| \to \infty$, uniformly over $\hat{\mathbf{r}} \in \mathcal{N}$,

$$a_{\mathbf{r}} = \left(G + O\left(s^{-1/2}\right)\right) \frac{1}{\sqrt{2\pi}} x^{-r} y^{-s} \sqrt{\frac{-y H_y}{s\, Q}} \tag{9.5.10}$$

evaluated at $\mathbf{z}(\hat{\mathbf{r}})$. The square root should be taken to be the principal square root of $(-yH_y)^3/Q$, divided by $-yH_y$.

Proof Applying Theorem 9.2.7, with $d = 2$, and using the bivariate notation $\mathbf{z} = (x, y)$, we see that the quantity $(2\pi r_k)^{(1-d)/2} G(\mathbf{z})/z_k H_k$ is equal to $\frac{G}{2\pi\, s} x^{-r} y^{-s}/(yH_y)$. Thus we need only to show that the reciprocal of the determinant of the Hessian matrix for ϕ_* at zero is equal to $(-yH_y)^3/Q$. Because $d - 1 = 1$, the matrix is one-dimensional. Hence we require

$$Q = (-yH_y)^3\, \phi_*''(0)\,, \tag{9.5.11}$$

which will show as well that the nonvanishing of Q is equivalent to the nonvanishing of $\phi''(0)$. Going back to the definition of ϕ_* in Equation 9.2.8, we see that $x = z_*^\circ e^{i\theta} = x_* e^{i\theta}$ so that $\phi_*(\theta) = \log g(x) + L$, where L is linear and

$d/d\theta = ix(d/dx)$. Thus

$$\begin{aligned}\frac{d^2}{d\theta^2}\phi_* &= ix\frac{d}{dx}\left(ix\frac{d}{dx}(\log g)\right)\\ &= -x\frac{d}{dx}\frac{x\,g'}{g}\\ &= -x\,\frac{g' + xg''}{g} + \frac{x^2(g')^2}{g^2}\,. \end{aligned} \tag{9.5.12}$$

The derivatives of g may be computed by implicitly differentiating the equation $H(x, g(x)) = 0$. Differentiating twice,

$$H_x + H_y\, g' = 0$$

$$H_{xx} + 2g'\, H_{xy} + g''\, H_y + (g')^2\, H_{yy} = 0\,.$$

Solving the first equation for g' gives $g' = -H_x/H_y$, after which solving the second equation for g'' gives

$$g'' = -\frac{1}{H_y}\left(H_{xx} - 2\frac{H_x}{H_y}H_{xy} + \frac{H_x^2}{H_y^2}H_{yy}\right).$$

Plugging this into (9.5.12) gives

$$\phi_*''(0) = \frac{x\left(xH_x^2H_y + g(x)\cdot[H_xH_y^2 + xH_{xx}H_y^2 - 2xH_xH_yH_{xy} + xH_x^2H_{yy}]\right)}{H_y^3\, g(x)^2}.$$

Multiplying by $(-yH_y)^3$ and evaluating at $\theta = 0$, $x = x_*$, $g(x) = y_*$ and comparing to the definition of Q in (9.5.9) establishes that $Q = (-yH_y)^3$ and finishes the proof. □

In the same way that Corollary 9.2.8 extends Theorem 9.2.7, we may extend Theorem 9.5.7 to finitely many critical points and arbitrary $\mathcal{V}_*$.

Corollary 9.5.8 *Assume the hypotheses of Theorem 9.5.7 but replace the assumption of strict minimality by the assumption that $\hat{\mathbf{r}}$ is nonflat and there is a finite nonempty set* **W** *of critical points for $h_{\hat{\mathbf{r}}}$ in $\mathcal{V}_*$. Then*

$$a_r = \sum_{(x,y)\in\mathbf{W}} \frac{G(x,y)}{\sqrt{2\pi}}\, x^{-r}y^{-s}\sqrt{\frac{-yH_y(x,y)}{s\,Q(x,y)}} + O\left(s^{-1}\left|x^{-r}y^{-s}\right|\right).$$

Example 9.5.9 (binomial coefficients continued once more) Having the formula (9.5.9), we may ignore the beginnings of the saddle-point computation in Examples 9.2.6 and 9.3.4 and plug $\mathbf{z}_* = (x_*, y_*) = \hat{\mathbf{r}} = (\frac{r}{r+s}, \frac{s}{r+s})$ from Example 8.3.1 directly into Theorem 9.5.7. With $G = 1$ and $yH_y = -y = -s/(r+s)$, we get $Q = xy(x+y)$. Evaluating at $(x, y) = (x_*, y_*)$ gives $x + y = 1$ and $Q = (rs)/(r+s)^2$. Then,

$$a_r \sim \left(\frac{r+s}{r}\right)^r \left(\frac{r+s}{s}\right)^s \sqrt{\frac{2\pi(r+s)}{(2\pi r)(2\pi s)}}.$$

This the usual approximation via Stirling's formula to the binomial coefficient $\binom{r+s}{s}$. ■

Example 9.5.10 (Delannoy numbers continued yet once more) Again, we go back to Example 2.4.12. We recall from (8.3.4) that

$$(x_*, y_*) = \left(\frac{\sqrt{r^2+s^2}-s}{r}, \frac{\sqrt{r^2+s^2}-r}{s}\right) \quad \text{or}$$

$$\left(\frac{-\sqrt{r^2+s^2}-s}{r}, \frac{-\sqrt{r^2+s^2}-r}{s}\right).$$

The first of these is a strictly minimal critical point. Substituting this point into Formula 9.5.9 for Q would lead to an ugly expression that Maple cannot simplify easily. Instead we solve directly for $-yH_y/Q$ by adding this into the critical point equations. Thus, after defining `H, Hx, Hy, Hxx, Hxy, Hyy` and `Q`, we compute

```
Groebner[Basis] ([s*x*(1+y)-r*y*(1+x),1-x-y-x*y,
                 W^2*Q+y*Hy], plex(x,y,W));
```

The first element of this is the elimination polynomial for the quantity $W := \sqrt{-yH_y/Q}$. Maple tells us this is

$$-s^2r + \left(-4\,r^3 - 4\,r^2s - 4\,s^2r - 4\,s^3\right)W^2 + \left(4\,s^2r + 4\,r^3\right)W^4.$$

After some simplification, denoting $\rho := \sqrt{r^2+s^2}$, this yields

$$a_{r,s} \sim \left(\frac{r}{\rho - s}\right)^r \left(\frac{s}{\rho - r}\right)^s \sqrt{\frac{1}{2\pi\rho}\frac{\sqrt{rs}}{r+s-\rho}}.$$

■

Example 9.5.11 (Chebyshev polynomials) Let $F(z, w) = 1/(1 - 2zw + w^2)$ be the generating function for Chebyshev polynomials of the second kind (Comtet, 1974), asymptotics for which are well known and easy to derive by other means. To use Corollary 9.5.8, first find the minimal points for the direction (r, s). They are $(i(\beta - \beta^{-1})/2, i\beta)$ for $\beta = \pm\sqrt{\frac{s-r}{s+r}}$. Computing $Q = 4a^2(1-a^2)$ and summing the two contributions then gives

$$a_{rs} \sim \sqrt{\frac{2}{\pi}}(-1)^{(s-r)/2}\left(\frac{2r}{\sqrt{s^2-r^2}}\right)^{-r}\left(\sqrt{\frac{s-r}{s+r}}\right)^{-s}\sqrt{\frac{s+r}{r(s-r)}},$$

when $r + s$ is even and zero otherwise. These asymptotics are uniform, as r/s varies over each compact subset of $(0, 1)$. ■

9.5.3 A Formula for Degenerate Points When $d = 2$

Example 9.4.8 gave a derivation by hand of bivariate smooth point asymptotics in a direction where the phase function ϕ vanished to order 4. The present derivation gives a general formula for arbitrary orders. For simplicity, we make two restrictions; namely, we assume the dominant singularities are minimal points, and we assume G is nonvanishing on at least one of these points. Both of these fail in Example 9.4.8, which is therefore not covered by our general result.

Fix $d = 2$ and $\hat{\boldsymbol{r}} \in \Xi$ and let (x_*, y_*) be a minimal point for $h = h_{\hat{\boldsymbol{r}}}$ on $\mathcal{V}$. Assume without loss of generality that $\partial H/\partial y(x_*, y_*) \neq 0$, and let $y = g(x)$ parametrize $\mathcal{V}$ near (x_*, y_*). Then (9.2.8) – (9.2.9) defines the functions ϕ and A on a neighborhood of zero in $\mathbb{R}$. Theorem 9.2.1 and Proposition 9.2.5 give

$$x_*^r y_*^s (a_{rs} - \chi) = O\left(e^{-\varepsilon s}\right),$$

where

$$\chi(r, s) = x_*^{-r} y_*^{-s} \frac{1}{2\pi} \int_{-\varepsilon}^{\varepsilon} e^{-s\phi(\theta)} A(\theta)\, d\theta\,. \tag{9.5.13}$$

Let $c = c_\kappa$ denote the leading coefficient of ϕ, i.e., $\phi(x) \sim c_\kappa x^\kappa$ as $x \to 0$. Define the quantity

$$\Phi_z^{(k)}(\boldsymbol{r}) := -\frac{\Gamma(1/\kappa)}{2\kappa\pi}(1 - \zeta)\frac{G(x_*, y_*)}{y_* \partial H/\partial y(x_*, y_*)} c^{-1/\kappa} s^{-1/k} x^{-r} y^{-s}\,, \tag{9.5.14}$$

where, as in Theorem 4.1.1 part (iii), $\zeta = -1$ if κ is even and $\zeta = \exp(\sigma i\pi/\kappa)$ if κ is odd.

Theorem 9.5.12 *Let $W \neq \emptyset$ denote the set of minimal points in direction $\hat{\boldsymbol{r}}$, and for each $(x, y) \in W$, let $\kappa(x, y) \geq 2$ denote the degree of vanishing of ϕ at (x, y). Then as $(r, s) \to \infty$ with the distance from (r, s) to the ray $\{t\hat{\boldsymbol{r}} : t \geq 0\}$ remaining bounded, there is an asymptotic series of the form*

$$a_{rs} \sim x_*^{-r} y_*^{-s} \sum_{j=0}^{\infty} \nu_j s^{(-1-j)/k}.$$

If $\sum_{z\in W} \Phi_z^{(k(z))}(\boldsymbol{r}) \neq 0$, then $\sum_{z\in W} \Phi_z^{(\kappa(z))}(\boldsymbol{r})$ is the leading term of the series.

Proof The asymptotic development is clear from (9.5.13) and Theorem 4.1.1. It remains to check that the leading term is given by (9.5.14). Starting from (9.5.13), fix $(x, y) \in W$, and use Theorem 4.1.1 with $\ell = 0$ to evaluate the leading term.

In the notation of Theorem 4.1.1, the leading term is

$$x^r y^s \chi \sim \frac{1}{2\pi} \int_{-\varepsilon}^{\varepsilon} A(x) e^{-s\phi(x)}\, dx$$

$$= \frac{1}{2\pi} I(s)$$

$$= \frac{1}{2\pi}(1-\zeta)C(\kappa, 0)A(0)(cs)^{-1/\kappa}.$$

Parametrizing $(x_1, x_2) = (x, y)$ by y means choosing coordinate $k = 2$, leading to the sign $(-1)^{k-1} = -1$ in $A(0) = -\dfrac{G(x_*, y_*)}{y_* \partial H/\partial y(x_*, y_*)}$. Recalling that $C(\kappa, 0) = \dfrac{\Gamma(1/\kappa)}{\kappa}$ and summing over $z \in W$ gives (9.5.14). □

Example 9.5.13 (cube root asymptotics) Let $F(x, y) = 1/(3 - 3x - y + x^2)$. The set $\mathcal{V}$ is parametrized via $y = g(x) := x^2 - 3x + 3$. The point $(1, 1) \in \mathcal{V}$ is the unique point where there is a degenerate critical point. Indeed, for directions above the diagonal, Theorem 9.5.7 may be used at the minimal points $\{(x, g(x)) : 0 < x < 1\}$, whereas each direction below the diagonal corresponds to a pair of complex minimal points fitting the hypotheses of Corollary 9.5.8. The result is that the coefficients decay exponentially at a rate that is uniform over compact subsets of directions not containing the diagonal.

The interesting behavior is on the diagonal. Here, the relevant critical point is $(1, 1)$ and the exponential term $\mathbf{1}^{-r}$ in the asymptotics drops out, yielding growth or decay i.e., at most polynomial. Computing $g''(0)$ via implicit differentiation gives

$$g''(x) = -3\frac{x(x^2 - 4x + 3)}{(x^2 - 3x + 3)^2}.$$

This vanishes when $x = 1$. Computing further, we find that g vanishes to order exactly 3 here, with $c := g'''(0)/3! = i$. Checking the signs gives $\zeta = -e^{i\pi/3}$, and therefore $i^{-1/3}(1 - \zeta) = e^{i\pi/6} + e^{-i\pi/6} = 2\cos(\pi/6) = \sqrt{3}$. Evaluating $A(0) = -G(1, 1)/yH_y(1, 1) = -1/(1 \cdot -1) = 1$, the result is a leading asymptotic term of

$$a_{r,r} \sim \frac{1}{2\pi} C(3, 0) i^{-1/3}(1 - \zeta) r^{-1/3} = \frac{\sqrt{3}\Gamma(1/3)}{6\pi} r^{-1/3}.$$

When $r = 100$, this gives $a_{100,100} \approx 0.053034$. The relative error between this and the true value of $0.052975\ldots$ is 0.111%. ■

Remark We have given a formula holding only very near the ray in direction $\hat{\mathbf{r}}$. Because the results for nondegenerate smooth points hold in a neighborhood of $\hat{\mathbf{r}}_*$, it remains to be seen whether asymptotics can be worked out that "bridge the gap," holding when the magnitude δ of $(\boldsymbol{r} - |\boldsymbol{r}|\hat{\boldsymbol{r}})$ satisfies $|\boldsymbol{r}| \gg \delta \gg 1$; see Section 13.4 for further discussion.

9.5.4 When the Pole has Multiplicity Greater than 1

As promised, we consider the asymptotics of the coefficients of $F = G/H$ when H is smooth but not squarefree. We require the following residue computation, which generalizes Proposition 9.3.3 to poles of higher order.

Lemma 9.5.14 *Let $dz_{\hat{k}}$ denote the $(d-1)$-form $dz_1 \wedge \cdots \wedge dz_{k-1} \wedge dz_{k+1} \wedge \cdots \wedge dz_d$. Wherever the functions Pz^{-r} and Q are analytic and Q is square free,*

$$\text{Res}\left(\frac{Pz^{-r}}{Q^\ell}\,dz\right) = \left[(-1)^{k-1}\binom{-r_k}{\ell-1}\frac{Pz^{-r}z_k^{-(\ell-1)}}{(\partial Q/\partial z_k)^\ell} + O\left(r_k^{\ell-2}\right)\right] dz_{\hat{k}}\,. \tag{9.5.15}$$

Proof We induct on ℓ. In the case $\ell = 1$, the right-hand side of (9.5.15) is

$$\left[(-1)^{k-1}\frac{Pz^{-r}}{(\partial Q/\partial z_k)} + O(r_k^{-1})\right] dz_{\hat{k}}\,.$$

Proposition 9.3.3, with $G = Pz^{-r}$, evaluates $\text{Res}(Pz^{-r}\,dz/Q)$ as exactly this, with no $O(r_k^{-1})$ term.

Now assume for induction that the lemma holds for $\ell - 1$. If η is meromorphic, then $\text{Res}(d\eta) = 0$. Apply this fact with $\eta = (-1)^{k-1}\dfrac{Pz^{-r}\,dz_{\hat{k}}}{(\ell-1)Q^{\ell-1}\partial Q/\partial z_k}$ to see that

$$\begin{aligned}
0 &= \text{Res}(d\eta) \\
&= \text{Res}\left[\frac{(\partial P/\partial z_k)z^{-r}}{(\ell-1)Q^{\ell-1}(\partial Q/\partial z_k)}dz + \frac{-r_k P z^{-r} z_k^{-1}}{(\ell-1)Q^{\ell-1}(\partial Q/\partial z_k)}dz\right. \\
&\qquad \left. - \frac{Pz^{-r}}{Q^\ell}dz - \frac{Pz^{-r}(\partial^2 Q/\partial z_k^2)}{(\ell-1)Q^{\ell-1}(\partial Q/\partial z_k)^2}dz\right].
\end{aligned}$$

The third term on the right is what we want. Isolating this yields

$$\text{Res}\left(\frac{Pz^{-r}}{Q^\ell}\,dz\right) = \text{Res}\left(\frac{-r_k P z^{-r} z_k^{-1}}{(\ell-1)Q^{\ell-1}(\partial Q/\partial z_k)}\,dz\right) + \text{Res}\left(\frac{Uz^{-r}}{Q^{\ell-1}}\,dz\right),$$

where U is an analytic function. Apply the induction hypothesis to the first residue on the right side with $\dfrac{-r_k}{\ell-1}\,\dfrac{P}{\partial Q/\partial z_k}$ in place of P and $\boldsymbol{r} + \delta_k$ in place of $\boldsymbol{r}$ to see that this residue is equal to

$$(-1)^{k-1}\left[\frac{-r_k}{\ell-1}\binom{-r_k-1}{\ell-2}\frac{(P/(\partial Q/\partial z_k))\,z^{-r-\delta_k}z_k^{-(\ell-2)}}{(\partial Q/\partial z_k)^{\ell-1}} + O\left(r_k^{\ell-3}\right)\right] dz_{\hat{k}}\,.$$

The induction hypothesis also shows that the second residue on the right side is $O\left(r_k^{\ell-2}\right)$. Combining powers of $(\partial Q/\partial z_k)$ and powers of z_k and simplifying $\frac{-r_k}{\ell-1}\binom{-r_k-1}{\ell-2}=\binom{-r_k}{\ell-1}$ gives the result. □

This enables the computation of asymptotics governed by quadratically nondegenerate smooth points when the denominator contains a power of order ℓ. If H factors as RQ^ℓ, wherever R does not vanish, we may replace G by G/R and assume without loss of generality that H is a perfect ℓ-power. The analogue of Theorem 9.3.2 for higher order poles is the following result.

Theorem 9.5.15 *Let $F = G/H^\ell$ and let z be a simple pole of $\mathcal{V}$ at which G is analytic. Define*

$$\Phi_z^{\text{non}}(\boldsymbol{r}) := (2\pi)^{(1-d)/2}\binom{-r_k}{\ell-1}(\det \mathcal{H}(\hat{\boldsymbol{r}}))^{-1/2}\frac{G(z)}{z_k^\ell(\partial H/\partial z_k)^\ell}. \tag{9.5.16}$$

Fix $\hat{\boldsymbol{r}}_*$ and suppose that c is the minimax height of the class C, and there is a set W such that $\sum_{z\in W} C(z)$ represents the projection of the class C to $H_d(\mathcal{M}, \mathcal{M}^{c-\varepsilon})$. Suppose that $h_{\boldsymbol{r}}$ is quadratically nondegenerate on W. Then in a neighborhood of $\hat{\boldsymbol{r}}_*$ there is an asymptotic expansion

$$a_{\boldsymbol{r}} = \sum_{j=\ell}^{\infty} \nu_j |\boldsymbol{r}|^{(1-d)/2+\ell-1} z^{-\boldsymbol{r}}$$

whose leading term is given by

$$\sum_{z\in W} \Phi_z^{\text{non}}(\boldsymbol{r}).$$

Proof Theorem 9.3.2 holds for any residue, and Lemma 9.3.6 concerns the topology of $\mathcal{V}$, not the power of H. Thus those two results remain valid in the present setup and show that

$$a_{\boldsymbol{r}} = \frac{1}{(2\pi i)^d}\sum_{z\in W}\int_{C(z)} \text{Res}\left(z^{-\boldsymbol{r}-\boldsymbol{1}}F(z)\,dz\right) + O\left(e^{(c-\varepsilon)|\boldsymbol{r}|}\right).$$

The only differences between this integral and the one in Theorem 9.3.7 are the constant factor $\binom{-r_k}{\ell-1}$ and the powers of ℓ on the factors z_k and $\partial H/\partial z_k$ in the denominator. These factors travel through the Fourier-Laplace integral, yielding the result. □

9.6 Limit Laws from Probability Theory

In this section we restrict our attention to the combinatorial case, namely $a_{\boldsymbol{r}} \geq 0$. Example 9.5.6 shows that one can have limit laws outside of this case with probabilistic interpretations. Nevertheless, the classical limit laws of probability theory, namely the laws of large numbers and central limit theorems, assume probabilities that are real and nonnegative.

In this case there is usually a combinatorial interpretation: $a_{\boldsymbol{r}}$ counts something indexed by $\boldsymbol{r}$. If there is a one-dimensional size parameter $\gamma(\boldsymbol{r})$, it often makes sense to ask about the typical behavior as the size parameter goes to infinity. The size parameter might be the sum $|\boldsymbol{r}| = r_1 + \cdots + r_d$ of the indices or the last index r_d. "Typical behavior" refers to a probabilistic interpretation.

A grand measure μ on $\mathbb{N}^d$ is defined by

$$\mu := \sum_{\boldsymbol{r}} a_{\boldsymbol{r}} \delta_{\boldsymbol{r}},$$

and its cross-sections are normalized to be probability measures:

$$\mu_k := \frac{1}{C_k} \sum_{\gamma(\boldsymbol{r})=k} a_{\boldsymbol{r}} \delta_{\boldsymbol{r}},$$

where $C_k := \sum_{\gamma(\boldsymbol{r})=k} a_{\boldsymbol{r}}$ is the normalizing constant.

One can ask for limit laws with varying degrees of subtlety. A weak law tells us that μ_k is concentrated on a region of diameter $o(k)$. More precisely, a weak law holds for the sequence $\{\mu_k\}$ with limit $\boldsymbol{m} \in (\mathbb{R}^+)^d$ if for any $\varepsilon > 0$,

$$\mu_k \left\{ \boldsymbol{r} : \left| \frac{\boldsymbol{r}}{k} - \boldsymbol{m} \right| > \varepsilon \right\} \to 0$$

as $k \to \infty$. More delicate is a central limit theorem. This gives information about the profile of μ_k in a region of diameter $O(k^{1/2})$ around $k\boldsymbol{m}$, namely that the shape of the distribution approaches a Gaussian . An ordinary central limit theorem gives information about sets on the scale of $k^{1/2}$. In particular,

$$\mu_k \left\{ \boldsymbol{r} : \frac{\boldsymbol{r} - k\boldsymbol{m}}{k^{1/2}} \in A \right\} \to \Phi(A)$$

as $k \to \infty$, where A is any nice region and Φ is a multivariate normal distribution. Even better is a local central limit theorem (LCLT), which estimates μ_k at individual points. An LCLT tells us that $\mu_k(\boldsymbol{r}) \sim \mathfrak{n}(\boldsymbol{r})$, where $\mathfrak{n}$ is the density of a multivariate normal. In the case where the size parameter is r_d, for example, the normal density looks like

$$\mathfrak{n}(\boldsymbol{r}) = (2\pi\, r_d)^{(1-d)/2} |M|^{1/2} \exp\left[(\boldsymbol{r}^\circ - r_d \boldsymbol{v})^T \, M \, (\boldsymbol{r}^\circ - r_d \boldsymbol{v}) \right],$$

where $\boldsymbol{r}^\circ$ denotes $(r_1, \ldots, r_{d-1})$, the vector $\boldsymbol{v}$ is the mean of the Gaussian distribution, and the matrix M is the inverse covariance matrix. As we mentioned at the beginning of the chapter, the so-called Ornstein-Zernike behavior $a_{\boldsymbol{r}} \sim C(\hat{\boldsymbol{r}}) |\boldsymbol{r}|^{(1-d)/2} \boldsymbol{z}^{-\boldsymbol{r}}$ leads to Gaussian estimates for individual probabilities if the coordinates of $\boldsymbol{z}$ are nonnegative real numbers. Ornstein-Zernike asymptotics are precisely the conclusion of our asymptotic estimates in the case where there is a single smooth point and the Hessian matrix is nondegenerate. Therefore, all we should require for a weak law and LCLT is that asymptotics are governed by a single smooth point with nondegenerate Hessian matrix. An example of this occurs in Chayes and Chayes (1986), where the point to plane generating function for self-avoiding random walks is analyzed. There, Gaussian behavior (Theorem 6.3) is deduced from meromorphicity (Theorem 6.1) and nondegeneracy (another assertion of Theorem 6.3). The remainder of this section is devoted to the statement and proof of a weak law and local central limit theorem.

Weak Laws

For the sake of clarity, we consider the case where the size parameter is $|\boldsymbol{r}|$ and then indicate what changes are necessary to handle other size parameters. We consider the generating function $F = G/H$ and let $h(\boldsymbol{x}) := F(x, \dots, x)$ be the composition of F with the diagonal embedding. Confining the statement to ordinary power series rather than Laurent series also simplifies matters by ensuring finite support of the cross-sectional measures μ_k.

Theorem 9.6.1 (weak law for diagonal slices) *Let $F = G/H$ be an ordinary generating function with nonnegative coefficients. Let $\mathbf{1}$ denote the d-vector of all ones. Let x_0 be the least positive value of x for which $H(x\mathbf{1}) = 0$. Assume the following:*

- *x_0 is a strictly minimal simple zero of $x \mapsto H(x\mathbf{1})$.*
- *$G(x_0\mathbf{1}) \neq 0$.*

Then the sequence $\{\mu_k\}$ of probability measures defined, as previously, by

$$\mu_k := \frac{1}{C_k} \sum_{|\boldsymbol{r}|=k} a_{\boldsymbol{r}} \delta_{\boldsymbol{r}}$$

satisfies a weak law with limit $\boldsymbol{m}$, where

$$\boldsymbol{m} := -\nabla_{\log} H(x_0, \dots, x_0)$$

is the outward normal to $H \circ \exp$ at the point $(\log x_0)\mathbf{1}$.

Proof The component B of the complement of the amoeba of H that contains a translate of the negative orthant is the logarithmic domain of convergence of the ordinary power series $\sum_{\boldsymbol{r}} a_{\boldsymbol{r}} \boldsymbol{z}^{\boldsymbol{r}}$. Recall from Proposition 8.4.3 that for $\boldsymbol{u} \in \partial B$, the point $\exp(\boldsymbol{u})$ is a zero of H.

Because B is convex and contains a translate of the negative orthant, it follows that for any positive vector $\mathbf{1}$, the set of real λ such that $\lambda\mathbf{1} \in B$ is bounded from above (or else B would have to be all of $\mathbb{R}^d$). In particular, x_0 is well defined. Suppose the gradient of H vanishes at $x_0\mathbf{1}$. Then $H(x_0\mathbf{1} + \boldsymbol{y}) = O(|\boldsymbol{y}|^2)$, and it follows that $H(x\mathbf{1})$ has a multiple zero at $x = x_0$. This contradicts the hypotheses of the theorem, whence $\nabla H(x_0\mathbf{1}) \neq \mathbf{0}$. Equivalently, $\nabla(H \circ \exp)((\log x_0)\,\mathbf{1}) \neq \mathbf{0}$.

Once we know the real function $H \circ \exp$ has a zero at $(\log x_0)\mathbf{1}$ with nonvanishing gradient, we see that the zero set of $H \circ \exp$ in a neighborhood of $(\log x_0)\mathbf{1}$

is a smooth hypersurface normal to $\boldsymbol{m}$ and coinciding with ∂B. It follows from this that any hyperplane through $(\log x_0 \mathbf{1})$ other than the one normal to $\boldsymbol{m}$ intersects the interior of B. In particular, for any $\boldsymbol{r}$ not parallel to $\boldsymbol{m}$, the maximum value over $\boldsymbol{x} \in B$ of $\boldsymbol{r} \cdot \boldsymbol{x}$ is strictly greater than $|\boldsymbol{r}|(\log x_0)$. This implies $a_{\boldsymbol{r}} = O((x_0 + \varepsilon)^{-|\boldsymbol{r}|})$ for some positive ε whose choice is uniform as $\hat{\boldsymbol{r}}$ varies over any neighborhood not containing $\hat{\boldsymbol{m}}$.

Observe that the generating function $\sum_k C_k z^k$ is equal to $F(z\mathbf{1})$, which has a minimal simple pole at x_0. It follows that $C_k \sim c x_0^{-k}$ and hence that $\sum_{\mathcal{N}} a_{\boldsymbol{r}} = o(C_k)$ as $k \to \infty$ where $\mathcal{N}$ is any conic neighborhood not containing $\boldsymbol{m}$. This implies the weak law with limit $\boldsymbol{m}$. □

Example 9.6.2 (multinomial distribution) Let $p_1, \ldots, p_d$ be nonnegative numbers summing to 1, and let $F = 1/(1 - \sum_{j=1}^d p_i x_i)$ be the generating function for the multinomial distribution with parameters $\boldsymbol{p}$, whose coefficients $a_{\boldsymbol{r}}$ are given by

$$a_{\boldsymbol{r}} = \binom{|\boldsymbol{r}|}{r_1, \ldots, r_d} p_1^{r_1} \cdots p_d^{r_d} .$$

The denominator of F is $H(\boldsymbol{x}) = 1 - \boldsymbol{p} \cdot \boldsymbol{x}$ whence $H(x\mathbf{1}) = 1 - x$ regardless of $\boldsymbol{p}$, and the hypotheses of the theorem are satisfied with $x_0 = 1$. The weak law is therefore satisfied with $\boldsymbol{m} = \nabla_{\log} H(x_0 \mathbf{1}) = \boldsymbol{p}$. This tells us what we already knew from basic probability theory, namely that repeated rolls of a die with weights $\boldsymbol{p}$ yield a weak law with $\boldsymbol{m} = \boldsymbol{p}$. ■

Often combinatorial classes are enumerated in such a way that the size parameter is not the sum $|\boldsymbol{r}|$ of the indices but is just one of the indices, say r_d. The weak law is more or less the same in this case except that the probability measures $\{\mu_k\}$ no longer have finite support, and an added hypothesis is required.

Theorem 9.6.3 (weak law for coordinate slices) *Let $F = G/H = \sum_{\boldsymbol{r}} a_{\boldsymbol{r}} \boldsymbol{z}^{\boldsymbol{r}}$ be a Laurent series on a component B of the complement of* amoeba(H) *with nonnegative coefficients. Suppose there is a real number x_0 such that $(1, \ldots, 1, x) \in \overline{B}$ if and only if $x \leq x_0$. Assume the following:*

(i) x_0 is a strictly minimal simple zero of $x \mapsto H(1, \ldots, 1, x)$.
(ii) $G(1, \ldots, 1, x_0) \neq 0$.

Let $C_k := \sum_{r_d = k} a_{\boldsymbol{r}}$. Then $C_k < \infty$ for all k, and the sequence $\{\mu_k\}$ of probability measures defined by

$$\mu_k := \frac{1}{C_k} \sum_{r_d = k} a_{\boldsymbol{r}} \delta_{\boldsymbol{r}}$$

satisfies a weak law with limit $\boldsymbol{m}$, *where*

$$\boldsymbol{m} := -\nabla_{\log} H(1, \ldots, 1, x_0)$$

is the outward normal to $H \circ \exp$ *at the point* $(0, \ldots, 0, \log x_0)$.

Proof Arguing as in the proof of Theorem 9.6.1, we see again that $\exp(\boldsymbol{u}) \in \mathcal{V}$ for every $\boldsymbol{u} \in \partial B$. We see also that $\nabla(H \circ \exp)$ is nonvanishing at $(0, \ldots, 0, \log x_0)$. Because $(1, \ldots, 1, x) \in B$ for $x < x_0$, we have convergence of the sum

$$\sum_{r_d = k} a_r r_d^k = \sum_k C_k r_d^k,$$

and hence C_k takes on finite values, and its univariate generating function has radius of convergence x_0. Our hypotheses of a simple strictly minimal pole at x_0 imply $C_k \sim c x_0^{-k}$. As before, the theorem follows once we show that the total weight of μ_k is $o(x_0^{-k})$ on sets for which $\hat{\boldsymbol{r}}$ is bounded away from $\boldsymbol{m}$. The argument is finished in the same way as in Theorem 9.6.1, by showing that any hyperplane through $(0, \ldots, 0, \log x_0)$ other than the hyperplane normal to the dth coordinate plane must intersect the interior of B. □

Example 9.6.4 (IID sums) Let μ be a probability measure on a finite subset $E \subseteq \mathbb{Z}^{d-1}$. The spacetime generating function F for convolutions of μ is given by

$$F(\boldsymbol{x}) = \sum_{k \geq 0} \sum_{\boldsymbol{r} \in \mathbb{Z}^{d-1}} \mu^{(k)}(\boldsymbol{r}) \boldsymbol{x}^{(\boldsymbol{r}, k)} = \frac{1}{1 - x_d \phi_\mu(x_1, \ldots, x_{d-1})},$$

where ϕ_μ is the $(d-1)$-variable generating function for μ. Then $\mathbf{1}$ is a simple pole of H and is strictly minimal as long as ϕ is aperiodic. Directly, $\nabla_{\log} H(\mathbf{1}) = (\boldsymbol{m}, 1)$, where $\boldsymbol{m}$ is the mean vector of μ. Theorem 9.6.3 then proves the weak law of large numbers for sums of IID samples from μ. ■

Remark More generally, we may allow μ to be any measure on $\mathbb{Z}^{d-1}$ whose moment generating function is finite everywhere. The function ϕ_μ will no longer be polynomial but will be entire. This takes us out of the theory of amoebas of polynomials; however, all the facts that are required concerning logarithmic domains of convergence still hold. Because we have not developed theory and notation for analytic amoebas, we do not include a statement or proof of this result. The greatest generality for the weak law via this type of argument is achieved by weakening the hypothesis to finiteness of the moment generating function just in a neighborhood of the origin.

Central Limits

The first local central limit theorem we derive is a general asymptotic for the profile of the numbers $\{a_{\boldsymbol{r}} : r_d = k\}$ as $k \to \infty$. One can also derive such a result for other slices such as $\{a_{\boldsymbol{r}} : |\boldsymbol{r}| = k\}$. The derivations of these are similar, and we find it simplest to stick to the case where the size parameter is the last coordinate, as in a spacetime generating function for a stochastic process on $\mathbb{Z}^{d-1}$. We do, however, wish to weaken the hypotheses so as not always to assume that F is a rational function, but rather that it is meromorphic in a suitable domain. We therefore begin with a lemma giving a version of Ornstein-Zernike behavior in this more general setting.

The lemma is somewhat long because each conclusion allows us to continue to define notation for the next conclusion. We use the notation $\boldsymbol{e}_j$ for the jth elementary unit vector and recall the notation $\mathbf{T}(\boldsymbol{x})$ for the set of complex vectors z whose coordinate-wise log-modulus $(\log|z_1|, \ldots, \log|z_d|)$ is equal to $\boldsymbol{x}$. Let $S = \mathbb{R}^{d-1} \times \{1\}$ denote the slice of $\mathbb{R}^d$, where the last coordinate is equal to 1.

Lemma 9.6.5 *Let $F(z) = \sum_{r \in \mathbb{Z}^d \times \mathbb{N}} a_{\boldsymbol{r}} z^{\boldsymbol{r}}$ be a d-variate series with a logarithmic domain of convergence of convergence $B \subseteq \mathbb{R}^d$. Suppose that $\overline{B}$ intersects the negative e_d axis in the ray $(-\infty, t]$ for some real number t and that F is meromorphic on a neighborhood of the torus $\mathbf{T}(t\, \boldsymbol{e}_d)$. Let G/H represent F as a quotient of analytic functions near $z_* := (1, \ldots, 1, e^t)$. Suppose z_* is a simple pole of F and is the only pole of F on $\mathbf{T}(t\, \boldsymbol{e}_d)$. Then the logarithmic pole variety $\log \mathcal{V}$ of F is a smooth complex analytic hypersurface in a neighborhood of $t\, \boldsymbol{e}_d$. If H is a real function, then the intersection of $\log \mathcal{V}$ with $\mathbb{R}^d$ near $t\, \boldsymbol{e}_d$ is a smooth real hypersurface. Let $\boldsymbol{m}$ denote a vector normal to $\log \mathcal{V}$ at $t\, \boldsymbol{e}_d$ and scaled so that $m_d = 1$. Let g be the function parametrizing $\log \mathcal{V}$ as a graph of a function over the first $d-1$ coordinates near $\boldsymbol{x}^\circ = \mathbf{0}$, i.e., $(\boldsymbol{x}^\circ, g(\boldsymbol{x}^\circ)) \in \log \mathcal{V}$. If the Hessian matrix $\mathcal{H}$ for g is nonsingular at the origin, then as $\boldsymbol{r}$ varies over a neighborhood of $\boldsymbol{m}$ in S, the point $\boldsymbol{x}(\boldsymbol{r}) \in \log \mathcal{V}$ with normal vector $\boldsymbol{r}$ varies smoothly. In this case there is an Ornstein-Zernike estimate*

$$a_{\boldsymbol{r}} \sim (2\pi\, r_d)^{(1-d)/2} |\,\mathcal{H}\,|^{-1/2} \frac{G(z)}{z_d(\partial H/\partial z_d)(z)} \exp(-\boldsymbol{r} \cdot \boldsymbol{x}(\boldsymbol{r})) \,. \tag{9.6.1}$$

Proof The first conclusion is that $\log \mathcal{V}$ is a smooth complex analytic hypersurface near $t\, \boldsymbol{e}_d$. Because $\nabla(H \circ \exp)$ is nonvanishing, this is a consequence of the implicit function theorem. If H is real, then the intersection of $\log \mathcal{V}$ with $\mathbb{R}^d$ is a smooth real hypersurface, for the same reason. Next we check how $z(\boldsymbol{r})$ varies with $\boldsymbol{r}$. The map $\boldsymbol{x} \mapsto \nabla(H \circ \exp)(\boldsymbol{x})$ is a version of the Gauss map, where we map to S rather than to the unit ball. Nonsingularity of the Jacobian of this map ensures that its inverse is smooth. We have seen in Proposition 9.5.3 that nonsingularity of the Jacobian of the Gauss map is equivalent to nonsingularity of the Hessian $\mathcal{H}(g)$ and to nonvanishing of the Gaussian curvature of $\mathcal{V}$. By Remark 9.2.2, we have established the conditions for Theorem 9.2.1. The asymptotic formula (9.2.12), with C_0 determined by (9.2.11), together are (9.6.1). □

Theorem 9.6.6 (LCLT) *Let $F = G/H$ be a d-variate generating function satisfying the hypotheses of Lemma 9.6.5. Define the value t for which $t\, \boldsymbol{e}_d \in \partial B$, the mean vector $\boldsymbol{m}$ and the parametrizing function g all as in the lemma, and let*

$M := \mathcal{H}(\mathbf{0})$ denote the Hessian matrix of g at the origin. Suppose M is nonsingular and hence negative definite (it is always nonpositive definite). Denote the $(d-1)$-variate normal density with mean $k\boldsymbol{m}$ and covariance kM by

$$\mathfrak{n}_k(\boldsymbol{r}^\circ) := (2\pi k)^{(1-d)/2}|M|^{-1/2} \exp\left[\frac{1}{2k}(\boldsymbol{r}^\circ - k\boldsymbol{m})^T M^{-1}(\boldsymbol{r}^\circ - k\boldsymbol{m})\right]. \quad (9.6.2)$$

Then there is a constant c such that

$$a_{\boldsymbol{r}^\circ,k} \sim c\ \mathfrak{n}_k(\boldsymbol{r}^\circ)$$

as $k \to \infty$ with $|\boldsymbol{r}^\circ - k\boldsymbol{m}| = o(k^{2/3})$. It follows that

$$\sup_{r:r_d=k} k^{d/2}\, |a_{\boldsymbol{r}} - c\,\mathfrak{n}(\boldsymbol{r}^\circ)| \to 0$$

as $k \to \infty$.

Remark In fact, we may evaluate c by comparing (9.6.1) and (9.6.2), although an LCLT traditionally does not require knowledge of the normalizing constant.

Proof Comparing (9.6.1) and (9.6.2), we see that we need to show that for $\boldsymbol{r} \in S$, the rate function $\beta(\boldsymbol{r}) := -\boldsymbol{r} \cdot \boldsymbol{x}(\boldsymbol{r})$ satisfies

$$\beta(k\boldsymbol{m} + \boldsymbol{y}) = \frac{1}{2k}\boldsymbol{y}^T M^{-1}\boldsymbol{y} + C_k + o(1)$$

as $k \to \infty$ with $|\boldsymbol{y}| = o(k^{2/3})$. The rate function is homogeneous of degree 1, so this is the same as

$$\beta(\boldsymbol{m} + \boldsymbol{y}) = \frac{1}{2}\boldsymbol{y}^T M^{-1}\boldsymbol{y} + C'_k + o(k^{-1}), \quad (9.6.3)$$

where the new $\boldsymbol{y}$ is $1/k$ times the old $\boldsymbol{y}$ and is required to be $o(k^{-1/3})$.

The point $\boldsymbol{x}(\boldsymbol{r})$, being the support point for a hyperplane normal to $\boldsymbol{r}$, is the minimizing point for $-\boldsymbol{r} \cdot \boldsymbol{x}$ on $\overline{B}$. The rate function $\beta(\boldsymbol{r})$ is therefore the minimum value of $-\boldsymbol{r} \cdot \boldsymbol{x}$ on B. When $\boldsymbol{r} \in S$, we may write $\boldsymbol{r} = (\boldsymbol{r}^\circ, 1)$. Let us also write the general point $\boldsymbol{x} \in \log \mathcal{V}$ as $(\boldsymbol{x}^\circ, g(\boldsymbol{x}^\circ))$, where the function g is concave (because locally the logarithmic domain of convergence of convergence $\overline{B}$ is described by $\{(\boldsymbol{x}, u) : u \le g(\boldsymbol{x})\}$ and logarithmic domain of convergences of convergence are convex [Hörmander, 1990]). Thus

$$\beta(\boldsymbol{r}^\circ, 1) = \inf_{\boldsymbol{x}^\circ \in \mathbb{R}^{d-1}} \{-g(\boldsymbol{x}^\circ) - \boldsymbol{r}^\circ \cdot \boldsymbol{x}^\circ\},$$

which is the negative of the *convex conjugate* of the convex function $-g$. The convex conjugate of the quadratic form $\boldsymbol{x} \mapsto \boldsymbol{x}^T A\boldsymbol{x}$ represented by a positive definite matrix A is represented in the dual basis by the inverse matrix $\boldsymbol{r} \mapsto \boldsymbol{r}^T A^{-1}\boldsymbol{r}$. This fact can be found, for example, in Boyd and Vandenberghe (2004, Example 3.22). The quadratic Taylor expansion of the convex conjugate at a point $\boldsymbol{r}$ is determined by the quadratic Taylor expansion of the function (assuming this is nondegenerate) at the point where the minimum occurs.

The minimizing point for $\boldsymbol{r} = \boldsymbol{m}$ is at the origin, and the quadratic term in the expansion of $-g$ at the origin is the matrix $-M$ representing the Hessian of $-g$ at the origin. Therefore, the Taylor expansion of β about $\boldsymbol{m}$ on S is given by

$$\beta(\boldsymbol{m} + \boldsymbol{y}) = \beta(\boldsymbol{m}) + \frac{1}{2}\boldsymbol{y}^T M^{-1} \boldsymbol{y} + O(|\boldsymbol{y}|^3).$$

The condition $|\boldsymbol{y}| = o(k^{-1/3})$ is exactly what is needed for $O(|\boldsymbol{y}|^3)$ to be $o(k^{-1})$, and taking C'_k to be $\beta(\boldsymbol{m})$ (not depending on k after all) establishes (9.6.3) and the first conclusion.

Pick ν with $1/2 < \nu < 2/3$. When $|\boldsymbol{r} - k\boldsymbol{m}| \leq k^\nu$, the first conclusion implies that

$$|a_{\boldsymbol{r}} - c\, \mathfrak{n}(\boldsymbol{r})| = o(\mathfrak{n}(\boldsymbol{r})) = o(k^{-d/2}).$$

It remains to establish the second conclusion when $|\boldsymbol{r} - k\boldsymbol{m}| \geq k^\nu$. In this case, both terms are small separately. The term $\mathfrak{n}(\boldsymbol{r})$ is in fact bounded about by $\exp(-ck^{\nu-1/2})$ for some c. However, when $\boldsymbol{r} \in S$, the quantity $\boldsymbol{r} \cdot \boldsymbol{x}(\boldsymbol{r})$ differs from its value at $\boldsymbol{r} = \boldsymbol{m}$ by at least a constant multiple of $|\boldsymbol{r} - \boldsymbol{m}|^2$. In general, 1-homogeneity tells us that when $r_d = k$, $\boldsymbol{r} \cdot \boldsymbol{x}(\boldsymbol{r})$ differs from its value at $\boldsymbol{r} = k\boldsymbol{m}$ by at least a constant multiple of $k^{-1}|\boldsymbol{r} - k\boldsymbol{m}|^2$. When $|\boldsymbol{r} - k\boldsymbol{m}| \geq k^\nu$, this is of order at least $k^{2\nu-1}$, which is a positive power of k. Plugging this into (9.6.1) shows that $a_{\boldsymbol{r}}$ is also at most $\exp(-ck^{2\nu-1})$, which completes the proof. □

As an example, we derive the LCLT for sums of independent lattice random variables whose moment generating functions are everywhere finite.

Example 9.6.7 (classical LCLT) Let μ be an aperiodic probability distribution on $\mathbb{Z}^{d-1}$, and let $\mu^{(k)}$ denote the k-fold convolution of μ. Let F denote the spacetime generating function for the random walk with increments distributed as μ:

$$F(\boldsymbol{z}) := \sum_{\boldsymbol{r} \in \mathbb{Z}^{d-1} \times \mathbb{N}} \mu^{(k)}(\boldsymbol{r}^\circ) \boldsymbol{z}^{\boldsymbol{r}}.$$

The $(d-1)$-dimensional probability generating function ϕ for μ is defined by

$$\phi(\boldsymbol{z}) := \sum_{\boldsymbol{r} \in \mathbb{Z}^{d-1}} \mu(\boldsymbol{r}) \boldsymbol{z}^{\boldsymbol{r}}.$$

Finiteness of the moment generating function is equivalent to ϕ being an entire function. The spacetime generating function F is related to the moment generating function ϕ for the distribution μ via

$$F(\boldsymbol{z}) = \frac{1}{1 - z_d\, \phi(\boldsymbol{z}^\circ)}.$$

The pole surface is defined by $\{z_d = 1/\phi(\boldsymbol{z}^\circ)\}$. Evidently, this is globally the graph of the function g defined by

$$g(\boldsymbol{x}) = -\log \phi(\exp(\boldsymbol{x})).$$

Denote this surface by $\mathcal{V}$ and its intersection with $\mathbb{R}^d$ by $\mathcal{V}_{\mathbb{R}}$. Next we check that $\log \mathcal{V}_{\mathbb{R}}$ is the boundary of the logarithmic domain of convergence of convergence. Indeed, nonnegativity of the coefficients implies that $(\boldsymbol{x}^\circ, t)$ is in the interior of the domain of convergence when $t < g(\boldsymbol{x}^\circ)$, whereas no point $(\boldsymbol{x}^\circ, g(\boldsymbol{x}^\circ))$ is in the logarithmic domain because it is on $\log \mathcal{V}$. When z is on the torus $\mathbf{T}(\boldsymbol{x}, g(\boldsymbol{x}))$, aperiodicity implies that $\phi(z) \neq 0$, unless z is real. Therefore, each point of $\mathcal{V}_{\mathbb{R}}$ is the only point of $\mathcal{V}$ on its torus, and in particular, such a minimal point is strictly minimal.

We may evaluate $\phi(1, \dots, t) = \sum_{k \geq 0} t^k$, which is finite if and only if $t \leq 1$. Taking logs, $g(\mathbf{0}) = 0$. We begin computing derivatives. One might recognize the standard computation of moments from derivatives of the moment generating function; for completeness, we reproduce it here. For $1 \leq j \leq d-1$, we let ϕ_j denote the partial derivative of ϕ with respect to $z_j = \exp(x_j)$ compute

$$\begin{aligned}
\frac{\partial}{\partial x_j} g(\mathbf{0}) &= -\frac{\partial}{\partial x_j} (\log \circ \phi \circ \exp)(\mathbf{0}) \\
&= -\frac{(\partial/\partial x_j)(\phi \circ \exp)}{\phi \circ \exp}(\mathbf{0}) \\
&= -\frac{e^{x_j} \phi_j \circ \exp}{\phi \circ \exp}(\mathbf{0}) \\
&= \phi_j(\exp(\mathbf{0}))
\end{aligned}$$

because all values e^{x_j} are equal to 1 and $\phi(1, \dots, 1) = 1$. This last partial derivative evaluates to $\sum_{\boldsymbol{r}} r_j \mu(\boldsymbol{r}) z^{\boldsymbol{r} - \boldsymbol{e}_j}$, and evaluating at $(1, \dots, 1)$ gives $\sum_{\boldsymbol{r}} r_j \mu(\boldsymbol{r})$. Thus

$$\boldsymbol{m} := \nabla(\phi \circ \exp)(\mathbf{0}) = \sum_{\boldsymbol{r}} r_j \mu(\boldsymbol{r})$$

is the mean of the distribution μ.

Differentiating again, if $i \neq j$, we find that

$$\begin{aligned}
\frac{\partial}{\partial x_i} \frac{\partial}{\partial x_j} g(\mathbf{0}) &= -\left[\frac{e^{x_i} e^{x_j} \phi_{ij} \circ \exp}{\phi \circ \exp} - \frac{e^{x_i} e^{x_j} (\phi_i \circ \exp)(\phi_j \circ \exp)}{(\phi \circ \exp)^2} \right] (\mathbf{0}) \\
&= \phi_{ij}(1, \dots, 1) - m_i m_j
\end{aligned}$$

so that the (i, j) entry of the Hessian of g at the origin is indeed the covariance of the i and j coordinates under μ. A similar computation works for $i = j$ and establishes that the Hessian matrix of g at the origin is the covariance matrix for μ. Applying Theorem 9.6.6, we see that $\mu^{(k)}(\boldsymbol{r}^\circ)$ is asymptotically equal to $c\ \mathfrak{n}_k(\boldsymbol{r}^\circ)$. There is no need to compute c because we know $\sum_{\boldsymbol{r}^\circ} \mu^{(k)}(\boldsymbol{r}^\circ) = 1$. Thus $c = 1$, and we recover the LCLT:

Theorem 9.6.8 *If μ is an irreducible aperiodic probability measure on $\mathbb{Z}^d$ with moment generating function ϕ everywhere finite, then*

$$\mu^{(k)}(\boldsymbol{r}) \sim \mathfrak{n}_k(\boldsymbol{r}),$$

as $k \to \infty$ with $|\boldsymbol{r} - k\boldsymbol{m}| = o(k^{2/3})$, where $\mathfrak{n}_k$ is defined by (9.6.2), with $\boldsymbol{m}$ equal to the mean of μ and M equal to its covariance matrix. It follows that

$$\sup_{\boldsymbol{r}} k^{1/2} \left| a_{\boldsymbol{r}} - \mathfrak{n}(\boldsymbol{r}) \right| \to 0,$$

as $k \to \infty$. ■

Notes

Precursors to the two derivations of the saddle-residue integrals in Section 9.2 were the multivariate asymptotic results of Bender and Richmond (1983). Breaking the symmetry among the coordinates, they wrote

$$F(z) = \sum_{n=0}^{\infty} f_n(z^\circ) z_d^n$$

and then used the fact that f_n is sometimes asymptotic to an nth power $f_n \sim C \cdot g \cdot h^n$ to obtain Gaussian asymptotics when certain minimality conditions are satisfied near a smooth critical point. Their language is inherently one-dimensional, so geometric concepts such as smooth point did not arise explicitly. The results presented in this chapter were first obtained via coordinate methods in Pemantle and Wilson (2002). These methods are valid only when z is a minimal point. The residue version of these computations appeared in print first in Baryshnikov, Brady, et al. (2010).

Extending the validity of the coordinate version beyond the case of finite intersection of $\mathcal{V}$ with $T(\boldsymbol{x}_{\min})$ was done only recently in Baryshnikov and Pemantle (2011).

Another rewriting of the leading term of the basic nondegenerate smooth point asymptotic formula is given in Bena, Berkooz, de Boer, El-Showk, and Van den Bleeken (2012, Appendix B). The formula uses the first and second partial derivatives of H.

Example 9.5.6 has an interesting history. The pictures in Figure 9.3 were first produced by a graduate student, Wil Brady, in an attempt to produce rigorous computations verifying the limit shapes of feasible regions that were suspected from simulations. At that time, Theorem 9.5.4 was not known, and the fact that the fine structure of the two plots agreed was a big surprise. This led to the reformulation of formulae such as Theorem 9.5.7 in terms of curvature.

Exercises

9.1 (the residue functor)
Let Res be the residue map on meromorphic forms with simple poles on a smooth variety V as defined in Proposition A.5.1. Prove that Res is functorial, i.e., it commutes with bi-holomorphic changes of coordinate.

9.2 (higher order poles)
Prove an analogue of Formula 9.3.1 for poles of higher order. First, show that the residue of $P\,dz/z_1^n$ is given by

$$\operatorname{Res}\left(\frac{P\,dz}{z_1^n};p\right) = \frac{1}{(n-1)!}\left(\frac{\partial}{\partial z_1}\right)^{n-1} P\,dz_2 \wedge \cdots \wedge dz_d\,.$$

Then copy the proof of Proposition A.5.1 to extend this to residues of G/H^n, computing the formula explicitly for $n = 2$ and $n = 3$ (assuming that dH is nonvanishing).

9.3 (binomial coefficients)
Let $F = 1/(1 - x - y)$ be the generating function for the binomial coefficients. In analogy with Example 9.4.3, find the residue integral $\int_\sigma \omega$ of Theorem 9.3.2 that estimates coefficients of the binomial generating function. Evaluate this asymptotically to obtain an asymptotic formula for a_{rs}, and compare with what you obtain from Stirling's formula for $\binom{r+s}{r}$.

9.4 (change of geometry)
This exercise looks at Example 9.5.13 from the viewpoint of coalsecing saddles. Let $f(x, y) = x^2 - 3x + 3 - y$. In the example, asymptotics in the diagonal direction reveal a quadratic degeneracy. To see what a quadratic degeneracy means topologically, begin by computing the critical points in direction $\hat{r} := (r, 1 - r)$ as a function of r on the unit interval. There should usually be two critical points. At what value r_* of r is there a single critical point of multiplicity 2? For each r, determine the subset $\texttt{local}(r) \subseteq \texttt{critical}(r)$. Does this vary continuously in r? What happens at r_*?

9.5 (binomial convolution)
Let $F(x, y) = 1/(1 - x - y)^\ell$. Compute the asymptotics for $a_{rs}^{(\ell)}$, and find the relation between these and the asymptotics of the binomial coefficients $a_{rs}^{(1)} = \binom{r+s}{r,s}$. Verify this combinatorially by finding the exact value of $a_{rs}^{(\ell)}$. (Hint: When $\ell = 2$, the bivariate convolution of the binomial array with itself can be represented as divisions of $r + s$ ordered balls into r balls of one color and s of another, with a marker inserted somewhere dividing the balls into parts I and II.)

9.6 (cube root asymptotics, next term in expansion)
In Example 9.5.13, the ratio of the error term to the leading term when $r = 100$ is $0.00111\ldots$. This seems to indicate that the next nonvanishing asymptotic term is $r^{-m/3}$ for some m strictly greater than 2, perhaps as great as 5. Compute enough derivatives of A and ϕ at zero to determine the next nonvanishing asymptotic term for $a_{r,r}$.

10

Multiple Point Asymptotics

In this chapter we give asymptotics for directions controlled by a multiple point of $\mathcal{V}$. We have alluded to multiple points in previous chapters. Intuitively, a multiple point is one where the singular variety locally decomposes as a union of smooth varieties. As in the smooth point case, because $\mathcal{V}$ is defined explicitly as the zero-set of H, some care is required in the formal definition. If we replace H with H^2, for example, the geometry does not change, but the asymptotics do. Asymptotic expressions for the coefficients a_r will again be of the form

$$a_r \sim \sum_{z \in \mathtt{contrib}} \Phi_z(r),$$

where now the points of `contrib` are multiple points. Several of the expressions we will prove are as follows. In these expressions, $\mathbf{\Gamma}_\Psi$ is the logarithmic gradient matrix defined in Definition (10.2.1), the vector $\boldsymbol{m}$ is the exponent in the case that the divisors are not simple poles, and Q is the Hessian matrix of the logarithmic parametrization defined in (10.3.4) later.

$$\Phi_z(\boldsymbol{r}) = z^{-r} \frac{G(z)}{\det \mathbf{\Gamma}_\Psi} \tag{10.3.1}$$

$$\Phi_{x_0, y_0}(r, s) = x_0^{-r-1} y_0^{-s-1} \frac{G}{\sqrt{\left(\dfrac{\partial^2 H}{\partial x \partial y}\right)^2 - \dfrac{\partial^2 H}{\partial x^2}\dfrac{\partial^2 H}{\partial y^2}}}(x_0, y_0)\,. \tag{10.3.2}$$

$$\Phi_z(\boldsymbol{r}) = \frac{1}{(\boldsymbol{m}-\mathbf{1})!} \frac{z^{-r} G(z)}{\det \mathbf{\Gamma}_\Psi(z)} (\boldsymbol{r}\mathbf{\Gamma}_\Psi^{-1})^{\boldsymbol{m}-\mathbf{1}} \tag{10.3.3}$$

$$\Phi_z(\boldsymbol{r}) = z^{-r} \frac{(\boldsymbol{r}\mathbf{\Gamma}_\Psi^{-1})^{\boldsymbol{m}-\mathbf{1}}}{(\boldsymbol{m}-\mathbf{1})!} \frac{G(z)}{\det \mathbf{\Gamma}_\Psi} (2\pi|\boldsymbol{r}|)^{-(d-k)/2} \frac{1}{\sqrt{\det Q}}\,. \tag{10.3.5}$$

Multiple points generalize smooth points, so technically our formulae for Φ_z when z is a multiple point will generalize those of the previous chapter. The

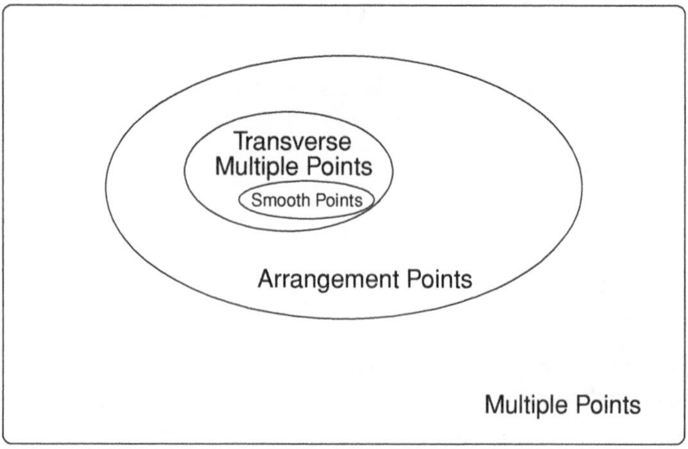

Figure 10.1 Four levels of geometric complexity.

problem of determining `contrib` is again a thorny issue, completely solved only in three cases: in addition to the case where $d = 2$ and the curve is smooth (see Section 9.4), we can identify `contrib` if there are minimal points or if H is a product of linear factors.

We first define and classify multiple points and develop some basic notation and taxonomy. This was not necessary for smooth points, but multiple points are substantially more complicated in their local geometry and topology. We then attack the integral from step (vi) of the general program. Geometrically, we define several levels of complexity, namely *transverse multiple points*, *arrangement points*, and (general) *multiple points*; see Figure 10.1. The analysis of the "base case" of transverse multiple points is carried out in two different ways, corresponding to the two approaches in Chapter 9. Each has its advantages. The residue approach has the advantage of naturality, making it easier for a sophisticated user to understand, the computations being of lesser complexity. Minimality is not required, although the problem of computing `contrib` is not always easy to solve in the absence of minimality. The coordinate-based surgical approach, which was historically first, is more elementary. It has the advantage of being able to compute asymptotics across the boundaries of cones where the leading term changes. It has the disadvantage of being restricted to minimal points. Due to the complexity of the integral formulae, the surgical approach is delayed until later in the chapter.

Having used either approach to compute $\Phi_z(\boldsymbol{r})$ for the base case, with square-free denominators and numerator equal to 1, we then build on this to compute $\Phi_z(\boldsymbol{r})$ in general when F is an arrangement point. We do not compute $\Phi_z(\boldsymbol{r})$ for general multiple points, although we consider some ad hoc examples in which the computation may be carried out.

10.1 Multiple Points

Geometric Classification

The classification of multiple points may be stated both geometrically and algebraically. We begin with the geometric classification.

Definition 10.1.1 (multiple point) *A point $z \in \mathcal{V}$ is said to be a* ***multiple point*** *if every sufficiently small neighborhood $U \cap \mathcal{V}$ of z in $\mathcal{V}$ is the union of finitely many smooth varieties $\mathcal{V}_1, \ldots, \mathcal{V}_n$.*

Throughout this chapter, we reserve the variable n for the number of smooth divisors. The variable k is used for the codimension. Thus any multiple point $z \in \mathcal{V}$ is an element of a unique stratum of $\mathcal{V}$, which is a smooth variety of complex dimension $d - k$. Later, when we allow the denominator of a generating function to contain factors with powers greater than 1, the powers corresponding to $\mathcal{V}_1, \ldots, \mathcal{V}_n$ are respectively denoted $m_1, \ldots, m_n$.

Definition 10.1.2 (transverse multiple point) *A point $z \in \mathcal{V}$ is said to be a* ***transverse multiple point*** *if every sufficiently small neighborhood $U \cap \mathcal{V}$ of z in $\mathcal{V}$ is the union of finitely many smooth varieties $\mathcal{V}_1, \ldots, \mathcal{V}_k$ containing z such that the normal vectors to $\mathcal{V}_j$ at z are linearly independent.*

Remark For transverse multiple points, the codimension k equals the number of sheets n. Note that a multiple point cannot be transverse according to our definition if $n > d$. The set of transverse multiple points includes the smooth points as a degenerate case ($k = 1$).

Example 10.1.3 (two lines in $\mathbb{C}^2$) The lowest degree example of a transverse multiple point is when $d = 2$ and H is the product of two linear factors. This is illustrated in Figure 10.2 in two different renderings. The specific example $H = \left(1 - \frac{1}{3}x - \frac{2}{3}y\right)\left(1 - \frac{2}{3}x - \frac{1}{3}y\right)$ was used to illustrate an amoeba in Figure 7.1(b).The combinatorial interpretation of the generating function $1/H$ and the asymptotics of its coefficients are worked out in Section 12.1.3. ■

Example 10.1.4 (figure eight) Let $F = 1/H$, where $H(x, y) = 19 - 20x - 20y + 5x^2 + 14xy + 5y^2 - 2x^2y - 2xy^2 + x^2y^2$. The real points of $\mathcal{V}$ are shown in Figure 10.3. At the point $(x, y) = (1, 1)$, the curve $\mathcal{V}$ intersects itself: intersecting $\mathcal{V}$ with a small neighborhood of $(1, 1)$ yields a union of two distinct segments of smooth curves, intersecting only at the point $(1, 1)$. ■

Hyperplane Arrangements

The arrangement points, the most general class of multiple point for which we have a general determination of Φ_z, borrow their name from the subject of hyperplane

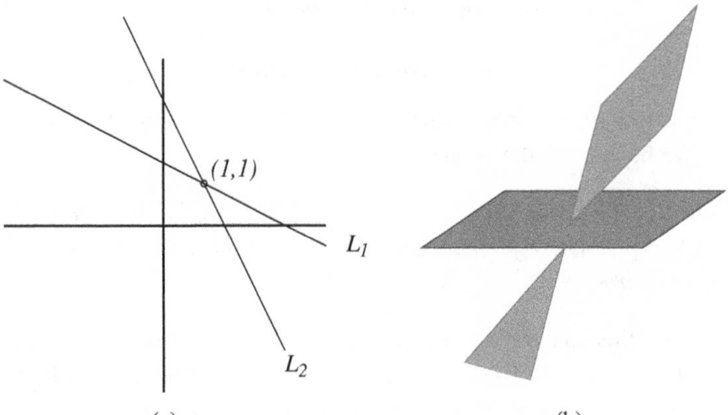

Figure 10.2 A transverse multiple point with $k = d = 2$ and both divisors linear, shown in two depictions.

arrangements. The relevant terminology is briefly summarized as follows. Let $\mathbb{K}$ be either the real or complex numbers. A hyperplane in $\mathbb{K}^d$ is an affine subspace $\{\boldsymbol{x} : \boldsymbol{x} \cdot \boldsymbol{v} = \boldsymbol{b}\}$ of codimension 1. A ***hyperplane arrangement*** is simply a finite collection of hyperplanes $P_1, \ldots, P_n$ in $\mathbb{K}^d$. A ***central hyperplane arrangement*** is one in which each hyperplane passes through the origin: thus $P_j = \{\boldsymbol{x} : \boldsymbol{x} \cdot \boldsymbol{v}_j = 0\}$ for some $\boldsymbol{v}_j \in \mathbb{K}^d$.

Each central hyperplane arrangement $\mathcal{A} = \{P_j : j \in A\}$ possesses a natural structure as a *matroid* $M(\mathcal{A})$, whose independent sets are the sets $\{P_j : j \in T\}$

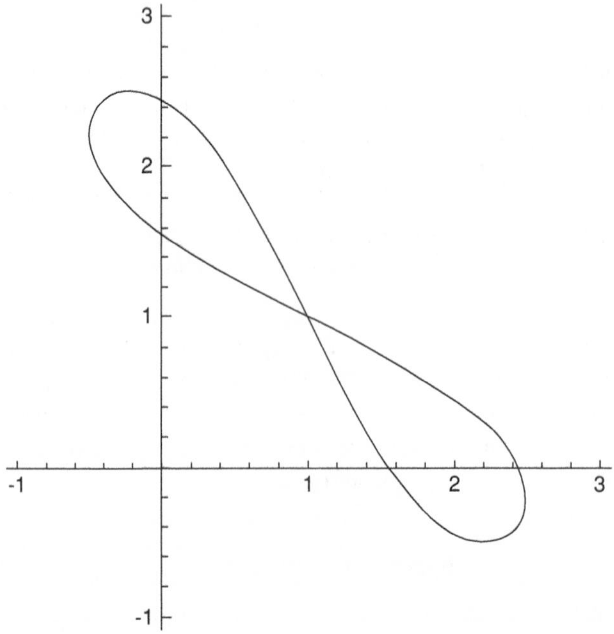

Figure 10.3 A local self-intersection.

such that $\text{codim}(\bigcap_{j \in T} P_j) = |T|$. Equivalently, these are the sets of hyperplanes whose normal vectors $\{\boldsymbol{v}_j : j \in T\}$ form a linearly independent collection. The set of ***circuits*** of $\mathcal{A}$, denoted by $\texttt{CIRC}$, is the set of (index sets of) minimal dependent sets. Thus $T \subseteq [n]$ is an element of $\texttt{CIRC}$ if $\{\boldsymbol{v}_j : j \in T\}$ is linearly dependent but no proper subset of $\{\boldsymbol{v}_j : j \in T\}$ is linearly dependent. The set of all intersections $L_T := \bigcap_{j \in T} P_j$ is called the ***lattice of flats*** of $\mathcal{A}$, as it possesses a natural lattice structure. The combinatorics of the set $\texttt{CIRC}$ can be axiomatized and many facts deduced; for a more complete discussion of matroids, see Björner, Las Vergnas, Sturmfels, White, and Ziegler (1999). We let $\overline{T}$ denote the set $\{j \in A : L_T \subseteq P_j\}$; in other words, $\overline{T}$ indexes the maximal set of hyperplanes having intersection L_T. This allows us to make an important definition.

Definition 10.1.5 (arrangement point) *A point $z \in \mathcal{V}$ is said to be an **arrangement point** of order n if every sufficiently small neighborhood $U \cap \mathcal{V}$ of z in $\mathcal{V}$ is the union of finitely many smooth varieties $\mathcal{V}_1, \ldots, \mathcal{V}_n$ containing z and having the additional property that the intersection lattice of the surfaces $\{\mathcal{V}_j\}$ coincides with the intersection lattice of their tangent planes. More formally, let $P_j := \tan_z(\mathcal{V}_j)$ denote the tangent plane to $\mathcal{V}_j$ at z, and let $\mathcal{A}$ be the central hyperplane arrangement $\{P_j : 1 \le j \le n\}$. For all $T \subseteq [n]$, if $\mathcal{V}_T$ denotes the intersection $\bigcap_{j \in T} \mathcal{V}_j$, then we require*

$$\mathcal{V}_T = \mathcal{V}_{\overline{T}}.$$

The simplest nontrivial example of an arrangement point i.e., not a transverse multiple point is when H is the product of linear divisors with $d = 2$ and $k = 3$. By definition, if H is any product of linear divisors then any point $z \in \mathcal{V}$ is an arrangement point, because the collection of surfaces $\{\mathcal{V}_j\}$ is a translation of the collection of tangent planes.

Example 10.1.6 (three curves in $\mathbb{C}^2$) Let $H := H_1 H_2 H_3 = (1 - x)(1 - y)(1 - xy)$. This is the generating function for the array $a_{rs} := \min\{r, s\}$. The three divisors are pairwise transverse, intersecting at the single point $(1, 1)$. The three have a common intersection there, which is perforce not transverse. This is illustrated in Figure 10.4. ■

To see why transversality is important, consider what can happen to the geometry in the absence of transversality. Suppose $H = (1 - z)(1 - z + (1 - y)^2 - (1 + x)^3)$. Then $\mathcal{V}$ is the union of two surfaces: the plane $\mathcal{V}_1 := \{z = 1\}$ and a surface $\mathcal{V}_2$ that is easily seen to be smooth. The two surfaces intersect in a translation by $(-1, 1, 1)$ of the curve $z = 0 = x^3 - y^2$, which has a cusp at the origin. In fact, with no transversality assumption, any variety is the intersection of smooth varieties. The transversality assumption serves to rule out geometrically complicated intersections, as shown by the following result.

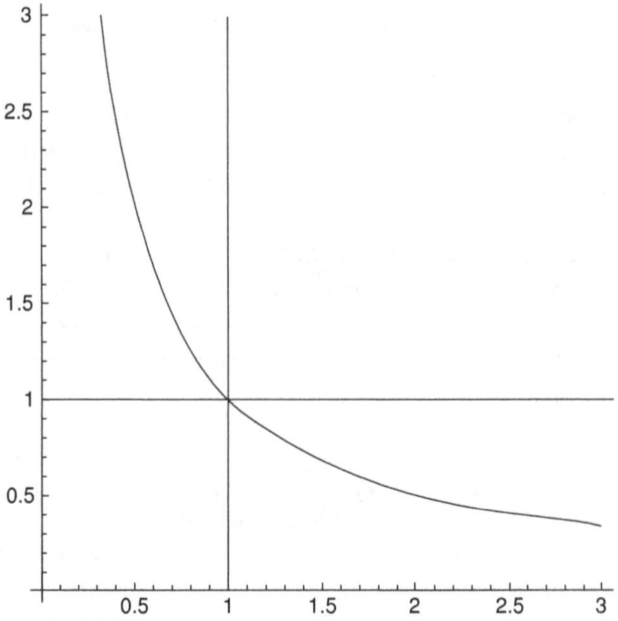

Figure 10.4 An arrangement point in dimension 2 where three curves intersect.

Proposition 10.1.7 *If z is an arrangement point of order n and $i \notin T \subseteq [n]$, then locally either $\mathcal{V}_T \subseteq \mathcal{V}_i$ or $\mathcal{V}_T$ intersects $\mathcal{V}_i$ transversely. Consequently, each intersection $\mathcal{V}_T$ is smooth, and its tangent space at z is the flat L_T.*

Proof We induct on the codimension of $\mathcal{V}_T$. When $\mathcal{V}_T$ has codimension 1, then $\mathcal{V}_T = \mathcal{V}_i$ for all $i \in \overline{T}$ with tangent space L_T at z. For $i \notin \overline{T}$, the hyperplane P_i is distinct from the tangent plane to T at z, and hence the intersection of $\mathcal{V}_i$ and $\mathcal{V}_T$ is transverse. Now suppose $\mathcal{V}_T$ has codimension $k \geq 2$ and $i \notin T$. By induction, T is smooth with tangent space L_T at z. If $i \in \overline{T}$, then there is nothing to prove, so assume $i \notin \overline{T}$ and let $T' := T \cup \{i\}$. Then the normal vector to $\mathcal{V}_i$ is not in $(L_T)^\perp$, and hence the surface $\mathcal{V}_i$ intersects $\mathcal{V}_T$ transversely. The transverse intersection of smooth varieties is smooth with tangent space given by the intersection of the tangent spaces, and hence $\mathcal{V}_{T'}$ is smooth with tangent space $L_{T'}$, completing the induction. □

Algebraic Classification

Our geometric classification of a point $\boldsymbol{p} \in \mathcal{V}$ is local, depending only on the geometry of $\mathcal{V}$ in a neighborhood of $\boldsymbol{p}$. To give equivalent algebraic definitions, we need to work in an appropriate local ring, which we now define. Let $\mathbf{R}_{\boldsymbol{p}}$ denote the ***local ring of germs of analytic functions*** at $\boldsymbol{p}$. By definition, this consists of all equivalence classes of analytic functions on neighborhoods of $\boldsymbol{p}$

under the relation of agreement on some neighborhood of $\mathbf{p}$. Because analytic functions are determined by their values in a neighborhood, all functions in such an equivalence class are analytic continuations of each other, making the situation somewhat simpler than for germs of smooth functions. Furthermore, because analytic functions have convergent power series, $\mathbf{R}_{\mathbf{p}}$ is isomorphic to $\mathbb{C}_{\mathbf{p}}\{\mathbf{z}\}$, the ring of formal power series about $\mathbf{p}$ that converge in some neighborhood of $\mathbf{p}$. The ring $\mathbf{R}_{\mathbf{p}}$ is a local ring whose unique maximal ideal is the germs of functions vanishing at $\mathbf{p}$. The local ring lies between the polynomial ring and the formal power series ring: $\mathbb{C}[\mathbf{z}] \subseteq \mathbf{R}_{\mathbf{p}} \subseteq \mathbb{C}_p[[\mathbf{z}]]$. It captures local geometry in the following sense. Suppose $H \in \mathbb{C}[\mathbf{z}]$ is square-free (generates a radical ideal) and is geometrically the union of smooth hypersurfaces $\mathcal{V}_j$ containing $\mathbf{z}$. Then there are $H_j \in \mathbf{R}_{\mathbf{p}}$ vanishing respectively on $\mathcal{V}_j$ such that $H = \prod_{j=1}^k H_j$ in $\mathbf{R}_{\mathbf{p}}$.

Example 10.1.8 (figure eight, continued) Let H be the polynomial from Example 10.1.4, whose zero set in $\mathbb{R}^2$ has the shape of a figure eight. The polynomial H is irreducible in $\mathbb{C}[x, y]$, but according to its geometry, it must factor in $\mathbf{R}_{(1,1)}$. It is not hard to compute the local parametrizations of the two branches, these being

$$y_1(x) = \frac{x^2 - (x-1)\sqrt{-4x^2+8x+5} - 7x + 10}{x^2 - 2x + 5}; \tag{10.1.1}$$

$$y_2(x) = \frac{x^2 + (x-1)\sqrt{-4x^2+8x+5} - 7x + 10}{x^2 - 2x + 5}.$$

The two branches have slopes -2 and $-1/2$ at the point $(1, 1)$. ■

Proposition 10.1.9 *The point $\mathbf{p} \in \mathcal{V}$ is a multiple point if and only if there is a factorization*

$$H = \prod_{j=1}^n H_j^{m_j} \tag{10.1.2}$$

in $\mathbf{R}_{\mathbf{p}}$ with $\nabla H_j(\mathbf{p}) \neq 0$ and $H_j(\mathbf{p}) = 0$. The point $\mathbf{p}$ is a transverse multiple point of order n if and only if in addition the gradient vectors $\{\nabla H_j(\mathbf{p}) : 1 \leq j \leq n\}$ are linearly independent.

When $m_j = 1$ for all j in the decomposition (10.1.2) with all H_j distinct, we say that H is ***square-free*** at $\mathbf{p}$.

To give an algebraic criterion for the multiple point $\mathbf{p}$ to be an arrangement point, let H_j be as in (10.1.2), and let ℓ_j denote the linear polynomial

$$\ell_j(\mathbf{z}) := \nabla H_j(\mathbf{p}) \cdot (\mathbf{z} - \mathbf{p}).$$

The leading term of H at $\mathbf{p}$ is given by

$$\texttt{hom}(H, \mathbf{p}) = \prod_{j=1}^n \ell_j^{m_j},$$

and therefore the zero set of $\mathtt{hom}(H, \boldsymbol{p})$, which is the algebraic tangent cone $\mathtt{algtan}_{\boldsymbol{p}}(H)$, equals the hyperplane arrangement $\mathcal{A}$ consisting of the planes tangent to some H_j at $\boldsymbol{p}$. We remark that this arrangement remains combinatorially identical as $\boldsymbol{p}$ varies over the stratum S of $\mathcal{V}$ containing $\boldsymbol{p}$. By definition, for $\boldsymbol{p}$ to be an arrangement point, such a product decomposition must hold, and in addition, the lattice of flats of $\mathcal{A}$ must be isomorphic to the intersection lattice of the local surfaces $\mathcal{V}_j := \mathcal{V}_{H_j}$. Repeated factors are allowed, so we may assume without loss of generality that each m_j is equal to 1, arriving at the following algebraic criterion.

Proposition 10.1.10 *The point $\boldsymbol{p} \in \mathcal{V}$ is an arrangement point if and only if both $\mathtt{hom}(H, \boldsymbol{p})$ and H factor into smooth factors (the former will be linear polynomials in $\mathbb{C}[\boldsymbol{z}]$, and the latter will be in $\mathbf{R}_{\boldsymbol{p}}$) and the two intersection lattices agree.*

Example 10.1.11 (figure eight, continued) Let $H(x, y)$ be the polynomial from Example 10.1.4 and let $\boldsymbol{p} = (1, 1)$. Taking the monomials of least degree of $H(1 + x, 1 + y)$ gives

$$\mathtt{hom}(H, \boldsymbol{p}) = 4x^2 + 10xy + 4y^2 .$$

Every homogeneous quadratic is the product of linear factors, which are distinct unless the *discriminant* vanishes – in this example, the discriminant equals 36, and so $\boldsymbol{p}$ is an arrangement point. It is in fact a transverse multiple point. ■

The simplest example of a multiple point i.e., not an arrangement point is the following product of two tangentially intersecting curves.

Example 10.1.12 (tangentially intersecting curves) Consider the generating function $F = \dfrac{1}{(2 - x - y)(1 - xy)}$, whose coefficients a_{ij} count the sums of normalized binomial coefficients $2^{-(i+j-2r)}\dbinom{i + j - 2r}{i - r}$ from $r = 0$ to the minimum of i and j. The two curves $2 - x - y = 0$ and $1 - xy = 0$ intersect tangentially at the point $(1, 1)$. The arrangement of tangent hyperplanes at $(1, 1)$ contains two copies of the hyperplane normal to the diagonal. Its lattice of flats is the Boolean lattice of order 1, whereas the lattice of intersections of the two curves is the Boolean lattice of order 2. Although this is not an arrangement point, we can compute the asymptotics of the coefficients, which we do in Example 10.4.14. ■

The following result, whose proof is left as an exercise, can be useful in detecting arrangement points.

Lemma 10.1.13 *Suppose $\boldsymbol{p}$ is a multiple point with local irreducible factors $H_1, \ldots, H_n$. Let J_T denote the ideal in $\mathbf{R}_{\boldsymbol{p}}$ generated by $\{H_i : i \in T\}$. The point $\boldsymbol{p}$ is an arrangement point if and only if for all $T \subseteq [n]$, the ideal J_T is radical. It follows that $J_T = J_{\overline{T}}$ for all T and that the codimension of this ideal is equal to the codimension of L_T.*

Computing Multiple Points

The foregoing discussion being of a theoretical nature, we pause to consider how the classification of multiple points might be computed. When given a polynomial H, one imagines being able to "look at its zero set" to see by inspection whether it is locally the union of smooth sheets, but of course we cannot do that when the dimension is high, the complexity is high, or when the procedure is automated. What, then, is an effective way to determine whether any of the singular points of H is in fact a multiple point?

The first problem is how to compute in the local ring at a singular point when such a point is known only as a solution to algebraic equations. Generically, a polynomial H will have no singularities at all. When H and its gradient vanish simultaneously, this will typically occur on a zero-dimensional variety, i.e., a finite, algebraic set of points. If there is only one point, then it is rational (see Exercise 6.4), and there is no difficulty making an explicit change of variables to translate the point to the origin. In general, decomposing the singular set into irreducible zero-dimensional varieties, we reduce to the case of a finite, algebraically irreducible finite algebraic set E defined by an ideal J. To localize at an element of E, let

$$J' := \langle J, H(z + \boldsymbol{w})\rangle$$

be the ideal generated by the equations for z to be a singular point of $\mathcal{V}$ together with the polynomial $H(z + \cdot)$ in the variables $\boldsymbol{w}$. Eliminating the z variables via a `plex` Gröbner basis computation yields an elimination ideal for the $\boldsymbol{w}$ variables. If this is a principal ideal $\langle h(\boldsymbol{w})\rangle$, then h is H translated so that the origin is a point of E; because J is irreducible, it does not matter which point of E.

Having successfully localized to a point of E, we examine the problem of classifying the local geometry. Here we can run into difficulties when trying a direct approach. We know no algorithm that will decide in $\mathbf{R}_p$ whether a point is multiple, or transverse, or an arrangement point, or compute the multiplicity. It seems that computation in $\mathbf{R}_p$ is essentially more difficult than in polynomial rings. Singularities can be classified only up to a weaker notion of equivalence than is useful for our purposes.

We do have a necessary condition and a sufficient condition for being a multiple point, and these are useful in practice. Without loss of generality, we assume that the point in question is the origin.

Proposition 10.1.14 *Let $H \in \mathbf{R} := \mathbf{R}_0$ and let $\tilde{h} = \mathtt{hom}(H)$. Let $\tilde{h} = h_1 \dots h_n$ be a factorization into homogeneous irreducible polynomials.*

- *If $\mathbf{0}$ is a multiple point of $\mathcal{V}$, then each h_i is linear.*
- *If each h_i is linear and all the h_i are distinct, then $\mathbf{0}$ is a multiple point of $\mathcal{V}$.*

Proof We know that $\mathbf{0}$ is a multiple point of $\mathcal{V}$ if and only if each irreducible factor of H in $\mathbf{R}$ has order of vanishing 1. This implies the first conclusion. If the second condition holds, then each irreducible factor of H in $\mathbf{R}$ vanishes to order 1, because the homogeneous part of H is the product of the distinct linear factors. □

Remark When the linear factors of $\tilde{h}$ are not distinct, we may or may not have a multiple point, because a product of two or more s_i may correspond to a leading term of some h_j. For example, consider two sheets that may or may not coincide in a neighborhood of $\mathbf{0}$, where we only have access to the factors as power series. We know no algorithm to decide whether the sheets do coincide and thus whether we are dealing with a repeated factor and a smooth point or distinct factors and a multiple point. We can rule out a transverse multiple point or an arrangement point easily via first-order information.

This situation is not as bad as it may seem. First, we shall see that asymptotic formulae for multiple points are such that each coefficient typically depends only on a finite number of derivatives of G and H, and so we have no need to answer the question definitively. Second, in many applications we are dealing with factors of algebraic functions that are themselves algebraic, in which case it is possible to distinguish the factors. Finally, when $d = 2$, the methods of Newton and Puiseux allow us to distinguish the factors in general.

Example 10.1.15 Let

$$H := 1 + xyz - \frac{1}{3}(x + y + z + xy + xz + yz)$$

be the denominator of the cube grove generating function from (8.6.6). The ideal generated by

$$\left\langle H, \frac{\partial H}{\partial x}, \frac{\partial H}{\partial y}, \frac{\partial H}{\partial z} \right\rangle$$

turns out to be $[x - 1, y - 1, z - 1]$; in other words, there is a single critical point at $(1, 1, 1)$. In this case, the polynomial h may be obtained directly as

$$h(u, v, w) = H(1 + u, 1 + v, 1 + w) = 2uv + 2uw + 2vw + 3uvw.$$

Had the solution $(1, 1, 1)$ not been explicitly computable, the same result could have been obtained by letting h be the elimination polynomial in (u, v, w) for the equations $H = H_x = H_y = H_z = 0$ along with $H(x + u, y + v, z + w)$. We remark that the simplification due to being able to represent the singular point explicitly is actually the rule rather than the exception: when a zero-dimensional variety consists of a single point, then this point is always rational (see Exercise 6.4).

Continuing, we let $\tilde{h} = 2(uv + uw + vw)$ be the leading homogeneous part of h. It is easily seen by writing down an attempted factorization that $\tilde{h}$ does not factor into linear factors, and hence h does not and $(1, 1, 1)$ is not a multiple point. The asymptotics for the coefficients of $F := 1/H$ are discussed in the next chapter, on cone points (see Example 11.3.3). ■

Remark Example 10.1.15 could be dealt with in a more sophisticated way, which may be useful for more complicated problems. If the complete factorization of $\tilde{h}$ into linear factors is not required, but only confirmation of its existence, a shortcut can be used. The set of homogeneous polynomials that do factor into linear factors is an algebraic variety that can be defined by explicit equations involving the coefficients of the polynomial. Various choices for these equations are possible and have appeared in the literature – see Briand (2010).

10.2 Iterated Residues

The theory of residues for simple poles on smooth complex hypersurfaces was given in Section 9.3. Key facts about the residue form of a meromorphic function $F = G/H$ are that it lives on the hypersurface $\mathcal{V}_H$ and that the integral of F is equal to the integral of its residue over an appropriately defined intersection class. In Section 9.5.4, the definition was extended to poles of order greater than 1. Computing the asymptotic contribution at a multiple point requires the extension of this theory to residues with respect to several hypersurfaces simultaneously. The quasi-local cycles whose integrals these compute are always a product of a k-torus with tori with a patch in $\mathbb{C}^{d-k}$. Rather than proving the existence of intersection classes in general, it is easier to prove what we need about iterated residues in this special case.

10.2.1 Transverse Multiple Points: Simple Poles

Let $\boldsymbol{p}$ be a square-free transverse multiple point for the meromorphic function F represented in $\mathbf{R}_{\boldsymbol{p}}$ as $G/\prod_{j=1}^{k} H_j$. Let $S := \bigcap_{j=1}^{k} \mathcal{V}_j$ denote the stratum of $\mathcal{V}$ containing $\boldsymbol{p}$. Recall that $\mathcal{M}$ has a local product structure $\tilde{N} \times S$ over some neighborhood U of $\boldsymbol{p}$ in $\mathbb{C}^d$. Because $\boldsymbol{p}$ is a transverse multiple point of order k, the normal link $\tilde{N}$ is homotopy equivalent to a k-torus. A convenient cycle representative for the generator of $H_k(\tilde{N})$ is a product of circles about $\boldsymbol{p}$ in each $\mathcal{V}_j$.

To make this more precise, we define a local coordinate system as follows. Choose a subset of $d - k$ coordinates that locally coordinatize S, i.e., for which the projection of S onto this $(d - k)$-dimensional subspace is nonsingular. The map $\Psi : U \to \mathbb{C}^d$ defined by

$$\Psi(\boldsymbol{z}) := \big(H_1(\boldsymbol{z}), \ldots, H_k(\boldsymbol{z}), z_{\pi(1)}, \ldots, z_{\pi(d-k)}\big) \tag{10.2.1}$$

is a bi-analytic change of coordinates taking $U \cap \mathcal{V}$ to a neighborhood of the origin in $\{\mathbf{0}\} \times \mathbb{C}^{d-k}$. Let $T_\varepsilon \subseteq \mathbb{C}^k \times \{\mathbf{0}\}$ denote the product of circles of radius ε in each of the first k coordinates. Taking ε small enough $T_\varepsilon \in \Psi[U]$, the cycle $T := \psi^{-1}[T_\varepsilon]$ will be a generator for $H_k(\tilde{N})$. Here and in what follows, we give U the local product structure that Ψ^{-1} induces from the product structure on $\mathbb{C}^d$; in particular, the normal slices $\tilde{N}$ are fibers of Ψ.

Definition 10.2.1 *For each $z \in S$, let $\mathbf{\Gamma}_\Psi(z)$ denote the* ***augmented lognormal matrix*** *whose first k rows are the lognormal vectors $\{\nabla_{\log} H_j(z) : 1 \le j \le k\}$ and whose last $d-k$ rows are the vectors $\{z_{\pi(j)} e_{\pi(j)} : 1 \le j \le d-k\}$. In other words, $\mathbf{\Gamma}_\Psi = DJ(\Psi)$ where D is the diagonal matrix with entries $z_1, \ldots, z_d$ and $J(\Psi)$ is the Jacobian matrix of the map Ψ.*

In the following construction of the iterated residue, we use the notation $\tilde{G}$ for $z^{-r-1}G$ to emphasize that the factor of z^{-r-1} plays no special role.

Theorem 10.2.2 *Let S be a stratum of transverse multiple points of the variety $\mathcal{V}$ locally decomposed as $\bigcup_{j=1}^k \mathcal{V}_j$ with $S = \bigcap_{j=1}^k \mathcal{V}_j$.*

(i) For any form $\eta = \tilde{G}\, dz / \prod_{j=1}^k H_j$ with simple poles on each $\mathcal{V}_j$, the restriction to S of any θ satisfying

$$dH_1 \wedge \cdots \wedge dH_k \wedge \theta = \tilde{G}\, dz \tag{10.2.2}$$

is independent of the particular solution θ.

(ii) Denoting this restriction by Res$(\eta; S)$, *for any $\mathbf{p} \in S$ there is a formula*

$$\operatorname{Res}\left(\frac{\tilde{G}\, dz}{\prod_{j=1}^k H_j}; S\right)(\mathbf{p}) := \frac{\tilde{G}(\mathbf{p})}{\det J(\Psi)(\mathbf{p})} dz_{\pi(1)} \wedge \cdots \wedge dz_{\pi(d-k)} . \tag{10.2.3}$$

(iii) Let σ be any $(d-k)$ chain in $S \cap U$. Then

$$\int_{T\times\sigma} \frac{\tilde{G}(z)\, dz}{\prod_{j=1}^k H_j(z)} = (2\pi i)^k \int_\sigma \operatorname{Res}\left(\frac{\tilde{G}\, dz}{\prod_{j=1}^k H_j}; S\right).$$

(iv) In particular,

$$\int_{T\times\sigma} \frac{z^{-r-1} G(z)\, dz}{\prod_{j=1}^k H_j(z)} = (2\pi i)^k \int_\sigma \frac{z^{-r} G(\mathbf{p})}{\det \mathbf{\Gamma}_\Psi(\mathbf{p})} dz_{\pi(1)} \wedge \cdots \wedge dz_{\pi(d-k)} . \tag{10.2.4}$$

Proof We first prove all four parts under the assumption that $H_j = z_j, 1 \le j \le k$. Set $\pi(i) = i_k$, $1 \le i \le d-k$. Clearly $\theta = \tilde{G}\, dz_{k+1} \wedge \cdots \wedge dz_d$ satisfies (10.2.2). Exactly as in the proof of Proposition A.5.1, the result of Exercise A.3 implies that $\iota^*\theta$ is well defined, yielding (i). The formula (10.2.3) is evident in this case: $J(\Psi)$ is the identity matrix, and hence (10.2.3) agrees with our choice of θ, proving (ii).

For (*iii*), write the left-hand side as an iterated integral

$$\int_{\sigma}\int_{\gamma_1}\cdots\int_{\gamma_k},$$

where γ_j is the circle of radius ε about the origin in coordinate j. Applying the univariate residue theorem to each of the k inner integrals in turn leaves the outer integral equal to

$$(2\pi i)^k \int_{\sigma} \tilde{G}(z)\, dz_{k+1} \wedge \cdots \wedge dz_d,$$

which proves (*iii*). Finally, (*iv*) follows from (*iii*) by plugging $\tilde{G} = z^{-r-1}G$ into (10.2.3) and absorbing one factor of each z_j in the denominator when going from $\det J(\Psi)$ to $\det \mathbf{\Gamma}_{\Psi}$.

For the general case, map by Ψ and use functoriality. The fact the Res is well defined and functorial follows exactly as in the proof of Proposition A.5.1. Applying the case already proved to the image space and pulling back by Ψ^{-1}, it remains only to observe that $\tilde{G}(\mathbf{0})$ pulls back to $\tilde{G}(\boldsymbol{p})$ and that $dz_{k+1} \wedge \cdots \wedge dz_d$ pulls back to $dz_{\pi(1)} \wedge \cdots \wedge dz_{\pi(d-k)} / \det J(\Psi)(\boldsymbol{p})$. □

Remarks

(i) The residue depends on H_j only via its gradient.
(ii) When the stratum S is a single point ($k = d$), the residue at $\boldsymbol{p}$ is just a number. In particular, it does not depend on $\boldsymbol{r}$.

Example 10.2.3 (two lines in $\mathbb{C}^2$, continued) Continuing Example 10.1.3, let $H := \left(1 - \frac{1}{3}x - \frac{2}{3}y\right)\left(1 - \frac{2}{3}x - \frac{1}{3}y\right)$, which has a transverse multiple point at $(x, y) = (1, 1)$. The gradients of the factors of H are $(1/3, 2/3)$ and $(2/3, 1/3)$. Whenever $z = \mathbf{1}$, the logarithmic gradients are the same as the gradients. The determinant of $\mathbf{\Gamma}_{\Psi}$ is therefore equal to $\pm 1/3$, the sign choice depending on the order in which we choose the factors. Up to sign, the iterated residue of $dx \wedge dy/H$ at $(1, 1)$ is the number 3. ■

Example 10.2.4 (queueing partition function) In this example, we take

$$F(x, y) = \frac{\exp(x + y)}{(1 - \rho_{11}x - \rho_{21}y)(1 - \rho_{12}x - \rho_{22}y)}$$

to be the partition generating function for a closed multiclass queueing network with one infinite server (see Pemantle and Wilson, 2008, Section 4.12 or Kogan, 2002, Equation [44]; or Bertozzi and McKenna, 1993, Equation [2.26] for details on queueing networks and their generating functions). Without loss of generality, for the most interesting case as discussed in Pemantle and Wilson (2008), we assume $\rho_{11} > \rho_{12}, \rho_{22} > \rho_{21}$ and $D := \rho_{11}\rho_{22} - \rho_{12}\rho_{21} > 0$. The two linear divisors intersect in the positive real quadrant at the point $(x_0, y_0) := \left(\frac{\rho_{22} - \rho_{21}}{D}, \frac{\rho_{11} - \rho_{12}}{D}\right)$. The residue of $F\, dx \wedge dy$ at (x_0, y_0), according to (10.2.3), is given by

$\exp(x_0 + y_0)/D$. In particular, if $\rho_{11} = \rho_{22} = 2/3$ and $\rho_{12} = \rho_{21} = 1/3$ then $x_0 = y_0 = 1$, $D = 1/3$, and the residue is the number $3e^2$. ■

Example 10.2.5 ($d = 3, k = 2$) Consider the generating function

$$F(x, y, z) := \frac{16}{(4 - 2x - y - z)(4 - x - 2y - z)}$$

whose coefficients satisfy a simple recurrence (see Pemantle and Wilson, 2004, Example 3.10). The divisors are two planes meeting at the complex line $S := \{(1, 1, 1) + \lambda(-1, -1, 3) \colon \lambda \in \mathbb{C}\}$. The gradients to the divisors are $(2, 1, 1)$ and $(1, 2, 1)$. At the point $(x_0, y_0, z_0) := (1, 1, 1)$, the logarithmic gradients are equal to the gradients. We may parametrize S by any of its three coordinates. Choosing the third coordinate, i.e., $\pi(1) = 3$, we obtain

$$\mathbf{\Gamma}_\Psi = \begin{bmatrix} 2 & 1 & 1 \\ 1 & 2 & 1 \\ 0 & 0 & 1 \end{bmatrix},$$

whence $\det \mathbf{\Gamma}_\Psi = 3$ and

$$\mathrm{Res}(F\, dx \wedge dy \wedge dz; S) = \frac{16}{3} dz\,.$$

Choosing one of the first two coordinates leads to an equivalent answer. The first two rows of $\mathbf{\Gamma}_\Psi$ are still $(2, 1, 1)$, $(1, 2, 1)$, whereas the third row becomes either $(1, 0, 0)$ or $(0, 1, 0)$. The determinants are (up to sign) equal to 1 now, instead of 3, which makes the residues equal to $16\,dx$ or $16\,dy$. These are both, up to sign, equal to $(16/3)\,dz$ on S. ■

10.2.2 Transverse Multiple Points: Higher Order Poles

In this section, we extend Theorem 10.2.2 to higher order poles. The method is somewhat analogous to the extension of smooth point analysis to higher order poles in Section 9.5.4. Let Ψ be the parametrization defined in (10.2.1), let E denote the corresponding set of indices $\{\pi(1), \ldots \pi(d - k)\}$, and let dz_E and z^E denote $dz_{\pi(1)} \wedge \cdots \wedge dz_{\pi(d-k)}$ and $\prod_{j=1}^{d-k} z_{\pi(j)}$ respectively.

Theorem 10.2.6 *Let S be a stratum of transverse multiple points of the variety $\mathcal{V}$ locally decomposed as $\bigcup_{j=1}^k \mathcal{V}_j$ with $S = \bigcap_{j=1}^k \mathcal{V}_j$. Let $\boldsymbol{m} = (m_1, \ldots, m_k)$ be a vector of positive integers, and let $P(\boldsymbol{r}) = P(\boldsymbol{r}, \boldsymbol{p})$ be the quantity defined in* (10.2.11) *and* (10.2.12).

(i) P is a polynomial in $\boldsymbol{r}$. The leading homogeneous part of P has degree $|\boldsymbol{m}| - k$ and is given by the following product:

$$\frac{1}{(\boldsymbol{m} - \mathbf{1})!} \frac{G(\boldsymbol{p})}{\det \mathbf{\Gamma}_\Psi(\boldsymbol{p})} (\boldsymbol{r}\mathbf{\Gamma}_\Psi^{-1})^{\boldsymbol{m}-\mathbf{1}}. \tag{10.2.5}$$

(ii) Let σ be any $(d-k)$ chain in $S \cap U$. Then

$$\int_{T\times\sigma} \frac{z^{-r-1} G(z)\, dz}{\prod_{j=1}^{k} H_j(z)^{m_j}} = (2\pi i)^k \int_{\sigma} P(\boldsymbol{r}, z) \frac{\boldsymbol{p}^{-r}}{z^E}\, dz_E\,. \tag{10.2.6}$$

(iii) The Res *operator defined by*

$$\mathrm{Res}\left(\frac{z^{-r-1} G\, dz}{\prod_{j=1}^{k} H_j^{m_j}}\,;\, S \right)(\boldsymbol{p}) := P(\boldsymbol{r}, \boldsymbol{p}) \frac{\boldsymbol{p}^{-r}}{z^E}\, dz_E \tag{10.2.7}$$

maps functorially under bi-analytic mappings.

Remark We use the following multi-index notation for the powers and the factorial. The factorial expands to $\prod_j m_j - 1)!$. The power $\boldsymbol{r}^{\boldsymbol{m}}$, where $\boldsymbol{r}$ has dimension d and $\boldsymbol{m}$ has dimension $k \le d$, is given by $\prod_{j=1}^{k} r_j^{m_j}$.

Example 10.2.7 To clarify some of the terms involved, we include an illustration with no particular combinatorial significance. Let a and b be positive integers and consider the function

$$F(x, y, z) := \frac{16}{(4 - 2x - y - z)^a\, (4 - x - 2y - z)^b}$$

generalizing the function in Example 10.2.5. Choosing again to parametrize S by the third coordinate, the previous residue of $(16/3)\, dz$ must be multiplied by $1/(\boldsymbol{m}-\boldsymbol{1})!$ and by $(\boldsymbol{r}\,\boldsymbol{\Gamma}_{\boldsymbol{\Psi}}^{-1})^{\boldsymbol{m}-\boldsymbol{1}}$. the product of $\boldsymbol{r} = (r, s, t)$ with $\boldsymbol{\Gamma}_{\boldsymbol{\Psi}}^{-1}$ is $\left(\frac{2r-s}{3}, \frac{2s-r}{3}, \frac{3t-r-s}{3}\right)$. The multi-exponent $(a-1, b-1)$ ignores the third of these, giving $\left(\boldsymbol{r}\,\boldsymbol{\Gamma}_{\boldsymbol{\Psi}}^{-1}\right)^{\boldsymbol{m}-\boldsymbol{1}} = \left(\frac{2r-s}{3}\right)^{a-1} \left(\frac{2s-r}{3}\right)^{b-1}$. Thus the leading term of P is equal to

$$\frac{16}{3\,(a-1)!\,(b-1)!} \left(\frac{2r-s}{3}\right)^{a-1} \left(\frac{2s-r}{3}\right)^{b-1},$$

and the residue at $(1, 1, 1)$ is this times dz/z. ■

Before embarking on the proof of Theorem 10.2.6, it is useful to remark on a difference between this theorem and the previous one; more details on cohomology may be found, if necessary, in Section A.3 of the appendixes. In Theorem 10.2.2, the simple residue was defined by the relation (10.2.2) and shown to satisfy the explicit equation (10.2.3). By contrast, the higher residue is defined by an explicit equation (10.2.7). This is because Theorem 10.2.6 works at the level of cohomology classes, satisfying the integral equation (10.2.6) rather than an equation on the level of individual forms.

More specifically, let Y denote the complement of $\mathcal{V}$ in the local coordinate neighborhood U. This is the product of the normal link $\tilde{N}$ with the space S.

For a transverse multiple point, the homology of $\tilde{N}$ is one-dimensional. The (complex) de Rham cohomology group $H^k(\tilde{N})$ is the space of equivalence classes of holomorphic k-forms on $\tilde{N}$ modulo the exact forms, these being the forms $d\eta$ where η is any holomorphic $(k-1)$-form. Cohomology is dual to homology via integration. Thus, if the k-forms η_1 and η_2 are equal in $H^k(\tilde{N})$, then their integrals are the same over the homology generator in $H_k(\tilde{N})$. We extend the notion of cohomological equivalence of forms to d forms on U by saying

$$[\eta_1 \wedge du_{k+1} \wedge \cdots \wedge du_d] = [\eta_2 \wedge du_{k+1} \wedge \cdots \wedge du_d]$$

if for all $\boldsymbol{p}$ in S, we have $[\eta_1] = [\eta_2]$ when restricted to $\tilde{N}(\boldsymbol{p})$. Two equivalent forms will then have the same integrals over products $T \times \sigma$, where T is a cycle and σ is any chain.

Proof (of Theorem 10.2.6)

As in the proof of Theorem 10.2.2, we tackle first the special case where $H_j = u_j$. It suffices to prove equality of the integrals over each $\tilde{N}(\boldsymbol{p})$. Accordingly, fix any index t with $1 \le t \le k$ and define a $(k-1)$-form η by

$$\eta := \frac{\tilde{G}\, d\boldsymbol{u}_{\hat{t}}}{\prod_{j=1}^{k} u_j^{m_j - \delta_{jt}}},$$

where $d\boldsymbol{u}_{\hat{t}}$ denotes the form $du_1 \wedge \cdots \wedge du_{t-1} \wedge du_{t+1} \wedge \cdots \wedge du_k$. By direct computation,

$$d\eta = \frac{(\partial/\partial u_t)\tilde{G}\, d\boldsymbol{u}}{\boldsymbol{u}^{\boldsymbol{m}-\delta_{jt}}} - \frac{(m_t - 1)\tilde{G}\, d\boldsymbol{u}}{\boldsymbol{u}^{\boldsymbol{m}}},$$

and the fact that $[d\eta] = 0$ implies that

$$\left[\frac{\tilde{G}\, d\boldsymbol{u}}{\boldsymbol{u}^{\boldsymbol{m}}}\right] = \frac{1}{m_t - 1}\left[\frac{(\partial/\partial u_t)\tilde{G}\, d\boldsymbol{u}}{\boldsymbol{u}^{\boldsymbol{m}-\delta_{jt}}}\right].$$

Applying this maneuver $m_t - 1$ times for each each $1 \le t \le k$ then yields

$$\left[\frac{\tilde{G}\, d\boldsymbol{u}}{\boldsymbol{u}^{\boldsymbol{m}}}\right] = \frac{1}{(\boldsymbol{m} - \boldsymbol{1})!}\left[\frac{(\partial/\partial \boldsymbol{u})^{\boldsymbol{m}-\boldsymbol{1}}\tilde{G}\, d\boldsymbol{u}}{\boldsymbol{u}^{\boldsymbol{1}}}\right]. \tag{10.2.8}$$

We put together what we have so far, using (10.2.8) in the first line, applying Theorem 10.2.2 in the second line to the function $(\partial/\partial \boldsymbol{u})^{\boldsymbol{m}-\boldsymbol{1}}\tilde{G}$ and all $H_j = u_j$, and using (10.2.3) in the third line:

$$\int_{T_\varepsilon} \frac{\tilde{G}(\boldsymbol{u})\,d\boldsymbol{u}}{\boldsymbol{u}^{\boldsymbol{m}}} = \frac{1}{(\boldsymbol{m}-\boldsymbol{1})!} \int_{T_\varepsilon} \frac{\left(\frac{\partial}{\partial \boldsymbol{u}}\right)^{\boldsymbol{m}-\boldsymbol{1}} \tilde{G}(\boldsymbol{u})\,d\boldsymbol{u}}{\boldsymbol{u}^{\boldsymbol{1}}}$$

$$= \frac{(2\pi i)^k}{(\boldsymbol{m}-\boldsymbol{1})!} \operatorname{Res}\left(\frac{\left(\frac{\partial}{\partial \boldsymbol{u}}\right)^{\boldsymbol{m}-\boldsymbol{1}} \tilde{G}(\boldsymbol{u})\,d\boldsymbol{u}}{\boldsymbol{u}^{\boldsymbol{1}}} ; \{\boldsymbol{p}\}\right)$$

$$= \frac{(2\pi i)^k}{(\boldsymbol{m}-\boldsymbol{1})!} \left(\frac{\partial}{\partial \boldsymbol{u}}\right)^{\boldsymbol{m}-\boldsymbol{1}} \tilde{G}(\boldsymbol{u}) . \tag{10.2.9}$$

We need some notation for the coordinates of Ψ^{-1}; because this is the map from $\boldsymbol{u}$ back to z, the most obvious "physics" notation is to denote $\Psi^{-1}(\boldsymbol{u})$ by $z(\boldsymbol{u}) = (z_1(\boldsymbol{u}), \ldots, z_d(\boldsymbol{u}))$. This notation is primarily for manipulating monomial powers such as $z(\boldsymbol{u})^{-\boldsymbol{r}-\boldsymbol{1}}$. To set up the final computation, we need to take

$$\tilde{G}(\boldsymbol{u}) = \frac{z(\boldsymbol{u})^{-\boldsymbol{r}-\boldsymbol{1}} G(\psi^{-1}(\boldsymbol{u}))}{J_\Psi(\Psi^{-1}(\boldsymbol{u}))} ,$$

which is chosen so that

$$\Psi^* \left(\frac{\tilde{G}(\boldsymbol{u})\,d\boldsymbol{u}}{\boldsymbol{u}^{\boldsymbol{m}}}\right) = \frac{z^{-\boldsymbol{r}-\boldsymbol{1}}\,dz}{\prod_{j=1}^k H(z)^{m_j}} .$$

We may now compute the higher order integral by mapping forward by Ψ, using (10.2.8) for this special case, then pulling back to the original space by Ψ^{-1}. In what follows, σ is any chain in S, and $d\boldsymbol{u}$ now denotes the d-form $du_1 \wedge \cdots du_d$.

$$\int_{T\times\sigma} \frac{z^{-\boldsymbol{r}-\boldsymbol{1}} G(z)\,dz}{\prod_{j=1}^k H_j^{m_j}}$$

$$= \int_{T_\varepsilon\times\Psi(\sigma)} \frac{\tilde{G}(\boldsymbol{u})\,d\boldsymbol{u}}{\boldsymbol{u}^{\boldsymbol{m}}}$$

$$= \frac{(2\pi i)^k}{(\boldsymbol{m}-\boldsymbol{1})!} \int_{\Psi(\sigma)} \left(\frac{\partial}{\partial \boldsymbol{u}}\right)^{\boldsymbol{m}-\boldsymbol{1}} \tilde{G}(\boldsymbol{u})\,du_{k+1} \wedge \cdots \wedge du_d$$

$$= \frac{(2\pi i)^k}{(\boldsymbol{m}-\boldsymbol{1})!} \int_{\Psi(\sigma)} \left(\frac{\partial}{\partial \boldsymbol{u}}\right)^{\boldsymbol{m}-\boldsymbol{1}} \frac{z(\boldsymbol{u})^{-\boldsymbol{r}-\boldsymbol{1}} G(\Psi^{-1}(\boldsymbol{u}))}{J_\Psi(\Psi^{-1}(\boldsymbol{u}))}\,du_{k+1} \wedge \cdots \wedge du_d$$

$$= \frac{(2\pi i)^k}{(\boldsymbol{m}-\boldsymbol{1})!} \int_\sigma \left(\frac{\partial}{\partial \boldsymbol{u}}\right)^{\boldsymbol{m}-\boldsymbol{1}} \left(\frac{z(\boldsymbol{u})^{-\boldsymbol{r}-\boldsymbol{1}} G(\Psi^{-1}(\boldsymbol{u}))}{J_\Psi(\Psi^{-1}(\boldsymbol{u}))}\right)\Bigg|_{\boldsymbol{u}=\Psi(z)} dz_E . \tag{10.2.10}$$

It looks (and is) quite messy to compute this order $|\boldsymbol{m}| - k$ mixed partial derivative. The following lemma skips the details while revealing the form of the result.

Lemma 10.2.8 *Let f, $f_1, \ldots, f_d$ be smooth functions of $\boldsymbol{u} \in \mathbb{C}^d$. Then*

$$\left(\frac{\partial}{\partial \boldsymbol{u}}\right)^{\boldsymbol{n}} f(\boldsymbol{u}) f_1(\boldsymbol{u})^{r_1} \cdots f_d(\boldsymbol{u})^{r_d} = f(\boldsymbol{u}) f_1(\boldsymbol{u})^{r_1} \cdots f_d(\boldsymbol{u})^{r_d} P(\boldsymbol{r}, \boldsymbol{u}),$$

where P depends polynomially on $\boldsymbol{r}$ and has degree $|\boldsymbol{n}|$. The leading term of P is $(\ell)^{\boldsymbol{n}} := \prod_{j=1}^{k} \ell_j^{n_j}$, where

$$\ell_j := \sum_i r_i \frac{\partial \log f_i}{\partial u_j}.$$

Proof We show by induction that for all $\boldsymbol{n}$,

$$\left(\frac{\partial}{\partial \boldsymbol{u}}\right)^{\boldsymbol{n}} f(\boldsymbol{u}) f_1(\boldsymbol{u})^{r_1} \cdots f_d(\boldsymbol{u})^{r_d} = f(\boldsymbol{u}) f_1(\boldsymbol{u})^{r_1} \cdots f_d(\boldsymbol{u})^{r_d} \left[(\ell)^{\boldsymbol{n}} + Q(\boldsymbol{r}, \boldsymbol{u})\right],$$

where Q is a polynomial of degree less than $|\boldsymbol{n}|$. When $\boldsymbol{n} = \boldsymbol{0}$, this holds with $Q = 0$. Assuming this holds for $\boldsymbol{n}$, differentiating with respect to u_j replaces $(\ell)^{\boldsymbol{n}} + Q$ by

$$\ell_j((\ell)^{\boldsymbol{n}} + Q) + \frac{\partial}{\partial u_j}((\ell)^{\boldsymbol{n}} + Q).$$

The terms other than $(\ell)^{\boldsymbol{n}+\delta_j}$ have degree at most $\boldsymbol{n}$, completing the induction. □

To continue the proof of Theorem 10.2.6, apply the lemma with $\boldsymbol{n} = \boldsymbol{m} - \boldsymbol{1}$, $f_i = z_i$, and $f = (\tilde{G}(\Psi^{-1}(\boldsymbol{u}))/J_\Psi(\Psi^{-1}(\boldsymbol{u}))$. The result is that

$$\left(\frac{\partial}{\partial \boldsymbol{u}}\right)^{\boldsymbol{m}-\boldsymbol{1}} \left(\frac{\boldsymbol{z}(\boldsymbol{u})^{-\boldsymbol{r}} G(\Psi^{-1}(\boldsymbol{u}))}{\prod_{j=1}^{d} z_j(\boldsymbol{u})\, J_\Psi(\Psi^{-1}(\boldsymbol{u}))}\right) = \frac{\boldsymbol{z}(\boldsymbol{u})^{-\boldsymbol{r}} G(\Psi^{-1}(\boldsymbol{u}))}{\prod_{j=1}^{d} z_j(\boldsymbol{u})\, J_\Psi(\Psi^{-1}(\boldsymbol{u}))} P(\boldsymbol{r}, \boldsymbol{p}), \tag{10.2.11}$$

where

$$P(\boldsymbol{r}, \boldsymbol{p}) = \left[\prod_{j=1}^{d} \left(\sum_i r_i \frac{\partial \log z_i}{\partial u_j}\right)^{m_j - 1} + Q(\boldsymbol{r}, \boldsymbol{p})\right] \tag{10.2.12}$$

with $\deg(Q) < |\boldsymbol{m}| - k$. Note that the exponent $-\boldsymbol{r} - \boldsymbol{1}$ in (10.2.10) has been replaced by $-\boldsymbol{r}$ due to incorporation of one power of each z_j into the denominator. The matrix $\boldsymbol{\Gamma}_\Psi$ is equal to $\left(\frac{\partial u_i}{\partial \log z_j}\right)$, whence $\frac{\partial \log z_i}{\partial u_j} = \left(\boldsymbol{\Gamma}_\Psi^{-1}\right)_{ij}$. Evaluating (10.2.12) at $\boldsymbol{u} = \Psi(\boldsymbol{p})$ and simplifying via $(\prod_{j \notin E} p_j) J_\Psi(\boldsymbol{p}) = \det \boldsymbol{\Gamma}_\Psi(\boldsymbol{p})$ yields

$$\left(\frac{\partial}{\partial \boldsymbol{u}}\right)^{\boldsymbol{m}-\boldsymbol{1}} \left(\frac{\boldsymbol{z}(\boldsymbol{u})^{-\boldsymbol{r}} G(\Psi^{-1}(\boldsymbol{u}))}{\prod_{j=1}^{d} z_j(\boldsymbol{u}) J_\Psi(\Psi^{-1}(\boldsymbol{u})}\right) = \frac{\boldsymbol{p}^{-\boldsymbol{r}} G(\boldsymbol{p})}{\det \boldsymbol{\Gamma}_\Psi(\boldsymbol{p})} \frac{P(\boldsymbol{r}, \boldsymbol{p})}{z^E}.$$

Plugging this into (10.2.10) gives (10.2.6). The construction of Res extends the definition in the case $H_j = u_j$ via a change of coordinates, which makes functoriality automatic. □

Remarks

(i) The leading term (10.2.5) depends on the divisors H_j only through their gradients.
(ii) When the stratum S is a single point ($k = d$), the residue at $\boldsymbol{p}$ is a 0-form, i.e., to say a number $P(\boldsymbol{r})$ varying polynomially with $\boldsymbol{r}$.

10.2.3 Arrangement Points

Next we consider residues of the class of forms $\tilde{G}\, dz / \prod_{j=1}^n H_j^{m_j}$ near an arrangement point $\boldsymbol{p} \in S = \bigcap_{j=1}^n \mathcal{V}_j$. Infact, all such residues may be expressed in terms of the residues we have already computed. This is in fact true at the level of forms and is accomplished via algebraic identities given in the following lemma. Recall that a circuit of the matroid $M(\mathcal{A})$ is a minimal dependent set and that the arrangement $\mathcal{A}$ of hyperplanes tangent to the surfaces $\mathcal{V}_j$ at $\boldsymbol{p}$ is independent of the particular choice of $\boldsymbol{p} \in S$.

Lemma 10.2.9 *Let C be any circuit of $M(\mathcal{A})$, where $\mathcal{A}$ is the arrangement of tangent hyperplanes to the varieties $\mathcal{V}_j$ at any point of S. There is a collection $\{g_i : i \in C\}$ of invertible elements of $\mathbf{R}_p$ such that $\sum_{i \in C} g_i H_i = 0$.*

Proof Fix any $i \in C$. The containment $\bigcap_{j \neq i} \mathcal{V}_j \subseteq \mathcal{V}_i$ is equivalent to H_i being in the radical of the ideal of $\mathbf{R}_p$ generated by $\{H_j : j \neq i\}$. This ideal is radical (see Lemma 10.1.13), whence $H_i = \sum_{j \neq i} g_j H_j$ for some elements g_j of $\mathbf{R}_p$. When $z \in S$, the gradient $\nabla(g_j H_j)$ is equal to $g_j\, \nabla(H_j)$. Taking the gradient of the equation $H_i = \sum_{j \neq i} g_j H_j$ shows that

$$\nabla H_i = \sum_{j \neq i} g_j \nabla H_j \,.$$

To see that the functions $\{g_j\}$ are all units, observe first that linear independence of the gradients on the right-hand side implies that the values $\{g_j(z)\}$ are uniquely determined when $z \in S$. Now compare these representations with two different choices, say i and t:

$$H_i - \sum_{j \neq i} g_j H_j = 0$$

$$H_t - \sum_{j \neq t} f_j H_j = 0$$

By uniqueness, the second of these must be f_i times the first at any point $z \in S$, which shows that $f_j g_t = 1$. Because t was arbitrary, we see that $g_j(z) \neq 0$ for all $j \neq i$ and all $z \in S$. Elements of $\mathbf{R}_p$ vanishing nowhere on S are invertible. □

Dividing through by $\prod_{j=1}^{k} H_j$ gives the identity we need. Letting H_{-i} denote the product $\prod_{j\neq i} H_j$ of all the divisors except H_i,

$$\frac{1}{H_{-i}} = \sum_{j\neq i} \frac{g_j}{H_{-j}}.$$

More generally, if $m_j > 0$ for all $j \neq i$ in some circuit C, then in monomial notation,

$$\frac{1}{H^{m}} = \sum_{t\neq j} \frac{g_j}{H^{m-\delta_j+\delta_i}}. \tag{10.2.13}$$

This identity replaces one term with many, but, if iterated in an organized manner, can be made to eliminate terms with linear dependence among the divisors. Define the support $\mathrm{supp}(g/\prod_{i\in S} H_i^{n_i})$ of a monomial fraction to be the set $T \subseteq S$ of indices of nonzero exponents (i.e., those i for which $n_i > 0$). The set of monomials with independent support spans the space of all monomials but is too big to be a basis. Consequently, the end result of repeated application of (10.2.13) is not well-defined. It is useful to choose a basis and a well defined reduction algorithm. One convenient choice is as follows. Say that a set B is a ***broken circuit*** if $B = C \setminus \{i\}$ for some circuit C with greatest element i.

Proposition 10.2.10 *Suppose the reduction procedure consists of repeatedly choosing a monomial whose support contains a broken circuit and replacing the left-hand side of* (10.2.13) *for that monomial by the right-hand side. Such a procedure must terminate in finite time in a set of monomials, none of whose supports contains a broken circuit. One cannot in general reduce further because the set of forms $\omega_T := dz/\prod_{j\in T} H_j$ is linearly independent.*

Proof The procedure replaces each monomial with monomials all of whose exponent vectors majorize the original monomial. Therefore, each monomial can only be replaced a finite number of times, according to its depth in the majorization order. Termination of the algorithm follows. Independence, in the case where H_j are linear, is a result of Orlik and Terao (1992, Theorems 3.43, 3.126, and 5.89). There, they show that the complex BC of bases of $\mathcal{A}$ containing no broken circuit indexes the Orlik-Solomon algebra, which is isomorphic to another algebra shown to be isomorphic to $H^*(Y)$. This implies independence in the case of linear divisors, from which the general case follows. □

Let $f = g/\prod_{j\in T} H_j^{m_j}$ be a monomial fraction with support set $T \in \mathrm{BC}$. Suppose that g vanishes on S. Then g is in the unique maximal ideal of the ring of analytic functions in a neighborhood of S, and it follows that $g = \sum_{i\in T} g_i H_i$. Making this substitution and canceling common factors replaces f by a sum of monomial fractions for which the degree of the denominator is at most $|\boldsymbol{m}| - 1$. Let $\mathcal{I}(f)$ denote the ideal $\langle H_j^{m_j} : j \in T\rangle$ of $\mathbf{R}_{\boldsymbol{p}}$. If $g \in \mathcal{I}$, then clearly one may cancel all the way down to a sum of monomial fractions whose support is a proper subset of T, and the converse also holds.

Putting together the algebraic reductions for linear dependences and vanishing numerators gives the following reduction scheme, which allows us to reduce to the transverse case.

Algorithm 10.2.11 (algebraic reduction) *Given* $F = G/\prod_{j=1}^{n} H_j^{m_j}$, *let* $\mathbf{p}$ *be an arrangement point in a stratum* S *of* $\mathcal{V}$, *and let* $\mathrm{BC}(F)$ *be the set of bases of* $\mathcal{A}$ *containing no broken circuit.*

Each stage of the following algorithm will have a sum of monomial fractions equal to F.

1. *If any monomial in the current sum has support containing a broken circuit, apply* (10.2.13). *Repeat until no longer possible.*
2. *Collect terms with the same denominator.*
3. *For each term* f, *check whether the numerator is in* $\mathcal{I}(f)$. *If there are such terms, choose one among them whose denominator has maximum degree, and replace it by a sum of terms with smaller support. Repeat until no longer possible.*
4. *For each term* f, *check whether its numerator vanishes on the stratum defined by the support of* f, *and if so, write it as the sum of terms with the same support but small degrees in the denominators. Repeat until no longer possible.*

Proposition 10.2.12 *The Algorithm 10.2.11 halts after finitely many steps. The result is an expression of* F *as the sum of monomial fractions with nonvanishing (at* $\mathbf{p}$*) numerators and supports that are subsets of* $\mathrm{BC}(F)$.

We end this section with a few examples of algebraic reduction.

Example 10.2.13 (three lines in $\mathbb{C}^2$) Let $H = H_1 H_2 H_3 = (3 - 2x - y)(3 - x - 2y)(2 - x - y)$. The variety $\mathcal{V}$ consists of three concurrent lines, as shown in Figure 10.5.

By examination, we easily find that $2H_1 + 2H_2 - 3H_3 = 0$. Dividing through by H,

$$\frac{1}{H_1 H_2} = \frac{2/3}{H_1 H_3} + \frac{2/3}{H_2 H_3}.$$

Starting with any monomial fraction $F := P/(H_1^a H_2^b H_3^c)$, this relation may be used in Algorithm 10.2.11 to eliminate any monomial fractions with both H_1 and H_2 in the denominator, resulting in a sum

$$F = \sum_j \frac{P_j}{H_1^j H_3^{a+b+c-j}} + \sum_j \frac{Q_j}{H_2^j H_3^{a+b+c-j}}.$$

Each summand has a transverse multiple point at $(1, 1)$. ■

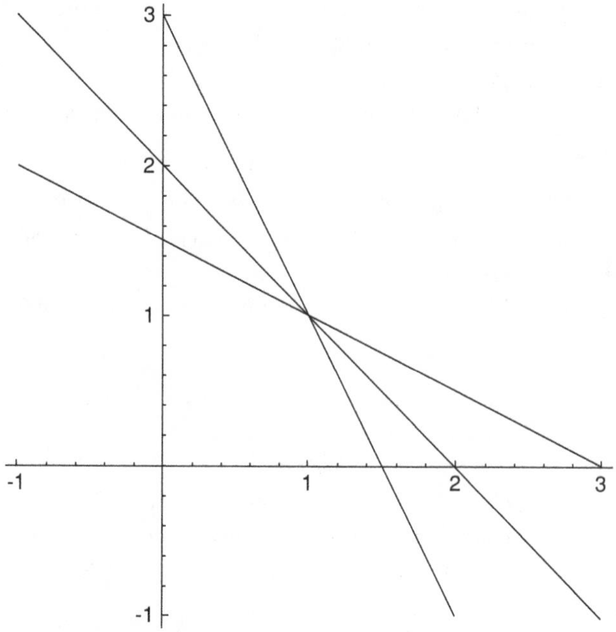

Figure 10.5 An arrangement point in dimension 2 with three linear divisors.

Example 10.2.14 (three curves in $\mathbb{C}^2$, continued) Let $H := H_1 H_2 H_3 = (1 - x)(1 - y)(1 - xy)$, as in Example 10.1.6. Localizing at the common intersection point $(1, 1)$, we know that each factor should be in the ideal generated by the others over $\mathbf{R}_{(1,1)}$. In fact, it is true over $\mathbb{C}[x, y]$:

$$H_3 = H_1 + H_2 - H_1 H_2 . \tag{10.2.14}$$

This implies a linear dependence over $\mathbf{R}_{(1,1)}$ among the terms $1/H_{-j}, 1 \leq j \leq 3$. The broken circuit is $\{1, 2\}$, so our standard procedure is to eliminate the term $1/H_{-3}$. Dividing (10.2.14) by H gives

$$\frac{1}{H_1 H_2} = \frac{1}{H_2 H_3} + \frac{1}{H_1 H_3} - \frac{1}{H_3} .$$

We may write the last term as $-H_2/(H_2 H_3)$, obtaining

$$\frac{1}{(1-x)(1-y)} = \frac{2-y}{(1-y)(1-xy)} + \frac{1}{(1-x)(1-xy)} .$$

This allows us to perform all reductions in Algorithm 10.2.11. For example, $F = 1/H$ reduces to

$$F_1 + F_2 = \frac{2-y}{(1-y)(1-xy)^2} + \frac{1}{(1-x)(1-xy)^2} .$$

Each of the generating functions F_1 and F_2 has a transverse multiple point at $(1, 1)$. ∎

Example 10.2.15 (integer solutions to linear equations) Let A be a $d \times m$ matrix of nonnegative integers, and for $\boldsymbol{r} \in \mathbb{N}^d$, let $a_{\boldsymbol{r}}$ denote the number of nonnegative integer solutions to $A\boldsymbol{x} = \boldsymbol{r}$. The generating function for the array $\{a_{\boldsymbol{r}}\}$ is given by

$$F(\boldsymbol{z}) := \sum_{\boldsymbol{r}} a_{\boldsymbol{r}} \boldsymbol{z}^{\boldsymbol{r}} = \prod_{j=1}^{m} \frac{1}{1 - \boldsymbol{z}^{Ae_j^T}},$$

where e_j is the jth elementary vector. This enumeration problem is discussed at length in De Loera and Sturmfels (2003) (see also Stanley, 1997, Section 4.6), in which they use the running example

$$A = \begin{bmatrix} 1 & 0 & 0 & 1 & 1 \\ 0 & 1 & 0 & 1 & 0 \\ 0 & 0 & 1 & 0 & 1 \end{bmatrix}.$$

In this case,

$$F(\boldsymbol{z}) = \frac{1}{H_1 H_2 H_3 H_4 H_5} = \frac{1}{(1-x)(1-y)(1-z)(1-xy)(1-xz)}.$$

The divisors are all binomials $1 - \boldsymbol{z}^{\boldsymbol{\alpha}}$, and all intersect at $(1, 1, 1)$. Their logarithmic gradients at $(1, 1, 1)$ are the respective columns of A. Every triple of columns of A except for $(1, 2, 4)$ and $(1, 3, 5)$ forms a linearly independent set. The circuits of the matroid are therefore these triples and the only quadruple not containing either triple, namely $(2, 3, 4, 5)$. The broken circuits are $(1, 2)$, $(1, 3)$, and $(2, 3, 4)$ and the bases containing no broken circuit are $(1, 4, 5)$, $(2, 3, 5)$, $(2, 4, 5)$, and $(3, 4, 5)$. By Proposition 10.2.10, a cohomology basis is given by

$$\left\{ \frac{1}{H_1 H_4 H_5}, \quad \frac{1}{H_2 H_3 H_5}, \quad \frac{1}{H_2 H_4 H_5}, \quad \frac{1}{H_3 H_4 H_5} \right\}.$$

To reduce F to the sum of terms whose support is in $\mathtt{BC}(F)$, we use relations expressing H_i in terms of $\{H_j : j \in C \setminus \{i\}\}$, where i is the greatest element in a circuit. A little scratch work uncovers these relations for the respective circuits $(1, 2, 4)$, $(1, 3, 5)$, and $(2, 3, 4, 5)$:

$$H_4 = H_1 + H_2 - H_1 H_2 \tag{10.2.15}$$

$$H_5 = H_1 + H_3 - H_1 H_3 \tag{10.2.16}$$

$$H_5 = -y^{-1} H_2 + y^{-1} H_3 + x H_4 . \tag{10.2.17}$$

Note that x, y, and y^{-1} are units in $\mathbf{R}_{(1,1,1)}$. The first relation (10.2.15) divided by $H H_4$ yields

$$\frac{1}{H} = \frac{1}{H_2 H_3 H_4^2 H_5} + \frac{1}{H_1 H_3 H_4^2 H_5} - \frac{1}{H_3 H_4^2 H_5}. \tag{10.2.18}$$

We are finished manipulating the third term of (10.2.18), as its support $\{3, 4, 5\}$ is in BC. The second term of (10.2.18), after an application of (10.2.16), becomes

$$\frac{1}{H_3 H_4^2 H_5^2} + \frac{1}{H_1 H_4^2 H_5^2} - \frac{1}{H_4^2 H_5^2}.$$

The first term of (10.2.18), after an application of (10.2.17), yields

$$\frac{-1/y}{H_3 H_4^2 H_5^2} + \frac{1/y}{H_2 H_4^2 H_5^2} + \frac{x}{H_2 H_3 H_4 H_5^2},$$

and using (10.2.17) once again on the last of these three terms replaces that term by

$$\frac{-x/y}{H_3 H_4 H_5^3} + \frac{x/y}{H_2 H_4 H_5^3} + \frac{x^2}{H_2 H_3 H_5^3}.$$

Putting this all together gives

$$\begin{aligned} F = &-\frac{1}{H_3 H_4^2 H_5} + \frac{1}{H_3 H_4^2 H_5^2} + \frac{1}{H_1 H_4^2 H_5^2} - \frac{1}{H_4^2 H_5^2} \\ &+ \frac{-1/y}{H_3 H_4^2 H_5^2} + \frac{1/y}{H_2 H_4^2 H_5^2} \\ &+ \frac{-x/y}{H_3 H_4 H_5^3} + \frac{x/y}{H_2 H_4 H_5^3} + \frac{x^2}{H_2 H_3 H_5^3}. \end{aligned}$$

■

Remark The dependence relation expressing some of the H_j in terms of the others, used in the previous few examples, is called a *Nullstellensatz certificate* and can be computed algorithmically. It is implemented, for example, in Singular via the command `lift`.

10.3 Formulae for Coefficients

This section consists of linking the computations of residues to the asymptotic evaluation of coefficients. This happens in two steps. First, residue theorems from Section 10.2 are applied to compute the leading term $\Phi_z(\boldsymbol{r})$ of the integral $\int_C z^{-r-1} F(z)\, dz$. Here C is the homology generator for $H_d(\mathcal{M}^{z,\mathrm{loc}})$ near a transverse multiple point z and is given in local coordinates by $T \times \sigma$, where T is a generator for $H_k(\tilde{N})$ and $\sigma \simeq \mathbb{R}^{d-k}$ is a patch over which the real part of $\hat{\boldsymbol{r}} \cdot \log(\cdot)$ is maximized at z. In this first step, we consider a number of sub-cases, depending on whether the intersection is complete ($k = d$) and whether the divisors are simple ($\boldsymbol{m} = \mathbf{1}$). The second step is to identify the subset `contrib` $\subseteq$ `critical` over which $\Phi_z(\boldsymbol{r})$ must be summed to give the asymptotics for the coefficients $a_{\boldsymbol{r}}$ of $F(z)$. This is the harder step and is done on an ad hoc basis outside of two general classes: minimal points and linear divisors. The proofs in this section are brief because all computations have already been carried out.

10.3.1 Asymptotics of Φ_z for Complete Intersections

Say that the transverse multiple point z is a ***complete intersection*** if $k = d$, i.e., if the stratum $\bigcap_j \mathcal{V}_j$ consists of the single point z (note that this is not the standard notion in algebraic geometry, which is more general). The simplest formula for Φ_z is when z is a complete intersection and each divisor is simple. As in Theorem 10.2.2, we let $\mathbf{\Gamma}_\Psi$ denote the matrix whose rows are the logarithmic gradients $\nabla_{\log} H_j(z)$.

Theorem 10.3.1 (square-free complete intersection) *Let $F = G/\prod_{j=1}^d H_j$ in $\mathbf{R}_z$, with each H_j square-free and all divisors intersecting transversely at z. Let $\mathbf{N}(z)$ denote the cone of positive linear combinations of rows of $\mathbf{\Gamma}_\Psi$. Suppose that G is holomorphic in a neighborhood of z and $G(z) \neq 0$. Then*

$$\frac{1}{(2\pi i)^d}\int_T z^{-r-1}F(z)\,dz \sim \Phi_z(r),$$

where $T = \Psi^{-1}(T_\varepsilon)$, with

$$\Phi_z(r) := z^{-r}\frac{G(z)}{\det \mathbf{\Gamma}_\Psi}. \tag{10.3.1}$$

The remainder term is of a lower exponential order, $\exp[|r|(\hat{r}\cdot\log z - \varepsilon)]$, *uniformly as $\hat{r}$ varies over compact subsets of the interior of* $\mathbf{N}(z)$.

Remark The sign of the integral depends on the orientation of T, which is determined by the order in which the divisors are listed.

Proof This is (10.2.4) of Theorem 10.2.2, setting $k = d$ so that the integral over σ drops out. □

This formula is quite explicit and easy to use, except for the fact that computing $\det \mathbf{\Gamma}_\Psi$ requires computation of the factorization of H. In fact, $\det \mathbf{\Gamma}_\Psi$ may be computed directly from the partial derivatives of H. We give the result in the special case $d = 2$ in which the Hessian of H suffices.

Corollary 10.3.2 (special case $d = 2$) *Assume the hypotheses of Theorem 10.3.1 with $d = 2$ and $z := (x_0, y_0)$. Then*

$$\Phi_{x_0,y_0}(r,s) = x_0^{-r-1}y_0^{-s-1}\frac{G}{\sqrt{\left(\frac{\partial^2 H}{\partial x\partial y}\right)^2 - \frac{\partial^2 H}{\partial x^2}\frac{\partial^2 H}{\partial y^2}}}(x_0,y_0). \tag{10.3.2}$$

Proof Write $H_1 = a(x - x_0) + b(y - y_0) + R$ and $H_2 = c(x - x_0) + d(y - y_0) + S$, where $|R| + |S| = O((x - x_0)^2 + (y - y_0)^2)$. Then $\det \mathbf{\Gamma}_\Psi = x_0 y_0(ad - bc)$. The second-order partial derivatives of H at (x_0, y_0) are given by $H_{xx} = (1/2)ac$, $H_{yy} = (1/2)bd$ and $H_{xy} = ad + bc$. The identity

$$(ad - bc)^2 = (ad + bc)^2 - (2ac)(2bd)$$

translates into

$$(ad - bc)^2 = H_{xy}^2 - H_{xx}H_{yy},$$

proving the result. □

The following result generalizes Theorem 10.3.1 to complete intersections with arbitrary exponents.

Theorem 10.3.3 (complete intersection) *Let $F = G/\prod_{j=1}^{d} H_j^{m_j}$ in $\mathbf{R}_z$ with each H_j square-free and all divisors intersecting transversely at z. Suppose that G is holomorphic in a neighborhood of z and $G(z) \neq 0$. Then*

$$\frac{1}{(2\pi i)^d} \int_T z^{-r-1} F(z)\, dz \sim \Phi_z(r)$$

with

$$\Phi_z(\boldsymbol{r}) := \frac{1}{(\boldsymbol{m}-\mathbf{1})!} \frac{z^{-\boldsymbol{r}} G(z)}{\det \boldsymbol{\Gamma}_\Psi(z)} (\boldsymbol{r}\boldsymbol{\Gamma}_\Psi^{-1})^{\boldsymbol{m}-\mathbf{1}}. \tag{10.3.3}$$

The remainder term is of a lower exponential order, $\exp[|\boldsymbol{r}|(\hat{\boldsymbol{r}} \cdot \log z - \varepsilon)]$, *uniformly as $\hat{\boldsymbol{r}}$ varies over compact subsets of the interior of* $\mathbf{N}(z)$.

Proof This is (10.2.5) and (10.2.6) of Theorem 10.2.6, with $k = d$ so that the integral over σ drops out. □

10.3.2 Asymptotics of Φ_z when $k < d$

When $k < d$, after computing the integral over the normal link, there is an integral over a $(d-k)$-chain in the stratum S containing $\boldsymbol{p}$. This integral is of the form $A(z)z^{-\boldsymbol{r}}$, where $A(z)$ is an integral over the normal link and varies smoothly with $z \in S$. The leading term of this integral is easy to write down except for a multiplicative constant. It is determined by two things: the value of A at the critical point $\boldsymbol{p}$ (provided this is nonzero) and the quadratic part of $-\boldsymbol{r} \cdot \log z$ on the stratum S. The contribution of the quadratic is conceptually simple, but its explicit computation requires a few more definitions. Recall the coordinate functions $z_1, \ldots, z_d$ of Ψ^{-1}. Letting E denote the set $\{\pi(1), \ldots, \pi(d-k)\}$, the coordinate functions z provide a parametrization of S by $\mathbb{C}^E$ embedded as a subset of $\mathbb{C}^d$. In other words, $\boldsymbol{g} := (g_j : j \notin E)$ parametrize S as a graph over a $(d-k)$-dimensional coordinate plane. Computing the quadratic requires logarithmic coordinates, which we may define for $j \notin E$ by

$$g_j(\boldsymbol{\theta}) := \log\left[z_j\left(\boldsymbol{p}\exp(i\boldsymbol{\theta})\right)\right].$$

The vector function $\boldsymbol{g}(\boldsymbol{\theta})$ parametrizes a neighborhood of $\log \boldsymbol{p}$ in $\log \mathcal{V}$ as $\boldsymbol{\theta}$ varies over a neighborhood of the origin in $\mathbb{C}^E$. By $\hat{\boldsymbol{r}} \cdot \boldsymbol{g}$, we denote the function

$\sum_{j \notin E} \hat{r}_j g_j$. The quadratic part of $\hat{\boldsymbol{r}} \cdot \boldsymbol{g}$ is the function defined by

$$Q(\boldsymbol{\theta}) := \sum_{i,j \in E} \frac{\partial^2(\hat{\boldsymbol{r}} \cdot \boldsymbol{g})}{\partial\theta_i \partial\theta_j}, \tag{10.3.4}$$

and we denote by $\det Q$ the determinant of the matrix whose (i, j)-entry is $\dfrac{\partial^2(\hat{\boldsymbol{r}} \cdot \boldsymbol{g})}{\partial\theta_i \partial\theta_j}$.

Theorem 10.3.4 (partial intersection) *Let $F = G / \prod_{j=1}^k H_j^{m_j}$ in $\mathbf{R}_z$, with each H_j square-free and all divisors intersecting transversely at a stratum S containing the point z. Suppose that G is holomorphic in a neighborhood of z and $G(z) \neq 0$. Let σ be a chain in S on which the real part of $-\hat{\boldsymbol{r}} \cdot \log(\cdot)$ is maximized at z. Then*

$$\frac{1}{(2\pi i)^d} \int_C z^{-r-1} F(z)\, dz \sim \Phi_z(\boldsymbol{r}),$$

where $C = \Psi^{-1}(T_\varepsilon) \times \sigma$ and

$$\Phi_z(\boldsymbol{r}) := z^{-r} \frac{(\boldsymbol{r}\boldsymbol{\Gamma}_\Psi^{-1})^{\boldsymbol{m}-\boldsymbol{1}}}{(\boldsymbol{m}-\boldsymbol{1})!} \frac{G(z)}{\det \boldsymbol{\Gamma}_\Psi} (2\pi|\boldsymbol{r}|)^{-(d-k)/2} \frac{1}{\sqrt{\det Q}}. \tag{10.3.5}$$

The remainder term is of order $|z^{-r}||\boldsymbol{r}|^{-(d-k)/2-1}$, uniformly as z varies over a compact subset of S over which G and $\det Q$ are nonvanishing, whereas $\hat{\boldsymbol{r}}$ varies over compact subsets of the interior of $\mathbf{N}(z)$.

Proof Fix $\boldsymbol{p} \in S$ and $\hat{\boldsymbol{r}}_*$ interior to $\mathbf{N}(\boldsymbol{p})$. Using (10.2.6) of Theorem 10.2.6 in the first line and changing coordinates via $z_j = \exp(i\theta_j)$ gives

$$\frac{1}{(2\pi i)^d} \int_C z^{-r-1} F(z)\, dz = \frac{1}{(2\pi i)^{d-k}} \int_\sigma z^{-r} P(\boldsymbol{r}, z) \frac{dz_E}{z^E}$$

$$= \frac{1}{(2\pi)^{d-k}} \int_N \exp(-|\boldsymbol{r}|\phi(\boldsymbol{\theta})) P(\boldsymbol{r}, z(\boldsymbol{\theta}))\, d\boldsymbol{\theta}.$$

Here the last integral is over a neighborhood $\mathcal{N}$ of the origin in $\mathbb{R}^{d-k}$ and $\phi(\boldsymbol{\theta}) := \hat{\boldsymbol{r}} \cdot \log z(\boldsymbol{\theta})$. This last integral is a standard multivariate saddle-point integral, which may be evaluated asymptotically via Theorem 5.1.2. Pulling out z^{-r} and applying the theorem with $\lambda = |\boldsymbol{r}|$ shows that the leading term of the integral is given by $(2\pi)^{(d-k)/2} z(\boldsymbol{0})^{-r} P(\boldsymbol{r}, z(\boldsymbol{0}))/\sqrt{\det \mathcal{H}}$, where $\mathcal{H}$ is the Hessian matrix for the phase function ϕ. Plugging in (10.2.5) for the leading term of $P(\boldsymbol{r}, z(\boldsymbol{0}))$ and collecting powers of 2π yields

$$(2\pi|\boldsymbol{r}|)^{-(d-k)/2} \frac{(\boldsymbol{r}\boldsymbol{\Gamma}_\Psi^{-1})^{(\boldsymbol{m}-\boldsymbol{1})}}{(\boldsymbol{m}-\boldsymbol{1})!} \frac{G(z)}{\det \boldsymbol{\Gamma}_\Psi(\boldsymbol{p})} z^{-r} (\det \mathcal{H})^{-1/2}.$$

The phase function ϕ is equal to $\hat{\boldsymbol{r}} \cdot \boldsymbol{g}$, which makes the Hessian matrix of ϕ with respect to $\boldsymbol{\theta}$ equal to Q. Equating $\det \mathcal{H}$ with $\det Q$ and reverting from $\boldsymbol{p}$ to z finishes the proof. □

10.3.3 *Computing* contrib

General results determining `contrib` are known only in two cases, namely minimal points and the case when all divisors are linear. Nevertheless, as the examples in this section indicate, it is possible to compute `contrib` on an ad hoc basis for many examples outside of these two general schema.

We begin with the case of linear divisors, $\ell_1, \ldots, \ell_n$. As usual, we let $\mathcal{M}$ denote the set where $\prod_{j=1}^d z_j \prod_{j=1}^n \ell_j$ is nonzero; furthermore, we denote $\mathcal{M}_{\mathbb{R}} := \mathcal{M} \cap \mathbb{R}^d$. Let L_S be a flat of codimension k in the intersection lattice of $\mathcal{A}$. Let $\sigma_S = \sigma_S(\hat{\boldsymbol{r}})$ denote the unique critical point for $h_{\hat{\boldsymbol{r}}}$ in L_S. Let $\mathcal{A}(S)$ denote the subarrangement of flats containing L_S, and let $\mathcal{M}_{\mathbb{R}}(S)$ denote the real complement of this subarrangement. Let B_S denote the unique component of $\mathcal{M}_{\mathbb{R}}(S)$ whose closure contains the origin. The function $h_{\hat{\boldsymbol{r}}}$ is minimized on B_S at σ_S if and only if $\hat{\boldsymbol{r}} \in \overline{\mathbf{N}(\sigma)}$.

Theorem 10.3.5 (linear divisors) *Assume $\hat{\boldsymbol{r}} \notin \partial\, \mathbf{N}(\sigma)$ for all critical points σ of $h_{\hat{\boldsymbol{r}}}$. Let W be the set of points σ_S for which $\hat{\boldsymbol{r}} \in \mathbf{N}(\sigma)$. Let $W' \subseteq W$ denote the set of $\sigma \in W$ on which h achieves its maximum on W. Then* `contrib` $= W'$. *In other words, $W' \neq \emptyset$, and if $c := h(\mathbf{z})$ for any/every $\mathbf{z} \in W'$, then T is homologous in $(\mathcal{M}, \mathcal{M}^{c-\varepsilon})$ to a sum $\sum_{\sigma \in W'} \alpha_\sigma$, where each α_σ is a nonzero element of the local homology group $(\mathcal{M}^{c-\varepsilon} \cup \mathcal{N}(\sigma), \mathcal{M}^{c-\varepsilon})$ for an arbitrarily small $\varepsilon > 0$ and neighborhood $\mathcal{N}(\sigma)$.*

Proof A sketch of the proof is as follows; full details appear in Baryshnikov and Pemantle (2013). When all the divisors are linear, it is possible to give actual cycles (rather than relative cycles) whose sum is the torus T. More specifically, for any real point $\boldsymbol{x} \notin \mathcal{M}_{\mathbb{R}}$, we let $\mathcal{F}(\boldsymbol{x})$ denote the "imaginary fiber" $\boldsymbol{x} + i\mathbb{R}^d$. This is not compact, but we may view it as a relative chain in $(\mathcal{M}, \mathcal{M}^{-\infty})$, where $M^{-\infty}$ denotes the set $\mathcal{M}^{\leq -c}$ for sufficiently large c. The class of $\mathcal{F}(\boldsymbol{x})$ depends only on the component of $\mathcal{M}_{\mathbb{R}}$ containing $\boldsymbol{x}$, and thus we may write $\mathcal{F}(B) := \mathcal{F}(\boldsymbol{x})$ for any $\boldsymbol{x}$ in the component B of $\mathcal{M}_{\mathbb{R}}$; the chain $\mathcal{F}(B)$ is null homologous if B is unbounded.

The homology of $(\mathcal{M}, \mathcal{M}^{-\infty})$ is generated by the cycles $\mathcal{F}(S)$. The torus T is the sum (with alternating orientations) of $\mathcal{F}(B)$ over the 2^d components B whose closure contains the origin. The generators for the local homology at σ_S may be written as sums of $\mathcal{F}(B)$ over the components B whose closure contains σ_S. Having written both T and each generator α as a sum of classes $\mathcal{F}(B)$, some combinatorial analysis establishes an identity in $H_d(\mathcal{M}, \mathcal{M}^{-\infty})$ of the form

$$T = \sum_S \alpha(S) .$$

Here, each $\alpha(S)$ is a sum of cycles $\mathcal{F}(B)$ (with some orientations) over components B whose closure contains σ_S and is a nonempty sum if and only if h is maximized on B_S at σ_S. Letting c be the height of the highest $\sigma \in W$ and projecting to $(\mathcal{M}, \mathcal{M}^{c-\varepsilon})$ kills all summands with $h(\sigma_S) < c$ and proves the theorem. $\square$

Proposition 10.3.6 (minimal points) *Let $F = G/H$ and $\mathcal{V} := \mathcal{V}_H$. Fix $\hat{\boldsymbol{r}}_*$ and let* $\mathtt{critical}(\hat{\boldsymbol{r}}_*)$ *denote the solution to the critical point equations in direction $\hat{\boldsymbol{r}}_*$. Fix a component B of the complement of the amoeba of H, and suppose the point $\boldsymbol{x}_{\min} \in \partial B$ is the unique minimizer of $h_{\hat{\boldsymbol{r}}_*}$ over $\overline{B}$. Assume that $\hat{\boldsymbol{r}}_* \notin \partial\, \mathbf{N}(\boldsymbol{z})$ for any critical point $\boldsymbol{z} \in \mathbf{T}(\boldsymbol{x}_{\min})$. Suppose further that the set W of critical points $\boldsymbol{z} \in \mathbf{T}(\boldsymbol{x}_{\min})$ for which $\hat{\boldsymbol{r}}_* \in \mathbf{N}(\boldsymbol{z})$ is nonempty. Then* $\mathtt{contrib}(\hat{\boldsymbol{r}}_*) = W$. *Furthermore, there is a neighborhood $\mathcal{N}$ of $\hat{\boldsymbol{r}}_*$ such that for $\hat{\boldsymbol{r}} \in \mathcal{N}$, the set* $\mathtt{contrib}(\hat{\boldsymbol{r}})$ *is the homotopy extension of* $\mathtt{contrib}(\hat{\boldsymbol{r}}_*)$ *as a subset of the continuously varying set* $\mathtt{critical}(\hat{\boldsymbol{r}})$.

Proof Suppose first that W is a singleton set $\{\boldsymbol{z}_*\}$. Suppose also that $\mathcal{V}$ is the union of transversely intersecting sheets $\mathcal{V}_j$ at the stratum containing $\boldsymbol{z}_*$. Because $\boldsymbol{z}_*$ is the only critical point in $\mathbf{T}(\boldsymbol{x}_{\min})$ with $\hat{\boldsymbol{r}}_* \in \mathbf{N}(\boldsymbol{z}_*)$, we know that either $\mathtt{contrib} = W$ or T is homologous to zero in $H_d(\mathcal{M}, \mathcal{M}^{c-\varepsilon})$. To rule out the latter, note that T may be taken as $\mathbf{T}(\boldsymbol{x})$ for any $\boldsymbol{x} \in B$. The local homology group at $\boldsymbol{z}$ is $H_d(\mathcal{M}^{\text{loc}}) := H_d(\mathcal{M}^{c-\varepsilon} \cup \mathcal{N}(\boldsymbol{z}_*), \mathcal{M}^{c-\varepsilon})$. With $\boldsymbol{x}$ sufficiently close to $\boldsymbol{x}_{\min}$, the projection of $\mathbf{T}(\boldsymbol{x})$ to $\mathcal{M}^{\text{loc}}$ is the same as the projection of $\mathcal{F}(\boldsymbol{x})$ to $\mathcal{M}^{\text{loc}}$. In $\mathcal{M}^{\text{loc}}$, the variety $\mathcal{V}$ is homotopic to its linearization $\mathtt{hom}(\mathcal{V}_j, \boldsymbol{z}_*)$ at $\boldsymbol{z}_*$, which is the union of the tangent planes to the divisors at $\boldsymbol{z}_*$. The relative cycle $\mathcal{F}(\boldsymbol{x})$ is nonzero in the local homology group if and only if it is nonzero in the local homology group once $\mathcal{V}$ has been replaced by $\mathtt{hom}(\mathcal{V}, \boldsymbol{z}_*)$. From Theorem 10.3.5, we know that this is the case because $\hat{\boldsymbol{r}}_* \in \mathbf{N}(\boldsymbol{z}_*)$.

To remove the hypothesis of a transverse intersection, use the local partial fraction decomposition to replace F by a sum of rational functions with transversely intersecting divisors. To remove the hypothesis that W is a singleton, use the decomposition of $(\mathcal{M}, \mathcal{M}^{c-\varepsilon})$ as a direct sum of local spaces $(\mathcal{M}^{c-\varepsilon} \cup \mathcal{N}(\boldsymbol{z}), \mathcal{M}^{c-\varepsilon})$. □

The following theorem is a direct corollary of Proposition 10.3.6 and Theorems 10.3.1–10.3.4.

Theorem 10.3.7 *Suppose $\boldsymbol{x}_{\min}$ is the unique minimizer for $h = h_{\hat{\boldsymbol{r}}_*}$ on the component ∂B of* $\mathtt{amoeba}(H)^c$ *and the set of critical points of h on $\mathbf{T}(\boldsymbol{x}_{\min})$ is a finite nonempty set of arrangement points. At each such point, $\boldsymbol{z}$, let $F = \sum_{\alpha \in W(\boldsymbol{z})} F_{\boldsymbol{z},\alpha}$ be the decomposition of F into monomial fractions whose supports are subsets of* $\mathtt{BC}(F)$ *and whose numerators do not vanish identically on the common intersection, S, of the divisors. Suppose that at least one of the monomials $F_{\boldsymbol{z},\alpha}$ has $\boldsymbol{r} \in \mathbf{N}(\boldsymbol{z})$ and $G \neq 0$ and that for all such monomials, the log-Hessian matrix $\mathcal{Q}$ is nonsingular. Then the coefficients $a_{\boldsymbol{r}}$ of the Laurent series for F over B are given asymptotically by*

$$a_{\boldsymbol{r}} \sim \sum_{\boldsymbol{z} \in \mathtt{contrib}} \; \sum_{\alpha \in W'(\boldsymbol{z})} \Phi_{\boldsymbol{z},\alpha}(\boldsymbol{r}),$$

where

(i) `contrib` *is the set of critical points for h on $\mathbf{T}(\boldsymbol{x}_{\min})$ for which at least one monomial has $\boldsymbol{r} \in \mathbf{N}(z)$.*
(ii) The set $W'(z) \subseteq W(z)$ selects monomials for which $\boldsymbol{r} \in \mathbf{N}(z)$ and $G \neq 0$.
(iii) Each expression $\Phi_{z,\alpha}$ is given by (10.3.5) *or its specializations to the case of complete intersection* (10.3.3) *or complete intersection with simple poles* (10.3.1).
(iv) If each monomial is a complete intersection with simple poles, then the remainder is exponentially smaller than the leading term of $a_{\boldsymbol{r}}$.

Examples

Example 10.3.8 (two lines in $\mathbb{C}^2$ continued) Continuing Examples 10.1.3 and 10.2.3, let $F = 1/H$, where $H := \left(1 - \frac{1}{3}x - \frac{2}{3}y\right)\left(1 - \frac{2}{3}x - \frac{1}{3}y\right)$. The point $(1, 1)$ is minimal. It is critical for every direction but only in `contrib` when $\boldsymbol{r}$ is in the cone between slope $1/2$ and slope 2. It is the only critical point on the unit torus, and therefore, by the residue computation in Example 10.2.3,

$$a_{rs} \sim \Phi_{(1,1)}(r, s) = 3\,.$$

If $r, s \to \infty$ with $1/2 + \varepsilon < r/s < 2 - \varepsilon$, then the error term will be $O\left(e^{-c(r+s)}\right)$ for some $c = c(\varepsilon) > 0$. ■

Example 10.3.9 (figure eight continued) Let $F = 1/H$ be the generating function from Example 10.1.4 whose pole set looks like a figure eight. The main work is showing that $(1, 1)$ is a strictly minimal point. Brute force is easiest here. The parametrization (10.1.1) shows that $\mathcal{V}$ does not enter the unit polydisk near the point $(1, 1)$. If this curve enters the unit polydisk anywhere else, then there is a point (x_0, y_0), with one coordinate having modulus 1 and the other having modulus less than 1. By symmetry, it suffices to check the points on $\mathcal{V}$ with $|x| = 1$ to see if ever $|y| \le 1$. Using a rational parametrization of the unit circle, such as $x(t) = (1 - t^2)/(1 + t^2) + i \cdot 2t/(1 + t^2)$, the square norm of $y(t)$ is an algebraic function of t. Checking zeros of the derivative of $|y(t)|^2$ and finding that $|y|^2 \ge 1$ there with equality only at $t = 0$ are easy computations and prove that $(1, 1)$ is strictly minimal.

The rest is easy. The cone $\mathbf{N}(1, 1)$ is the set of (r, s) lying between the logarithmic normals to the two branches of $\mathcal{V}$; the lognormal directions were computed in Example 10.1.8 to be the rays with slopes 2 and $1/2$. Theorem 10.3.7 implies that $a_{\boldsymbol{r}} \sim \Phi_{(1,1)}(\boldsymbol{r})$. The point $(1, 1)$ is a square-free complete intersection, so we may apply Corollary 10.3.2. The values of x_0, y_0, and $G(x_0, y_0)$ are all equal to 1. Evaluating the determinant gives $H_{xx} = H_{yy} = 8$, $H_{xy} = 10$, which puts $100 - 64$

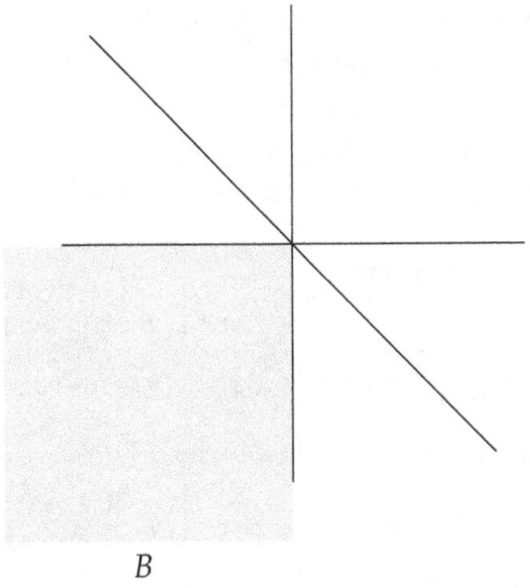

Figure 10.6 The normal cone to the component B at $(1, 1)$ is the positive quadrant.

under the radical, leading to $\Phi_{(1,1)} = 1/6$. Therefore,

$$a_{rs} = \frac{1}{6} + R,$$

where R is a remainder converging to zero exponentially as $(r, s) \to \infty$, with r/s remaining in a compact subset of $(1/2, 2)$. Numerically, $a_{30,30} \approx 0.1652$ for a numeric error of 0.8%. The relative error for $a_{60,60}$ has decreased by a factor of 40, corroborating a faster than polynomial decay. ■

Example 10.3.10 (three curves continued: the minimum function) Let $F = 1/H = 1/H_1 H_2 H_3 = 1/[(1 - x)(1 - y)(1 - xy)]$, as in Examples 10.1.6 and 10.2.14. We already know that $a_{rs} = \min\{r, s\}$, but let us see how to obtain this from Theorems 10.3.3 and 10.3.7 and Algorithm 10.2.11. Because each factor of H is a binomial, the amoeba $\texttt{amoeba}(H)$ is somewhat degenerate: it is the union of three lines, namely the x-axis, the y-axis, and the line $x + y = 0$. The ordinary power series for F corresponds to the component B of $\texttt{amoeba}(H)^c$, which is the negative quadrant, as shown in Figure 10.6.

For any $\mathbf{r}$ in the strictly positive quadrant, $(0, 0)$ is unique minimizer for $h_{\hat{\mathbf{r}}}$ on B. As long as $\mathbf{r}$ is not the diagonal direction, the only critical point for $h_{\mathbf{r}}$ on $\mathbf{T}(0, 0)$ is the point $(1, 1)$. The decomposition at this point

$$F = F_1 + F_2 = \frac{2 - y}{(1 - y)(1 - xy)^2} + \frac{1}{(1 - x)(1 - xy)^2}$$

was worked out in Example 10.2.14. The stratum S is the singleton $\{(1, 1)\}$, and neither the numerator of F_1 nor F_2 vanishes there. The hypotheses of Theorem 10.3.7 are satisfied with $\texttt{contrib} = \{(1, 1)\}$.

For F_1, the normal cone $\mathbf{N}_1(1, 1)$ is the cone of directions between the diagonal and the y-axis. For F_2, the normal cone $\mathbf{N}_2$ is the cone of directions between the diagonal and the x-axis. Therefore, for vectors $\boldsymbol{r}$ above the diagonal, $\boldsymbol{r} \in F_\alpha$ if and only if $\alpha = 1$, whereas for vectors $\boldsymbol{r}$ below the diagonal, $\boldsymbol{r} \in F_\alpha$ if and only if $\alpha = 2$. Therefore, uniformly over compact subcones of the region above the diagonal, $a_{\boldsymbol{r}} \sim \Phi_{(1,1),1}(\boldsymbol{r})$, whereas uniformly over compact subcones of the region below the diagonal, $a_{\boldsymbol{r}} \sim \Phi_{(1,1),2}(\boldsymbol{r})$, where $\Phi_{(1,1),\alpha}$ is computed from Theorem 10.3.3 for F_α.

Computing $\Phi_{(1,1),1}$ first, we see that $G(1,1) = 1$. Also, $\boldsymbol{m} = (1,2)$, so $(\boldsymbol{m} - \boldsymbol{1})! = 1$. The logarithmic gradient matrix for F_1 is the same as the gradient matrix, namely $\begin{bmatrix} 0 & 1 \\ 1 & 1 \end{bmatrix}$ and its inverse is $\begin{bmatrix} -1 & 1 \\ 1 & 0 \end{bmatrix}$. Thus $(r, s)\boldsymbol{\Gamma}_\Psi^{-1} = (s - r, r)$ and $(\boldsymbol{r}\boldsymbol{\Gamma}_\Psi^{-1})^{\boldsymbol{m}-\boldsymbol{1}} = r$. Putting this together with a similar computation for F_2 gives

$$\Phi_{(1,1),1}(r, s) = r$$

$$\Phi_{(1,1),2}(r, s) = s,$$

and plugging into Theorem 10.3.7 gives the asymptotics $a_{r,s} \sim r$ when $r < s$ and $a_{r,s} \sim s$ when $r > s$. In other words, $a_{r,s} \sim \min\{r, s\}$ for all r, s, up to an exponential term whose decay rate is bounded from below when r/s is bounded away from 0, 1, and ∞. Although we know the asymptotics to be exact once 1 is added, we cannot expect the theorem to give us uniformity across the diagonal because the function $\min\{r, s\}$ has a non-analyticity here (in fact is not differentiable). ■

10.4 The Base Case via Surgery

The previous two sections used multivariate residues to compute asymptotics for multiple points. We now go back and obtain the same results using only univariate residue theory. For any $z \in \mathbb{C}^d$, let $z^\circ \in \mathbb{C}^{d-1}$ denote the first $d - 1$ coordinates of z. Standing assumptions on the function F and the point z_* throughout this section are as follows.

(1) The point z_* is a strictly minimal point (later we allow finitely minimal points) of $\mathcal{V}$, and all its coordinates are nonzero.
(2) The function F is meromorphic in a neighborhood of the polydisk $\mathbf{D}(z_*) := \{\boldsymbol{w} : |w_j| \le |(z_j)_*|, 1 \le j \le d\}$.
(3) In a neighborhood of z_*, the function F may be written as

$$\frac{\psi}{\prod_{j=1}^n (1 - z_d v_j(z^\circ))}, \tag{10.4.1}$$

where $\psi(z_*) \ne 0$ and each v_j is smooth and satisfies $v_j(z_*^\circ) = 1/(z_d)_*$.

Remarks

(i) The functions v_j are inverses to the root functions u_j in the usual Weierstass factorization $\psi / \prod_j (z_d - u_j(z^\circ))$.
(ii) We allow repetitions among the inverse root functions $\{v_j\}$.

(iii) The value of ψ at z_* is easily seen to satisfy

$$\psi(z_*) = \frac{G(z_*)}{\prod_j \left(-z_d \partial H_j / \partial z_d\right)\Big|_{z=z_*}}. \tag{10.4.2}$$

The Residue Sum

Letting $\mathcal{N}$ be a neighborhood of z_*° on which (10.4.1) holds, we repeat the argument for Theorem 9.2.1 to prove the following result in which C_1 is a circle of radius less than $(z_*)_d - \varepsilon$ and C_2 is a circle of radius $(z_*)_d + \varepsilon$.

Lemma 10.4.1 *Let*

$$\chi := \int_{\mathcal{N}} \int_{C_1 - C_2} z^{-r} F(z) \frac{dz_d}{z_d} \frac{dz^\circ}{z^\circ},$$

where $\int_{C_1 - C_2}$ *denotes the difference of the two integrals. Then*

$$\left| z_*^r \right| \; \left| a_r - (2\pi i)^{-d} \chi \right| \to 0 \tag{10.4.3}$$

exponentially as $|r| \to \infty$ *with* $\hat{r} \to \hat{r}_*$.

Proof The details having been done once before, we sketch the proof. Let T° be the product of circles of radius $|(z_*)_j|$ in coordinate j for $1 \le j \le d-1$. Choose $\mathcal{N}$ small enough so that for $z^\circ \in \mathcal{N}$, no pole of $F(z^\circ, \cdot)$ lies on C_2. Then

$$\begin{aligned} (2\pi i)^d a_r &= \int_{T^\circ} \int_{C_1} z^{-r} F(z) \frac{dz_d}{z_d} \frac{dz^\circ}{z^\circ} \\ &= R_1 + \int_{\mathcal{N}} \int_{C_1} z^{-r} F(z) \frac{dz_d}{z_d} \frac{dz^\circ}{z^\circ} \\ &= R_1 + R_2 + \chi, \end{aligned}$$

where

$$R_1 := \int_{\mathcal{N}^c} \int_{C_1} z^{-r} F(z) \frac{dz_d}{z_d} \frac{dz^\circ}{z^\circ};$$

$$R_2 := \int_{\mathcal{N}} \int_{C_2} z^{-r} F(z) \frac{dz_d}{z_d} \frac{dz^\circ}{z^\circ}.$$

The quantity $z_*^r R_1$ is exponentially small because $F(z^\circ, \cdot)$ is analytic in a disk of radius at least $|(z_*)_d| + \varepsilon$ for $z^\circ \notin \mathcal{N}$. The quantity $z_*^r R_2$ is exponentially small because the integrand is exponentially smaller than z_*^{-r}. □

Let z° be any point for which $u_1(z^\circ), \ldots, u_n(z^\circ)$ are all distinct, and define the negative residue sum R at z° by

$$R(z^\circ) := -\sum_{j=1}^{n} u_j(z^\circ)^{-r_d} \operatorname{Res}\left(\frac{F(z^\circ, \cdot)}{z_d}; u_j(z^\circ)\right). \tag{10.4.4}$$

Note that $R(z^\circ)$ also depends on $\boldsymbol{r}$ via r_d.

Lemma 10.4.2 *Suppose that the set Λ of values of $z^\circ \in \mathcal{N}$ for which $\{u_1, \ldots, u_n\}$ are not distinct has measure zero. Then*

$$\chi = 2\pi i \int_{\mathcal{N}} (z^\circ)^{-r^\circ} R(z^\circ) \frac{dz^\circ}{z^\circ} . \tag{10.4.5}$$

Proof By choice of C_2 and $\mathcal{N}$, the poles of $F(z^\circ, \cdot)$ lying between C_1 and C_2 are enumerated by $u_j(z^\circ)$, $1 \le j \le n$. By the univariate residue theorem, when $z^\circ \notin \Lambda$,

$$\int_{C_1 - C_2} z^{-r} F(z) \frac{dz_d}{z_d} = 2\pi i \, (z^\circ)^{-r^\circ} R(z^\circ) . \tag{10.4.6}$$

When Λ has measure zero, the outer integral is unaffected by the behavior of the inner integral on Λ. Integrating (10.4.6) against dz°/z° over $\mathcal{N}$ results in (10.4.5). Note that we do not need to assume integrability because this follows from the existence of the Cauchy integral together with Lemma 10.4.1. □

Evaluating (10.4.5) would be easy if R were of the form $Ae^{\lambda\phi}$, because then we could use Fourier-Laplace integral theory directly (e.g., Theorem 5.1.2). The n summands composing R are indeed of this form, but each on its own is not integrable. The solution is to use an integral representation from DeVore and Lorentz (1993) i.e., based on representing R as a symmetric rational sum.

Let Δ_{n-1} denote the standard $(n-1)$-simplex in $\mathbb{R}^n$, defined by

$$\Delta_{n-1} := \{\mathbf{t} \in (\mathbb{R}^+)^n : \sum_{j=1}^{n} t_j = 1\}.$$

This is parametrized by its projection to the first $n - 1$ coordinates,

$$\pi \Delta_{n-1} = \{\mathbf{t} \in (\mathbb{R}^+)^{n-1} : \sum_{j=1}^{n-1} t_j \le 1\}.$$

We write ι for the map $\iota(\mathbf{t}) := (t_1, \ldots, t_{n-1}, 1 - \sum_{j=1}^{n-1} t_j)$ that inverts the projection, so that $\iota(\mathbf{t}) \cdot \boldsymbol{v}$ is the convex combination

$$\iota(\mathbf{t}) \cdot \boldsymbol{v} = t_1 v_1 + \cdots + t_{n-1} v_{n-1} + \left(1 - \sum_{j=1}^{n-1} t_j\right) v_n .$$

Lemma 10.4.3 *Let f be an analytic function of one complex variable, and let $v_1, \ldots, v_n \in \mathbb{C}$ be distinct. Then*

$$\sum_{j=1}^{n} \frac{f(v_j)}{\prod_{r \ne j} (v_j - v_r)} = \int_{\pi \Delta_{n-1}} f^{(n-1)} (\iota(\mathbf{t}) \cdot \boldsymbol{v}) \, d\mathbf{t} . \tag{10.4.7}$$

Applying this with $f(v) := v^{r_d+n-1}\psi(z^\circ, 1/v)$ *to the inverse roots* $v_j(z^\circ)$ *yields*

$$R(z^\circ) = \int_{\pi\Delta_{n-1}} f^{(n-1)}(\iota(\mathbf{t}) \cdot \boldsymbol{v})\, d\mathbf{t}. \tag{10.4.8}$$

Proof The identity (10.4.7) is proved in DeVore and Lorentz (1993, p. 121, Eqs 7.7 and 7.12). The rest follows from the form of the summand in (10.4.4). Specifically,

$$u_j(z^\circ)^{-r_d}\,\mathrm{Res}\left(\frac{F(z^\circ,\cdot)}{z_d}\,;\, u_j(z^\circ)\right) = \frac{v_j(z^\circ)^{r_d+n-1}\psi(z^\circ, 1/v_j(z^\circ))}{\prod_{r\neq j}[v_r(z^\circ) - v_j(z^\circ)]}$$

$$= \frac{f(v_j(z^\circ))}{\prod_{r\neq j}[v_r(z^\circ) - v_j(z^\circ)]},$$

and applying (10.4.7) gives (10.4.8). □

Theorem 10.4.4 *The standing hypotheses (1)–(3) imply the estimate*

$$a_r \sim \left(\frac{1}{2\pi i}\right)^{d-1} \int_N (z^\circ)^{-r^\circ} \int_{\pi\Delta_{n-1}} h\,[\iota(\mathbf{t}) \cdot \boldsymbol{v}(z^\circ)]\; d\mathbf{t}\,\frac{dz^\circ}{z^\circ}, \tag{10.4.9}$$

where

$$h(y) := \frac{d^{n-1}}{dy^{n-1}}\left(y^{r_d+n-1}\psi\left(z^\circ, \frac{1}{y}\right)\right).$$

Proof In the case when $v_1, \ldots, v_n$ are distinct except on a set of measure zero, this follows directly from Lemmas 10.4.1–10.4.3. For the general case, let v_j^s be functions approaching v_j as $s \to 0$, such that v_j^s are distinct for s in a punctured neighborhood of 0, and let $g^s(z) = z^{-r_d-1}\psi(z)/\prod_{j=1}^n(z_d - 1/v_j^s(z^\circ))$, so that $g^s \to g$ as well. The sum of the residues of g^s may be computed by integrating g^s around the circle $|z_d - (z_*)_d| = \varepsilon$, and because $g^s \to g$, this sum approaches the sum of the residues of g. Because the right-hand side of (10.4.9) is continuous in the variables v_j, this proves the general case. □

Fourier-Laplace Integral Formulae

Until this point, the geometry of $\mathcal{V}$ has played a limited role. We used minimality of z_* to perform surgery and reduce the integral to a neighborhood of z° (Lemma 10.4.1). We used the fact that the number of local roots of $H(z^\circ, \cdot) = 0$ is a constant, n, but i.e., true independent of any assumptions and so are the results through Lemma 10.4.2. Finally, the representation as $\int_N \int_{\pi\Delta_{n-1}}$ relied on the labeling of the roots valid throughout a neighborhood of z°, in other words, lack of monodromy. Theorem 10.4.4 is therefore valid for any multiple point. Further restrictions become necessary as we try to evaluate the integrand of (10.4.9) and apply the Fourier-Laplace integral machinery of Chapter 5. We remark that because the region of integration is not a manifold, we will be using the more general result Theorem 5.4.8 rather than Theorem 5.1.2.

As in the smooth point case, we change from $\mathbf{z}$ to $\boldsymbol{\theta}$ via $z_j = (z_*)_j e^{i\theta_j}$, converting the outer integral to an integral over a neighborhood $\mathcal{N}'$ of the origin in $\mathbb{R}^{d-1}$. The following are expressions for the resulting phase function, amplitude function, and a constant that will help to reduce the notation. For $0 \le j \le n-1$, $\boldsymbol{\theta} \in \mathcal{N}'$ and $t \in \mathbb{R}^{n-1}$, define

$$p_j(s) := \frac{(n-1)!\Gamma(s+n)}{j!(n-1-j)!\Gamma(s+j+1)}$$

$$\phi(\boldsymbol{\theta}, t) := \frac{i\mathbf{r}^\circ \boldsymbol{\theta}}{s} - \log\left(\frac{\iota(\mathbf{t})\boldsymbol{v}(\mathbf{z}_*^\circ e^{i\boldsymbol{\theta}})}{\boldsymbol{v}(\mathbf{z}_*^\circ)}\right)$$

$$A_j(\boldsymbol{\theta}) := \left(\frac{d}{dy}\right)^j \psi(\mathbf{z}_*^\circ e^{i\boldsymbol{\theta}}, 1/y)\Bigg|_{y=\boldsymbol{v}(\mathbf{z}_*^\circ e^{i\boldsymbol{\theta}})}.$$

Observe that $p_j(s)$ is a constant multiple of a falling factorial with $n-j-1$ terms and therefore has degree $n-j-1$ in s.

Lemma 10.4.5 (reduction to Fourier-Laplace integral) *The right side of* (10.4.9) *in Theorem 10.4.4 may be rewritten as*

$$\chi = \left(\frac{1}{2\pi}\right)^{d-1} \mathbf{z}_*^{-\mathbf{r}} \sum_{j=0}^{n-1} p_j(r_d) \int_{\mathcal{E}} e^{-r_d\phi(\boldsymbol{\theta},t)} A_j(\boldsymbol{\theta})\, d(\boldsymbol{\theta}) \times d\mathbf{t}, \tag{10.4.10}$$

where $\mathcal{E} := \mathcal{N} \times \pi\Delta_{n-1}$ *and* $d\boldsymbol{\theta} \times d\mathbf{t}$ *denotes Lebesgue measure in each of the coordinates. Furthermore,* $\operatorname{Re}\phi \ge 0$ *on* $\mathcal{E}$.

Proof Applying Leibniz' rule for differentiating the function f in (10.4.8) yields

$$\begin{aligned} f^{(n-1)}(y) &= \sum_{j=0}^{n-1} \binom{n-1}{j} \left(\frac{d}{dy}\right)^{n-1-j} y^{r_d+n-1} \left(\frac{d}{dy}\right)^j \psi(\mathbf{z}_*^\circ e^{i\boldsymbol{\theta}}, 1/y) \\ &= \sum_{j=0}^{n-1} \frac{(n-1)!(r_d+n-1)!}{j!(n-1-j)!(r_d+j)!} y^{r_d+j} \left(\frac{d}{dy}\right)^j \psi(\mathbf{z}_*^\circ e^{i\boldsymbol{\theta}}, 1/y) \\ &= y^{r_d} \sum_{j=0}^{n-1} p_j(r_d) y^j \left(\frac{d}{dy}\right)^j \psi(\mathbf{z}_*^\circ e^{i\boldsymbol{\theta}}, 1/y). \end{aligned}$$

Plugging this into (10.4.8) and using the definitions of A_j, ϕ, and p_j yields the stated formula.

By strict minimality of $\mathbf{z}_*$, for each j the modulus of $v_j(\mathbf{z}^\circ)$ achieves its maximum only when $\mathbf{z} = \mathbf{z}_*$. Thus each convex combination of $v_j(\mathbf{z}^\circ)$ with $\mathbf{z} \ne \mathbf{z}_*$ has modulus less than $|v_j(\mathbf{z}_*^\circ)|$. This shows that $\operatorname{Re}\phi \ge 0$ on $\mathcal{E}$. □

Remark We cannot conclude that $\mathrm{Re}\,\phi$ is strictly positive except at the origin. In fact, the origin can be connected to the boundary of $\mathcal{E}$ by a set on which $\mathrm{Re}\,\phi$ vanishes, which is why we need the generality of stratified spaces and cannot simply reduce to a sub-neighborhood i.e., a manifold.

Suppose that $\hat{\boldsymbol{r}}_*$ is on the boundary of a set for which the hypotheses of Lemma 10.4.5 hold. Even if the lemma holds for $\hat{\boldsymbol{r}}_*$, because it does not hold for a neighborhood of $\hat{\boldsymbol{r}}_*$, the conclusion for integer vectors $\boldsymbol{r}$ with $\boldsymbol{r}/|\boldsymbol{r}| = \hat{\boldsymbol{r}}_*$ may be vacuous. To bolster the argument in such a case, we record an easy corollary.

Corollary 10.4.6 *Fix $\hat{\mathbf{r}}_*$ and let $\mathbf{r} = \lambda\hat{\mathbf{r}}_* + \boldsymbol{\alpha}$. If the standing hypotheses (1)–(3) hold for $\hat{\mathbf{r}}_*$ and if $\mathbf{r} \to \infty$ with $|\boldsymbol{\alpha}| = O(1)$, then* (10.4.9) *remains valid.*

Proof Dependence on $\boldsymbol{r}$ is via the factor of $-\boldsymbol{r} \cdot \log \boldsymbol{z}$ in the phase function ϕ. When $\boldsymbol{r} = \lambda\hat{\boldsymbol{r}}$, this is $\lambda h_{\hat{\boldsymbol{r}}}(\boldsymbol{z})$ and is in the form $e^{-c\lambda}$ required for the Fourier-Laplace integral. If instead $\boldsymbol{r} = \lambda\hat{\boldsymbol{r}} + \boldsymbol{\alpha}$, then one has the identical phase term but a factor of $\boldsymbol{z}^{-\boldsymbol{\alpha}}$ in the amplitude term $A(\boldsymbol{z})$. This is integrated over a bounded $\boldsymbol{z}$-neighborhood. If $\boldsymbol{\alpha}$ is bounded as well, this amplitude term and its derivatives are uniformly bounded. Thus the remainder terms change by a bounded factor, and the asymptotic expansion is still valid. Pulling the $\boldsymbol{z}^{-\boldsymbol{\alpha}}$ term back out of the integral and combining it with the $e^{-\lambda h(\hat{\boldsymbol{r}})}$ term reconstitutes the $\boldsymbol{z}^{-\boldsymbol{r}}$ term of (10.4.9), proving that this conclusion still holds. □

Asymptotics for $a_{\boldsymbol{r}}$ in directions controlled by multiple points can now be computed using Lemma 10.4.5, provided that we can compute asymptotics of the relevant Fourier-Laplace integrals. The easiest situation technically, and the only one for which anything approaching a complete theory exists, occurs when there is a single critical point, which is in the interior of $\mathcal{E}$ and which is quadratically nondegenerate. We shall consider only this case, and this is why our standing hypotheses are required.

We first need to understand the behavior of ϕ. First,

$$\phi(\mathbf{0}, \alpha(t)) = 0 \quad \text{for all } t \in \Delta_{n-1},$$

as is easily seen from the definition. However, ϕ does not vanish to order 2 whenever $\boldsymbol{\theta} = \mathbf{0}$, unlike the case when we analyzed smooth points. We must also ensure that no other critical points arise, so we first determine the critical point set exactly.

Let $\boldsymbol{\Gamma}$ denote the lognormal matrix. In the previous analysis, everything was reduced to the case of $k \le d$ sheets meeting transversely; in that case, $n = k$ and $\boldsymbol{\Gamma}$ is the first k rows of the augmented lognormal matrix $\boldsymbol{\Gamma}_\Psi$ from Section 10.2.1. In general, $\boldsymbol{\Gamma}$ will be a $n \times d$ matrix. Let $\boldsymbol{C}$ denote the matrix resulting when the rows of $\boldsymbol{\Gamma}$ are normalized to have all 1's in the last column. On the hypersurface

$\mathcal{V}_i := \{H_i = 0\}$, the logarithmic coordinates of points satisfy a familiar implicit derivative identity:

$$\frac{\partial \log z_d}{\partial \log z_j} = -\frac{\partial H_i / \partial \log z_j}{\partial H_i / \partial \log z_d}.$$

Parametrizing the hypersurface $\mathcal{V}_i$ by z°, the quantity $-\log z_d$ is $\log v_i(z^\circ)$. Thus $\partial \log v_i / \partial \log z_j = (\partial H_i / \partial \log z_j)/(\partial H_i / \partial \log z_d)$, which implies

$$\boldsymbol{C}_{ij} := \frac{\boldsymbol{\Gamma}_{ij}}{\boldsymbol{\Gamma}_{id}} = \frac{z_j}{z_d}\frac{\partial v_i}{\partial z_j}(\boldsymbol{z}_*).$$

Proposition 10.4.7 (critical points of ϕ) *Fix a vector $\boldsymbol{r} \in \mathbf{N}(\boldsymbol{z}_*)$. The set of critical points of ϕ in direction $\boldsymbol{r}$ on $\mathcal{E}$ is the set of points $(\mathbf{0}, \mathbf{t})$ with $\mathbf{t} \in \mathcal{S}(\boldsymbol{r})$, where*

$$\mathcal{S}(\boldsymbol{r}) := \{\mathbf{t} \in \pi\, \Delta_{n-1} \mid \iota(\mathbf{t})\boldsymbol{C}(\boldsymbol{z}_*) = \boldsymbol{r}/r_k\}$$

is the set of convex combinations that, when applied to the rows of $\boldsymbol{C}$, produce the vector $\boldsymbol{r}/r_d$.

Proof When $\boldsymbol{\theta} \neq \mathbf{0}$, $\operatorname{Re}\phi > 0$, from which it follows that all critical points of ϕ are of the form $(\mathbf{0}, \mathbf{t})$ for $\mathbf{t} \in \Delta_{n-1}$. In fact, ϕ is somewhat degenerate: $\phi(\mathbf{0}, \mathbf{t}) = 0$ for all $\mathbf{t} \in \Delta$, so not only does the real part of ϕ vanish when $\boldsymbol{\theta} = \mathbf{0}$, but also the $\mathbf{t}$-gradient of ϕ vanishes there. We compute the $\boldsymbol{\theta}$-derivatives at $\boldsymbol{\theta} = \mathbf{0}$ as follows. For $j \neq d$,

$$\frac{\partial \phi}{\partial \theta_j} = i\left(\frac{r_j}{r_d} - \frac{z_j}{\iota(\mathbf{t})\boldsymbol{v}(\boldsymbol{z}_*^\circ)}\iota(\mathbf{t})\left(\frac{\partial}{\partial z_j}\right)\boldsymbol{v}(\boldsymbol{z}^\circ)\right)\Bigg|_{\boldsymbol{z}=\boldsymbol{z}_*}.$$

Recalling the definition of $\boldsymbol{C}$ and noting that $z_d = v_j(\boldsymbol{z}_*^\circ)$ for all j, and hence that $\iota(\mathbf{t})\boldsymbol{v}(\boldsymbol{z}_*^\circ) = z_d$, we see that the preceding $\boldsymbol{\theta}$-derivatives vanish simultaneously if and only if $\boldsymbol{r}/r_d = \iota(\mathbf{t})\boldsymbol{C}(\boldsymbol{z}_*)$, which finishes the proof. □

This proposition may look ugly, but we can see what the result says in some simple situations. Suppose first that $\boldsymbol{z}_*$ is a transverse multiple point and is square-free. Then the rows of $\boldsymbol{C}$ are linearly independent, so there is at most one point in $\mathcal{S}$. Conversely, the normal cone $\mathbf{N}(\boldsymbol{z}_*)$ is the set of convex combinations of the lognormal vectors, so there is at least one point in $\mathcal{S}$; thus $\mathcal{S}$ is a singleton $\{\mathbf{t}\}$. If $\boldsymbol{r}$ is interior to $\mathbf{N}(\boldsymbol{z}_*)$, then $\mathbf{t}$ is in the interior of $\pi\,\Delta$. Suppose next that the two sheets $\mathcal{V}_1$ and $\mathcal{V}_2$ are tangent. Then the first two rows of $\boldsymbol{C}$ are equal, and any solution $\mathbf{t}$ leads to a line of solutions $(t_1 - s, t_2 + s, t_3, \ldots, t_n)$. For example, if $d = n = 2$, then $\mathbf{N}(\boldsymbol{z}_*)$ is a singleton and $\mathcal{S}$ is the whole unit interval.

Proposition 10.4.8 (Hessian of ϕ in the singleton case) *Suppose that $\mathcal{S}$ consists of the single point $(\mathbf{0}, \mathbf{t}_*)$. Then the Hessian of $\phi(\boldsymbol{\theta}, \mathbf{t})$ at this point has the block form*

$$\mathcal{H}(\boldsymbol{z}_*, t_*) = \begin{pmatrix} Q(\boldsymbol{z}_*, t_*) & -i\boldsymbol{C}^\circ(\boldsymbol{z}_*)^T \\ -i\boldsymbol{C}^\circ(\boldsymbol{z}_*) & 0 \end{pmatrix}. \tag{10.4.11}$$

In this decomposition:

- *The* $(d-1)\times(d-1)$ *block* $Q(z_*, t_*)$ *is the Hessian of the restriction of* ϕ *to the* $\boldsymbol{\theta}$*-directions.*
- *The zero block has dimensions* $(n-1)\times(n-1)$*.*
- *The* $(n-1)\times(d-1)$ *matrix* $\boldsymbol{C}^\circ$ *is* $\boldsymbol{C}$ *with the last row and column stripped off.*

Remark The last column, corresponding to the dth coordinate, is stripped off because this is a function of the others in the $\boldsymbol{\theta}$ parametrization; the last row, corresponding to the nth sheet, is stripped off because our parametrization of the simplex is by the first $(n-1)$ coordinates.

Proof Constancy of ϕ in the $\mathbf{t}$-directions at $\boldsymbol{\theta}=\mathbf{0}$ shows that the second partials in those directions vanish, giving the upper-left block of zeros. Computing $(\partial/\partial\theta_j)\phi$ up to a constant gives

$$-\frac{i}{\iota(\mathbf{t})\boldsymbol{v}(z_*^\circ)}\iota(\mathbf{t})z_j\frac{\partial}{\partial z_j}\boldsymbol{v}(z_*^\circ),$$

and because $\iota(\mathbf{t})\boldsymbol{v}(z_*^\circ)$ is constant when $\boldsymbol{\theta}=\mathbf{0}$, differentiating in the $\mathbf{t}$ directions recovers the blocks $-i\boldsymbol{C}^\circ(z_*)$ and $-i\boldsymbol{C}^\circ(z_*)^T$. The second partials in the $\boldsymbol{\theta}$ directions are of course unchanged. □

We first present the main asymptotic result in the case $q \le d$. Note that this is consistent with the smooth point result in Theorem 9.2.7.

Theorem 10.4.9 (transverse square-free multiple point) *Suppose that* $n \le d$ *and let* z_* *be a strictly minimal transverse square-free multiple point of* $\mathcal{V}$ *such that* $G(z_*) \ne \mathbf{0}$*. For* $\boldsymbol{r} \in \mathbf{N}(z_*)$*, let* $\mathbf{t}_*(\boldsymbol{r})$ *be the unique point of* $\mathcal{S}(\boldsymbol{r})$*.*

(i) There is an asymptotic series

$$a_{\boldsymbol{r}} \sim (2\pi)^{(n-d)/2} z_*^{-\boldsymbol{r}} \sum_{l=0}^{\infty} C_\ell(\boldsymbol{r}) r_k^{(n-d)/2-\ell}. \tag{10.4.12}$$

(ii) The estimate is uniform as long as $\hat{\boldsymbol{r}}$ *is restricted to a compact subset of* $\mathbf{N}(z_*)$*.*

(iii) If $\hat{\boldsymbol{r}}$ *is in the interior of* $\mathbf{N}(z_*)$ *and* $G(z_*) \ne 0$*, the leading coefficient of* (10.4.12) *is given by*

$$C_0 = (\det \mathcal{H}(z_*, \mathbf{t}_*(\boldsymbol{r})))^{-1/2} \frac{G(z_*)}{\prod_j \left(-z_d \partial H_j/\partial z_d\right)\Big|_{z=z_*}}, \tag{10.4.13}$$

where the square root of the determinant is the product of the principal square roots of the eigenvalues of $\mathcal{H}(z_*\,;\,\mathbf{t}_*(\boldsymbol{r}))$*.*

Proof Because $\mathcal{S}$ is a singleton, we may apply Theorem 5.4.8 to the integral over $\mathcal{E}$ in each summand of (10.4.10). The resulting quantity has an asymptotic series in decreasing powers of r_d with leading term $cr_d^{1-(n_d)/2}$. The polynomials p_j decrease in degree from $n-1$ to zero. Integration may be carried out term by term because the remainders are uniformly one power of r_d lower as long as $\hat{\boldsymbol{r}}$ is restricted to a compact subset of $\mathbf{N}(z_*)$. Already this is sufficient to give the form of (10.4.12).

The leading term is evidently obtained by setting $\ell = j = 0$. Equation (10.4.10) then gives

$$\chi = (2\pi)^{1-d} z_*^{-r} p_0(r_d) \int_{\mathcal{E}} \exp\left(-r_d \phi(\boldsymbol{\theta}, \mathbf{t})\right) A_0(\boldsymbol{\theta})\, d\boldsymbol{\theta} d\mathbf{t}\,.$$

Evaluating $A_0(\mathbf{0}) = \psi(z_*^\circ, (z_d)_*) = G(z_*)/\prod_{j=1}^n (-z_d \partial H_j/\partial z_d)(z_*)$, and observing that p_0 is monic shows the $j = 0$ term of (10.4.10) to be asymptotically $cr_d^{-(d-n)}$, where

$$c = (2\pi)^{(n+d)/2-1} (\det \mathcal{H}(z_*, \mathbf{t}_*(\boldsymbol{r}))^{-1/2} \frac{G(z_*)}{\prod_{j=1}^n (-z_d \partial H_j/\partial z_d)(z_*)}\,.$$

Multiplying by $(2\pi)^{1-d} z_*^{-r}$ completes the computation of the leading term of (10.4.10) and establishes (10.4.13). □

Remark The degenerate case where there is only one sheet ($n = 1$) corresponds to a smooth point. Because $n - 1 = 0$, the Hessian $\mathcal{H}(z_*, \mathbf{t}_*(\boldsymbol{r}))$ reduces to the Hessian $\mathcal{H}(\hat{\boldsymbol{r}})$ in the $\boldsymbol{\theta}$ directions that appears in Theorem 9.2.7. Also, the fraction at the end of (10.4.13) reduces to the analogous quantity with a single factor from Theorem 9.2.7, making it clear that the present formula is a generalization of the one in that theorem. Note that when $G(z_*) = 0$, the leading coefficient involves derivatives of G. See Section 13.3 for more details.

Example 10.4.10 We analyze the rational function $1/(H_1 H_2)$ from Example 8.3.3. Each point in the line of intersection is square-free. Given the index (r, s, t), we know that the point

$$(x_0, y_0, z_0) := \left(\frac{4(r+s)}{3(r+s+t)}, \frac{4(r+s)}{3(r+s+t)}, \frac{4t}{r+s+t}\right)$$

lies on this line and is critical for the direction in question. In fact, it is the only point of $\mathcal{V}$ contributing to asymptotics in this direction. Of course, this point determines asymptotics in other directions too, because the normal cone at this point is two-dimensional. Direct computation using Theorem 10.4.9 is possible.

For example, when $(x_0, y_0, z_0) = (1, 1, 1)$, the vector (r, s, t) lies in the normal cone $\mathbf{N}$ if and only if $r + s = 3t$. Further specializing to the directions where $r = s$, we obtain, for example, the first-order approximation $a_{3t,3t,2t} \sim (48\pi t)^{-1/2}$. This differs by less than 0.3% from the true value when $t = 30$. ■

Comparison to Method of Iterated Residues

Some results are harder to glean from the surgical approach. For example, in the case of a transverse intersection with $n = d$, where there are d sheets intersecting in a single point, Theorem 10.3.3 gives asymptotics of cz_*^{-r}, where c is given by a relatively transparent determinant and the remainder is exponentially small. The surgical method obscures the computation of the constant and does not prove exponential decrease of the remainder.

However, the surgical method handles complicated geometries of the stationary set better, in principle, than does the method of iterated residues. One example is when $\hat{\boldsymbol{r}}$ is on a *facet* of $\mathbf{N}(z_*)$, i.e., a face of the cone whose dimension is one less than the dimension of the whole cone. Because this case is relatively easy to analyze, we state it as a corollary.

Corollary 10.4.11 *Assume the hypotheses of Theorem 10.4.9, but with* $\hat{\mathbf{r}}$ *on a facet of* $\mathbf{N}(z_*)$ *instead of being in the interior. Then if* $\boldsymbol{r} \to \infty$ *with* $\boldsymbol{r} - \hat{\boldsymbol{r}}\mathbb{R}^+$ *bounded, the formulae of Theorem 10.4.9 hold, except with an extra factor of* $1/2$ *in* (10.4.13).

Proof The condition that $\hat{\boldsymbol{r}}$ is on a facet of $\mathbf{N}(z_*)$ is equivalent to the condition that $\iota(\mathbf{t}(\boldsymbol{r}))$ is on a facet of Δ_{n-1}. The inner integral in (10.4.10) is local, the main contribution being in an ε-ball about $\mathbf{t}(\boldsymbol{r})$ for any $\varepsilon > 0$; when $\mathbf{t}$ is on a facet, this integral is over a halfspace neighborhood rather than a ball neighborhood. The leading term in such an integral is precisely half of the leading term for a ball neighborhood. After introducing a factor of $1/2$ in the inner integral of (10.4.10), the new hypotheses leave everything else unchanged. This proves the result when $\boldsymbol{r}$ is a precise multiple of $\hat{\boldsymbol{r}}$. When $\boldsymbol{r} - |\boldsymbol{r}|\hat{\boldsymbol{r}}_*$ is bounded, for some $\hat{\boldsymbol{r}}_*$, an invocation of Corollary 10.4.6 shows that (10.4.9) and hence (10.4.10) remain valid, introducing the factor of $1/2$ and completing the proof. □

Example 10.4.12 (two-lines, case of boundary directions) Suppose $F = 1/H$, where $H := \left(1 - \frac{1}{3}x - \frac{2}{3}y\right)\left(1 - \frac{2}{3}x - \frac{1}{3}y\right)$ as in Examples 10.1.3, 10.2.3, and 10.3.8. We saw previously that $a_{rs} \sim 3$ when $r, s \to \infty$ with r/s in a compact subset of $[1/2, 2]$. By Corollary 10.4.11, the coefficients at slope $1/2$ are asymptotically half of the coefficients on the interior of the cone $1/2 < r/s < 2$, thus

$$a_{2s,s} \sim 3/2$$

as $s \to \infty$. This holds in fact for any a_{rs} with $r - 2s = O(1)$. For discussion of what happens when $r \gg r - 2s \gg 1$, see Section 13.4. ■

Aside from the case of boundary directions, the surgical method also beats iterated residues when dealing with large stationary sets. The general such integral is not worked out in Chapter 5, but the case of two curves leads, as we will see, to a two-dimensional integral with a one-dimensional stationary set, which can be computed without great difficulty. To handle more complicated cases, the work of Varchenko (1977) is an excellent guide.

Proposition 10.4.13 (two curves intersecting tangentially) *Suppose that $d = 2$ and that $F = G/H$ has a multiple point at $(1, 1)$ with $n = 2$ and two tangentially intersecting curves $\mathcal{V}_j$ with common lognormal direction $(r_*, 1)$. Let u_1 and u_2 denote the parametrizations of the two curves through the point; i.e., near the point $(1, 1)$, the set $\mathcal{V}_j$ coincides with the set $\{(x, u_j(x))\}$ for $j = 1, 2$. Let $v_j = 1/u_j$ denote the inverse roots and let $g_j := \log v_j(e^{i\theta})$ be the parametrization of the inverse roots in log space. Let $\kappa_j := g_j''(0) \geq 0$ denote the quadratic coefficients of g_j, so that*

$$g_j(\theta) = i r_* \theta - \frac{\kappa_j}{2}\theta^2 + O(\theta)^3 .$$

Then as $(r, s) \to \infty$ with $r = r_* s + O(1)$,

$$a_{rs} \sim \frac{G(1,1)}{\sqrt{2\pi}} \frac{2}{\sqrt{\kappa_1} + \sqrt{\kappa_2}} s^{1/2} .$$

Proof By Lemma 10.4.5, and its extension via Corollary 10.4.6 to $(r, s) = \lambda(\hat{r}, \hat{s}) + O(1)$,

$$a_{rs} \sim (2\pi)^{-1} s \int_{\mathcal{E}} e^{-s\phi(\theta,t)} A_0(\theta)\, d\theta\, dt .$$

Here, $\mathcal{E} = \mathcal{N} \times \pi\Delta_1$, where $\mathcal{N}$ is a neighborhood $(-\delta, \delta)$ of zero in $\mathbb{R}^1$ and $\pi\Delta_1$ is the unit interval; also we have used that fact that $p_0(s) = s + 1 \sim s$. The amplitude A_0 is equal to $G/(xy)$. In particular, $A_0(1, 1) = G(1, 1)$.

The phase function is the convex combination

$$-\phi(\theta, t) = t g_1(\theta) + (1 - t) g_2(\theta) .$$

On $(-\delta, \delta) \times [0, 1]$, the phase function ϕ has nonnegative real part, vanishing on the line segment $\mathcal{S} := \{0\} \times [0, 1]$. As opposed to previous cases we have considered, this time the entire line segment is critical for ϕ. This takes us beyond what was developed in Chapter 5, so we do it by hand. Integrating $e^{-s\phi(\theta,t)}\, d\theta$ over $(-\delta, \delta)$ for fixed t gives

$$(2\pi s \kappa_t)^{-1/2} + O\left(s^{-3/2}\right),$$

where $\kappa_t = t\kappa_1 + (1 - t)\kappa_2$ is the quadratic term of $\phi(\cdot, t)$. The $O(s^{-3/2})$ term is uniformly $o(s^{-1/2})$; therefore, the expression for the leading term may be integrated over [0, 1]. Using the change of variables $y = \kappa_t$ leads to

$$a_{rs} \sim \frac{G(1,1)\sqrt{s}}{\sqrt{2\pi}} \int_0^1 \kappa_t^{-1/2}\, dt$$
$$= \frac{G(1,1)\sqrt{s}}{\sqrt{2\pi}} \int_{\kappa_2}^{\kappa_1} y^{-1/2} \frac{dy}{\kappa_1 - \kappa_2}$$
$$= \frac{G(1,1)\sqrt{s}}{\sqrt{2\pi}} \frac{2}{\sqrt{\kappa_1} + \sqrt{\kappa_2}} .$$

The substitution $y = \kappa_t, dy = (\kappa_1 - \kappa_2)\, dt$ is valid only when $\kappa_1 \neq \kappa_2$, but the resulting expression for $\int_0^1 \kappa_t^{-1/2}$ is valid for all $\kappa_1, \kappa_2 \geq 0$ with $\kappa_1 + \kappa_2 > 0$. □

Example 10.4.14 (partial sums of normalized binomials) Recall the generating function $F = \dfrac{2}{(2 - x - y)(1 - xy)}$ from Example 10.1.12. The diagonal coefficients are the normalized binomial sums $\sum_{j=0}^{s} P(2j, j)$, this sum yielding the expected number of returns of a simple random walk to zero by time $2s$. Computing the logarithmic parametrizations gives $g_1(\theta) = -\log(2 - e^{i\theta})$ and $g_2(\theta) = i\theta$. This gives $\kappa_1 = 2$ and $\kappa_2 = 0$. Plugging in $G(1, 1) = 2$ gives

$$a_{s,s} \sim \frac{2}{\sqrt{2\pi}} \frac{2}{0 + \sqrt{2}} s^{1/2} = 2\sqrt{\frac{s}{\pi}} .$$

It is easy to check this because the diagonal of F has the generating function $f(z) := (1 - z)^{-3/2}$, whose coefficients are easily seen to be asymptotic to $2\sqrt{s/\pi}$. Note that in the tangential case, there is no longer exponential convergence. For example, $a_{100,100} \approx 11.326$, which differs from $2\sqrt{100/\pi} \approx 11.284$ by about 0.4%; the relative error is $O(s^{-1})$ but is not exponentially small. ■

Notes

The residues in this chapter were first computed via coordinate methods in Pemantle and Wilson (2002) and Pemantle and Wilson (2004). These methods are valid only when z is a minimal point. The computation for multiple points in the case of linear factors H_j was generalized in an unpublished manuscript of Baryshnikov and Pemantle. All these computations for multiple points are based on the somewhat well-known theory of iterated residues. More information on iterated residues can be found in Aĭzenberg and Yuzhakov (1983). The application of these techniques to asymptotics of multivariate generating functions first appeared in Bertozzi and McKenna (1993) in an analysis of queueing models.

The computation of the residue from an explicit factorization via Algorithm 10.2.11 is effective, assuming the factors H_j are in a nice class of functions such as polynomials or algebraic functions, but the efficiency of such algorithms is a topic of ongoing study (see, e.g., De Loera and Sturmfels, 2003).

The fact that the iterated residue of a rational function is a polynomial, and its consequences for generating function asymptotics, are discussed in Pemantle (2000), where conditions are given for the coefficients to be piecewise polynomial functions of the index $\boldsymbol{r}$. A well-known example of piecewise polynomial behavior is the enumeration of integer solutions to linear equations as in De Loera and Sturmfels (2003). The decomposition of the positive orthant into regions (chambers) in which the counts vary polynomially are objects of classical study, as are the counts themselves; one example is the enumeration of vertices in the *Birkhoff polytope*, where even the leading term asymptotic was found only recently by Canfield and McKay (2009). Although the computation of the polynomials is effective, the best way to compute it is a subject of considerable interest, as is discussed in De Loera and Sturmfels (2003).

Exercises

10.1 (criterion for arrangement point)
Prove Lemma 10.1.13.

10.2 (local geometry in dimension 2)
Suppose that $d = 2$ and let $\boldsymbol{p}$ be a homogeneous point of $\mathcal{V}$ of degree k. Then $\boldsymbol{p}$ is a multiple point, or it is a cusp all of whose tangents are equal.

10.3 (unimodular counting)
Example 10.2.15 considered the generating function

$$F(z) = \frac{1}{H_1 H_2 H_3 H_4 H_5} = \frac{1}{(1-x)(1-y)(1-z)(1-xy)(1-xz)},$$

which is a running example in De Loera and Sturmfels (2003) and counts nonnegative integer solutions to $A\boldsymbol{x} = \boldsymbol{r}$ with

$$A = \begin{bmatrix} 1 & 0 & 0 & 1 & 1 \\ 0 & 1 & 0 & 1 & 0 \\ 0 & 0 & 1 & 0 & 1 \end{bmatrix}.$$

Algorithm 10.2.11 was used in the example to decompose F into the sum of nine terms supported in BC.

(i) Use Theorem 10.3.3 to compute $\Phi_{(1,1)}(\boldsymbol{r})$ for each of these nine terms.
(ii) Use this result along with Theorem 10.3.7 to write asymptotic estimates for $\boldsymbol{r}$ in a set of conic regions, the union of whose closures is the positive orthant.

11
Cone Point Asymptotics

Given $F = G/H = \sum_r a_r z^r$ convergent on a component B of the complement of $\mathtt{amoeba}(H)$ and a nonflat direction $\hat{r}_*$, we return to the question of whether $\overline{\beta}(\hat{r}_*) = \beta^*(\hat{r}_*)$. In this chapter, we nearly settle this question. We also give asymptotics in the case where the contributing critical point is a so-called cone point. By its definition, the class of cone points contains smooth and multiple points, also allowing certain more complicated singularities. Therefore, the methods of this chapter apply to smooth and multiple points. In those cases the more elementary approach, described in the previous two chapters, is easier, although these serve as useful examples of the more general theory of cone points.

This chapter is not entirely self-contained. As mentioned in Section 1.4, the development of these results in Baryshnikov and Pemantle (2011) already condenses a fair amount of background from Atiyah, Bott, and Gårding (1970). We give a complete rendition of the geometric results, which extend the foundation built in Chapter 7 to the context of hyperbolic polynomials, but largely quote without proof the generalized function theory from Baryshnikov and Pemantle (2011, Chapter 6) necessary for rigorous justification of certain Fourier transforms. The first section, containing the geometric results, is quite long and is divided into a number of subsections: using cones to construct deformations, cones of hyperbolicity (homogeneous case), cones of hyperbolicity (general case), strong and weak hyperbolicity, semi-continuity, proof of Theorem 8.4.2, and a coda in which a projective version of the deformation is constructed.

Before proceeding with a barrage of definitions and propositions, we outline intuitively what will happen. Let $\boldsymbol{x}_{\min}$ be the minimizing point for $\hat{r}_*$. We recall what was promised in the not yet proved Theorem 8.4.2. Trivially, the cycle of integration can be deformed to a location arbitrarily near $\mathbf{T}(\boldsymbol{x}_{\min})$. An obstruction to going further will lie on $\mathbf{T}(\boldsymbol{x}_{\min})$. Let $\mathcal{V}_1 := \mathcal{V} \cap \mathbf{T}(\boldsymbol{x}_{\min})$. We check for critical points on $\mathcal{V}_1$. It was claimed that, homologically, the only obstacles to further lowering the cycle of integration are critical points and that even a critical point $z \in \mathcal{V}_1$ is not an obstacle unless $\hat{r}_*$ lies in the cone $\mathbf{N}(z)$. We prove this by finding a loglinear deformation: for each such critical point $z = \exp(\boldsymbol{x}_{\min} + i\boldsymbol{y})$, we find a vector $\boldsymbol{v}$ and deform $\exp(\boldsymbol{x} + i\boldsymbol{y})$ along the path $\{\exp(\boldsymbol{x} + t\boldsymbol{v} + i\boldsymbol{y}) : t \geq 0\}$,

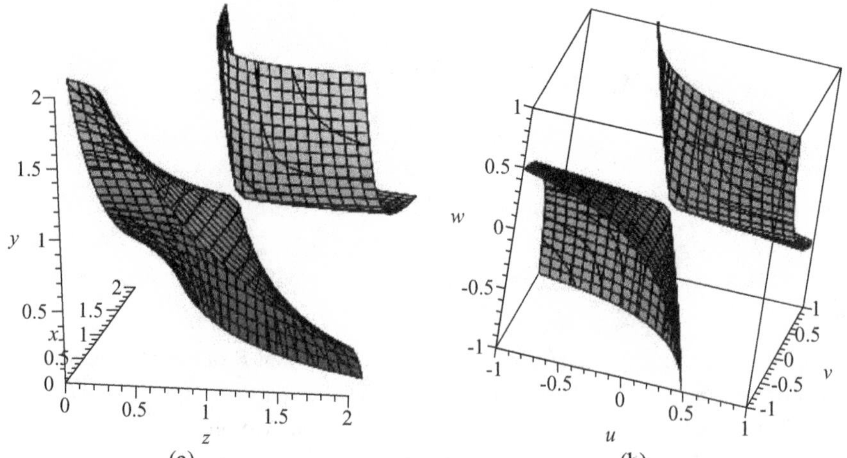

(a) (b)

Figure 11.1 The variety $\mathcal{V}$ in ordinary and logarithmic coordinates.

where $\boldsymbol{x}$ is near $\boldsymbol{x}_{\min}$ (of course, we cannot take $\boldsymbol{x} = \boldsymbol{x}_{\min}$ without intersecting $\mathcal{V}$). We then use the theory of hyperbolic polynomials to show that this vector $\boldsymbol{v}(\boldsymbol{y})$ may be chosen continuously in $\boldsymbol{y}$, resulting in a homotopy $\{\exp(\boldsymbol{x} + t\boldsymbol{v}(\boldsymbol{y}) + i\,\boldsymbol{y}) : 0 \le t \le L\}$, which pushes the cycle down below height h_*.

11.1 Cones, Hyperbolicity, and Deformations

In this section, we prove Theorem 8.4.2. Throughout this section, we fix $F = G/H = \sum_r a_r z^r$ convergent on the component B of the complement of $\mathtt{amoeba}(H)$ and a minimizing point $\boldsymbol{x}_{\min}$ for a nonflat direction $\hat{\boldsymbol{r}}_*$. As usual, $\mathcal{M}$ denotes the set $(\mathbb{C} \setminus \{0\})^d \setminus \mathcal{V}$ and $\mathbf{T}_{\text{flat}} := (\mathbb{R}/(2\pi\mathbb{Z}))^d$.

The main work of Section 11.1 is the construction of a deformation that moves the chain $\mathbf{T}(\boldsymbol{x}_{\min})$ to a chain C_δ lying below height h_* except in a δ-neighborhood of each $z \in \mathtt{local}$. As $\delta \to 0$, in a neighborhood of z, the chain C_δ will approach a chain $\overline{C}(z)$, which is a cone over z (in logarithmic coordinates) and over which $\mathrm{Re}(\hat{\boldsymbol{r}}_* \cdot z')$ is $\Theta(|z - z'|)$. A few pictures provide some intuition. In Figure 11.1, an instance is shown where $\mathcal{V}$ has a quadratic point at $(1, 1, 1)$. The surface shown is the real part of $\mathcal{V}$. At $(1, 1, 1)$, the surface is tangent to a cone whose axis is in the main diagonal. Of course we could only draw the real part. Changing to logarithmic coordinates gives the lower picture, which is the real part of the zero set of $\log \mathcal{V}$ of $f := H \circ \exp$. The tangent cone is a circular cone, which we did not include in the figure because it nearly coincides with the zero set of f. Figure 11.2 is in logarithmic coordinates. The cone in the picture is the tangent cone to $\log \mathcal{V}$ at the point $\boldsymbol{x}_{\min}$. The point $\boldsymbol{u} \in \mathbb{R}^d$ is in the set B, which is tangent to the leftward opening half cone. The dashed line represents the linear 3-space in which the real part of each point is $\boldsymbol{u}$ and the imaginary part is arbitrary. This

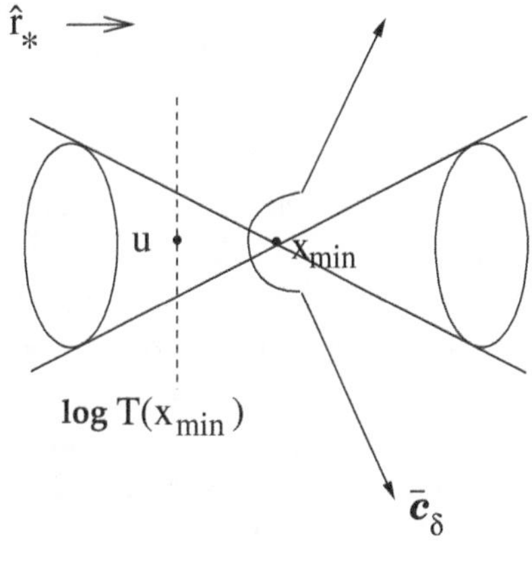

Figure 11.2 $\hat{\boldsymbol{r}}_*$ points to the right; the dotted plane is the chain $\mathbf{T}(\boldsymbol{u})$; the solid chain is $\overline{c}_\delta$; on the right is $\overline{c}$.

chain, $\boldsymbol{u} + i\mathbb{R}^d$, is the logarithm of $\mathbf{T}(\boldsymbol{x}_{\min})$. This chain can be deformed to the chain $\overline{C}_\delta = \exp[\overline{c}_\delta]$, which is depicted by a solid one-dimensional sketch (it is actually a 3-chain). Taking δ to zero gives the chain $\overline{c}$ shown in the lower picture.

11.1.1 How to Use Cones to Construct Deformations

Let $\mathbf{W} := \{\boldsymbol{y} \in \mathbf{T}_{\text{flat}} : \exp(\boldsymbol{x}_{\min} + i\boldsymbol{y}) \in \texttt{local}(\hat{\boldsymbol{r}}_*)\}$ denote the set of arguments of locally oriented critical points in direction $\hat{\boldsymbol{r}}_*$. We use the following result when `local`, hence $\mathbf{W}$, is empty.

Lemma 11.1.1 (deformation defined by cones) *Suppose that for each $\boldsymbol{y} \in \mathbf{T}_{\text{flat}}$, there is a convex cone $Z(\boldsymbol{y})$ with the following properties.*

(i) There is a fixed nonzero $\boldsymbol{u}$ contained in every $Z(\boldsymbol{y})$.
(ii) The set $\{\exp(\boldsymbol{x}_{\min} + \boldsymbol{x} + i\boldsymbol{y}) : \boldsymbol{x} \in Z(\boldsymbol{y}), |\boldsymbol{x}| \le \varepsilon\}$ is disjoint from $\mathcal{V}$ for sufficiently small $\varepsilon > 0$.
(iii) Each cone $Z(\boldsymbol{y})$ contains a vector $\boldsymbol{v}(\boldsymbol{y})$ whose dot product with $\hat{\boldsymbol{r}}_*$ has strictly positive real part.
(iv) The cones $\{Z(\boldsymbol{y}) : \boldsymbol{y} \in \mathbf{T}_{\text{flat}}\}$ vary semicontinuously in $\boldsymbol{y}$ in the sense that if $\boldsymbol{y}_n \to \boldsymbol{y}$, then

$$Z(\boldsymbol{y}) \subseteq \liminf_n Z(\boldsymbol{y}_n) .$$

Then the torus $\mathbf{T}(\boldsymbol{x}_{\min} + \boldsymbol{u})$ is homotopic in $\mathcal{M}$ to a cycle C_* whose maximum height is less than $h_* := h_{\hat{\boldsymbol{r}}_*}(\boldsymbol{x}_{\min})$.

Proof We claim first that for each $\boldsymbol{y} \in \mathbf{T}_{\text{flat}}$, there is a neighborhood $\mathcal{N}(\boldsymbol{y})$ of $\boldsymbol{y}$ in $\mathbf{T}_{\text{flat}}$ such that $\boldsymbol{v}(\boldsymbol{y}) \in Z(\boldsymbol{y}')$ when $\boldsymbol{y}' \in \mathcal{N}(\boldsymbol{y})$. This follows directly from semicontinuity. By compactness we may cover $\mathbf{T}_{\text{flat}}$ with finitely many neighborhoods $\{\mathcal{N}(\boldsymbol{y}_j) : 1 \le j \le m\}$. Let $\{\psi_j : 1 \le j \le m\}$ be a partition of unity subordinate to this finite cover. For $\boldsymbol{y} \in \mathbf{T}_{\text{flat}}$ define

$$u(\boldsymbol{y}) := \sum_{j=1}^{m} \psi(\boldsymbol{y})\boldsymbol{v}(\boldsymbol{y}_j).$$

For each j such that $\boldsymbol{y} \in \mathcal{N}(\boldsymbol{y}_j)$, the vector $\boldsymbol{v}(\boldsymbol{y}_j)$ is in the cone $Z(\boldsymbol{y})$. The vector $u(\boldsymbol{y})$ is a convex combination of these vectors, and therefore, by convexity of the cone $Z(\boldsymbol{y})$, we see that $u(\boldsymbol{y}) \in Z(\boldsymbol{y})$. By linearity of the dot product with $\hat{\boldsymbol{r}}_*$, we see also that $u(\boldsymbol{y}) \cdot \hat{\boldsymbol{r}}_* > 0$. Again, using convexity of $Z(\boldsymbol{y})$, we see that the line segment between $\boldsymbol{u}$ and $u(\boldsymbol{y})$ is in $Z(\boldsymbol{y})$.

We now have enough to define a homotopy. Replacing $\boldsymbol{u}$ and the function $u(\boldsymbol{y})$ by small scalar multiples if necessary, we can assume without loss of generality that $\boldsymbol{u}$, $|u(\boldsymbol{y})| < \varepsilon$, where ε is as in hypothesis *(ii)* of the Lemma. Define

$$\phi_t(\boldsymbol{y}) := \exp[i\boldsymbol{y} + \boldsymbol{x}_{\min} + (1-t)\boldsymbol{u} + tu(\boldsymbol{y})]. \qquad (11.1.1)$$

Clearly $\phi : \mathbf{T}_{\text{flat}} \times [0, 1]$ is continuous, with $\phi_0[\mathbf{T}_{\text{flat}}]$ being the chain $\mathbf{T}(\boldsymbol{u})$. Because each line segment from $\boldsymbol{u}$ to $u(\boldsymbol{y})$ is in $Z(\boldsymbol{y})$ and $\exp(i\boldsymbol{y} + \boldsymbol{x}_{\min} + \boldsymbol{v}) \notin \mathcal{V}$ for any $\boldsymbol{v}$ in any $Z(\boldsymbol{y})$, we see that the homotopy ϕ avoids $\mathcal{V}$. The homotopy ϕ deforms $\mathbf{T}(\boldsymbol{u})$ to the cycle $\phi_1[\mathbf{T}_{\text{flat}}]$. The height of a point $\phi_1(\boldsymbol{y})$ is given by $-\hat{\boldsymbol{r}}_* \cdot \text{Re}\{\boldsymbol{x}_{\min} + u(\boldsymbol{y})\}$. We have seen that $\hat{\boldsymbol{r}}_* \cdot u(\boldsymbol{y})$ is strictly positive, which finishes the proof. □

For later use, we record an extension.

Corollary 11.1.2 (deformation to `local`) *Assume the hypotheses of Lemma 11.1.1, weakened by allowing (iii) to fail on a finite set* $\mathbf{W}$. *Then for any* $\delta > 0$, *the cycle* $\mathbf{T}(\boldsymbol{x}_{\min} + \boldsymbol{u})$ *is homotopic to a cycle* C_δ *whose maximum height is less than* h_* *except on an* δ*-neighborhood of* $\exp(\boldsymbol{x}_{\min} + i\,\mathbf{W})$.

Proof The neighborhoods $\mathcal{N}(\boldsymbol{y})$ in the previous proof will not exist for $\boldsymbol{y} \in \mathbf{W}$, and none of the other neighborhoods will contain any point of $\mathbf{W}$. Instead, for $\boldsymbol{y} \in \mathbf{W}$, we define $\mathcal{N}(\boldsymbol{y})$ to be the log of the δ ball centered at $\exp(\boldsymbol{x}_{\min} + i\boldsymbol{y})$ and set $\boldsymbol{v}(\boldsymbol{y}) = \mathbf{0}$. Completing to an open cover and using the same construction as before produces the modified cycle C_δ, which satisfies the modified conclusion. □

Remark The conclusions of Lemma 11.1.1 and Corollary 11.1.2 may be strengthened to hold for all $\hat{\boldsymbol{r}}$ in some neighborhood of $\hat{\boldsymbol{r}}_*$. This is immediate from the finite cover and the strict inequalities.

11.1.2 Cones of Hyperbolicity in the Homogeneous Case

The proof of part *(ii)* of Theorem 8.4.2, and the main work of this section, consists of verifying the hypotheses of Lemma 11.1.1. We begin with a definition.

Definition 11.1.3 (hyperbolicity) *Say that a homogeneous complex polynomial* A *of degree* $m \geq 1$ *is* ***hyperbolic*** *in direction* $\boldsymbol{v} \in \mathbb{R}^d$ *if* $A(\boldsymbol{v}) \neq 0$ *and for all* $\boldsymbol{x} \in \mathbb{R}^d$, *the polynomial* $t \mapsto A(\boldsymbol{x} + t\boldsymbol{v})$ *has only real roots.*

Remark Seemingly weaker than but equivalent to hyperbolicity is the condition that $A(\boldsymbol{v} + i\boldsymbol{y}) \neq 0$ for all $\boldsymbol{y} \in \mathbb{R}^d$.

The set of $\boldsymbol{v}$ for which A is hyperbolic in direction $\boldsymbol{v}$ is an open set whose components are convex cones in $\mathbb{R}^d$ and are components of the complement of the zero set of A in $\mathbb{R}^d$. These are called ***cones of hyperbolicity*** for the homogeneous polynomial A. Denote by $\mathbf{K}^{\boldsymbol{v}}(A)$ the cone of hyperbolicity of A containing a given $\boldsymbol{v}$. Some multiple of A is positive on $\mathbf{K}^{\boldsymbol{v}}(A)$ and vanishing on $\partial\,\mathbf{K}^{\boldsymbol{v}}(A)$, and for $\boldsymbol{x} \in \mathbf{K}^{\boldsymbol{v}}(A)$, the roots of $A(\boldsymbol{x} + t\boldsymbol{v})$ will all be negative. These properties are proved, among other places, in Gülen (1997, Theorem 3.1).

Example 11.1.4 (linear function) Suppose A is real linear, thus $A(\boldsymbol{x}) = \boldsymbol{v} \cdot \boldsymbol{x}$. Then the cones of hyperbolicity are the halfspaces $\{\boldsymbol{x} : \boldsymbol{v} \cdot \boldsymbol{x} > 0\}$ and $\{\boldsymbol{x} : \boldsymbol{v} \cdot \boldsymbol{x} < 0\}$. ■

Example 11.1.5 (quadratic function) Suppose A is the ***standard Lorentzian quadratic***, i.e., $A(\boldsymbol{x}) = x_1^2 - \sum_{j=2}^d x_j^2$. Then there are two cones of hyperbolicity B_+ and B_-, defined respectively by

$$x_1 > \left(\sum_{j=2}^d x_j^2\right)^{1/2}$$

$$x_1 < -\left(\sum_{j=2}^d x_j^2\right)^{1/2} .$$

These are the so-called positive and negative time-like cones. ■

The following proposition and definition define a family of cones $\{\mathbf{K}^{A,B}(\boldsymbol{x})\}_{\boldsymbol{x}\in\mathbb{R}^d}$, which will be used to establish the semicontinuity hypothesis in Lemma 11.1.1.

Proposition 11.1.6 (first semicontinuity result) *Let A be any hyperbolic homogeneous polynomial, and let m be its degree. Fix $\boldsymbol{x}$ with $A(\boldsymbol{x}) = 0$ and let $\tilde{A} := \mathrm{hom}(A, \boldsymbol{x})$ denote the leading homogeneous part of A at $\boldsymbol{x}$. If A is hyperbolic in direction $\boldsymbol{u}$, then $\tilde{A}$ is also hyperbolic in direction $\boldsymbol{u}$. Consequently, if B is any cone of hyperbolicity for A, then there is some cone of hyperbolicity for $\tilde{A}$ that contains B.*

Proof This follows from the conclusion (3.45) of Atiyah, Bott, and Gårding (1970, Lemma 3.42). Because the development there is long and complicated, we give here a short, self-contained proof, provided by J. Borcea (personal communication). If P is a polynomial whose degree at zero is k, we may recover its leading homogeneous part $\mathrm{hom}(P)$ by

$$\mathrm{hom}(P)(\boldsymbol{y}) = \lim_{\lambda\to\infty} \lambda^k P(\lambda^{-1}\boldsymbol{y}) .$$

The limit is uniform as $\boldsymbol{y}$ varies over compact sets. Indeed, monomials of degree k are invariant under the scaling on the right-hand side, whereas monomials of degree $k+j$ scale by λ^{-j}, uniformly over compact sets.

Apply this with $P(\cdot) = A(\boldsymbol{x} + \cdot)$ and $\boldsymbol{y} + t\boldsymbol{u}$ in place of $\boldsymbol{y}$ to see that for fixed $\boldsymbol{x}$, $\boldsymbol{y}$, and $\boldsymbol{u}$,

$$\tilde{A}(\boldsymbol{y} + t\boldsymbol{u}) = \lim_{\lambda \to \infty} \lambda^k A(\boldsymbol{x} + \lambda^{-1}(\boldsymbol{y} + t\boldsymbol{u}))$$

uniformly as t varies over compact sub-intervals of $\mathbb{R}$. Because A is hyperbolic in direction $\boldsymbol{u}$, for any fixed λ, all the zeros of this polynomial in t are real. Hurwitz's theorem on the continuity of zeros. Conway (1978, Corollary 2.6) says that a limit, uniform on bounded intervals, of polynomials having all real 0's will either have all real 0's or vanish identically. The limit, $\tilde{A}(\boldsymbol{y} + t\boldsymbol{u})$, has degree $k \geq 1$; it does not vanish identically, and therefore it has all real 0's. This shows $\tilde{A}$ to be hyperbolic in direction $\boldsymbol{u}$. □

Definition 11.1.7 (family of cones in the homogeneous case) *Let A be a hyperbolic homogeneous polynomial, and let B be a cone of hyperbolicity for A. If $A(\boldsymbol{x}) = 0$, define*

$$\mathbf{K}^{A,B}(\boldsymbol{x})$$

to be the cone of hyperbolicity of $\mathtt{hom}(A, \boldsymbol{x})$ *containing B, whose existence we have just proved. If $A(\boldsymbol{x}) \neq 0$, we define $\mathbf{K}^{A,B}(\boldsymbol{x})$ to be all of $\mathbb{R}^d$.*

Remark For $\boldsymbol{x} \neq 0$, the cone $\mathbf{K}^{A,B}$ depends only on the projective vector $\widehat{\mathbf{x}} := \boldsymbol{x}/|\boldsymbol{x}|$.

Example 11.1.8 As an example of hyperbolicity in a homogeneous quadratic, let $S = x_1^2 - x_2^2 - \cdots - x_d^2$ be the standard Lorentzian quadratic. In Example 11.1.5 we saw that the two cones of hyperbolicity at the origin are the time-like cones B_+ and B_-. Let $B := B_+$. For $\boldsymbol{x} \in \partial B$, if $\boldsymbol{x} \neq 0$, then $\mathbf{K}^{S,B}(\boldsymbol{x})$ is the tangent halfspace $\{\boldsymbol{x} + \boldsymbol{y} : \boldsymbol{y} \cdot \nabla S \geq 0\}$. If $\boldsymbol{x} = \mathbf{0}$, then $\mathbf{K}^{S,B}(\boldsymbol{x}) = B$. By definition, a ***Lorentzian quadratic*** is obtained from S by a real linear transformation. We see therefore that for any Lorentzian quadratic, the boundary of the cone of hyperbolicity is the algebraic tangent cone, and the cones of hyperbolicity at these points are halfspaces whose boundaries are support hyperplanes to the time-like cones. ■

11.1.3 Cones of Hyperbolicity in the General Case

Although the definition of hyperbolicity in the homogeneous case is valid for complex polynomials and involves complex roots, the role played by the real subspace is essential. In this section we generalize these notions beyond homogeneous polynomials in two ways. We define cones of hyperbolicity $\mathbf{K}^{f,B}(\boldsymbol{z})$ for polynomials f that are not necessarily homogeneous (Definition 11.1.9), and we prove versions of semicontinuity for this family of cones (Lemma 11.1.15). We cannot do this in full generality but do so under the following assumptions.

- The function f is $H \circ \exp$ for some Laurent polynomial H.
- The point z is equal to $x + iy$ for some x on the boundary of a component B of the complement of $\mathtt{amoeba}(H)$.

We do not know whether $\mathbf{K}^{f,B}(z)$ may be defined when z is not on the boundary of the amoeba of H, or when f is not a log-Laurent polynomial, in such a way that semicontinuity results still hold.

The following definitions of $\mathbf{K}(z)$ and $\mathbf{N}(z)$ pay off our debt from Section 8.4 and ultimately enable us to prove Theorem 8.4.2.

Definition 11.1.9 (hyperbolicity and normal cones at a point of $\mathcal{V}_f$) *Let H be a Laurent polynomial, B a component of $\mathbb{R}^d \setminus \mathtt{amoeba}(H)$, and $z = \exp(x + iy) \in \mathcal{V}_f$ with $x \in \partial B$. We let $f := H \circ \exp$ and define*

$$\mathbf{K}(z) := \mathbf{K}^{f,B}(z) := \mathbf{K}^u(\mathtt{hom}(f, x + iy)), \tag{11.1.2}$$

the (open) cone of hyperbolicity of $A := \mathtt{hom}(f, x + iy)$ that contains B. The existence of this cone is guaranteed by Proposition 11.1.14.

Remarks

(i) Defining $\mathbf{K}^{f,B}$ as $\mathbf{K}^{A,B}$ appears almost trivial; however, the greater difficulty of the analogous properties in this case convinces us that this extension is quite nontrivial.
(ii) We may extend the definition of $\mathbf{K}(\cdot)$ to all of $\mathbf{T}(x)$ by taking $\mathbf{K}(z) = \mathbb{R}^d$ when $f(z) \neq 0$.
(iii) When we wish to stress the dependence on f and/or B, we write $\mathbf{K}^{f,B}(z)$.
(iv) When $z = \exp(x + iy)$ and x is understood, we sometimes write $\mathbf{K}^{f,B}(y)$.
(v) If f is homogeneous, then this definition agrees with Definition 11.1.7.

Before going further, we give a few examples.

Example 11.1.10 (cones of hyperbolicity at smooth points) Suppose z is a minimal smooth point of $\mathcal{V}_f$. Then $\overline{f}$ is linear, as in Example 11.1.4. Writing $\overline{f}(x) = v \cdot x$, the assumption of minimality implies that the vector v is a complex scalar multiple of a real vector. The cones of hyperbolicity at z are halfspaces normal to v. ■

Example 11.1.11 (cones of hyperbolicity at multiple points) Suppose z is a multiple point of $\mathcal{V}_f$. Then $\overline{f}$ is a product of linear functions, i.e., a central hyperplane arrangement. If $f = H \circ \exp$ and z is a minimal point of $\mathcal{V}_H$, then the normal vectors to the factors of $\overline{f}$ are scalar multiples of real vectors. The real parts of these complex are real hyperplanes, and these divide $\mathbb{R}^d$ into projective cones. Each of these is a cone of hyperbolicity for $\overline{f}$. ■

Example 11.1.12 (cones of hyperbolicity at quadratic cone points) Let $f = \overline{f} = x_1^2 - \sum_{j=2}^d x_j^2$ be the standard Lorentzian quadratic. The polynomial f is homogeneous whence at the origin, $f = \mathtt{hom}(f)$ and the two cones of hyperbolicity are the positive and negative light cone $x_1 > (\sum_{j=2}^d x_j^2)^{1/2}$ and $x_1 < -(\sum_{j=2}^d x_j^2)^{1/2}$. Everywhere in $\mathbb{R}^d$ other than the origin, f is smooth and the cones are halfspaces as before. ■

11.1.4 Strong and Weak Hyperbolicity

To finish the job, we need to prove the existence result Proposition 11.1.14 and the semicontinuity result Lemma 11.1.15. This involves defining notions of strong and weak hyperbolicity that we will not be needing for any other purposes. These definitions are somewhat less natural than those that have preceded, and readers not interested in these developments may safely skip to the proof of Theorem 8.4.2. The following definitions may be found in Baryshnikov and Pemantle (2011, Section 2.4).

Definition 11.1.13 (strong and weak hyperbolicity) *Let $f : \mathbb{C}^d \to \mathbb{C}$ vanish at $\mathbf{z}$ and be holomorphic in a neighborhood of $\mathbf{z}$. We say that f is **strongly hyperbolic** at $\mathbf{z}$ in direction of the unit vector $\hat{\mathbf{v}}$ if there is an $\varepsilon > 0$ such that $f(\mathbf{z} + t\mathbf{v}' + i\mathbf{u}) \neq 0$ for all real $0 < t < \varepsilon$, all $\mathbf{v}'$ at distance at most ε from $\hat{\mathbf{v}}$, and all $\mathbf{u} \in \mathbb{R}^d$ of magnitude at most ε. In this case we may say that f is strongly hyperbolic at $\mathbf{z}$ in direction $\hat{\mathbf{v}}$ with radius ε. Say that f is **weakly hyperbolic** in direction $\mathbf{v}$ if for every $M > 0$, there is an $\varepsilon > 0$ such that $f(\mathbf{z} + t\mathbf{v} + i\mathbf{u}) \neq 0$ for all real $0 < t|\mathbf{v}| < \varepsilon$, and for all $\mathbf{u} \in \mathbb{R}^d$ of magnitude at most ε additionally satisfying $|\mathbf{u}|/(t|\mathbf{v}|) \leq M$.*

Strong hyperbolicity implies weak hyperbolicity, and weak hyperbolicity is equivalent to hyperbolicity of the homogeneous part. The first of these is obvious and the second is easy, and is proved as Proposition 2.11 of Baryshnikov and Pemantle (2011) (see Exercise 11.1).

Proposition 11.1.14 *Let H be a Laurent polynomial in d-variables. Suppose that B is a component of $\mathtt{amoeba}(H)$ and $\mathbf{x} \in \partial B$, so that $f := H \circ \exp$ vanishes at some point $\mathbf{x} + i\mathbf{y}$. Let $\overline{f} := \mathtt{hom}(f, \mathbf{x} + i\mathbf{y})$ denote the leading homogeneous part of $f(\mathbf{x} + i\mathbf{y} + \cdot)$. Then f is strongly hyperbolic at $\mathbf{x} + i\mathbf{y}$, some complex scalar multiple of $\overline{f}$ is real and hyperbolic, and some cone of hyperbolicity $\mathbf{K}^{\mathbf{u}}(\overline{f})$ contains $\mathtt{tan}_{\mathbf{x}}(B)$.*

Proof Strong hyperbolicity of f in any direction $\mathbf{u} \in \mathtt{tan}_{\mathbf{x}}(B)$ follows from the definition of the amoeba. Strong hyperbolicity is stronger than weak hyperbolicity, and hence hyperbolicity of $\overline{f}$ in direction $\mathbf{u}$ follows from Exercise 11.1. The vector $\mathbf{u} \in \mathtt{tan}_{\mathbf{x}}(B)$ is arbitrary, whence $\mathbf{K}^{\mathbf{u}}(\overline{f}) \supseteq \mathtt{tan}_{\mathbf{x}}(B)$. To see that some multiple of $\overline{f}$ is real, let $\mathbf{u}$ be any real vector in $\mathtt{tan}_{\mathbf{x}}(B)$, let m denote the degree of $\overline{f}$, and let γ denote the coefficient of the z^m term of $A(\overline{f}\mathbf{u} + \mathbf{y})$. Then γ is the degree m

coefficient of $\overline{f}(zu)$, and hence is nonzero and does not depend on y. For any fixed y, the fact that $\overline{f}(zu + y)$ has all real roots implies that the monic polynomial $\gamma^{-1}\overline{f}(zu + y)$ has all real coefficients. □

11.1.5 Semicontinuity

Some of the intuition regarding the role played by hyperbolicity in fulfilling hypothesis (iv) of Lemma 11.1.1 is as follows. Near any point in any stratum of any complex algebraic variety, there are one or more cones contained in the complement of the variety. Hyperbolicity may be thought of as a kind of orientability for families of such cones, ensuring a consistent choice of "inward tangent cone," namely the one containing B. In case it seems obvious that such a choice should be possible, see Exercise 11.2. This should provide some intuition regarding why polynomials should be hyperbolic near points on the boundary of the domain of convergence: the meaning of "inward" is clear there. Hyperbolicity is also used to ensure the cones are convex.

Still aiming at hypothesis (*iv*) of Lemma 11.1.1, we quote and briefly outline proofs of the following semicontinuity results from Baryshnikov and Pemantle (2011, Theorem 2.14, Corollary 2.15), where they are attributed to Atiyah, Bott, and Gårding (1970, Lemma 3.22) and Gårding (1950, Theorem H 5.4.4).

Lemma 11.1.15 (semicontinuity)

(i) Let A be a homogeneous polynomial and B a cone of hyperbolicity for A. Then the cone $\mathbf{K}^{A,B}(y)$ is semicontinuous in y.

(ii) Let $f = H \circ \exp$ for some Laurent polynomial H. Let B be a component of the complement of amoeba(H) *and let $x \in \partial B$. Then $\mathbf{K}^{f,B}(z)$ is semicontinuous as z varies over $\mathbf{T}(x)$.*

(iii) Let f, H, B, x be as in (ii). The following mixed semicontinuity result holds:

$$K^{A,B}(\widehat{y}) \subseteq \liminf K^{f,B}(y_n)$$

as $y_n \to 0$ with $y_n/|y_n| \to \widehat{y}$.

Remarks

(i) The first conclusion is nearly a specialization of the second conclusion to homogeneous functions, except that not every homogeneous function is the homogenization of a Laurent polynomial composed with the exponential function.

(ii) In particular, when the homogeneous part A of f does not vanish at $\widehat{x}$, then the final conclusion of the lemma implies that $f(w) \neq 0$ if w is sufficiently small and $w/|w|$ is sufficiently close to $\widehat{x}$.

Proof (sketch) We first establish the following result on strong hyperbolicity. Suppose an analytic function f is strongly hyperbolic in the direction v at the point $z = x = iy$. Let $A :=$ hom$(\overline{f}, z)$. If $u \in \mathbf{K}^{v}(f)$, then f is strongly hyperbolic in direction $t\boldsymbol{v} + (1-t)\boldsymbol{u}$ for any $t \in [0,1]$. The first two conclusions follow relatively easily from this, and the third with a little work. The proofs are a little over a page in Baryshnikov and Pemantle (2011). □

Example 11.1.16 As an example of semicontinuity, we look at a multiple point z. The homogeneous part $\overline{f} := \mathtt{hom}(f, z)$ is the product of linear factors $\sum b_j z_j$. The cones of hyperbolicity are projective cones that are the components of $\mathbb{R}^d$ when the hyperplanes on which the linear factors vanish are removed. Suppose $f = H \circ \exp$ where H has only multiple point singularities. Let $z' \to z$ in $\mathcal{V}_H$ while remaining in a single stratum S. Then z is in either S or ∂S. In the latter case, the partition near z of $\mathbb{R}^d$ into projective cones is finer at z that at the points z'. In this case, the semicontinuity is strict. ■

11.1.6 Proof of Theorem 8.4.2

Proof Taking the conclusions of this theorem in reverse order, the third requires no proof. To prove the second conclusion, we apply Lemma 11.1.1. For every $\boldsymbol{y} \in \mathbf{T}_{\text{flat}}$, we let

$$Z(\boldsymbol{y}) := \mathbf{K}^{f,B}(\exp(\boldsymbol{x}_{\min} + i\boldsymbol{y})) = \mathbf{K}^{\boldsymbol{u}}(\mathtt{hom}(f, \boldsymbol{x}_{\min} + i\boldsymbol{y})),$$

where $\boldsymbol{u}$ is any vector in $\mathtt{tan}_{\boldsymbol{x}_{\min}}(B)$. By definition, $\boldsymbol{u} \in Z(\boldsymbol{y})$ for all $\boldsymbol{y}$, so the hypothesis (i) of Lemma 11.1.1 is satisfied. By definition of cones of hyperbolicity, hypothesis *(ii)* is satisfied. Because `local` is empty, each point $z = \boldsymbol{x}_{\min} + i\boldsymbol{y}$ is not in `local`. By definition of `local`, this means that $\hat{\boldsymbol{r}}_* \notin \mathbf{N}(z)$. Because $\mathbf{N}(z)$ is defined as the dual to $\mathbf{K}(z) = \mathbf{K}^{f,B}(z)$, it is immediate that $\hat{\boldsymbol{r}}_* \notin \mathbf{N}(z)$ implies $\hat{\boldsymbol{r}}_* \cdot \boldsymbol{v} > 0$ for some $\boldsymbol{v} \in Z(\boldsymbol{y})$. This establishes hypothesis *(iii)*. Finally, the semicontinuity hypothesis *(iv)* is conclusion *(ii)* of Lemma 11.1.15. The conclusion of the lemma is the existence of the cycle C_*, which is the second conclusion of the theorem.

To prove the first conclusion, we show the contrapositive. Suppose that z is not a critical point of the function $h_{\hat{\boldsymbol{r}}_*}$ on $\mathcal{V}$. Letting $z = \exp(\boldsymbol{x}_{\min} + i\boldsymbol{y})$, we see that $\boldsymbol{x}_{\min} + i\boldsymbol{y}$ is not critical for $f = H \circ \exp$ on the logarithmic space, and letting S denote the stratum of $\log \mathcal{V}$ containing $\boldsymbol{x}_{\min} + i\boldsymbol{y}$, we see that the differential of $h_{\hat{\boldsymbol{r}}_*} \circ \exp$ is not identically zero on S. This differential is just the dot product with $\hat{\boldsymbol{r}}_*$.

The linear space $T := T_{\boldsymbol{x}_{\min}+i\boldsymbol{y}}(S)$ is what Atiyah, Bott, and Gårding (1970) call a *lineality* for the function $\overline{f} = \mathtt{hom}(f, \boldsymbol{x}_{\min} + i\boldsymbol{y})$, meaning that $\overline{f}(\boldsymbol{w} + \boldsymbol{w}') = \overline{f}(\boldsymbol{w})$ for any $\boldsymbol{w}' \in T$ and any $\boldsymbol{w} \in \mathbb{C}^d$. This is shown, for example, in the proof of Baryshnikov and Pemantle (2011, Proposition 2.22). By Proposition 11.1.14, the function $\overline{f}$ has a cone of hyperbolicity $\mathbf{K}^{f,B}(z)$ containing B. By Atiyah, Bott, and Gårding (1970, Lemma 3.52), the real part of the linear space T is in the *edge* of the cone $\mathbf{K}^{f,B}(z)$, meaning that translations by elements of T map the cone into itself. Any real hyperplane not containing the edge of a cone intersects the interior of the cone. Our initial assumption that z is not a critical point implies that the real hyperplane $\{x : \hat{r}_* \cdot x = 0\}$ does not contain T; therefore, there is some vector $v \in \mathbf{K}^{f,B}(x + iy)$ with $\hat{r}_* \cdot v > 0$. This implies $\hat{r}_* \notin \mathbf{N}(z)$, finishing the proof of conclusion (i) and hence of Theorem 8.4.2. □

11.1.7 Projective Deformations

Suppose that $z = \exp(x) \in \texttt{local}$ and that the vector $\hat{r}_*$ is in the interior of the dual cone $\mathbf{N}(z)$ (in particular, the interior must be nonempty). Let $q := \texttt{hom}(H, z)$ be the homogeneous part of H at z, and suppose that for every $z' = \exp(x + iy)$ in a neighborhood of z, the cone of hyperbolicity of q at z', denoted $\mathbf{K}(z')$, contains a vector $v(z')$ whose dot product with $\hat{r}_*$ is strictly positive. We may then carry out the construction in the proof of Theorem 8.4.2 in a more deliberate manner, as follows.

Replacing H by q, for each y on the unit sphere, we choose a vector $v(y) \in \mathbf{K}(\exp(x + iy))$ for which $v \cdot \hat{r}_* > 0$. Each $v(y)$ works for all y' in some neighborhood of y and $\hat{r}$ in a neighborhood of $\hat{r}_*$, so, as before, using a finite open cover of the sphere by these neighborhoods and a partition of unity, we may choose $v(y)$ continuously. For $y \neq 0$, we let $v_0(y)$ denote the 0-homogeneous extension $v(y/|y|)$. By conclusion (iii) of Lemma 11.1.15, there is some $\varepsilon > 0$ such that for $|y| < \varepsilon$, the cone of hyperbolicity for H, namely $\mathbf{K}^{H,B}(\exp(x + iy))$, contains $v(y)$. Now extend v to a continuous 1-homogeneous function by $v(y) = |y| v_0(y)$ for $y \neq \mathbf{0}$ and $v(\mathbf{0}) = \mathbf{0}$.

Using this on the ε-ball about z and extending via partitions of unity to the remainder of $\mathbf{T}(x)$ gives a vector field i.e., projective in a neighborhood of z and satisfies hypotheses (i)–(iv) of Lemma 11.1.1, except that (iii) is not satisfied at the origin, because there $v = 0$. The deformation (11.1.1) provides a homotopy in $\mathcal{M}$ to a chain $\overline{C}$ that, when intersected with an ε-ball about z, is the exponential image of a cone; the homotopy stays in $\mathcal{M}$ at times less than 1 and at time 1 except at the point z. We denote this piece of $\overline{C}$ by $\overline{C}(z)$.

We sum up what we have constructed in the following lemma, which includes the routine extension of piecing these together for all $z \in \texttt{local}$ via a partition of unity. Following Baryshnikov and Pemantle (2011), with q as previously defined, we say that the direction $\hat{r}_*$ is ***non-obstructed*** for the point z if, for every $z' = \exp(x + iy)$ in a neighborhood of z, the cone of hyperbolicity of q at z' contains a vector $v(z')$ whose dot product with $\hat{r}_*$ is strictly positive.

Lemma 11.1.17 *Suppose that* $z = \exp(x + iy) \in \texttt{local}$ *and that* $\mathbf{N}(z)$ *has non-empty interior. Suppose the vector* $\hat{r}_*$ *is interior to* $\mathbf{N}(z)$ *and is not obstructed. Then there is a homotopy* $\Phi : \mathbf{T}_{\text{flat}} \times [0, 1] \to \mathbb{C}^d$ *taking* $\mathbf{T}(x)$ *to a chain* C *such that*

(i) $\Phi(z, t) \in \mathcal{M}$ *except when* $t = 1$ *and* $z \in \texttt{local}$, *in which case* $\Phi(z, 1) = z$.

(ii) There is a $\delta > 0$ *such that for each* $z \in$ `local`, *restricting the homotopy to a* δ*-neighborhood of* z *results in a chain* $\overline{C}_\delta(z) = \exp[\overline{c}_\delta(y)]$ *where* $\overline{c}_\delta(y)$ *is an intersection of a neighborhood of* $x + iy$ *with an affine cone* $\{x + iy + \lambda S : \lambda \geq 0\}$ *for some compact set S.*

(iii)

$$\inf\{\operatorname{Re}\{r \cdot w\} : w \in S\} > 0 .$$

(iv) It follows from the first two properties that

$$h(w) > c|w - z|$$

on $\overline{C}_\delta(z)$ *for some* $c > 0$.

11.2 Evaluating the Asymptotics

We return now to the case where the set `local` of locally oriented critical points is non-empty. Let $\hat{r}_*$ be a nonflat direction with minimizing point $x_{\min}$. Using the results of the previous section, the Cauchy integral can be localized to `local`. To spell this out, use Lemma 11.1.17 to deform the cycle $\mathbf{T}(x)$, over which we know the Cauchy integral (1.3.1) is valid, to a union of cycles $C_\delta(z)$ over $z \in$ `local`, together with a leftover cycle bounded below height h_*. The contributions to the Cauchy integral are negligible outside of the fragments $C_\delta(z)$. Applying Cauchy's integral formula to F, changing coordinates via the exponential map and breaking into pieces local to each $w \in \mathbf{W}$ gives the following corollary to Lemma 11.1.17.

Corollary 11.2.1 *Under the hypotheses of Lemma 11.1.17,*

$$(2\pi)^d z^r a_r = \sum_{w \in \mathbf{W}(r)} \int_{\overline{C}_\delta(w)} e^{-r \cdot z'} f(z')\, dz' + O(e^{-c|r|})$$

for some $c > 0$.

11.2.1 Approximating by Canonical Functions

We are left with the task of asymptotically evaluating the Cauchy integral over chains $C_\delta(z)$. When asymptotically evaluating integrals, the first step is always to approximate the amplitude via a series. The summands should have a canonical form i.e., particularly easy to integrate. For instance, in Chapters 4 and 5, saddle integrals with general amplitudes are reduced to those whose amplitudes are monomials. Another classical case of this is the Flajolet-Odlyzko transfer theorems. Cauchy integrals of the particular functions $(z - z_0)^\alpha$, $(\log(z - z_0))^\beta$, and $(\log\log(z - z_0))^\gamma$ are evaluated exactly near a point z_0. Using these results, the Cauchy integral can be evaluated for any function that can be approximated by a function of the canonical form or by a series of such functions. The key lemma in doing this is the big-O lemma, which tells us that functions differing by a small quantity will have integrals differing by a correspondingly small quantity.

In this section we treat the class of functions of the form $z^m/A(z)$, where m is a vector of nonnegative integers and A is a homogeneous function. The class of homogeneous functions is still quite large. In fact we won't make much of a dent in it: we consider the simplest cases not handled by previous chapters. Having already mastered smooth and multiple points, we consider points of degree 2, near which the leading homogeneous term is a Lorentzian quadratic. The article by Baryshnikov and Pemantle (2011) handles a product of an arbitrary power of such a quadratic with one or more smooth factors, which is all we attempt here.

We begin by a lemma approximating general functions by functions of the canonical form $z^m/A(z)$. This is proved in Baryshnikov and Pemantle (2011, Lemma 2.24).

Lemma 11.2.2 (straightening lemma) *Suppose that $q(x) = \tilde{q}(x) + R(x)$, where $\tilde{q}$ is a homogeneous polynomial of degree α and R is analytic in a neighborhood of the origin with $R(x) = O(|x|^{\alpha+1})$. Let K be any closed cone on which $\tilde{q}$ does not vanish. Then on the intersection of K with some neighborhood of the origin, the function q does not vanish, and there is a convergent expansion*

$$q(x)^{-s} = \sum_{n=0}^{\infty} \tilde{q}(x)^{-s-n} \left[\sum_{|m| \geq n(\alpha+1)} c(m,n) x^m \right] . \tag{11.2.1}$$

Furthermore,

$$q(x)^{-s} - \sum_{|m|-\alpha n < N} c(m,n) x^m \tilde{q}(x)^{-s-n} = O\left(|x|^{-\alpha s+N}\right) . \tag{11.2.2}$$

Proof Let $R(x) = \sum_{|m| \geq h+1} b(m) x^m$ be a power series for R absolutely convergent in some ball B_ε centered at the origin. Let

$$M := \frac{\sup_{|x| \in B_\varepsilon} \sum |b(m)||x|^m}{\inf_{|x| \in \partial B_\varepsilon \cap K} \tilde{q}(x)} .$$

Then by homogeneity,

$$\sum_m \frac{|b(m) x^m|}{|\tilde{q}(x)|} \leq 1/2$$

on the $\varepsilon/(2M)$ ball. The binomial expansion $(1+u)^{-s} = \sum_{n \geq 0} \binom{-s}{n} u^n$ converges for $|u| < 1$ and in particular for $|u| = 1/2$. Therefore, plugging in $\sum_m b(m) x^m / \tilde{q}(x)$ in for u yields a series

$$\left(1 + \frac{R(x)}{\tilde{q}(x)}\right)^{-s} = \sum_{n \geq 0} \binom{-s}{n} \left(\sum_m b(m) \frac{x^m}{\tilde{q}(x)}\right)^n$$

that converges on $B_{\varepsilon/(2M)} \cap K$. Multiply through by $\tilde{q}^{-s}$ to get (11.2.1). Convergence on any neighborhood of the origin implies the estimate (11.2.2). □

11.2.2 Big-O Estimate

Next we state the big-O lemma. Recall from Definition 7.1.1 the degree of vanishing of an analytic function Q at a point z, denoted $\deg(Q, z)$, which we take to be zero if $Q(z) \neq 0$. Define the degree of vanishing of the real power Q^s to equal $s \deg(Q, z)$. We remark that if Q is a Laurent polynomial, then a branch of Q^s may be defined on the domain of convergence $\operatorname{Re}\log^{-1}[B]$ component B of any Laurent series for F. We define the degree of $H := \prod_{j=1}^k Q_j^{s_j}$ by

$$\deg(H, z) = \sum_{j=1}^{k} s_j \deg(Q_j, z),$$

and we abuse notation slightly by letting

$$\texttt{amoeba}(H) := \texttt{amoeba}\left(\prod_{j=1}^{k} Q_j\right) = \bigcap_{j=1}^{k} \texttt{amoeba}(Q_j).$$

Lemma 11.2.3 (multivariate big-O lemma) *Let $Q_1, \ldots, Q_k$ be Laurent polynomials. Let $s_1, \ldots, s_k$ be real numbers that are not negative integers, let $F = G/H$ where G is analytic and $H = \prod_{j=1}^k Q_j^{s_j}$, and let $f := F \circ \exp$. Let B be a component of the complement of $\texttt{amoeba}(Q)$, and let a_r be the coefficients such that $F(z) = \sum_r a_r z^r$ on B. Let $x \in \partial B$ and fix $w \in \mathbf{W}$. Let $z = \exp(x + iw)$, and let $\overline{C}_\delta(z) = \exp[\overline{c}_\delta(w)]$ be a chain satisfying the conclusion of Lemma 11.1.17. Let $\hat{\mathbf{r}}_*$ be any non-obstructed vector in interior to the dual cone $\mathbf{N}(z)$. Then the following estimates hold uniformly as $\mathbf{r}$ varies over a conical neighborhood of $\hat{\mathbf{r}}_*$.*

(i) If $\phi(z)$ is any function i.e., $O(|z|^\beta)$ at z and $\beta + d > 0$, then

$$|z^r| \int_{\overline{c}(w)} \exp(-\mathbf{r} \cdot z)\phi(z)\, dz = O(|\mathbf{r}|)^{-d-\beta}. \tag{11.2.3}$$

(ii) The same estimate holds for the chain $\overline{c}_\delta(w)$ in place of $\overline{c}(w)$.

(iii) Let $D := \deg(F, z) = \deg(G, z) - \sum_{j=1}^k s_j \deg(Q_j, z)$. Then for any bounded function ψ,

$$|z^r| \int_{\overline{c}_\delta(w)} \exp(-\mathbf{r} \cdot z)\psi(z) f(z)\, dz = O\left(|\mathbf{r}|^{-d-D}\right).$$

(iv) $|z|^r a_r = O(|\mathbf{r}|^{-d-D_*})$ where $D_* := \min_{z \in \texttt{local}} \deg(F, z)$.

Proof The cone $\overline{c}(w)$ is a subset of an infinite cone $\bigcup_{\lambda \geq 0} \lambda S$. We may decompose $dz = t^{d-1}\, dt \wedge dS$, where dS is a finite measure on S. It follows from conclusion (*iv*) of Lemma 11.1.17 that there is a $\theta > 0$ for which $\operatorname{Re}\{\mathbf{r} \cdot y\} \geq \theta |\mathbf{r}|$ on S.

Thus

$$
\begin{aligned}
|z^r| \left| \int_{\overline{c}(w)} \exp(-r \cdot z)\phi(z)\,dz \right| &\le \int_0^\infty \left(\int_S C\, e^{-\theta\lambda|r|}\lambda^\beta\, dS \right) \lambda^{d-1}\, d\lambda \\
&\le \int_0^\infty C' e^{-\theta\lambda|r|}\lambda^{\beta+d-1}\, d\lambda \\
&= O(|r|)^{-d-\beta},
\end{aligned}
$$

proving (i).

The chains $\overline{c}_\delta(w)$ are all homotopic in $\mathcal{M}$ to each other. For any fixed r, therefore, the integral in (ii) is independent of δ. We have seen that $\beta + d > 0$ implies absolute integrability on $\overline{c}(w)$. The same estimates imply that the integral over the intersection of $\overline{c}_\delta(w)$ with an ε-neighborhood of w goes to zero as $\varepsilon \to 0$ uniformly in δ. This implies convergence of the integrals in (ii) to the integral in (i), and because the integrals in (ii) are all the same, they are all equal to the integral in (i). This proves the second conclusion. The third conclusion follows from the first with $\phi(z) = \psi(z) f(z)$ and from the estimate $f(z) = O(|z|)^D$ on $\overline{c}_\delta(w)$, which is a consequence of Lemma 11.2.2 with $q = f$. The fourth conclusion follows from the second and Corollary 11.2.1. □

11.2.3 Fourier Transforms

Up to this point in the chapter, $F = \prod Q_j^{s_j}$ has been an arbitrary product of powers of Laurent polynomials. We now specialize to $k = 1$ and Q, a polynomial whose leading term is a Lorentzian quadratic. The summands in Corollary 11.2.1 are evidently Fourier transforms, which, as we see shortly, are classically known. Recall that the standard Lorentzian quadratic $S(y) := y_1^2 - \sum_{j=2}^d y_j^2$ (see Example 11.1.12) is in fact equivalent to any other Lorentzian quadratic q in the sense that there is a real linear map M such that $q = S \circ M^{-1}$. The Fourier transform of a Lorentzian quadratic is known (see Riesz (1949), Atiyah, Bott, and Gårding (1970), and Baryshnikov and Pemantle (2011)), and in fact it is known for any power S^{-s} as long as $s \neq 0, d-1$. In the formula that follows, S^* denotes the dual quadratic, which has an identical formula $r_1{}^2 - \sum_{j=2}^d r_j^2$, whereas the dual $q^*(r)$ is equal to $S^*(M^* r)$, where M^* is the adjoint of the linear map M.

Proposition 11.2.4 (Fourier transform of a Lorentzian quadratic) *Let s be any real number other than 0 or $d/2 - 1$. The generalized Fourier transform of S^{-s} is given by*

$$
\widehat{S^{-s}}(r) = e^{i\pi s} \frac{S^*(r)^{s-d/2}}{2^{2s-1}\pi^{(d-2)/2}\Gamma(s)\Gamma(s+1-d/2)}.
$$

More generally, for any monomial $\boldsymbol{x}^{\boldsymbol{m}}$ and any Lorentzian quadratic q, the Fourier transform of $\boldsymbol{x}^{\boldsymbol{m}}\, q^{-s}$ is given by

$$\widehat{\boldsymbol{x}^{\boldsymbol{m}} q^{-s}}(\boldsymbol{r}) = e^{i\pi s}\, i^{|\boldsymbol{m}|} \frac{|M|(\partial/\partial \boldsymbol{r})^{\boldsymbol{m}} q^*(\boldsymbol{r})^{s-d/2}}{2^{2s-1}\pi^{(d-2)/2}\Gamma(s)\Gamma(s+1-d/2)}. \tag{11.2.4}$$

There is a catch here. The function $\boldsymbol{x}^{\boldsymbol{m}} q^{-s}$ will fail to be integrable at infinity if the homogeneous degree $|\boldsymbol{m}| - 2s$ is $-d$ or more. Also, the integral defining the Fourier transform blows up at the origin if $|\boldsymbol{m}| - 2s \leq -d$. One of these is bound to happen. Proposition 11.2.4 is stated in terms of *generalized functions*. These generalized functions are defined as limits of actual functions on $\boldsymbol{u} + i\mathbb{R}^d$ as $\boldsymbol{u} \to 0$ in B, their integrals over noncompact sets are defined by weak limits of compact integrals, and their Fourier transforms are defined not by direct integration against $e^{i\boldsymbol{r}\cdot\boldsymbol{x}}$ but by their integrals against (classical) Fourier transforms of smooth, compactly supported functions. For further details of generalized functions, we refer to Gel'fand and Shilov (1964) or the summary in Baryshnikov and Pemantle (2011).

Fortunately we do not have to worry about these subtleties here due to the following result, proved by Baryshnikov and Pemantle (2011, Lemma 6.3). The proof is not trivial, involving the right choice of insertions of compactly supported functions and truncation estimates.

Lemma 11.2.5 (Fourier magic) *The generalized Fourier transform $\widehat{S^{-s}}(\boldsymbol{r})$ correctly computes the integral over the chain $\overline{\mathcal{C}}_\delta(\boldsymbol{w})$ of* $\exp(-\boldsymbol{r}\cdot\boldsymbol{x})S^{-s}(\boldsymbol{x})$. *More generally, the same is true of $\boldsymbol{x}^{\boldsymbol{m}} q^{-s}$ and of $F \circ \exp(\boldsymbol{x})$ when F is a Laurent polynomial and $z = \exp[\boldsymbol{x}_{\min} + i\boldsymbol{w}]$ is in* `local`.

11.2.4 Main Result on Asymptotics of Coefficients

We are now ready to state and prove a general result on asymptotics of coefficients. Let $H = \prod_{j=1}^k Q_j^{s_j}$ be a product of powers of Laurent polynomials and let B be a component of the complement of $\mathtt{amoeba}(H)$. Let $F = G/H = \sum_{\boldsymbol{r}} a_{\boldsymbol{r}} \boldsymbol{z}^{\boldsymbol{r}}$ on B with $\boldsymbol{x}_{\min} \in \partial B$. Suppose there is a direction $\hat{\boldsymbol{r}}_*$ in $\mathtt{normal}_{\boldsymbol{x}_{\min}}(B)$, and a point $z = \boldsymbol{x}_{\min} + i\boldsymbol{y} \in \mathtt{local}(\hat{\boldsymbol{r}}_*)$ such that: *(i)* $\hat{\boldsymbol{r}}_*$ is in the interior of $\mathbf{N}(z)$; *(ii)* $\hat{\boldsymbol{r}}_*$ is nonobstructed for z; and *(iii)* $\mathtt{local}(\hat{\boldsymbol{r}}_*)$ is the singleton $\{z\}$. Let $q_j := Q_j \circ \exp$ and $\tilde{q}_j = \mathtt{hom}(q_j, \boldsymbol{x}_{\min} + i\boldsymbol{y})$. The straightening Lemma 11.2.2 allows us to develop each $q_j^{-s_j}$ as a sum of terms of the form $c(\boldsymbol{m}, n, j)\boldsymbol{x}^{\boldsymbol{m}} q_j^{-s-n}$. Multiplying these series and then multiplying by the Laurent polynomial G gives

$$f := F \circ \exp = \sum c(\boldsymbol{n}, \boldsymbol{m})\boldsymbol{x}^{\boldsymbol{m}} Q^{-\boldsymbol{s}-\boldsymbol{n}}, \tag{11.2.5}$$

where $Q^{-\boldsymbol{s}-\boldsymbol{n}} = \prod_{j=1}^k q_j^{-s_j-n_j}$ and $\boldsymbol{n}$ is a nonnegative integer vector. This sum contains only terms whose degrees are the degree of H or greater and contains only finitely many terms of degree less than any constant.

Theorem 11.2.6 *Let $F, G, H, Q_j, q_j, \boldsymbol{r}_*, \boldsymbol{z}$ and the expansion (11.2.5) be as previously and let $\boldsymbol{b} := (b_1, \ldots, b_k)$ be the sequence of degrees of $q_1, \ldots, q_k$. Let $\chi_{\boldsymbol{n},\boldsymbol{m}}$ be the generalized Fourier transform of $\boldsymbol{x}^{\boldsymbol{m}} Q^{-\boldsymbol{s}-\boldsymbol{n}}$. Then there is an asymptotic development*

$$a_{\boldsymbol{r}} = (2\pi)^{-d} \boldsymbol{z}^{-\boldsymbol{r}} \sum_{\boldsymbol{n},\boldsymbol{m}} c(\boldsymbol{n}, \boldsymbol{m}) \chi_{\boldsymbol{n},\boldsymbol{m}}(\boldsymbol{r}) \tag{11.2.6}$$

valid when $\boldsymbol{r} \to \infty$ with $\hat{\boldsymbol{r}}$ in some neighborhood of $\hat{\boldsymbol{r}}_*$. When $\boldsymbol{n} \cdot \boldsymbol{b} + |\boldsymbol{m}| > -\boldsymbol{s} \cdot \boldsymbol{b} - d$, the remainder term is $O(|\boldsymbol{r}|^{|\boldsymbol{m}|+\boldsymbol{s}\cdot\boldsymbol{b}-\boldsymbol{n}\cdot\boldsymbol{b}-d})$.

If $\mathtt{local}(\hat{\boldsymbol{r}}_*)$ has cardinality greater than 1 and $\hat{\boldsymbol{r}}_*$ is interior to $\mathbf{N}(\boldsymbol{z})$ and non-obstructed for every $\boldsymbol{z} \in \mathtt{local}(\hat{\boldsymbol{r}}_*)$, then the series on the right-hand side of (11.2.6) can be summed over $\boldsymbol{z} \in \mathtt{local}(\hat{\boldsymbol{r}}_*)$ to give an asymptotic series for $a_{\boldsymbol{r}}$.

Proof We have seen in Corollary 11.2.1 that, up to a term $O(e^{-\boldsymbol{r}\cdot\boldsymbol{z}-c|\boldsymbol{r}|})$ of lower exponential order, $a_{\boldsymbol{r}}$ is a sum of integrals over chains $\overline{\mathcal{C}}_\delta(\boldsymbol{y})$ of Fourier integrands $e^{-\boldsymbol{r}\cdot\boldsymbol{z}'} f(\boldsymbol{z}')\, d\boldsymbol{z}'$. We assume without loss of generality that $\mathtt{local} = \{\boldsymbol{z}\} = \{\exp(\boldsymbol{x}_{\min} + i\,\boldsymbol{y})\}$, the case of cardinality greater than 1 following by exactly the same argument.

Expand f via the series (11.2.5), ordered by increasing homogeneous degree. Equation (11.2.2) of Lemma 11.2.2 shows that the series is a true asymptotic development in the sense that the remainders beginning with a term of a given homogeneous degree β are $O(|\boldsymbol{x}|^\beta)$ near $\boldsymbol{x} + i\,\boldsymbol{y}$ on any closed cone avoiding $\log \mathcal{V}$.

The big-O lemma tells us we can integrate $e^{-\boldsymbol{r}\cdot\boldsymbol{z}'} f(\boldsymbol{z}')\, d\boldsymbol{z}'$ term by term over $\overline{\mathcal{C}}_\delta$ over all terms of homogeneous degree less than β, and as long as $\beta > -d$, the remainder of the integral will be $O(|\boldsymbol{r}|^{-\beta-d})$. By Fourier magic (Lemma 11.2.5), each integral over $\overline{\mathcal{C}}$ is given by its generalized Fourier transform, $\chi_{\boldsymbol{n},\boldsymbol{m}}$. This proves the theorem in the case that $\beta > -d$. Finally, if $\beta \le -d$, we observe that the generalized Fourier transform of a homogeneous function of degree α is always homogeneous of degree $-\alpha - d$. Therefore, letting α run over all degrees of terms that are in the interval $[\beta, -d]$, the remainder is expressed as the sum of finitely many terms of type $O(|\boldsymbol{r}|^{-\alpha-d})$ together with a remainder i.e., at most $O(1)$. This establishes that the remainder is $O(|\boldsymbol{r}|^{-\beta-d})$ and finishes the proof. □

11.3 Examples and Consequences

In practice, few computations of asymptotics in any setting compute beyond the leading term. The material in this chapter is quite new, most of it drawing on only one published work, whence previously worked examples are limited. We give three here.

Example 11.3.1 (power of a cone) Let $F = 1/H = 1/Q^\beta$, where $Q(x, y, z) := (1-x)(1-y) + (1-x)(1-z) + (1-y)(1-z)$. In logarithmic coordinates

(u, v, w), the leading homogeneous term of Q at the singular point $(0, 0, 0)$ of $\log \mathcal{V}$ is the second elementary symmetric function $q := uv + uw + vw$; this is not surprising because Q is this elementary symmetric function applied to $(1 - x, 1 - y, 1 - z)$. The ordinary Taylor series for Q corresponds to the component of $\texttt{amoeba}(Q)^c$ containing the negative diagonal. The normal cone is the cone of all (r, s, t) such that $r^2 + s^2 + t^2 < 2(rs + rt + st)$. This cone is contained in the positive orthant, and its boundary is tangent to the coordinate hyperplanes $\{r = 0\}$, $\{s = 0\}$ and $\{t = 0\}$ at the face diagonals $\lambda \cdot (1, 1, 0)$, $\lambda \cdot (1, 0, 1)$ and $\lambda \cdot (0, 1, 1)$. Thus $\texttt{normal}_{(0,0,0)}(B)$ is the symmetric cone inscribed in the positive orthant. In the study of a discretized time-dependent wave equation in two spatial dimensions, Friedrichs and Lewy (see Askey and Gasper (1972, page 327)) required the coefficients of Q^{-s} to be nonnegative in the special case $\beta = 1$. Szegö (1933) showed this for $\beta \geq 1/2$. A necessary condition for nonnegativity is that the asymptotic estimate be nonnegative. We now compute this asymptotic estimate.

Substituting the result of Proposition 11.2.4 into Theorem 11.2.6, taking β to be greater than $1/2$, we compute that the dual to q is $2(rs + st + rt) - (r^2 + s^2 + t^2)$ and arrive at an estimate for the leading term:

$$a_r \sim \frac{4^{1-\beta}}{\sqrt{\pi}\Gamma(\beta)\Gamma(\beta - 1/2)}(2rs + 2rt + 2st - r^2 - s^2 - t^2)^{\beta-3/2} .$$

■

Stringing together several facts we have accumulated concerning amoebas, tangent cones, and hyperbolicity leads to a useful one-sided bound. Given a Laurent polynomial H and a component B of the complement of $\texttt{amoeba}(H)$, we know from Proposition 11.1.14 that any $\mathbf{z} = \exp(\mathbf{x} + i\,\mathbf{y}) \in \partial B \cap \mathcal{V}$ has a cone of hyperbolicity containing $\tan_{\mathbf{x}}(B)$. From the remarks following Definition 11.1.3, we know that this cone K is convex and is a component of the complement of the zero set of $A := \texttt{hom}(H \circ \exp, \mathbf{x})$ in $\mathbb{R}^d$. The homogeneous polynomial A vanishes on the boundary of the cone K. Dualizing, we see that the algebraic dual A^* to A vanishes on the boundary of the dual cone $\mathbf{N}(\mathbf{z})$. It follows that $\mathbf{N}(\mathbf{z})$ is a subset of L, where is the largest subset of the halfspace dual to $\mathbf{u}$ i.e., bounded by A^*, the algebraic dual to $A := \texttt{hom}(H \circ \exp, \mathbf{x})$. From this, it follows that for any $\mathbf{r} \notin L$, the set $\texttt{local}(\mathbf{r})$ is empty, whence by Theorem 8.4.2, the exponential rate $\overline{\beta}(\hat{\mathbf{r}})$ is strictly less than $\beta^*(\hat{\mathbf{r}}) = -\hat{\mathbf{r}} \cdot \mathbf{x}$. We have therefore proved the following theorem.

Theorem 11.3.2 *Let B be a component of the complement of* $\texttt{amoeba}(H)$*, let H vanish at $\mathbf{z} = \exp(\mathbf{x} + i\,\mathbf{y})$, and let $\mathbf{u}$ be any element of* $\tan_{\mathbf{x}}(B)$. *Then $\overline{\beta}(\hat{\mathbf{r}}) < -\hat{\mathbf{r}} \cdot \mathbf{x}$ for any $\hat{\mathbf{r}}$ outside the closure of L, where L is the largest subset of the halfspace dual to $\mathbf{u}$ i.e., bounded by A^*, the algebraic dual to $A :=$* $\texttt{hom}(H \circ \exp, \mathbf{x})$.

In many examples, $\mathbf{x}$ is the origin. This simplifies the interpretation of Theorem 11.3.2 because then for any $\mathbf{r} \in \texttt{normal}(\mathbf{x})$, the quantity $\beta^*(\mathbf{r})$ vanishes; because $\overline{\beta} < \beta^*$, the coefficients decay exponentially in a conic neighborhood of $\mathbf{r}$.

Example 11.3.3 (cube groves) Recall the cube grove creation generating function

$$F(x, y, z) = \frac{1}{1 + xyz - (1/3)(x + y + z + xy + xz + yz)}$$

from (8.6.6). Let $\boldsymbol{x} = \boldsymbol{0}$ and let B be the component of $\mathtt{amoeba}(H)^c$ containing the negative orthant. Then

$$A := \hom(H \circ \exp) = 2xy + 2xz + 2yz$$

is twice the second elementary symmetric function. This quadratic form is represented by the matrix

$$M := \begin{bmatrix} 0 & 1 & 1 \\ 1 & 0 & 1 \\ 1 & 1 & 0 \end{bmatrix},$$

whereas the dual is represented by the matrix

$$M^{-1} := \frac{1}{2} \begin{bmatrix} -1 & 1 & 1 \\ 1 & -1 & 1 \\ 1 & 1 & -1 \end{bmatrix}.$$

Thus

$$A^*(r, s, t) = rs + rt + st - \frac{1}{2}\left(r^2 + s^2 + t^2\right)$$

(compare the discussion at the end of Section 7.1). The zero set of A^* is a circular cone ∂L tangent to the three bounding planes of the positive orthant at the diagonals $\{x = y, z = 0\}$, $\{x = z, y = 0\}$, and $\{y = y, x = 0\}$ and bounding a solid cone L. It follows from Theorem 11.3.2 that $a_{\boldsymbol{r}}$ decays exponentially as $\boldsymbol{r} \to \infty$ in any closed cone disjoint from L. ■

This is the "easy" direction, in the sense that it is guaranteed by Theorem 8.4.2, to which only the computation of the dual cone need be added. Nevertheless, this computation and its counterpart for orientation probabilities (where the denominator has an extra factor of $1 - z$) are the main results in the work that introduced cube groves (Petersen and Speyer, 2005). The analysis here is considerably simpler. The main reason is that hyperbolicity results reduce geometric questions in complex codimension 1 to the corresponding analyses in real codimension 1, where one can use connectivity and natural orientations. The machinery of algebraic duals and Theorem 11.3.2 combine to make it almost automatic to show exponential decay outside a set of directions whose boundary is the algebraic dual.

The "hard" direction of the computation of $\overline{\beta}$ is to show that there is no exponential decay when $\boldsymbol{r}$ is in the interior of L. It is hard because to do so, we typically need to evaluate the integral near the point $(1, 1, 1) \in \mathtt{local}$ and show that it is indeed a quantity not decaying exponentially. For this, we use

Theorem 11.2.6. Luckily, the one Fourier transform we have computed is the Lorentzian quadratic, which is precisely what we need here. Proposition 11.2.4 identifies the Fourier transform as a scalar multiple of the dual quadratic. There are no obstructed directions. We therefore arrive at the following asymptotic result for cube grove creation rates.

Corollary 11.3.4 *The creation rates* $\{a_{r,s,t}\}$ *satisfy*

$$a_{r,s,t} \sim \frac{1}{\pi}\left[rs + rt + st - \frac{1}{2}(r^2 + s^2 + t^2)\right]^{-1/2}$$

as $(r, s, t) \to \infty$ *within a closed subcone in the interior of* L.

We may of course apply Theorem 11.3.2 to cases in which the Fourier transform has not been computed or for which the computation is too difficult to include here. As an example, we consider the generating function for a random tilings of the so-called ***Aztec Diamond*** and a related tiling ensemble known as the ***fortress*** or ***diabolo*** tiling ensemble.

Example 11.3.5 (Aztec Diamond) The Aztec Diamond has a rich history of study dating back nearly twenty years. We do not discuss this, nor even pause to define the combinatorial objects, but skip directly to the generating function

$$F(x, y, z) = \frac{z/2}{(1 - yz)Q},$$

where $Q(x, y, z) = (1 - (x + x^{-1} + y + y^{-1})z/2 + z^2)$ is a Laurent polynomial. The singular variety $\mathcal{V}_Q$ is smooth except at $\pm(1, 1, 1)$, where it is represented by a Lorentzian quadratic. Its homogenization at the origin is the circular cone $2z^2 - (x^2 + y^2)$, and the cone of hyperbolicity containing the negative z-axis is the cone $B_- := \{(x, y, z) : z < -\sqrt{(x^2 + y^2)/2}\}$. This corresponds to Laurent series with no negative powers of z, which is the Laurent expansion with combinatorial significance. Its dual is given by

$$(B_-)^* = \left\{(r, s, t) : r^2 + s^2 \le \frac{1}{2}t^2\right\}.$$

The other factor $q := (1 - yz)$ is smooth at $\pm(1, 1, 1)$ and is in fact already linear when put in logarithmic coordinates: if $(x, y, z) = \exp(u, v, w)$ then $\{yz = 1\}$ becomes $\{v + w = 0\}$. The cones of hyperbolicity of q are the two halfspaces $\mathcal{H}_- := \{v + w < 0\}$ and $\mathcal{H}_+ := \{v + w > 0\}$, the former containing the negative z-axis. The amoeba of a product is the intersection of the amoebas of the factors, hence $B = B_- \cap \mathcal{H}_-$. Dualizing, B^* is equal to the convex hull of $(B_-)^* \cup (\mathcal{H}_-)^*$. Projectively, $(B_-)^*$ is the cone over the circle $\{\hat{r}^2 + \hat{s}^2 \le 1/2\}$, whereas $(\mathcal{H}_-)^*$ is the single point $(0, 1)$. The convex hull of the union is the teardrop shape shown in Figure 11.3. We then have the following consequence of Theorem 11.3.2. ■

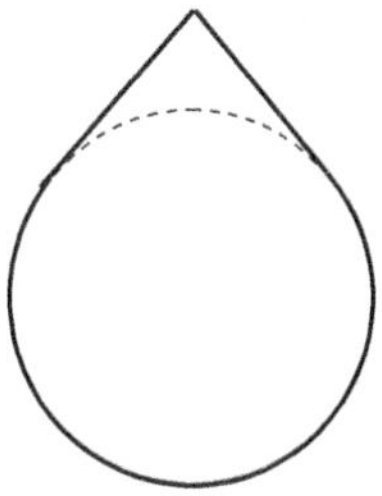

Figure 11.3 Teardrop-shaped region for Example 11.3.5.

Corollary 11.3.6 *Outside of the teardrop-shaped region* $\texttt{hull}((B_-) * \cup\{(0, 1)\})$, *the Aztec Diamond placement probabilities decay exponentially.*

Remark A similar result to Corollary 11.3.6 holds for the cube grove probability generating function, which is the same as the cube grove creation function except for an added factor of $(1 - z)$ in the denominator. Again, the dual is a cone over the convex hull of a circle together with a point outside the circle, which is again a teardrop.

Example 11.3.7 (fortress) The fortress tiling ensemble has a generating function of the form $G/(H_1 \cdots H_k \cdot Q)$, where H_i are all smooth at the point $(1, 1, 1)$ and Q is a nondegenerate quartic. The homogeneous part of $Q \circ \exp$ at $(0, 0, 0)$ is given by

$$A(x, y, z) := 200z^2 \left(2z^2 - x^2 - y^2\right) + 9\left(x^2 - y^2\right)^2$$

(see, e.g., Du, Gessel, Ionescu, and Propp, 2011). The zero set of A is a cone over the curve

$$400 - 200x^2 - 200y^2 + 9\left(x^2 - y^2\right)^2 = 0$$

depicted in Figure 11.4. The Fourier transform of A will be computed in forthcoming work of Baryshnikov and Pemantle. Without this, however, we can still prove a one-sided result, namely that the coefficients a_{rst} decay exponentially outside the algebraic dual curve. The Maple command

```
Basis([r - diff(A,x), s - diff(A,y), t - diff(A,z), A],
plex(x,y,z,r,s,t));
```

produces a Gröbner basis whose first entry is the algebraic dual

$$A^* := 729\,t^8 - 13608\,t^6s^2 - 22896\,s^4t^4 + 64000\,s^6t^2 + 102400\,s^8 - 13608\,r^2t^6$$
$$+ 412992\,s^2t^4r^2 - 1104000\,s^4r^2t^2 + 870400\,s^6r^2 - 22896\,r^4t^4 - 1104000\,r^4s^2t^2$$
$$+ 2054400\,s^4r^4 + 64000\,r^6t^2 + 870400\,r^6s^2 + 102400\,r^8\,.$$

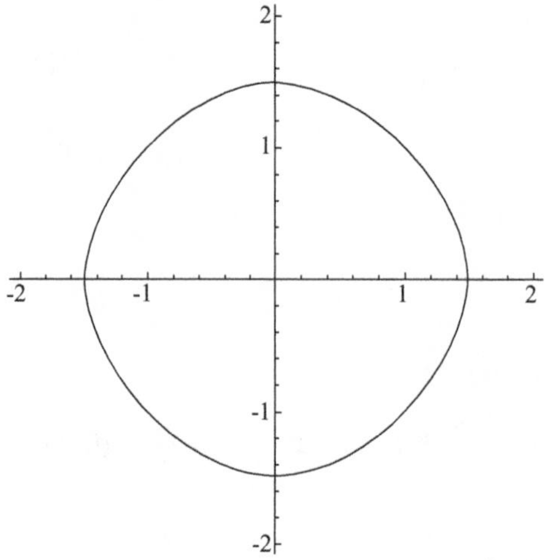

Figure 11.4 Cross-section of the homogeneous part of the fortress generating function.

This projective curve is a cone over the octic affine curve

$$q^* := 729 - 13608\,s^2 - 22896\,s^4 + 64000\,s^6 + 102400\,s^8 - 13608\,r^2 + 412992\,s^2r^2 - 1104000\,s^4r^2 + 870400\,s^6r^2 - 22896\,r^4 - 1104000\,r^4s^2 + 2054400\,s^4r^4 + 64000\,r^6 + 870400\,r^6s^2 + 102400\,r^8$$

whose zero set is shown in Figure 11.5. The real part of this octic curve has two components: there is a concave aster-shaped region inside a nearly circular region. The dual cone $\mathbf{N}(1, 1, 1)$ must be contained within the outer curve, leading to the following "octic circle" result, conjectured by Cohn and Pemantle in 1998 and proved when Kenyon and Okounkov (2007) obtained asymptotics for this ensemble. ■

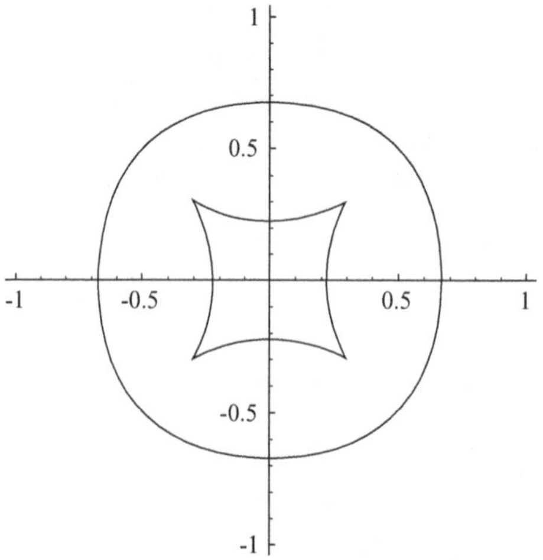

Figure 11.5 Fortress dual curve.

Theorem 11.3.8 *Let K be the cone over the region bounded by the outer dual curve q^*. Then uniformly over closed cones disjoint from the closure of K, the coefficients of the fortress generating function decay exponentially.*

Coda: A Tricky Integral

In fact, the Fourier transforms for the Aztec Diamond and the cube grove probability generating functions are explicitly computable. The computation requires six pages in Baryshnikov and Pemantle (2011), which follow four pages of preparation, more than we are willing to include here. The result, however, can be understood intuitively as follows. The Fourier transform of a linear function is a Heaviside function, i.e., the delta function of a ray $\{t\hat{\boldsymbol{u}} : t \geq 0\}$. The Fourier transform of a product is the convolution of the transforms of the factors, whence the Fourier transform of QL, if L is linear, is given by

$$\int_0^\infty \hat{Q}(\boldsymbol{r} - t\hat{\boldsymbol{u}})\, dt\,.$$

Here $\hat{Q}$ is supported on a cone (the *Paley-Wiener theorem*), and for any $\boldsymbol{r}$, the integrand vanishes for sufficiently large t. This leads to a simple integral, but rigorizing the intuition appears daunting. Instead, the integral was computed in a roundabout but rigorous manner in Baryshnikov and Pemantle (2011), leading to inverse trigonometric functions. We quote the following results. We refer to Baryshnikov and Pemantle (2011, Sections 4.1 and 4.2) for further details.

Theorem 11.3.9 (Baryshnikov and Pemantle, 2011, Theorems 4.1 and 4.2) *The Taylor coefficients a_{rst} of* $\dfrac{z/2}{(1-yz)(1-(x+x^{-1}+y+y^{-1})/2+z^2)}$ *are given asymptotically by*

$$a_{rst} \sim (1+(-1)^{i+j+n+1})\frac{1}{2\pi}\arctan\left(\frac{\sqrt{t^2-2r^2-2s^2}}{t-2s}\right).$$

The Taylor coefficients b_{rst} of $\dfrac{2z^2}{(1-z)(1+xyz-(x+y+z+xy+xz+yz)/3}$ *are given asymptotically by*

$$b_{rst} \sim \frac{1}{\pi}\arctan\left(\frac{\sqrt{2(rs+rt+st)-(r^2+s^2+t^2)}}{r+s-t}\right).$$

Notes

The material in this section is largely taken from Baryshnikov and Pemantle (2011). The use of hyperbolicity in this context is much older. Hyperbolicity, as defined here, arose first in the context of partial differential equations. If f is a complex polynomial, let $f(D)$ denote the corresponding linear partial differential operator obtained by replacing each x_j by $\partial/\partial x_j$. For example, if f is the standard Lorentzian quadratic $x_1^2 - \sum_{j=2}^{d} x_j^2$, then $f(D)$ is the wave operator. Gårding set out to investigate when the equation

$$f(D)u = g$$

with g supported on a halfspace has a solution supported in the same halfspace. When $f(D)$ is the wave operator, this is true and the solution is unique. It turns out that the class of homogeneous polynomials f for which this is true is precisely characterized as the hyperbolic homogeneous polynomials.

The concept of hyperbolicity was exploited in the study of lacunas in Atiyah, Bott, and Gårding (1970) to construct the deformations we have borrowed in this chapter. Later, the property of hyperbolicity turned up in algebraic combinatorics under the name of the ***real root property***. Polynomials with this property are called ***real stable***. Real stability is linked to a wide range of theorems and conjectures, including the van der Waerden conjecture. Its importance seems to stem from the closure of the class of real stable polynomials under a great many algebraic and combinatorial operations. A recent survey is Wagner (2011). For polynomials that are generating functions of joint distributions of binary variables, the property implies a number of negative dependence properties, and this has resulted in the solution of a number of outstanding conjectures in the theory of negative dependence (Borcea, Branden, and Liggett, 2009).

There is another possible approach to asymptotics governed by points other than smooth and multiple points, namely resolution of singularities. A resolution at a singular point z is a change of variables that is one-to-one away from z and after which the local geometry at z is a ***normal crossing***, i.e., one or more smooth, transversely intersecting sheets. Resolution of singularities is effective (Bierstone and Milman, 1997). Unfortunately, the phase function becomes highly degenerate, which complicates the integral substantially, and this approach has not yet been made to work in general. However, see Section 13.2 for some early steps.

Exercises

11.1 (weak hyperbolicity suffices)

Let f be analytic with $f(z) = 0$ and let $A = \mathtt{hom}(f, z)$. Prove that A is hyperbolic in direction $\boldsymbol{u}$ if and only if f is weakly hyperbolic in direction $\boldsymbol{u}$ at z. This was asserted in Section 11.1 and was proved as Proposition 2.11 of Baryshnikov and Pemantle (2011).

11.2 (tangent cones at cubic point)

Let $f(x, y, z) := xy + z^3$. First compute a stratification of the zero set $\mathcal{V}$ of f. The tangent cones to $\mathcal{V}$ at (x, y, z) vary continuously as (x, y, z) moves within a stratum. Describe these. Then, prove or disprove: there exists a semicontinuous choice of tangent cone $K(x, y, z)$ in some neighborhood of the origin.

11.3 (explicitly constructing the vector field)
Let $f(x, y, z) = z^2 - x^2 - y^2$ be the standard Lorentzian quadratic and let $B = \{(x, y, z) : z < -\sqrt{x^2 + y^2}\}$ be the cone of hyperbolicity for f containing the downward direction. Let $\hat{\boldsymbol{r}}_* = e_3$, the elementary vector in the positive z-direction. Find a projective vector field $\boldsymbol{v}(\boldsymbol{y})$ such that for all $\boldsymbol{y} \neq \boldsymbol{0}$, the function $f(i\boldsymbol{y} + \cdot)$ does not vanish on $t\boldsymbol{u} + (1 - t)\boldsymbol{v}(\boldsymbol{y})$ for $t \in [0, 1]$. You may use the proof of Lemma 11.1.17 or try to do this by drawing it.

12

Worked Examples

We have used a few simple running examples to illustrate our results so far, most notably binomial coefficients and Delannoy numbers. We now present a selection of examples of gradually increasing complexity. A wealth of worked examples is available in Pemantle and Wilson (2008) along with an accompanying website (linked from this book's website) containing Maple worksheets. A selection of that material is reproduced in Section 12.1. Our emphasis here is on how to work out the math; for the combinatorial significance or for more fully worked examples, we recommend consulting Pemantle and Wilson (2008). The subsequent two sections concern a schema of applications known as Riordan arrays. Some general results are worked out in Section 12.2 and applied to specific examples. Section 12.3 concerns Lagrange inversion, viewed as an application of the asymptotics of Riordan arrays. Section 12.4 discusses the transfer matrix method and works examples arising from this method. The latter part of this section is devoted to quantum random walks, which are a rich class of transfer matrix examples.

12.1 Four Independent Examples

We begin with two examples in which the contributing critical points are smooth, followed by an example in which it is a multiple point.

12.1.1 Horizontally Convex Polyominoes

The function

$$F(x, y) = \sum_{r,s} a_{rs} x^r y^s = \frac{xy(1-x)^3}{(1-x)^4 - xy(1 - x - x^2 + x^3 + x^2 y)} \tag{12.1.1}$$

counts ***horizontally convex polyominoes*** (HCPs) by total size r and number of rows s. The coefficients are nonnegative integers and are nonzero exactly when $r \geq s$. Definitions, origins (Pólya, 1969), and citations (Wilf, 1994; Odlyzko, 1995; Stanley, 1997) are given in Pemantle and Wilson (2008, Section 4.6). Letting H denote the denominator of F, it is easy to check that $\nabla H \neq 0$ except at the point

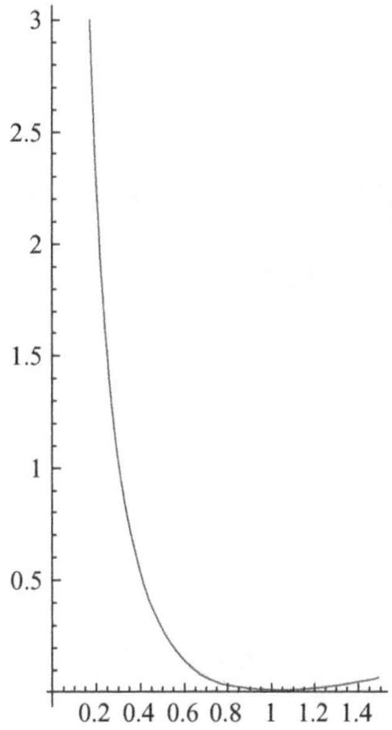

Figure 12.1 Minimal points of $\mathcal{V}$ in the positive real quadrant.

(1, 0) and therefore that all minimal critical points are smooth. By nonnegativity of the coefficients, there is a component of the graph of H in the first quadrant consisting of minimal points; this component is shown in Figure 12.1.

As $\hat{r}$ varies over $\Xi = \{(\hat{r}, \hat{s}) : 0 < \hat{s} < 1/2\}$ from the horizontal to the diagonal, the point $\texttt{contrib}_{\hat{r}}$ moves along this graph from $(1, 0)$ to $(0, \infty)$. The numerator $G := xy(1 - x)^3$ is obviously nonvanishing on this component; therefore, the complete asymptotics may be deduced from Theorem 9.5.7 provided that the quantity Q defined in the theorem is also nonvanishing on the component. Using Gröbner bases, we find that the solutions to $Q = H = 0$ are at $(1, 0)$ and at complex locations where $xH_x/(yH_y)$ is not real.

We will check (later) that each point on the component is strictly minimal. It follows that the asymptotics for a_{rs} are uniform as s/r varies over a compact subset of the interval (0, 1) and given by

$$a_{rs} \sim Cx^{-r}y^{-s}r^{-1/2} .$$

Algebraic methods may then be used to determine x, y, and C as explicit functions of $\lambda := s/r$, giving asymptotics for the number of HCPs that are uniform as long as s/r remains in a compact sum-interval of (0, 1).

Before embarking on this computation, here is a less involved computation that gives a weak law. We consider the distribution of number of rows for a fixed size n as $n \to \infty$. Let $h(x) = H(x, 1) = 1 - 5x + 7x^2 - 4x^3$ and let $x_0 \approx 0.311957$ be the root of h with minimum modulus. Theorem 9.6.3, with the roles of x and y switched, then shows that the number of rows k of a uniformly chosen HCP of total size n satisfies $k/n \to m$ in probability, where $m = yH_y/(xH_x)$ evaluated at

$(x_0, 1)$. This computation may be carried out in $\mathbb{Q}[x]/\langle h(x)\rangle$, leading to

$$\frac{1}{m} = \frac{4(5 - 14x_0 + 12x_0^2)}{5 - 9x_0 + 11x_0^2} = \frac{1}{47}(147 - 246x_0 + 344x_0^2) \approx 2.207\,.$$

We conclude that for large n, the quantity n/k converges in probability to $2.207\ldots$; in other words, the average row size converges to a little over 2.2.

The computation to verify strict minimality and compute x, y, and C as functions of $\lambda := s/r$ uses the following Maple code, which is a useful fragment to keep on hand.

```
Hx := diff(H,x): Hy := diff(H,y): X:=x*Hx: Y:=y*Hy:
Hxx := diff(Hx,x): Hxy := diff(Hx,y): Hyy := diff(Hy,y):
Q := -X^2*Y-X*Y^2-x^2*Y^2*Hxx-X^2*y^2*Hyy+2*X*Y*x*y*Hxy:
L := [H,lambda*X-Y]:
gb := Basis(L, plex(y,x)):
```

The value of gb returned by Maple is a Gröbner basis for (x, y) in terms of λ. In this case, the elimination polynomial for x is $(1 - x)^5\beta$ where

$$\beta(x) := (1 + \lambda)x^4 + 4(1 + \lambda)^2x^3 + 10(\lambda^2 + \lambda - 1)x^2 + 4(2k - 1)^2x + (1 - \lambda)(1 - 2\lambda)$$

is irreducible for generic λ. The second polynomial in gb gives y as a linear function of x. Because β has degree four, each $\lambda \in (0, 1)$ corresponds to four critical points. Precisely one is minimal, this being the one on the component in Figure 12.1. To check this for any given λ is easy. To check it simultaneously for all λ requires a subdivision of the component into finitely many pieces on which minimality cannot change. Homotopy methods for this are discussed in DeVries (2011).

Having performed these checks, we may use Theorem 9.5.7 to conclude that

$$C = \frac{xy(1 - x)^3}{\sqrt{2\pi}}\sqrt{\frac{y\left(-x\left(1 - x - x^2 + x^3 + x^2y\right) - x^3y\right)}{Q}}\,.$$

The minimal polynomial for $\sqrt{2\pi}C$ turns out to have degree 8 for generic λ and may be found using computer algebra. Alternatively, if x and y are given as floating point numbers, we may approximate C directly. Simplifying algebraically as much as possible before making any floating point computations will reduce numerical error. For example, suppose we are interested in counting HCPs with $\lambda = 1/2$. The minimal polynomials for x and y simplify to $3x^2 + 18x - 5$ and $75y^2 - 288y + 256$. The critical point is the quadratic point

$$(x_0, y_0) := \left(\sqrt{\frac{32}{3}} - 3, \frac{48 - \sqrt{512}}{25}\right) \approx (0.265986, 1.397442)\,.$$

A floating point computation gives $a_{n,n/2} \sim (0.237305\ldots)(3.18034\ldots)^n n^{-1/2}$. For $n = 60$, the relative error in this first-order approximation is about 1.5%. The

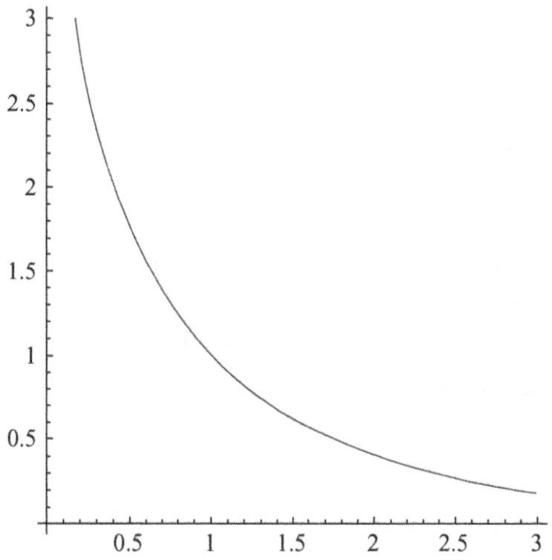

Figure 12.2 Zero-set of $(xe^y - ye^x)/(x - y)$ in the first quadrant.

exponential growth rate of 3.18034... is only a tiny bit less than the exponential growth rate of 3.20557 for all HCPs. This is because the exponential growth rate varies quadratically around its maximum.

12.1.2 Symmetric Eulerian Numbers

The symmetric Eulerian numbers $A(r, s)$ (Comtet, 1974, page 246) count the number of permutations of the set $[r + s + 1] := \{1, 2, \ldots, r + s + 1\}$ with precisely r descents. Their exponential generating function is symmetric in r and s and is given in Goulden and Jackson (2004, page 2.4.21) as

$$F(x, y) = \frac{e^x - e^y}{xe^y - ye^x} = \sum_{r,s} \frac{A(r, s)}{r!s!} =: \sum_{r,s} a_{rs} x^r y^s . \qquad (12.1.2)$$

To represent numerator and denominator as analytic functions with no common divisor, factor $(x - y)$ out of both top and bottom: $F(x, y) = G(x, y)/H(x, y)$, where $G = (e^x - e^y)/(x - y)$ and $H = (xe^y - ye^x)/(x - y)$. We know a_{rs} to be nonnegative; therefore, we need only check the graph of H in the positive quadrant, where we find it is monotone decreasing and asymptotic to both axes; see Figure 12.2.

We check easily that Q is nonvanishing here, discovering also that Q reaches its minimum value of $e^3/12$ at the point $(1, 1)$. We check also that $\mathcal{V}$ is smooth (the gradient of H is nonvanishing on $\mathcal{V}$). By symmetry (or L'Hôpital's rule), $\mathbf{L}(1, 1)$ is the line with slope 1. At any point other that $(1, 1)$ we may compute $\mathbf{L}$ directly. We find that $\mathbf{L}(x, y)$ is the line through $(\alpha, 1 - \alpha)$, where

$$\alpha = \frac{x\partial H/\partial x}{x\partial H/\partial x + y\partial H/\partial y}.$$

On $\mathcal{V}$ this expression for α simplifies greatly, to $(1-x)/(y-x)$. Solving the equations

$$\frac{1-x}{y-x} = \alpha\,; \qquad (12.1.3)$$
$$xe^y = ye^x$$

for x and y in terms of α therefore computes $\boldsymbol{x}_{\min}$ for the generic direction $\hat{\boldsymbol{r}} = (\alpha, 1-\alpha)$. It is clear there is a unique positive real solution. To apply Theorem 9.5.7, it is necessary to check that the torus $\mathbf{T}(\boldsymbol{x}_{\min})$ contains no other critical points. This is easy to do for any given $\hat{\boldsymbol{r}}$. The result, recalling that F is an exponential generating function, is that the symmetric Eulerian number $A(r, s)$ is asymptotically estimated by

$$A(r,s) \sim C_\alpha (r+s)^{-1/2} r!s!\gamma^{r+s},$$

where

$$\gamma = x^{-\alpha} y^{-(1-\alpha)}$$

and C_α is a messy constant determined by (9.5.9) and (9.5.10) in Theorem 9.5.7.

12.1.3 Number of Successes

An independent sequence of random numbers uniform on $[0, 1]$ is used to generate biased coin-flips: if p is the probability of heads, then a number $x \leq p$ means heads and $x > p$ means tails. The coins will be biased so that $p = 2/3$ for the first n flips, and $p = 1/3$ thereafter. A player desires to get r heads and s tails and is allowed to choose n. On average, how many choices of $n \leq r + s$ will be winning choices?

The probability that n is a winning choice for the player is precisely

$$\sum_{a+b=n} \binom{n}{a} (2/3)^a (1/3)^b \binom{r+s-n}{r-a} (1/3)^{r-a} (2/3)^{s-b}\,.$$

Let a_{rs} be this expression summed over n. The array $\{a_{rs}\}_{r,s\geq 0}$ is just the convolution of the arrays $\binom{r+s}{r}(2/3)^r(1/3)^s$ and $\binom{r+s}{r}(1/3)^r(2/3)^s$, so the generating function $F(z, y) := \sum a_{rs} x^r y^s$ is the product

$$F(x,y) = \frac{G(x,y)}{H(x,y)} = \frac{1}{\left(1 - \frac{1}{3}x - \frac{2}{3}y\right)\left(1 - \frac{2}{3}x - \frac{1}{3}y\right)}\,.$$

Although the signs in one factor are different, the denominator has the same amoeba as in Figure 8.4. However, the graph of $\mathcal{V}$ in real space now has a double point at $(1, 1)$, shown in on the right of Figure 12.3, instead of at the nonminimal point $(-3, 3)$, as shown on the left of the figure.

The dual cones $\mathbf{N}(\boldsymbol{x})$ are all rays except at $(1, 1)$, where the dual cone occupies everything between the rays of slope 2 and $1/2$. See Figure 12.4.

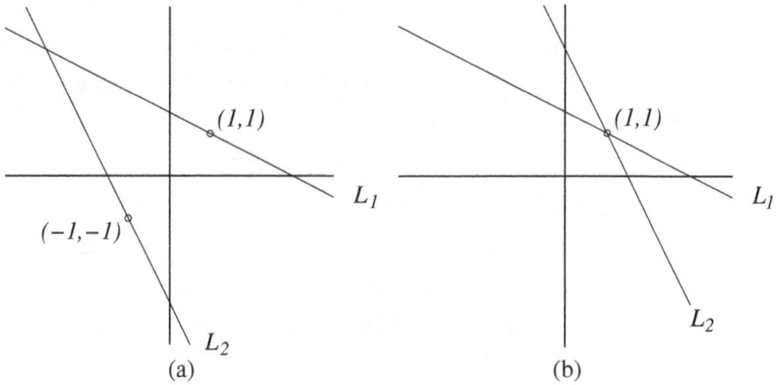

Figure 12.3 Comparison with Example 8.4.1.

Applying Theorem 10.3.7 and Corollary 10.3.2 with $G \equiv 1$ and $\det \mathcal{H} = -1/9$, we see that $a_{rs} = 3$ plus a correction that is exponentially small as $r, s \to \infty$ with $r/(r+s)$ staying in any subinterval of $(1/3, 2/3)$. A purely combinatorial analysis of the sum may be carried out to yield the leading term, 3, but says nothing about the correction terms. The diagonal extraction method of Hautus and Klarner (1971) yields very precise information for $r = s$ but nothing more general in the region $1/3 < r/(r+s) < 2/3$.

12.1.4 One-Dimensional Quantum Walk

In Section 9.5 the notion of a quantum random walk was introduced, and its spacetime generating function was given. Letting $\psi(r, n)$ denote the amplitude for the random walk to be at location r at time n,

$$F(z) := \sum_{r,n} \psi(r, n) x^r y^n = (I - yMU)^{-1} .$$

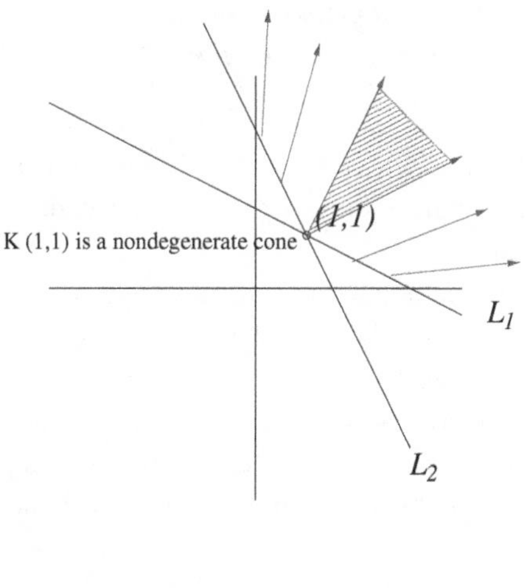

Figure 12.4 $\mathbf{N}(x, y)$ is a ray except when $(x, y) = (1, 1)$.

Here U is a $k \times k$ unitary matrix and M is a $k \times k$ diagonal matrix whose entries x^a run through k possible steps (a, b) of the walk. It is shown in Bressler and Pemantle (2007) that when $k = 2$, there is no loss of generality in taking U to be the real matrix

$$U_c := \begin{bmatrix} c & \sqrt{1-c^2} \\ \sqrt{1-c^2} & -c \end{bmatrix}.$$

There is also no loss of generality in taking the entries of M to be 1 and x, meaning that the walk either stays where it is or moves one to the right (the other natural choice of moving by ± 1 is avoided because it introduces periodicity). The universal spacetime generating function for two-dimensional quantum walks is therefore given by

$$F_c(x, y) = \frac{G_c}{H_c} = \frac{G_c}{1 - cy + cxy - xy^2}.$$

The numerator depends on initial chiralities (one of k hidden states of the walk) and plays no special role. For example, when $k = 2$ and starting and ending chiralities are both in state 2, $G_c = 1 - cy$.

At time n, the possible locations for the walk are $0, 1, \ldots, n$. It turns out that the feasible velocity set for the walk is the interval $J_c := [(1-c)/2, (1+c)/2]$. This means that as $r, n \to \infty$ with $r/n \to \lambda$, the amplitudes $p(r, n)$ decay exponentially if $\lambda \notin J_c$ and are of order $n^{-1/2}$ if λ is in the interior of J_c. This was proved in several places, of which Carteret, Ismail, and Richmond (2003) is perhaps the most complete. The following specific results for $P(r, s) := |\psi(r, s)|^2$ with starting and ending state 2 were given in Bressler and Pemantle (2007).

Theorem 12.1.1 (spacetime asymptotics for one-dimensional quantum walk) *Let $\lambda := \frac{r}{s}$. Then there is a real phase function $\rho(r, s)$ such that*

$$P(r, s) = \frac{2}{\pi} \frac{\lambda\sqrt{1-c^2}}{(1-\lambda)s\sqrt{-((1-c^2) - 4\lambda + 4\lambda^2)}} \cos^2(\rho(r, s)) + O\left(s^{-3/2}\right)$$

uniformly as λ varies over any compact subset of the interior of J_c. Conversely, if λ varies over a compact subset of the complement of J_c, then $P(r, s) \to 0$ exponentially.

The variation of probabilities in the feasible region for $c = 1/2$ is illustrated in Figure 12.5. Qualitatively similar results hold for the other starting and ending chiralities and for combinations of chiralities.

To prove this, one first computes the critical points for $\mathcal{V}$. The variety $\mathcal{V}$ is a quadratic curve in $\mathbb{C}^2$ so it is not hard to analyze. We compute it for $c = \sqrt{1/2}$; the computation is nearly identical for other values of $c \in (0, 1)$.

As mentioned in Section 9.5, the denominator $H = 1 - \sqrt{1/2}(1 - x)y - xy^2$ satisfies the torality hypothesis (9.2.1). This means we can expect $\mathcal{V}$ to intersect the unit torus in a curve of dimension one. Indeed, the intersection $\mathcal{V}_1$ of $\mathcal{V}$ with the

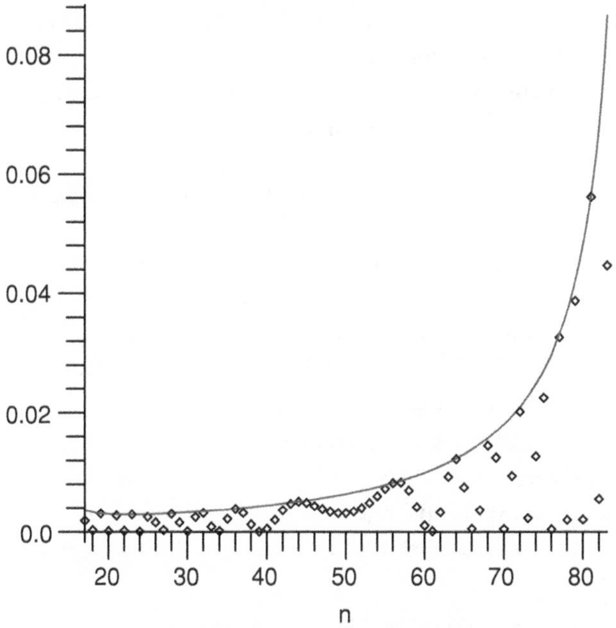

Figure 12.5 $c = 1/2$: time $n = 100$ probabilities starting and ending in state 2, and their upper envelope obtained by dropping the $\cos^2(\rho)$ term.

unit torus is a topological circle $x = (cy - 1)/(cy - y^2)$ winding twice around the torus in the x direction and once in the y direction. The logarithmic Gauss map is a smooth map on this circle with two extreme values, $(1 - c)/2$ and $(1 + c)/2$, and no other critical points. Therefore, for each λ in the interior of J_c, there are precisely two points (x_1, y_1) and (x_2, y_2) in $\mathcal{V}_1$ with $\nabla_{\log} H(x_j, y_j) \parallel (\lambda, 1)$. Therefore, for directions λ interior to J_c, Corollary 9.5.8 with $(s = n,\ G = 1 - \sqrt{1/2}y$, and $|x_j| = |y_j| = 1$ gives

$$\psi(r, s) = \sum_{j=1}^{2} \frac{1 - cy_j}{\sqrt{2\pi}} x_j^{-r} y_j^{-n} \sqrt{\frac{-y_j H_y}{n\, Q}}(x_j, y_j) + O\left(n^{-1}\right),$$

where Q is given by (9.5.9). The two critical points (x_1, y_1) and (x_2, y_2) are conjugate, which makes the sum twice the real part:

$$\psi(r, s) = 2\,\mathrm{Re}\left\{\frac{1 - cy_j}{\sqrt{2\pi}} x_j^{-r} y_j^{-n} \sqrt{\frac{-y_j H_y}{n\, Q}}(x_j, y_j)\right\} + O\left(n^{-1}\right).$$

Defining ρ to be the argument of the expression in braces, writing (x, y) for either one of the two points, and taking the square modulus,

$$P(r, s) = \frac{2}{\pi} \cos^2(\rho) \left|(1 - cy)^2 \frac{-yH_y}{n\, Q(x, y)}\right| + O\left(n^{-3/2}\right).$$

The rest of the case $\lambda \in J_c$ is computer algebra. Set $w := (1-cy)^2 \frac{-yH_y}{n\,Q(x,y)}$, plug in the minimal polynomial for Q, and introduce a variable $z := 1/(nQ)$. The elimination polynomial in w for the ideal generated by $H(x, y)$, $xH_x - \lambda y H_y$, $1 - nzQ$ and $w + (1-cy)^2 y H_y z$ turns out to be

$$\lambda^2(1-c^2) + 4\left(\frac{(1+c)}{2} - \lambda\right)\left(\lambda - \frac{(1-c)}{2}\right)(1-\lambda)^2 w^2.$$

In this computation λ and c may be treated as indeterminates. Solving, we find

$$|w| = \frac{\sqrt{1-c^2}\lambda}{(1-\lambda)s\sqrt{-((1-c^2)-4\lambda+4\lambda^2)}}.$$

This completes the verification of the asymptotics for $\lambda \in J_c$.

Exponential decay for $\lambda \notin J_c$ follows most easily as a disjunction. Either $(0, 0)$ is not the minimzing point for direction $(\lambda, 1)$, in which case $\beta^* < 0$ and β^* is an upper bound for the exponential rate $\overline{\beta}$, or $(0, 0)$ is the minimizing point, in which case by Theorem 8.4.2, because `local` is empty, we have $\overline{\beta} < \beta^* = 0$.

12.2 Powers, Quasi-Powers, and Riordan Arrays

Given a polynomial or power series $v(z)$, we may wish to estimate $[z^r]v(z)^k$, i.e., the $\boldsymbol{r}$ coefficient of a large power of a given function $v(z)$. Clearly, this is equal to the coefficient of $z^r w^k$ of the generating function

$$F(z, w) := \frac{1}{1 - wv(z)}. \tag{12.2.1}$$

One place where such a function arises is in the enumeration of a combinatorial class whose objects are strings built from given blocks. Let $v(x) := \sum_{n=1}^{\infty} a_n x^n$ count the number a_n of blocks of size n. Then the generating function (12.2.1) counts objects of a given size by the number of blocks in the object.

Example 12.2.1 A long sequence of 0's and 1's may be divided into blocks by repeatedly stripping off the unique initial string that is a leaf of T, a given ***prefix tree***. Lempel-Ziv coding, for instance, does this but with an evolving prefix tree. When $v(x)$ is the generating function for the number a_n of leaves of T at depth n, then $1/(1 - yv(x))$ generates the numbers a_{rs} of strings of length r made of s blocks (the final block must be complete). ■

Under suitable hypotheses, as we see later, large powers lead to Gaussian behavior. This well-known fact was noted, for instance, in Bender and Richmond (1983). There, they also observed that this behavior is robust enough to hold not only for exact powers but also for ***quasi-powers***, meaning sequences of functions $\{f_n\}$ satisfying

$$f_n(z) \sim C_n g(z) \cdot h(z)^n \tag{12.2.2}$$

uniformly as z ranges over a certain polydisk. Gaussian behavior of coefficients of quasi-powers is the basis for the GF-sequence method developed by Bender, Richmond, and others in a series of publications including Bender (1973), Bender and Richmond (1983), Gao and Richmond (1992), and Bender and Richmond (1999); see also the work of H.-K. Hwang extending this to the algebraic-logarithmic class: Hwang (1996), Hwang (1998a), and Hwang (1998b). They give conditions under which a multivariate generating function

$$F(z_1, \ldots, z_d, w) = \sum_{n=0}^{\infty} f_n(z) w^n \tag{12.2.3}$$

is a quasi-power in the sense of (12.2.2). They then show that g and h are analytic in a Camembert region (recall Figure 3.1), and if h has a unique dominant singularity where the boundary of the region intersects the positive real axis, and if the quadratic part of h is nondegenerate there, then a Gaussian limit holds for the rescaled coefficient array of f_n as $n \to \infty$.

Our approach to the estimation of coefficients of large powers is via the multivariate generating function (12.2.1). This approach might seem too narrow to cover quasi-powers, where a closed form of the generating function $F = \sum_{n=0}^{\infty} F_n w^n$ may not be known. It should be noted, however, that the quasi-power behavior of $\{f_n\}$ in (12.2.3) in the multivariate central limit theorems of Bender and Richmond (1983) stem from knowledge of the $(d+1)$-variate generating function F, and this allows us to try the multivariate approach. Before working out results in this case, we note two more applications where powers and quasi-powers naturally appear. A third application, namely Lagrange inversion, is taken up in the next section.

One obvious place where coefficients of exact powers arise is in sums of independent random variables. Let $v(z) = \sum_r a_r z^r$ be the probability generating function for a distribution on $\mathbb{Z}^d$, i.e., $a_r = \Pr(X_j = r)$ where $\{X_j\}$ are a family of independent, identically distributed lattice-valued random variables. Then $v(z)^n$ is the probability generating function for the partial sum $S_n := \sum_{j=1}^{n} X_j$, and hence

$$\mathbb{P}(S_n = r) = [z^r] v(z)^n .$$

Indeed, this was the subject of Theorem 9.6.8.

One further combinatorial application of quasi-powers is in the analysis of Riordan arrays. A ***Riordan array*** is defined to be an array $\{a_{nk} : n, k \geq 0\}$ whose generating function $F(x, y) := \sum_{n,k\geq 0} a_{nk} x^n y^k$ satisfies

$$F(x, y) = \frac{\phi(x)}{1 - y v(x)} . \tag{12.2.4}$$

Here, ϕ and v are functions with $v(0) = 0$ and $\phi(0) \neq 0$. If in addition $v'(0) \neq 0$, the array is called a ***proper Riordan array***. Just as (12.2.1) represents sums of independent, identically distributed random variables when v is a probability generating function, the format (12.2.4) is of a ***delayed renewal sum*** (see, e.g., Durrett (2004, Section 3.4), where an initial summand X_0 may be added i.e., distributed differently from the others.

Riordan arrays have been widely studied. In addition to enumerating a great number of combinatorial classes, Riordan arrays also behave in an interesting way under matrix multiplication (note that the condition $v(0) = 0$ implies $a_{nk} = 0$ for $k < n$, and, by triangularity of the infinite array, that multiplication in the Riordan group is well defined). Surveys of the Riordan group and its combinatorial applications may be found in Shapiro et al. (1991), and Sprugnoli (1994).

When ϕ and v are known explicitly, the analysis of a Riordan array is straightforward. Riordan arrays, however, are often specified by parameters other than ϕ and v. Most commonly, this arises when one has an explicit recursion of the form

$$a_{n,k+1} = \sum_{s=1}^{k-n} c_s a_{n+s,k}.$$

Typically, the generating function $A(t) := \sum_{j=1}^{\infty} c_j t^j$ is known and is fairly simple, but the function $v(x)$ in (12.2.4) is known only implicitly through the Lagrange inversion equation (Rogers, 1978, Equation 6)

$$v(x) = xA(v(x)).$$

We are concerned here with the application of multivariate generating function technology. To avoid a long digression, we stick to the case where ϕ and v are explicitly known. For the case where the array is described in terms of A, we refer to Wilson (2005), where the necessary properties of v are computed from A without computing v itself, and multivariate asymptotics theorems such as Theorem 9.2.7 and 9.6.3 are applied to obtain the asymptotics of the array.

We remark that it is not important here to require $v(0) = 0$. For example, neither the binomial coefficient nor Delannoy number examples satisfies that condition. We therefore drop this hypothesis and consider generalized Riordan arrays that satisfy (12.2.4) but may have $v(0) \neq 0$.

Begin by computing the quantities

$$\mu(v; x) := \frac{xv'(x)}{v(x)}; \tag{12.2.5}$$

$$\sigma^2(v; x) := \frac{x^2 v''(x)}{v(x)} + \mu(v; x) - \mu(v; x)^2 = x \frac{d\mu(v; x)}{dx}. \tag{12.2.6}$$

It is readily established that any $(x, y) \in \mathcal{V}$ is a smooth point, whence the dual cone is the normal to the surface, which is the projective direction given by

$$\mathbf{N}(x, y) = \overline{(\mu(v; x), 1)}.$$

In other words, $(x, 1/v(x)) \in \texttt{critical}_{(r,s)}$ if and only if $s\mu(v; x) = r$. Furthermore, this relation holds between x and r/s, then the function Q in Theorem 9.5.7 is given by

$$Q\left(x, \frac{1}{v(x)}\right) = \sigma^2(v; x).$$

Provided that ϕ and σ^2 are nonzero at a minimal point, the leading term of its asymptotic contribution in (9.5.10) then becomes

$$a_{rs} \sim x^{-r} v(x)^s \frac{\phi(x, 1/v(x))}{\sqrt{2\pi s \sigma^2(v; x)}}, \tag{12.2.7}$$

where $\mu(v; x) = r/s$. The notations μ and σ^2 are of course drawn from probability theory. These quantities are always nonnegative when v has nonnegative coefficients. To relate this to the limit theorems in Section 9.6, observe that setting $x = 1$ gives

$$\mu(v; 1) = \frac{v'(1)}{v(1)}; \tag{12.2.8}$$

$$\sigma^2(v; 1) = \frac{vv'' - (v')^2 + vv'}{v^2}(1). \tag{12.2.9}$$

Thus, when hypotheses (i) and (ii) of Theorem 9.6.3 on v and ϕ, respectively, are satisfied, a WLLN will hold with mean $\boldsymbol{m} = \mu(v; 1)$. Of course we see here that $\mu(v; 1)$ is simply the mean of the renormalized distribution on the nonnegative integers with probability generating function v. Similarly, $\sigma^2(v; 1)$ is the variance of the renormalized distribution. The quadratic form in the exponent of (9.6.2) in Theorem 9.6.6 is given by $(s - \mu(v; 1)r)^2/(2k\sigma^2(v; 1))$, whence a local central limit theorem holds with variance $\sigma^2(v; 1)$. An analysis of the relation between r/s and x (see Pemantle and Wilson, 2008, Section 4.3) then leads to the following result.

Theorem 12.2.2 (Pemantle and Wilson, 2008, Proposition 4.2) *Let $(v(x), \phi(x))$ determine a generalized Riordan array. Suppose that $v(x)$ has radius of convergence $R \in (0, \infty]$ and is aperiodic with nonnegative coefficients and that ϕ has radius of convergence at least R. The function $\mu(v; x)$ is strictly increasing on the interval $J := (A, B)$, where $A := \mu(x, 0)$ and $B := \mu(v; R)$ are defined as one-sided limits. Uniformly as r/s varies over compact subsets of J,*

$$a_{rs} \sim v(x)^s x^{-r} s^{-1/2} \sum_{k=0}^{\infty} b_k(r/s) s^{-k} \tag{12.2.10}$$

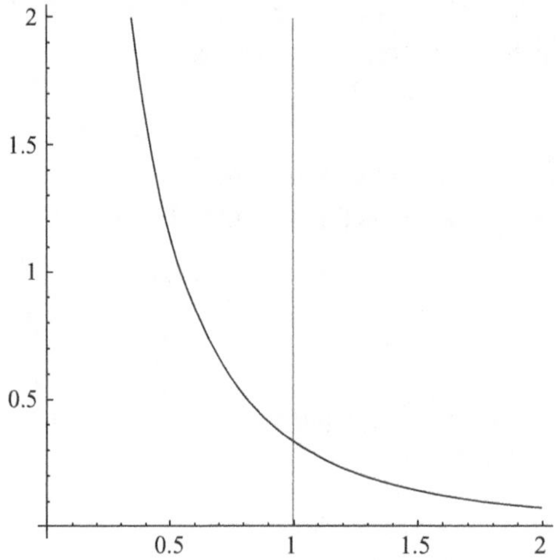

Figure 12.6 $\mathcal{V}$ in the case $d = 3$.

where x *is the unique positive real solution to* $\mu(v; x) = r/s$ *and* $b_0 = \frac{\phi(x)}{\sqrt{2\pi\sigma^2(v;x)}} \neq 0$.

Remarks If v has coefficients of mixed sign, more complicated behavior can occur.

The condition on the radius of convergence of ϕ is satisfied in most applications. One way in which it may fail is when F is a product of more than one factor. We now illustrate this possibility with an example that is taken from Pemantle and Wilson (2002, Section 4.4).

Example 12.2.3 (maximum number of distinct subsequences) Flaxman, Harrow, and Sorkin (2004) considered strings over an alphabet of size d, which we take to be $\{1, 2, \ldots, d\}$ for convenience. They examined strings of length n that contain as many distinct subsequences (not necessarily contiguous) of length k as possible. Let a_{nk} denote the maximum number of distinct subsequences of length k that can be found in a single string of length n. Initial segments $S|_n$ of the infinite string S consisting of repeated blocks of the string $12\cdots d$ turn out always to be maximizers; i.e., $S|_n$ has exactly a_{nk} distinct subsequences of length k. The generating function for $\{a_{nk}\}$ is given by Flaxman, Harrow, and Sorkin (2004, Equation 7):

$$F(x, y) = \sum_{n,k} a_{nk} x^n y^k = \frac{1}{1 - x - xy(1 - x^d)}.$$

This is of Riordan type with $\phi(x) = (1 - x)^{-1}$ and $v(x) = x + x^2 + \cdots + x^d$.

Assume for nontriviality that $d \geq 2$. The singular variety $\mathcal{V}$ is the union of the line $x = 1$ and the smooth curve $y = 1/v(x)$ and they meet transversely at the double point $(1, 1/d)$; see Figure 12.6 for an illustration of this when $d = 3$. This is a case where the radius of convergence of ϕ is less than the radius of convergence of

v, the former being 1 and the latter being infinite. We have $\mu(v;x) = 1/(1-x) - dx^d/(1-x^d) = (1+2x+3x^2+\cdots+dx^{d-1})/(1+x+x^2+\cdots+x^{d-1})$. As x increases from 0 to 1 (the radius of convergence of ϕ, which is the value of x at the double point), μ increases from 1 to $(d+1)/2$. Thus when $\lambda := n/k$ remains in a compact sub-interval of $(1, \frac{d+1}{2})$, the Gaussian asymptotics of equation (12.2.10) hold.

To compute these in terms of λ, we solve for x in

$$\mu(v;x) = \lambda := \frac{n}{k} \tag{12.2.11}$$

and plug this into (12.2.7). One can do this numerically, but in the case where v is a polynomial, one can do better with computational algebraic techniques. The quantity σ^2 is algebraic, in the same degree-$(d-1)$ extension of the rationals that contains the value of x that solves (12.2.11). We look therefore for a polynomial with coefficients in $\mathbb{Q}(\lambda)$, of degree $d-1$, which annihilates the σ^2 in Theorem 12.2.2. We illustrate with the case $d=3$, although this procedure is completely general and will work any time v is a polynomial.

Plugging in the expression (12.2.5) for $\mu(v;x)$ in (12.2.11) and clearing denominators gives a polynomial equation for x:

$$x\frac{dv}{dx} - \lambda v = 0.$$

In our example,

$$x\left(1+2x+3x^2\right) - \lambda\left(x+x^2+x^3\right) = 0\,. \tag{12.2.12}$$

We now need to evaluate

$$\sigma^2(v;x) = x\frac{d\mu}{dx} = \frac{x\left(1+4x+x^2\right)}{\left(1+x+x^2\right)^2} \tag{12.2.13}$$

at the value x that solves (12.2.12).

To do this we compute a Gröbner basis of the ideal in $\mathbb{Q}(\lambda)[x, S]$ generated by $\mu(v;x) - \lambda$ and $\sigma^2(v;x) - S$ (after clearing denominators). The commands

```
p1:=(1+2*x+3*x^2)-lambda*(1+x+x^2):
p2:=x*(1+4*x+x^2)-S*(1+x+x^2)^2:
Basis([p1, p2], plex(x, S));
```

produce the elimination polynomial

$$p(S;\lambda) = 3S^2 + (6\lambda^2 - 24\lambda + 16)S + 3\lambda^4 - 24\lambda^3 + 65\lambda^2 - 68\lambda + 24,$$

which is easily checked to be irreducible (e.g., using Maple's `factor` command) and hence is generically the minimal polynomial for σ^2.

It is easy to choose the right branch of the curve. The variety given by $p(S,\lambda) = 0$ is easily checked to be smooth (there are no solutions to $p = \partial p/\partial S = \partial\lambda = 0$). Thus the two branches given by solving $p(S,\lambda)$ for S do not meet. Because σ^2

is a continuous function of λ on the interval $[1,2]$, σ^2 must stay on one of the branches. It remains only to identify which one. However, this is easy, because $\lim_{\lambda\to 1^+}\sigma^2 = \lim_{x\to 0^+}\sigma^2(v;x) = 0$. Thus σ^2 is given by the value of S on the branch passing through $S = 0$, $\lambda = 1$. We see, for example, that σ^2 increases from 0 to $2/3$ as λ goes from 1 to 2.

To finish describing the asymptotics, we first note that values of λ greater than d are uninteresting. It is obvious that any prefix of S of length at least dk will allow all possible k-subsequences to occur. Thus $a_{nk} = d^k$ when $\lambda \geq d$. We already know that as $\lambda := n/k \to (d+1)/2$ from below, the asymptotics are Gaussian and the exponential growth rate approaches d. For $\lambda \geq (d+1)/2$, we use Proposition 8.4.3 to see that for each such λ, there is a minimal point in the positive quadrant controlling asymptotics in direction λ. The only minimal point of $\mathcal{V}$ we have not yet used is the double point $(1, 1/d)$. It is readily computed that this cone has extreme rays corresponding to $\lambda = (d+1)/2$ and $\lambda = \infty$, and thus asymptotics in the interior of the cone will be supplied by the double point. Using Theorem 10.3.7 and Corollary 10.3.2, we obtain $a_{\lambda k,k} \sim d^k$. ■

12.3 Lagrange Inversion

Suppose that a univariate generating function $f(z)$ satisfies the functional equation $f(z) = zv(f(z))$ for some function v analytic at the origin and not vanishing there. Such functions often arise, among other places, in graph and tree enumeration problems. If v is a polynomial, then f is algebraic. Asymptotics for coefficients of univariate algebraic generating functions may be obtained automatically via an algorithm due to Chabaud (see Flajolet and Sedgewick, 2009, page VII.36, and Chabaud, 2002). Even in this case, the work may be cumbersome or the implementation may not halt. In some of these cases, and in all cases where v is not a polynomial, Lagrange inversion is the obvious device to use. One common formulation, for example, Goulden and Jackson (2004, Theorem 1.2.4), is as follows; we supply the short proof via change of variables, not because it is not well known, but because of the danger that the reader will stumble upon the more common and less illuminating formal power series proof.

Proposition 12.3.1 (***Lagrange inversion formula***) *If* $f(z) = zv(f(z))$ *with* v *analytic and* $v(0) \neq 0$, *then*

$$[z^n]f(z) = \frac{1}{n}\left[y^{n-1}\right] v(y)^n , \tag{12.3.1}$$

where $[y^n]$ *denotes the coefficient of* y^n.

Proof Change variables to $y = f(z)$ so that the implicit equation for f then implies $z = y/v(y)$. We see that $dz = dy[1/v(y) - yv'(y)/v(y)^2]$. The Cauchy integral computing $(2\pi i)[z^n]f(z)$, namely $\int z^{-n-1} f(z)\, dz$ around the origin, then becomes

$$\int \left[\left(\frac{v(y)}{y} \right)^n - \left(\frac{v(y)}{y} \right)^{n-1} v'(y) \right] dy ,$$

around the origin in the y-plane. To see that this is equal to $\frac{1}{n} \int \left(\frac{v(y)}{y} \right)^n dy$, observe that the difference between the integrands is an exact differential

$$d\left[\frac{y}{n} \left(\frac{v(y)}{y} \right)^n \right] = \left(\frac{v(y)}{y} \right)^{n-1} v'(y) - \frac{n-1}{n} \left(\frac{v(y)}{y} \right)^n .$$

□

Using multivariate asymptotic analysis to estimate the right-hand side of (12.3.1), we consider the generating function

$$\frac{1}{1 - xv(y)} = \sum_{n=0}^{\infty} x^n v(y)^n ,$$

which generates the powers of v. The $x^n y^{n-1}$ term of this is the same as the y^{n-1} term of of $v(y)^n$. In other words,

$$[z^n]f(z) = \frac{1}{n} [x^n y^n] \frac{y}{1 - xv(y)} . \tag{12.3.2}$$

This formula holds at the level of formal power series and, if v has a nonzero radius of convergence, at the level of analytic functions.

Using (12.3.2), results have been derived that give (univariate) asymptotics for $[z^n]f(z)$ in terms of the power series coefficients of v. For example, it follows from Theorem VI.6 of Flajolet and Sedgewick (2009) that

$$[z^n]f(z) \sim \frac{1}{\sqrt{2\pi v''(y_0)/v(y_0)}} n^{-3/2} v'(y_0)^n , \tag{12.3.3}$$

where y_0 is the least $y > 0$ such that the tangent line to v at $(y, v(y))$ passes through the origin and we suppose that $\psi'(y_0) \neq 0$; a geometric interpretation of y_0 is that it is where the secant from the origin coincides with the tangent to the graph of ϕ; see Figure 12.7.

For a fixed power k, one may also easily obtain the estimate

$$[z^n]f(z)^k \sim \frac{k}{n} \frac{y_0^{k-1}}{\sqrt{2\pi n v''(y_0)/v(y_0)}} v'(y_0)^n .$$

Using multivariate methods, we may derive bivariate asymptotics for $[z^n]f(z)^k$ as $k, n \to \infty$ uniformly as $\lambda =: k/n$ varies over compact subsets of $(0, 1)$. The following result is from Wilson (2005).

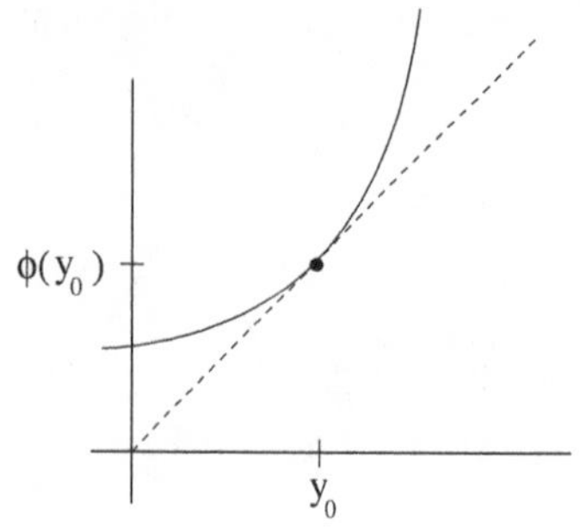

Figure 12.7 Interpretation of y_0.

Proposition 12.3.2 *Let v be analytic and nonvanishing at the origin, with nonnegative coefficients, aperiodic and of order at least 2 at infinity. Let f be the positive series satisfying $f(z) = zv(f(z))$ and define μ and σ^2 by equations (12.2.5) and (12.2.6), respectively. For each n, k, set $\lambda = k/n$ and let x_λ be the positive real solution of the equation $\mu(v; x) = (1 - \lambda)$. Then*

$$[z^n]f(z)^k \sim \lambda v(x_\lambda)^n x_\lambda^{k-n} \frac{1}{\sqrt{2\pi n \sigma^2(v; x_\lambda)}} = \lambda(1-\lambda)^{-n} v'(x_\lambda)^n \frac{x_\lambda^k}{\sqrt{2\pi n \sigma^2(v; x_\lambda)}}. \tag{12.3.4}$$

The asymptotic approximation holds uniformly provided that λ lies in a compact subset of (0, 1).

Proof First, we recall a classical extension of Proposition 12.3.1 (the proof is left as Exercise 12.1). If ψ is analytic, then

$$[z^n]\psi(f(z)) = \frac{1}{n}\left[y^{n-1}\right]\psi'(y)v(y)^n. \tag{12.3.5}$$

When $\psi(y) = y^k$, this becomes

$$\begin{aligned}[z^n]f(z)^k &= \frac{k}{n}\left[y^{n-k}\right]v(y)^n \\ &= \frac{k}{n}[x^n y^{n-k}]\frac{1}{1 - xv(y)}.\end{aligned} \tag{12.3.6}$$

This represents the coefficients of the powers of f as a Riordan array determined by the (known) function v. We may therefore apply Theorem 12.2.2 with $\phi \equiv 1$. To obtain the format of Theorem 12.2.2, we reverse the roles of x and y. We apply Theorem 12.2.2 to the Riordan array $\{a_{rs}\}$ with generating function $1/(1 - yv(x))$. Denoting $x := x_\lambda$, we conclude finally that

$$[z]^n f(z)^k = \frac{k}{n} a_{n-k,k}$$

$$\sim \lambda v(x)^n x^{k-n} n^{-1/2} \frac{1}{\sqrt{2\pi\sigma^2(v; x)}}.$$

□

Example 12.3.3 (trees with restricted offspring sizes) Let $\mathcal{W}$ be the class of unlabeled plane trees, with the restriction that the number of children of each node must lie in a prescribed finite subset $\Omega \subseteq \mathbb{N}$. The generating function $f(z)$ counting such trees by their total number of nodes satisfies

$$f(z) = z\omega(f(z)),$$

where $\omega(z) := \sum_{k\in\Omega} y^k$ (see Flajolet and Sedgewick, 2009, Section VII.3). For example, unary-binary trees are defined by $\Omega = \{0, 1, 2\}$. Take $v(z) = \omega(z)$. In the case of unary-binary trees, $v(z) = 1 + z + z^2$. This leads to

$$\mu(v; z) = \frac{2z^2}{z^2 + z + 1}$$

$$\sigma^2(v; z) = \frac{z^4 + z^3 + 2z^2 - z - 1}{(z^2 + z + 1)^2}.$$

Proposition 12.3.2 applies, but we may use the simpler (12.3.3). Computing $y_0 = 1$, $v'(y_0) = 3$, and $v''(y_0) = 2$, we find that the number of unary-binary trees with n nodes is asymptotically given by $a_n \sim (4\pi/3)^{-1/2} n^{-3/2} 3^n$. ■

12.4 Transfer Matrices

The univariate transfer matrix method was discussed in Chapter 2 and is easily extended to multivariate generating functions. It allows us to count multiplicatively weighted paths. If M is a matrix indexed by a finite set S, define the weight $w(\boldsymbol{x})$ of the path $x_0, \ldots, x_n$ to be $\prod_{r=1}^n M_{x_{r-1},x_r}$. The sum of weights of all paths of length n from i to j is given by $(M^n)_{i,j}$, and this leads to the generating function by length $F(z) = (I - zM)^{-1}$. Suppose now that we wish to break down the count according to additive integer valued functions $v_1, \ldots, v_d$. We then replace the matrix M by the matrix $M_{\boldsymbol{v}}$ whose (i, j)-coefficient is $M_{i,j}\boldsymbol{y}^{\boldsymbol{v}}$. The corresponding multivariate generating function

$$F_{\boldsymbol{v}}(\boldsymbol{y}, z)_{i,j} = \sum_n \sum_{i=x_0,\ldots,x_n=j} w(\boldsymbol{x}) y_1^{v_1(\boldsymbol{x})} \cdots y_d^{v_d(\boldsymbol{x})}$$

is given by the (i, j)-element of the matrix $(I - zM_{\boldsymbol{v}})^{-1}$. We consider two examples. The first is a simple example from Pemantle and Wilson (2008), which models message passing. The second is a more complicated example involving a lattice quantum random walk in $\mathbb{Z}^d$.

Message Passing

Let G be the graph on $K + L + 2$ vertices, which is the union of two complete graphs of sizes $K + 1$ and $L + 1$ with a loop at every vertex and one edge $\overline{xy}$ between them. Paths on this graph correspond perhaps to a message or task being passed around two workgroups, with communication between the workgroups not allowed except between the bosses. If we sample uniformly among paths of length n, how much time does the message spend, say, among the common members of group 1 (excluding the boss)?

Let the vertices 1, 2, 3 and 4 denote, respectively, the Group 1 members, Group 1 boss, Group 2 boss, and Group 2 members. Every time the message moves to vertex 1 it can do so in K ways, and every move to vertex 4 can be done in L ways. The generating function counting paths by time spent among the common members of each workgroup and by total length is $(I - A)^{-1}$, where

$$A = \begin{bmatrix} Kuz & Kuz & 0 & 0 \\ z & z & z & 0 \\ 0 & z & z & z \\ 0 & 0 & Lvz & Lvz \end{bmatrix}$$

and z counts the length of the path. The entries of $(I - A)^{-1}$ are rational functions with common denominator $\det(I - A) = H(Ku, Lv, z)$, where

$$H(u, v, z) = uz^2 + uz^2v - uz - uz^4v + z^2v - 2z - zv + 1 + z^3v + uz^3.$$

The monomials whose z-degree is n give the probability law of time spent in the two groups for a path sampled uniformly from all paths of length n. By Theorem 9.6.3, these "slices" satisfy a weak law, provided that we check that the minimal modulus root of $H(K, L, z)$ is simple and the numerator of the generating function is nonvanishing there. Checking where H and H_z have a common factor, we find an algebraic condition on u and v that appears to have no positive solutions. As long as (K, L) is not a solution, Theorem 9.6.3 then shows that the times spent in Group 1 and Group 2 as a portion of the length n converge to

$$\boldsymbol{m} := -\nabla_{\log} H(K, L, z_0),$$

where z_0 is the minimal modulus root of $H(K, L, \cdot)$. By symmetry, the portion spent in each group is the same, and it is given by $uH_u/(zH_z)$ evaluated at (K, L, z_0). Plugging in $K = L = 1$, for example, we see that $H(1, 1, z) = 1 - 4z + 3z^2 + 2z^3 - z^4$. This leads to $z_0 \approx 0.381966$ and a proportion of approximately $z - 1/4 = 0.1381966$ of the time spent in Group 1. If bosses and employees had equal access to communication, then by symmetry the portion would have been $1/4$, so the effect of communicating through bosses only was to reduce the time each message spends with each non-boss by a factor of nearly 2. This effect is more marked when the workgroups have different sizes. Increasing the size of the second group to 2, we plug in $K = 1, L = 2$ and find that $z_0 \approx 0.311108$ and that the fraction of time spent in Group 1 has plummeted to just under 0.024.

Quantum Random Walk

As we have seen in (9.5.8), the spacetime generating function for a quantum walk may be computed via the transfer matrix method, yielding

$$F(\mathbf{z}) := \sum_{\mathbf{r},n} (\mathbf{z}^\circ)^{\mathbf{r}} z_{d+1}^n ,$$

where $\mathbf{z}^\circ = (z_1, \ldots, z_d)$ are d space variables and z_{d+1} is a time variable. The spacetime generating function is rational and is given by

$$F(\mathbf{z}) := (I - z_{d+1} M U)^{-1},$$

where M is the diagonal matrix whose (j, j)-entry is $(\mathbf{z}^\circ)^{\boldsymbol{v}^{(j)}}$. The common denominator of the entries F_{ij} is $H := \det(I - z_{d+1} MU)$. We saw that a general consequence was that the feasible velocity region R is the image of the logarithmic Gauss map. The limit law for the amplitudes, up to an oscillatory term, may be written in terms of the Gaussian curvature. Here, we work an example to see how the computations are carried out.

The example worked here is a family of walks for which the unitary matrix is given by

$$U = S(p) = \begin{pmatrix} \frac{\sqrt{p}}{\sqrt{2}} & \frac{\sqrt{p}}{\sqrt{2}} & \frac{\sqrt{1-p}}{\sqrt{2}} & \frac{\sqrt{1-p}}{\sqrt{2}} \\ -\frac{\sqrt{p}}{\sqrt{2}} & \frac{\sqrt{p}}{\sqrt{2}} & -\frac{\sqrt{1-p}}{\sqrt{2}} & \frac{\sqrt{1-p}}{\sqrt{2}} \\ \frac{\sqrt{1-p}}{\sqrt{2}} & -\frac{\sqrt{1-p}}{\sqrt{2}} & -\frac{\sqrt{p}}{\sqrt{2}} & \frac{\sqrt{p}}{\sqrt{2}} \\ -\frac{\sqrt{1-p}}{\sqrt{2}} & -\frac{\sqrt{1-p}}{\sqrt{2}} & \frac{\sqrt{p}}{\sqrt{2}} & \frac{\sqrt{p}}{\sqrt{2}} \end{pmatrix}.$$

The four vectors $\boldsymbol{v}^{(1)}$, $\boldsymbol{v}^{(2)}$, $\boldsymbol{v}^{(3)}$, and $\boldsymbol{v}^{(4)}$ are the four nearest neighbor unit vectors. This QRW is analyzed in Baryshnikov, Brady, Bressler, and Pemantle (2010, Section 4.1). The first step is to show that $\mathcal{V}_1 := \mathcal{V}_H \cap \mathbf{T}(\mathbf{0})$ is smooth.

Checking Smoothness. We verify smoothness by checking that the gradient of H never vanishes on three-dimensional unit torus $\mathbf{T}(\mathbf{0})$. It helps to allow Maple to simplify expressions involving p, but it can only do this if we reparametrize by $\alpha := \sqrt{2p}$ so as to obtain polynomial dependence. This gives

$$\begin{aligned} H &= (x^2y^2 + y^2 - x^2 - 1 + 2xyz^2)z^2 - 2xy \\ &\quad - \alpha z(xy^2 - y - x + z^2y - z^2x + z^2xy^2 + z^2x^2y - x^2y). \end{aligned}$$

The Maple command `Basis([H, Hx, Hy, Hz], plex(x, y, z, α))` then gives a Gröbner Basis with first term

$$z\alpha^2(\alpha^2 - 1)(\alpha^2 - 2) = 2zp(2p - 1)(2p - 2).$$

The only way this can vanish when $x, y, z \neq 0$ and $0 < p < 1$ is to have $p = 1/2$. In this case $\alpha = 1$, and the Gröbner basis for the ideal where $(H, \nabla H) = \mathbf{0}$

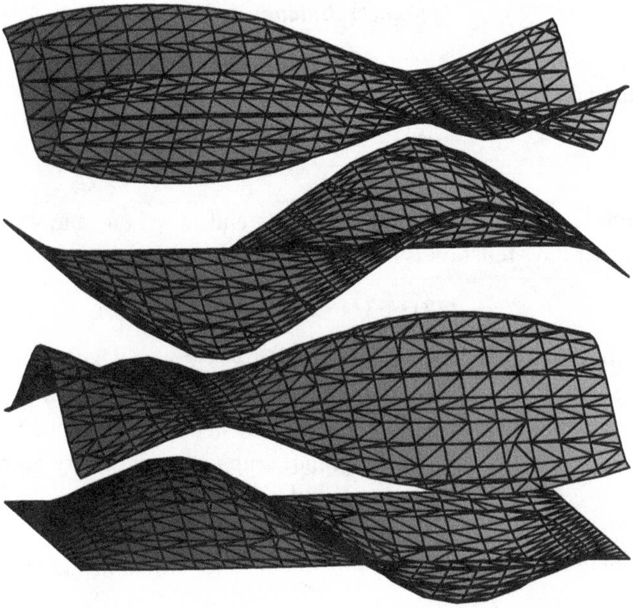

Figure 12.8 The four tori comprising $\mathcal{V}_1$ for $S(1/2)$.

is $(-z + z^5, z^3 + 2y - z, -z - z^3 + 2x)$. The first of these vanishes on the unit circle for $z = \pm 1, \pm i$. However, for $z = \pm 1$, the second vanishes only when $y = 0$, and for $z = \pm i$, the third vanishes only when $x = 0$. Thus ∇H does not vanish on T_3.

Parametrization of $\mathcal{V}_1$. Next we check that the projection of $\mathcal{V}_1$ by $(x, y, z) \to (x, y)$ is a smooth fourfold cover of the unit torus in $\mathbb{C}^2$. From the fact that H is degree 4 in z, it follows that there are at most four, and generically four points of $\mathcal{V}$ projecting to each point of (x, y), with torality guaranteeing that all four points are unit complex numbers as long as x and y are. By the implicit function theorem, if $\partial H/\partial z$ is nonvanishing, then no solutions can ever coalesce. We know that ∇H never vanishes, so if $\partial H/\partial z$ vanishes, then the point (x, y, z) contributes to nonvanishing asymptotics in the direction $(r, s, 0)$ for some $(r, s) \neq (0, 0)$. This is ruled out from our knowledge of the generating function because the velocity of a QRW is at most the longest step, in this case 1.

There are many four-fold covers of the two-torus, but in this case some trigonometry shows that $\mathcal{V}_1$ is in fact the union of four two-tori, each mapping diffeomorphically to the two-torus under the logarithmic Gauss map. Figure 12.8 shows the four components for the parameter value $p = 1/2$ by graphing z as a function of x and y with the torus depicted as the unit cube with wraparound boundary conditions. For details, we refer to Baryshnikov, Brady, Bressler, and Pemantle (2010, Section 4.1).

Final Result

Having shown that $\mathcal{V}_1$ is smooth, we fix states i and j in $\{1, 2, 3, 4\}$ such that the numerator G of F_{ij} is nonvanishing. The following result follows directly from Corollary 9.5.5.

Theorem 12.4.1 *For the $S(p)$ walk with $0 < p < 1$, and states i and j such that the numerator G of ij does not vanish on W, for each $\hat{\mathbf{r}}$ in the image of the logarithmic Gauss map on $\mathcal{V}_1$, let W be the set of four pre-images of $\mathbf{r}$ in $\mathcal{V}_1$. Then as $\mathbf{r} \to \infty$ with $\hat{\mathbf{r}}$ in a compact subset of R,*

$$a_{\mathbf{r}} = (-1)^{\delta} \frac{1}{2\pi |\mathbf{r}|} \sum_{z \in \mathbf{W}} z^{-\mathbf{r}} \frac{G(z)}{|\nabla_{\log} H(z)|} \frac{1}{\sqrt{|\mathcal{K}(z)|}} e^{-i\pi\tau(z)/4} + O\left(|\mathbf{r}|^{-3/2}\right), \tag{12.4.1}$$

where $\delta = 1$ if $\nabla_{\log} H$ is a negative multiple of $\hat{\mathbf{r}}$ (to account for the absolute value in Corollary 9.5.5) and zero otherwise.

Notes

The material in Sections 12.2 and 12.3 is largely taken from Pemantle and Wilson (2008, Section 4.3), as is the message passing example from Section 12.4.

The idea for Exercise 12.2 comes from Noble (2010).

Exercises

12.1 (general Lagrange inversion)
Use the same change of variables as in the proof of Proposition 12.3.1 and the exact differential $d\left[\frac{\psi(y)}{n}\left(\frac{v(y)}{y}\right)^n\right]$ to prove the more general Lagrange inversion formula (12.3.5).

12.2 (a binomial sum)
For each $n \in \mathbb{N}$, define $\mu(m, n) = \sum_{k=0}^{n} (-1)^k \binom{n}{k}\binom{2m}{k}$. For example, the values for $m = n \le 10$ are $1, -1, -1, 8, -17, -1, 116, -344, 239, 1709, -7001$. Compute the leading term asymptotic approximation for $\mu(n, n)$. (Hint: First replace -1 by z, multiply by $x^m y^n$, and sum over k, m, n to obtain a trivariate generating function. Set $z = -1$ to obtain a bivariate generating function of Riordan type.)

13

Extensions

13.1 The Diagonal Method

We recall the *diagonal method* from Section 2.4. When used for asymptotic coefficient extraction from a bivariate generating function F, it consists of two steps:

(i) Find a closed form or defining equation for the diagonal generating function $\operatorname{diag} F$.
(ii) Apply univariate singularity analysis to $\operatorname{diag} F$ to compute the asymptotics.

However, this method is extremely limited in scope, working well only for the main diagonal in two variables. On the positive side, it does allow the determination of the entire diagonal generating function, not just asymptotics of the coefficients. In this section we give more details on the limitations of this method. Although they have been alluded to in Section 2.4, our experience has shown that the results in this book are better appreciated if the reader has a thorough understanding of why such results are not as easily obtainable via the older diagonal method machinery.

The first step usually works well for computing main diagonal asymptotics in two variables, as already seen in the case of Delannoy numbers in Example 2.4.12. The second can be often carried out in this case using standard univariate asymptotics. For example, if F is bivariate rational, then the diagonal is algebraic, and the transfer theorems such as Theorem 3.4.2 can be applied. The full procedure is demonstrated in the following example.

Example 13.1.1 (zigzag-free binary words; Munarini and Zagaglia Salvi, 2002/2004) The bivariate generating function

$$F(x, y) = \sum_{m,n} F_{mn} x^m y^n = \frac{1 + xy + x^2y^2}{1 - x - y + xy - x^2y^2}$$

counts the number of words over a binary alphabet, $\{0, 1\}$ say, that have m zeros and n ones and do not contain zigzags, i.e., the subwords 010 and 101. The main diagonal coefficients F_{nn}, then, count zigzag-free binary words with an equal number of 0's and 1's.

To compute the asymptotics of F_{nn} using the standard diagonal method, we proceed as follows. Because $F(x, y)$ is rational and holomorphic in a neighborhood of the origin, for fixed x small enough $F(x/t, t)$ will be rational and holomorphic as a function of t in some annulus about $t = 0$. Thus in that annulus it can be represented by a Laurent series whose constant term is $[t^0]F(x/t, t) = \sum_{n\geq 0} F_{nn}x^n$, the series we want. By Cauchy's integral formula and the residue theorem, for some circle γ_x about $t = 0$

$$\begin{aligned} G(x) := \operatorname{diag} F(x) &= \sum_n F_{nn}x^n \\ &= [t^0]F(x/t, t) \\ &= \frac{1}{2\pi i}\int_{\gamma_x} \frac{F(x/t, t)}{t}dt \\ &= \sum_k \operatorname{Res}(F(x/t, t)/t; t = s_k), \end{aligned}$$

where the s_k are the "small" singularities of $F(x/t, t)/t$, i.e., the ones satisfying $\lim_{x\to 0} s_k(x) = 0$. Because F is rational, these singularities are poles and algebraic functions of x, so that the residue sum, the diagonal generating function G, is also an algebraic function of x.

In particular,

$$F(x/t, t)/t = \frac{1 + x + x^2}{-t^2 + (1 + x - x^2)t - x},$$

which has a single simple pole approaching zero as x approaches zero, namely, $s = \frac{1}{2}(1 + x - x^2 - \sqrt{1 - 2x - x^2 - 2x^3 + x^4})$. Thus

$$G(x) = \lim_{t\to s} F(x/t, t)/t = \sqrt{\frac{x^2 + x + 1}{x^2 - 3x + 1}}.$$

The singularity of G closest to the origin is $\omega := (3 - \sqrt{5})/2$, and its reciprocal is the exponential growth order of the coefficients of G. To determine the leading subexponential factor, we note that

$$G(x) \sim \frac{1}{\sqrt{1 - \frac{x}{\omega}}}\left(\sqrt{\frac{x^2 + x + 1}{-\omega\left(x - \frac{3+\sqrt{5}}{2}\right)}}\right)_{x=\omega} = \left(1 - \frac{x}{\omega}\right)^{-1/2}\frac{2}{\sqrt{\sqrt{5}}}$$

as $x \to \omega$, so that

$$F_{nn} \sim \omega^{-n}\frac{2}{\sqrt{\sqrt{5}}}\frac{n^{1/2-1}}{\Gamma(1/2)} = \left(\frac{2}{3 - \sqrt{5}}\right)^n \frac{2}{\sqrt{\sqrt{5}\pi n}}$$

by Theorem 3.4.2. ■

However, the standard diagonal method encounters major problems even in two variables once we leave the main diagonal, as illustrated in the next example adapted from Stanley (1999, Section 6.3). Recall that we computed asymptotics for this case in Example 9.5.10.

Example 13.1.2 (Delannoy numbers off the main diagonal) The bivariate generating function

$$F(x, y) = \frac{1}{1 - x - y - xy}$$

as usual encodes the Delannoy numbers. To compute the asymptotics of the general diagonal coefficients $F_{an,bn}$ using the standard diagonal method, we fix $x > 0$ small enough and observe that

$$\begin{aligned} G(x) &:= \operatorname{diag} F(x) := \sum_n F_{an,bn} x^n \\ &= [t^0] F(x^{1/a}/t^b, t^a) \\ &= \frac{1}{2\pi i} \int_{\gamma_x} \frac{F(x^{1/a}/t^b, t^a)}{t} dt \\ &= \frac{1}{2\pi i} \int_{\gamma_x} \frac{t^{b-1}}{t^b - x^{1/a} - t^{a+b} - x^{1/a} t^a} dt. \end{aligned}$$

Because $t^b - x^{1/a} - t^{a+b} - x^{1/a}t^a$ has a zero of multiplicity b at $t = 0$ when $x = 0$, it follows that $F(x^{1/a}/t^b, t^a)/t$ has a single pole s of order b satisfying $\lim_{x\to 0} s(x) = 0$. Thus

$$\begin{aligned} G(x) &= \operatorname{Res}(F(x^{1/a}/t^b, t^a)/t; t = s(x)) \\ &= \lim_{t\to s(x)} \frac{1}{(b-1)!} \partial_t^{b-1} \left((t - s(x))^b F(x^{1/a}/t^b, t^a)/t \right), \end{aligned}$$

where ∂_t is the derivative with respect to t. By using Leibniz's rule for the iterated derivative of a product, we could express this limit in terms of $s(x)$ and $g(t) := (t - s(x))^{-b}(t^b - x^{1/a} - t^{a+b} - x^{1/a}t^a)$ and use it to find an algebraic equation satisfied by G. However, even with the help of a computer algebra system, this seems unlikely for general a and b.

The complexity of the computation of G grows with $a + b$, and representing G explicitly, and even implicitly, seems difficult. Thus the first step of the diagonal method fails even for one of the simplest examples, once we move off the main diagonal. Note that Exercise 2.5 asks for the solution in the simple case $a = 2$, $b = 1$, which can be done by hand. ■

Even the main diagonal poses problems in three or more variables, as illustrated by the next example.

Example 13.1.3 (trinomial coefficients) The trivariate generating function

$$F(x, y, z) = \frac{1}{1 - x - y - z}$$

counts the the number of words F_{lmn} over a ternary alphabet, $\{0, 1, 2\}$ say, that have l 0's, m 1's, and n 2's. A simple combinatorial argument shows that $F_{lmn} = \binom{l+m+n}{l,m,n}$, to which one could apply Stirling's formula to derive the asymptotics.

To compute the asymptotics of the main diagonal coefficients F_{nnn} using the standard diagonal method instead, we iterate the contour integration process. First

$$\begin{aligned}\sum_n F_{m,n,n} x^m y^n &= [t^0] F(x, y/t, t) \\ &= \frac{1}{2\pi i} \int_{\gamma_x} \frac{F(x, y/t, t)}{t} dt \\ &= \frac{1}{2\pi i} \int_{\gamma_x} \frac{1}{-t^2 + (1 - x)t - y} dt,\end{aligned}$$

which has a single simple pole approaching zero as y and x approach zero, namely, $s = \frac{1}{2}(1 + x - \sqrt{1 - 2x + x^2 - 4y})$. The residue is then

$$G(x, y) := \frac{1}{\sqrt{1 - 2x + x^2 - 4y}}.$$

Second, because

$$\lim_{x \to 0} G(x/t, t) = \lim_{x \to 0} \frac{1}{\sqrt{1 - 2x/t + x^2/t^2 - 4t}} = \frac{1}{\sqrt{1 - 4t}},$$

$G(x/t, t)$ is a holomorphic function of t in some annulus about $t = 0$. Thus computing

$$\frac{1}{2\pi i} \int_{\gamma_x} \frac{G(x/t, t)}{t} dt = \frac{1}{2\pi i} \int_{\gamma_x} \frac{1}{t\sqrt{1 - 2x/t + x^2/t^2 - 4t}} dt$$

will give us the diagonal generating function. Proceeding, we employ a computer algebra system and find that two singularities of $G(x/t, t)/t$ approach 0 as x approaches 0, namely

$$-\frac{1}{24} p(x) + \frac{q(x)}{p(x)} + \frac{1}{12} + \frac{1}{2} i\sqrt{3}\left(\frac{1}{12} p(x) + \frac{q(x)}{p(x)}\right)$$

and its conjugate, where $p(x) = (-36x + 216x^2 + 1 + 24\sqrt{-3x^3 + 81x^4})^{1/3}$ and $q(x) = x - \frac{1}{24}$. Although both of these singularities are algebraic in x, we could not compute the sum of their residues in reasonable time with a computer algebra system.

The problem is that the residue sum, i.e., the diagonal generating function $H(x)$, say, is not algebraic: as mentioned previously, by means other than the diagonal method, we know that $H(x) = \sum \binom{3n}{n,n,n} x^n$, which is not algebraic (see Stanley, 1999, Exercise 6/3). Although H is D-finite, it is not trivial to find its defining differential equation by hand (although fairly straightforward with a computer algebra package). We obtain the linear differential equation $27x(x-1)H''(x)+(54x-1)H'(x)+6H(x)=0$, which has a solution in terms of Gauss's hypergeometric function. Moreover, even given the defining differential equation, we still require a general theory of univariate singularity analysis of D-finite functions to compute the asymptotics. This theory has not been worked out in general, although much is known from the theory of linear differential equations. Indeed (Flajolet and Sedgewick, 2009, pages 521–522), certain aspects of it, such as the so-called connection problem, might not be computable, although they can be carried out in an ad hoc manner for many problems. ■

When the diagonal method succeeds, it does yield an explicit generating function for diag F. However, if asymptotics are the main goal, then the methods explained in earlier chapters are much more powerful. For our preceding examples, we can easily obtain the leading term asymptotics using smooth point methods. We can also determine the higher order terms using the methods of Section 13.3.

The converse procedure, namely that of writing an algebraic function as the diagonal of a rational one, is covered in Section 13.2.

13.2 Algebraic Generating Functions

Algebraic generating functions in one variable can be well analyzed by the transfer methods discussed in Section 3.4 and often by the use of Lagrange inversion as described in Section 12.3. In fact, the theory as described in Flajolet and Sedgewick (2009, Chapter VII) is very detailed and algorithmic. However, in several variables, the situation is, as we by now should expect, much more complicated. Of course, for the purposes of asymptotics, we can sometimes ignore the fact that the generating function is algebraic, because the contributing points to the asymptotics might be smooth, for example. The more difficult cases occur when the asymptotics are determined by an algebraic singularity rather than a smooth or multiple point.

The basic setup is as follows. Let $P(w, \mathbf{z})$ be a complex polynomial in $d+1$ variables $w, z_1, \ldots, z_d$, such that $P(0, \mathbf{0}) = 0$. There are in general several solutions for w as a function of $\mathbf{z}$ in a sufficiently small punctured neighborhood of $\mathbf{0}$, but not all need be analytic there. Assuming that we have an analytic branch at the origin, we aim to compute asymptotics of the coefficients of its power series expansion as done for rational and meromorphic functions in previous chapters.

We know from Section 2.4 that the diagonal of a complex rational function in two variables is necessarily algebraic, but that for three or more variables, the analogous result is not true. However, we may still hope for the converse result, which will express every algebraic function as a diagonal of a rational one.

In fact there are several known results of this type. We say that an algebraic function f analytic near $\mathbf{0} \in \mathbb{C}^d$ *lifts* from dimension d to dimension m, where

$m > d$, if the power series expansion of f near the origin is expressible as a diagonal of a rational function F on $\mathbb{C}^m$. Safonov (1987) showed that every univariate algebraic function lifts to dimension 2, and gave an explicit way to find the rational function in terms of the defining polynomial P. Safonov (2000) also showed that lifting from dimension 2 to dimension 3 is not possible in general, but showed how to lift from dimension d to dimension $d+1$ using a more general notion of diagonal. We give some (but not all) details of these results later.

These results lead to an obvious idea for computing asymptotics of algebraic generating functions, because we can reduce to the rational case at the cost of increasing the number of variables and then use results of previous chapters. It turns out that there are several difficulties with this approach. It is clear that in general we cannot be satisfied with the leading term asymptotics, at least if the lifting increases dimension by the minimum amount. This is because the leading order does not always match up correctly. For example, the transfer theorems of Flajolet and Odlyzko lead eventually to the fact that a univariate algebraic irrational function must yield asymptotics of the form $a_n \sim C\alpha^n n^\beta$, where $\beta = b + 1/k$ for some integer b and integer $k \geq 2$. However, when $d = 2$, the smooth point formulae of Chapter 9 give such a representation with $\beta = -1$. We deal with the computation of higher order terms in Section 13.3.

There is a more serious problem, which is that the lifting procedure may not preserve positivity of coefficients. This makes the determination of contributing points in the lifted generating function much harder, because minimal points are not guaranteed to contribute. The Morse-theoretic approach of Section 9.4, as displayed in Example 9.4.8 but extended to more variables, seems to us the best way to make further progress. Of course, there is no guarantee that the lifted function will have a contributing singularity of a type that we can deal with using current technology, but in our experience this has not proven to be the major difficulty.

Despite the somewhat negative preceding comments, in many cases it is quite feasible to apply the obvious lifting method and obtain useful results. We now give an idea of the constructions involved in lifting. We first consider the simplest case, where there is a single analytic branch at the origin.

Lemma 13.2.1 (Safonov, 2000, Lemma 1) *Suppose that f is an algebraic function as described previously and that in some neighborhood of $\mathbf{0}$, there is a factorization $P(w, z) = (w - f(z))^k u(w, z)$, where $u(0, \mathbf{0}) \neq 0$ and $k \geq 1$ is an integer.*

Then f is the elementary diagonal of the rational function F given by

$$F(z_0, z) = \frac{z_0^2 P_1(z_0, z_0 z_1, z_2, \ldots, z_d)}{k P(z_0, z_0 z_1, z_2, \ldots, z_d)},$$

where by P_1 we mean the partial derivative of P with respect to the variable w.

Proof The verification makes use of the following well-known consequence of Cauchy's residue theorem. Let h be a function of one variable analytic at 0, and let g be an analytic function. Then the integral of $g(z)f'(z)/f(z)$ around a circle of sufficiently small radius centered at 0 equals the sum over all residues r of $g(r)$ times the multiplicity of the pole r.

We apply this with $f(w) = P(w, z)$ and $g(w) = w$ for fixed z. Because there is a single pole, namely at $w = f(z)$, and it has multiplicity k, the integral will equal $f(z)$.

Let F be the desired lifting and define $\tilde{F}(w, z) = F(w, z_1/w, z_2, \ldots, z_d)$. Then as in Theorem 2.4.11, for a suitable contour of integration

$$f(z_1, \ldots, z_d) = \frac{1}{2\pi i} \int \tilde{R}(w, z) \frac{dw}{w} = \sum \operatorname{Res}(\tilde{F}/w). \tag{13.2.1}$$

Thus choosing $\tilde{F}(w, z) = w^2 P_w(w, z)/P(w, z)$ yields the result we want, and this is equivalent to the stated formula for the rational function R. □

The simplest interesting example, which we now check, occurs in the enumeration of binary trees by external nodes.

Example 13.2.2 Consider the shifted Catalan number generating function $f(x) = \sum_n a_n x^n = (1 - \sqrt{1-4x})/2$ with minimal polynomial $P(x, y) = y^2 - y + x$. Lemma 13.2.1 yields the lifting $F(y, z) = y(1-2y)/(1-x-y)$. Asymptotics of the univariate generating function are therefore given by computing asymptotics in the diagonal direction for F. The variety $\mathcal{V}$ is globally smooth and the critical point equation yields the single solution $(1/2, 1/2)$, which is clearly strictly minimal. This gives the correct exponential rate of $\log 4$, but the first term in the basic smooth point expansion is zero, precluding easy use of the basic smooth point results. We give further details in Example 13.3.6. ■

Example 13.2.3 The Narayana numbers given by the explicit formula

$$a_{rs} = \frac{1}{r}\binom{r}{s}\binom{r}{s-1}$$

are a refinement of the Catalan numbers in Flajolet and Sedgewick (2009, Example III.13), which enumerate, for example, rooted ordered trees by edges and leaves. Their generating function is

$$f(x, y) = \frac{1}{2}\left(1 + x(y-1) - \sqrt{1 - 2x(y+1) + x^2(y-1)^2}\right).$$

The defining polynomial is

$$P(w, x, y) = w^2 - w\,[1 + x(y-1)] + xy,$$

which factors as $[w - f(x, y)]\left[w - \overline{f}(x, y)\right]$, where $\overline{f}$ denotes the algebraic conjugate of f, obtained by changing the sign in front of the square root. Using

Lemma 13.2.1, we obtain

$$F(u,x,y) = \frac{u(1-2u-ux(1-y))}{1-u-xy-ux(1-y)}.$$

Note that when $y = 1$, we obtain exactly the lifting of the Catalan generating function: i.e., in this case at least, lifting commutes with the specialization $y = 1$. By construction $a_{rs} = b_{rrs}$, where b_{rst} is the generic Maclaurin coefficient of F. For example, to derive asymptotics when $r = \alpha n, s = \beta n$, we consider the direction determined by (α, α, β). The critical point equations yield $u = \beta/\alpha, x = (\alpha-\beta)^2/\alpha\beta, y = \beta^2/(\alpha-\beta)^2$. Thus the exponential rate is $2\alpha \log \alpha - 2\beta \log \beta - 2(\alpha-\beta)\log(\alpha-\beta)$ and the smooth point formulae of Chapter 9 coupled with the higher order expansions in Section 13.3 can be used to derive greater detail. In particular, the maximum exponential rate is $2 \log 2$, achieved when $\alpha/\beta = 2$. ■

When there is more than one analytic branch going through the origin, we must distinguish them somehow. This is precisely the issue of resolution of singularities. Safonov gives the following example, which shows that we cannot expect the result of Lemma 13.2.1 to hold in general.

Example 13.2.4 (Safonov, 2000, Example 2) Let $f(x,y) = x\sqrt{1-x-y}$. Suppose that f occurs as the elementary diagonal of a rational function $F(x_0, x, y)$. Extracting Maclaurin coefficients yields

$$[x]f(x,y) = \frac{\partial f}{\partial x}(0,y) = \sqrt{1-y}$$

$$[x]f(x,y) = \frac{\partial F}{\partial x}(0,0,y),$$

and thus $\sqrt{1-y}$ is rational, a contradiction. ■

Safonov (2000) used explicit methods to resolve the singularity and derived a strong result. Recall that a *unimodular matrix* is one with integer entries whose determinant is ± 1.

Definition 13.2.5 *Let $F(\mathbf{z}) = \sum_{\mathbf{r}} a_{\mathbf{r}} \mathbf{z}^{\mathbf{r}}$ be a formal power series in $d+1$ variables, and M a unimodular $d \times d$ matrix with nonnegative entries. The* M-diagonal *of F is the formal power series in d variables given by $\sum_{\mathbf{r}^\circ} b_{\mathbf{r}^\circ} \mathbf{z}^{\circ \mathbf{r}^\circ}$, where $b_{\mathbf{r}^\circ} = a_{s_1, \mathbf{s}}$ and $\mathbf{s} = \mathbf{r}^\circ M$.*

Example 13.2.6 For the function f given in Example 13.2.4, it turns out (from the algorithm in the proof of Theorem 13.2.7) that with

$$F(x_0, x_1, x_2) = \frac{x_0 x_1 (2 + x_1 + x_1 x_2 + 2x_0^2 + 3x_0)}{2 + x_0 + x_1 + x_1 x_2} = \sum_{r_0, r_1, r_2} a_{r_0 r_1 r_2} x_0^{r_0} x_1^{r_1} x_2^{r_2},$$

the matrix M is given by $M = \left(\begin{smallmatrix} 1 & 0 \\ 1 & 1 \end{smallmatrix}\right)$ and $f(x_1, x_2) = \sum_{r_1, r_2} a_{r_1+r_2, r_1+r_2, r_2} x_1^{r_1} x_2^{r_2}$.

Note that asymptotics of the (r, s) coefficient of f can be given by asymptotics of $a_{r+s,r+s,s}$ and that if (r, s) lies on a fixed ray, then so does $(r + s, r + s, s)$. ■

Theorem 13.2.7 (Safonov, 2000, Theorem 1) *Let f be an algebraic function in d variables. Then there is a unimodular matrix M with positive entries and a rational function F in $d + 1$ variables such that f is the M-diagonal of F.*

Proof We omit the details, referring to the original publication. The basic idea is to apply a sequence of *blowups* to partially resolve the singularity until we arrive at a case to which Lemma 13.2.1 applies. These changes of variable are all monomial substitutions of the form $z_i \mapsto z_i z_j$, which yields the unimodular matrix. □

Safonov's approach is not the only way to lift an algebraic function to a rational one, but we do not know of a simpler one i.e., computationally effective. See the Notes to this chapter for more details. It seems that there is still plenty of scope for advances in this area.

13.3 Higher-Order Asymptotics

We have seen several formulae for asymptotic expansions in previous chapters. Each of these gives the general form of the expansion, but an explicit formula is given only for the leading term. There are several important classes of applications that require precise information about higher order terms (we give three such examples next). We know from our asymptotic formulae in previous chapters that whenever G vanishes at a contributing point, the order 0 coefficient is zero, and the relatively compact formulae we have obtained for the leading term are not valid.

Example 13.3.1 Let $F(x, y) = (x - y)/(1 - x - y)$ as in Example 8.1.2. The (r, r)-coefficient is zero for all r. Thus all coefficients in the basic smooth point asymptotic formula must be zero when we look in the principal diagonal direction $(1, 1)$, corresponding to the strictly minimal smooth point $(1/2, 1/2)$. ■

All our explicit computations ultimately reduce to an expansion for a Fourier-Laplace integral. The only truly explicit form that we know is presented in Hörmander (1983) and proved using smooth methods. We quote the result here (translated to our own notation) and refer to the original for the proof.

Lemma 13.3.2 (Hörmander, 1983, Theorem 7.7.5) *Let $X \subseteq \mathbb{R}^d$ be an open neighborhood of $\mathbf{0} \in \mathbb{R}^d$ and ϕ, A be smooth functions on X such that* $\operatorname{Re}\phi \geq 0$ *on X. Furthermore, suppose that on* $\operatorname{supp} A$, *ϕ has a unique stationary point at $\mathbf{0}$ that $\phi(\mathbf{0}) = 0$. Finally, suppose that the Hessian $\mathcal{H}$ of ϕ at $\mathbf{0}$ is nonsingular.*

Then for each positive integer M, letting $N = M + \lceil d/2 \rceil$,

$$\left| \int_X \phi(t) e^{-\lambda \phi(t)} dt - [\det(\lambda \mathcal{H}/2\pi)]^{-1/2} \sum_{k<M} \lambda^{-k} L_k(A\phi) \right|$$

$$\leq C(\phi) \lambda^{-M} \sum_{|\beta| \leq 2N} \sup |\mathcal{D}^\beta A| \qquad (\lambda > 0).$$

Here $C(\phi)$ is bounded when ϕ stays in a bounded set in C^{3N+1}, and $|t|/|\phi'(t)|$ has a uniform bound. With

$$\underline{\phi}(t) := \phi(t) - \frac{1}{2}\langle t, \mathcal{H} t \rangle,$$

which vanishes to order three at $\mathbf{0}$,

$$L_k(A\phi) = (-1)^k \sum_{l \leq 2k} \frac{\mathcal{D}^{l+k}(A\underline{\phi}^l)(\mathbf{0})}{2^{l+k} l!(l+k)!}, \tag{13.3.1}$$

where $\mathcal{D}$ is the differential operator $\langle \nabla, \mathcal{H}^{-1} \nabla \rangle = \sum_{i,j} \mathcal{H}_{ij}^{-1} \partial_i \partial_j$.

In every term of $L_k(A\phi)$ the total number of derivatives of A and of $\mathcal{H}$ is at most $2k$.

Rewriting this in our notation, we obtain the following refinement of Theorem 5.4.8. The extension to finitely many critical points as in that theorem is trivial to do but harder to state.

Corollary 13.3.3 (full expansion of F-L integral) *Suppose there is a single critical point in Theorem 5.4.8 and that $\phi = 0$ there. Then $c_k = \det \mathcal{H}^{-1/2} L_k(A\phi)$ for each $k \geq 0$, where L_k is as defined in* (13.3.1).

Proof The proof of Theorem 5.4.8 shows that the contribution from the boundary of the domain of integration may be ignored – we may localize to a neighborhood of the critical point i.e., diffeomorphic to an open ball in $\mathbb{R}^d$. In other words, we can without loss of generality multiply by a bump function: replace A by $A\alpha$ for some compactly supported smooth function α whose value is 1 on that neighborhood. The result now follows directly, by Lemma 13.3.2. □

Remark Note that $L_0(A\phi) = A(\mathbf{0})$, as expected.

We can now apply the full expansion to our analyses of previous chapters. A computer algebra system is necessary in practice. Similar computations can be made with multiple points – we refer to Raichev and Wilson (2012b) for full details. See the Notes to this chapter for details of our computer algebra implementation.

Example 13.3.4 We compute the first two terms of the asymptotic expansion of the Delannoy numbers $F_{4n,3n}$ as $n \to \infty$. The critical points of $\mathcal{V}$ in this direction are $(-2, -3)$ and $(1/2, 1/3)$. Both points are smooth, and the second point, which we denote by c, is strictly minimal. Combining the reduction to Fourier-Laplace integral in Chapter 9 with the detailed formulae of Corollary 13.3.3, we obtain

$$F_{4n,3n} = 432^n \left(a_1 n^{-1/2} + a_2 n^{-3/2} + O(n^{-5/2})\right),$$

as $n \to \infty$, where

$$a_1 = \frac{\sqrt{2}\sqrt{3}\sqrt{5}}{10\sqrt{\pi}} \approx 0.3090193616$$

$$a_2 = -\frac{\sqrt{2}\sqrt{3}\sqrt{5}}{288\sqrt{\pi}} \approx -0.01072983895.$$

Comparing this approximation with the actual values of $F_{n\alpha}$ for small n (using 10-digit floating-point arithmetic), we get the following table.

n	1	2
$F_{n\alpha}c^{n\alpha}$	0.2986111111	0.2147687329
$a_1 n^{-1/2}$	0.3090193616	0.2185096861
$a_1 n^{-1/2} + a_2 n^{-3/2}$	0.2982895227	0.2147161152
1-term rel. error	−0.03485553660	−0.01741851906
2-term rel. error	0.001076947318	0.0002449968910

4	8	16
0.1531773658	0.1087821882	0.07708745613
0.1545096808	0.1092548431	0.07725484041
0.1531684509	0.1087806467	0.07708718667
−0.008697858486	−0.004344965359	−0.002171355521
0.00005819932814	0.00001417078449	0.000003495529179

For an arbitrary direction α, the two-term asymptotic expansion of $F_{n\alpha}$ is just as easy to compute symbolically in α. The corresponding constants c_1, c_2, a_1, a_2 are square roots of rational functions of α_1, α_2, and $\sqrt{\alpha_1^2 + \alpha_2^2}$. The exact formulas are somewhat long, so we omit them. ■

The previous example gave one obvious application for higher order terms in our asymptotic approximations, namely improved accuracy for smaller values of $|\mathbf{r}|$. The question of the optimal order of truncation of the asymptotic series goes beyond our scope here. See the Notes to this chapter for more details.

Another reason why we may need higher order terms is because of cancellation of terms when we combine asymptotic expansions of related functions. The following example gives a nice illustration.

Example 13.3.5 Consider the $(d+1)$-variate function

$$W(x_1, \ldots, x_d, y) = \frac{A(x)}{1 - yB(x)},$$

where $A(x) = 1/[1 - \sum_{j=1}^{d} x_j/(x_j+1)]$, $B(x) = 1 - (1 - e_1(x))A(x)$, and $e_1(x) = \sum_{i=j}^{d} x_j$. Using the symbolic method, it is not very difficult to show that W counts words over a d-ary alphabet X, where x_j marks occurrences of letter j of X and y marks occurrences of ***snaps***, nonoverlapping pairs of duplicate letters. Here $A(x)$ counts snapless words over X, which are simply Smirnov words as described in Example 2.2.10.

The diagonal coefficient $[x_1^n \dots x_d^n, y^s]W(x, y)$ is then the number of words with n occurrences of each letter and s snaps. We compute the expected number and variance of snaps as $n \to \infty$ as follows. Let ψ denote the random variable counting snaps.

$$
\begin{aligned}
\mathbb{E}(\psi) &= \frac{[x^{n\alpha}]\frac{\partial W}{\partial y}(x,1)}{[x^{n\alpha}]W(x,1)} = \frac{[x^{n\alpha}]A(x)^{-1}B(x)(1-e_1(x))^{-2}}{[x^{n\alpha}](1-e_1(x))^{-1}} \\
\mathbb{E}(\psi^2) &= \frac{[x^{n\alpha}]\left(\frac{\partial^2 W}{\partial y^2}(x,1) + \frac{\partial W}{\partial y}(x,1)\right)}{[x^{n\alpha}]W(x,1)} \\
&= \frac{[x^{n\alpha}]A(x)^{-2}B(x)(B(x)+1)(1-e_1(x))^{-3}}{[x^{n\alpha}](1-e_1(x))^{-1}} \\
\mathbb{V}(\psi) &= \mathbb{E}(\psi^2) - \mathbb{E}(\psi)^2,
\end{aligned}
$$

Let $H(x) = 1 - e_1(x)$. Then the only critical point of $\mathcal{V}$ is $c := (1/d, \dots, 1/d)$, which is strictly minimal. Applying the previously derived full expansions (combined with the extensions to repeated factors in Chapter 9) to $F_1(x) := W(x, 1)$, $F_2(x) := \partial W/\partial y(x, 1)$, and $F_3(x) := \partial^2 W/\partial y^2(x, 1) + \partial W/\partial y(x, 1)$ yields

$$
\begin{aligned}
\mathbb{E}(\psi) &= \frac{\frac{3\sqrt{3}}{8\pi} - \frac{61\sqrt{3}}{192\pi}n^{-1} + O(n^{-2})}{\frac{\sqrt{3}}{2\pi}n^{-1} - \frac{\sqrt{3}}{9\pi}n^{-2} + O(n^{-3})} \\
&= (3/4)n - 15/32 + O(n^{-1}) \\
\mathbb{E}(\psi^2) &= \frac{\frac{9\sqrt{3}}{32\pi}n - \frac{35\sqrt{3}}{128\pi} + O(n^{-1})}{\frac{\sqrt{3}}{2\pi}n^{-1} - \frac{\sqrt{3}}{9\pi}n^{-2} + O(n^{-3})} \\
&= (9/16)n^2 - (27/64)n + O(1) \\
\mathbb{V}(\psi) &= (9/32)n + O(1).
\end{aligned}
$$

Comparing these approximations with the actual values for small n (using 10-digit floating-point arithmetic), we obtain the following table.

n	1	2	4	8
$\mathbb{E}(\psi)$	0	1.000000000	2.509090909	5.520560294
$(3/4)n$	0.7500000000	1.500000000	3	6
$(3/4)n - 15/32$	0.2812500000	1.031250000	2.531250000	5.531250000
one-term relative error	undefined	0.5000000000	0.1956521740	0.08684620409
two-term relative error	undefined	0.03125000000	0.008831521776	0.001936344398
$\mathbb{E}(\psi^2)$	0	1.800000000	7.496103896	32.79620569
$(9/16)n^2$	0.5625000000	2.250000000	9	36
$(9/16)n^2 - (27/64)n$	0.1406250000	1.406250000	7.312500000	32.62500000
one-term relative error	undefined	0.2500000000	0.2006237006	0.09768795635
two-term relative error	undefined	0.2187500000	0.02449324323	0.005220289555
$\mathcal{V}(\psi)$	0	0.800000000	1.200566706	2.31961973
$(9/32)n$	0.2812500000	0.5625000000	1.125000000	2.250000000
relative error	undefined	0.2968750000	0.06294253008	0.03001342380

■

Our final application of higher order expansions comes from the lifting method described in Section 13.2. Exercise 13.4 shows that this will be necessary whenever we use Safonov's procedure.

Example 13.3.6 In Example 13.2.2 we saw that the shifted Catalan number generating function lifts to $w(1-2w)/(1-z-w)$. The asymptotics in direction $(1,1)$ are controlled by the point $(1/2, 1/2)$, and the numerator vanishes there. Using the first two terms of the expansion as earlier gives the correct first-order asymptotic $a_n \sim 4^{n-1}/\sqrt{\pi n^3}$. ■

13.4 Phase Transitions

Our asymptotic approximations often hold uniformly for all $\hat{r}$ in a cone of directions, corresponding to contributing points at which the local geometry does not change. For example, the basic smooth point results hold for all directions for which the Hessian is nonsingular. In all our results, exactly which term of the approximation is the leading term depends only on the order of vanishing of G at the contributing point(s). In terms of Fourier-Laplace integrals, the phase controls the asymptotic scale, and the amplitude controls the leading term. Note that the polynomial correction terms are at issue here – the exponential order is determined by the location of the contributing point, not the local geometry.

When the local geometry does change, the results obviously cannot continue to be valid. In three dimensions, this behavior arises, for example, in quantum random walks. The logarithmic Gauss map maps a 2-torus to the simply connected subset Ξ of the plane. Such a map must have entire curves on which it folds over itself and some points of greater degeneracy, where such curves meet or fold on themselves.

Consider the situation of Example 9.5.13. We have asymptotics on the diagonal $r = s$ in the scale $s^{-1/3}$, and also asymptotics in other directions in the scale $s^{-1/2}$, uniform as long as we remain bounded away from the main diagonal. Without some kind of result to bridge the gap, we cannot, for instance, conclude that

$$\limsup |\boldsymbol{r}|^{1/3} a_{\boldsymbol{r}} = \frac{\sqrt{3}\,\Gamma(1/3)}{6\pi} \tag{13.4.1}$$

or even that

$$\limsup \log a_{\boldsymbol{r}} / \log |\boldsymbol{r}| = -1/3 \,. \tag{13.4.2}$$

Indeed such asymptotics can be worked out. A combinatorial generating function with this behavior has been discussed in Banderier, Flajolet, Schaeffer, and Soria (2001) under the name of "Airy phenomena." This name is due to the fact that in the rescaled window $s = \lambda r + O(r^{1/3})$, the leading term converges to an Airy function. A start on a general formulation of such asymptotics in dimension $d = 2$ was made by Lladser in Lladser (2003). For instance, (13.4.2) follows from Lladser (2003, Corollary 6.12). Lladser (2006) also shows that when there is a change of degree of the amplitude and the phase does not change degree, we can derive a uniform formula for the coefficients in the expansion. Clearly, this work is only a first step and only in dimension 2, and much more work is desirable.

A similar issue for trivariate functions arises in the analysis of spacetime generating functions for two-dimensional quantum random walks. The logarithmic Gauss map maps a 2-torus to the simply connected subset Ξ of the plane. Such a map must have entire curves on which it folds over itself and some points of greater degeneracy where such curves meet or fold on themselves. Further work on asymptotics in such regimes would be desirable. In particular, just understanding the power laws of the intensities near such points would represent a step forward.

13.5 Conclusion

In this book we have focused on deriving algorithmic methods for computing asymptotic approximations to coefficients of multivariate generating functions of meromorphic type. Such methods have many applications in various fields via their application to enumerative combinatorics and probability. For example, the basic results for smooth and multiple points have so far been used by authors to study problems arising in dynamical systems, bioinformatics, number theory, and statistical physics, and queueing theory is also a rich source of examples. We expect the number of applications to grow steadily.

Although many (most?) applied problems can be tackled by smooth point analysis, there are many interesting problems that involve much more complicated local geometry, such as the tiling models discussed in Section 11.3. To tackle more complex models, we will need to deal with these geometries. The results of Chapter 11 show that these analyses will be nontrivial in general.

From the standpoint of mathematical analysis, it seems to us that the tools required to extend our results already exist, and it is just a matter of mastering them. Problems for which minimal points control asymptotics usually sidestep complicated topology, and the Morse-theoretic intuition behind our results can often be ignored in such cases by casual users seeking to solve a specific problem. However, when dealing with contributing points that are not necessarily minimal, substantial topological difficulties can arise as we have already seen in Section 9.4, and the resolution of singularities approach in Section 13.2 leads directly to such situations, even when all coefficients in the original problem are positive. We believe that to make further progress in this area, substantial work in the Morse-theoretic framework will be required.

The formulae given in terms of derived quantities after various changes of variable are often rather useful and lead to a complete algorithm for computation (provided the contributing points can be identified, as discussed earlier). However, for some purposes, not least geometric insight, it is desirable to have coordinate-free formulae expressed in terms of the original data. We have presented one such interpretation in terms of Gaussian curvature in Section 9.5. Analogous formulae for higher order coefficients and for nonsmooth points would be very welcome.

This book is certainly not the last word on the subject, but rather an invitation to join the authors in further development of this research area, which combines beauty, utility, and tractability to a high degree and which has given us considerable enjoyment over many years.

Notes

The case $k = 1$ of Lemma 13.2.1 was observed in Furstenberg (1967). Denef and Lipshitz (1987) described how to lift a d-variate algebraic function defined over any field (in fact, more general rings) to a $2d$-variate rational function, using the complete diagonal. They also lift to dimension $d + 1$ using a specific M-diagonal. However, to our knowledge these constructions are not computationally effective. The proofs rely on existence of a generator for a certain type of ring extension. More details can be found in Adamczewski and Bell (2012), which also gives a different, computationally effective procedure for lifting.

There are several older methods for computing asymptotics of D-finite generating functions, often used for computing diagonal asymptotics in the combinatorial literature. These include the "method of Frobenius" and methods arising from the work of Birkhoff and Trjitzinsky (Wimp and Zeilberger, 1985). The latter in particular is still controversial after more than 80 years, as well as computationally intensive. These methods all suffer from the apparent undecidability of the general connection problem, although in concrete cases, extra information can help to solve that problem.

There are many multivariate generalizations of the Lagrange Inversion Formula, but we know of none that are useful for our purposes.

The examples in this chapter are taken from Raichev and Wilson (2007), Raichev and Wilson (2008), Raichev and Wilson (2012a), and Raichev and Wilson

(2012b). Links to computer algebra implementations can be found on the website for this book.

Exercise 13.2 is suggested by a recent line of work on rook walks. See, for example, Kauers and Zeilberger (2011).

An interesting question is when to stop computing higher order terms in an asymptotic expansion, if our goal is numerical accuracy for fixed $\boldsymbol{r}$. Roughly speaking, in most applications one should stop at the smallest term (recall that asymptotic expansions usually diverge as the number of terms tends to infinity). This often leads to error terms that are in fact exponentially small ("superasymptotics"). One can even go beyond this ("hyperasymptotics," "asymptotics beyond all orders"). The books by Paris and Kaminski (2001) and Paris (2011) give a good introduction, and applications to integrals arising from coefficient extraction are treated in Delabaere and Howls (2002).

Exercises

13.1 (other "diagonals")

Let $\boldsymbol{\alpha} \in \mathbf{N}^d$ represent a direction (with all coefficients nonzero and relatively prime). Show how to express $\sum_{\boldsymbol{r}=n\boldsymbol{\alpha}} a_{\boldsymbol{r}} z^n$ using the complete diagonal extraction operation. Apply this to find the generating function for the Delannoy numbers restricted to the ray defined by (2, 1) and compare with your solution to Exercise 2.5. Hint: roots of unity.

13.2 (rook walks)

Let $a_{\boldsymbol{r}}$ be the number of ways in which a chess rook can move from the origin to $\boldsymbol{r}$ by moves that increase one coordinate and do not decrease any other. The generating function is given by methods of Chapter 2 as

$$\sum_{\boldsymbol{r}} a_{\boldsymbol{r}} \boldsymbol{z}^{\boldsymbol{r}} = \frac{1}{1 - \sum_{i=1}^{d} \frac{z_i}{1-z_i}} = \frac{\prod_i (1 - z_i)}{\sum_{j=0}^{d} (-1)^j (j+1) e_j(\boldsymbol{z})},$$

where e_j is the elementary symmetric polynomial of degree j.

(i) For $d = 2$, use the diagonal method to find the diagonal generating function, and then derive a linear recurrence relation for a_{rr}, for example, by the method outlined in Proposition 2.4.2. What happens when you try this for $d = 3$?

(ii) Compute the first-order asymptotic approximation to $a_{\boldsymbol{r}}$ for $d = 3$.

(iii) Use a computer algebra system, or write your own program, to compute $a_{\boldsymbol{r}}$ exactly, for values of d up to 10. Compare with the first-order asymptotic when $\boldsymbol{r} = (100, \dots, 100)$.

(iv) Can you compute the next term in the expansion? If so, how much better is the accuracy of the 2-term asymptotic approximation when $d = 3$ and $\boldsymbol{r} = (100, 100, 100)$?

13.3 (pairs of compositions with the same number of parts)

Derive the bivariate generating function $\sum_{rs} a_{rs} x^r y^s$ for the number of ordered pairs of ordered sequences of integers with parts in a fixed set $A \subseteq \mathbb{N}$, the first summing to r and the second to s, and each having the same number of parts. Compute the asymptotics of the coefficents on the leading diagonal. Compare the results and methods with those in Banderier and Hitczenko (2012).

13.4 (Safonov's vanishing numerator)
It is expected (as explained in Section 13.2) that if we lift an algebraic function A to a rational one R, the numerator of R must vanish at a contributing critical point. Show that this in fact always happens when R is produced using the method of Lemma 13.2.1.

13.5 (assembly trees)
Vince and Bona (2012) define the concept of *assembly tree* of a graph and show that the generating function for the number of assembly trees of the complete bipartite graph K_{rs} is given by

$$\sum_{rs} a_{rs} x^r y^s = 1 - \sqrt{(1-x)^2 + (1-y)^2 - 1}.$$

Use the lifting method to compute the asymptotics of the diagonal coefficient a_{rr}. Compare with the result in the cited article and the methods used there.

13.6 (new and old leaves)
The Narayana numbers can be further refined by considering different types of leaves in a rooted ordered tree. Chen, Deutsch, and Elizalde (2006) call a leaf of such a tree *old* if it is the leftmost child of its parent and *young* otherwise. They enumerate such trees according to the number of old leaves, number of young leaves, and number of edges, finding the algebraic equation

$$G(x, y, z) = 1 + \frac{z(G(x, y, z) - 1 + x)}{1 - z(G(x, y, z) - 1 + y)}.$$

Use the lifting procedure to express this as a diagonal of a 4-variable rational function R. How exactly do we obtain the Narayana generating function from G? Does the lifting procedure commute with this operation?

13.7 (alternative way of computing higher order terms [Pemantle and Wilson, 2010])
This exercise outlines an alternative way of computing higher order terms in the basic Fourier-Laplace integral. Suppose for simplicity that we have a single critical point of the phase ϕ at $\mathbf{0}$ and the amplitude A vanishes outside the closure of a neighborhood of $\mathbf{0}$. Let S be the standard quadratic $S(z) = \sum_i z_i^2$. Corollary 5.2.3 gives an exact formula for the higher order terms in this case.

(i) Show that when $\phi = S$, the differential operator

$$\sum_{|r|=k} \frac{\partial_1^{2r_1} \cdots \partial_d^{2r_d}}{4^k r_1! \cdots r_d!}$$

when applied to A and evaluated at $\mathbf{0}$, gives the coefficient c_k from Theorem 5.4.8.

(ii) The Morse Lemma states there is a change of variables $S = \phi \circ \psi$, where S is the standard quadratic $S(z) = \sum_i z_i^2$. Note that this changes the amplitude function A to $(A \circ \psi) \det \psi'$. Apply the result of (i), and solving a triangular system to compute the derivatives of ψ at $\mathbf{0}$, we can derive the desired result.
Carry this out in the case $d = 1, k = 1$ and derive an explicit formula. Check it against the formula given in Corollary 13.3.3.

13.8 (binomial transition)
Consider the binomial coefficient generating function $(1 - x - y)^{-1}$, and compute first-order asymptotics for the coefficient a_{rs}, where $s/r \to 0$ as $r, s \to \infty$. How many different cases are there in the analysis?

PART IV

Appendixes

Appendix A
Integration on Manifolds

In this first appendix, we formally construct the apparatus for integrating differential forms on real and complex manifolds. Roughly speaking, this allows us to fully understand the first two of the five steps listed at the end of Section 1.3. These two steps are the replacement of a_r by an integral that is a variant of (1.3.1):

$$a_r = \left(\frac{1}{2\pi i}\right)^d \int_C z^{-r-1} F(z)\, dz, \tag{1.3.1'}$$

where C is any cycle homologous to T in the domain $\mathcal{M}$ of holomorphy of the integrand. To make sense of this, we must define differential forms and their integrals, state the multivariate Cauchy formula, construct the singular homology of the domain $\mathcal{M}$, and connect the singular homology of $\mathcal{M}$ to the integration of exact forms over cycles in $\mathcal{M}$. The final two sections on intersection classes and residue forms concern the residue integral theorem, a Stokes-type theorem generalizing the familiar univariate residue integral theorem. The first half of this appendix is well known and contained in a standard graduate course; this brief review is included for ease of reference; the second half contains material beyond the usual canon. The main topics of the appendix are as follows.

- **Section A.1.** Real differential forms and their integrals over chains are defined, leading to Stokes' Theorem (Theorem A.1.1).
- **Section A.2.** Complex forms are defined and Cauchy's integral formula (Theorem A.2.4) is stated.
- **Section A.3.** Singular homology is constructed, the long exact homology sequence is derived (Theorem A.3.2), and the relation between homology and integration is stated (Theorem A.3.8). This section also contains discussions of the homology of pairs and the Künneth product formula.
- **Section A.4.** The intersection class is defined.
- **Section A.5.** The residue form is defined and a Stokes-type theorem is proved for its integral over the intersection cycle.

A.1 Differential Forms in $\mathbb{R}^n$

Manifolds

The notion of a manifold is undoubtedly familiar, but there are several different formalizations. The manifolds relevant to us are submanifolds of Euclidean space, so we use definitions specific to $\mathbb{R}^n$. We define a d-manifold to be a subset $\mathcal{M}$ of $\mathbb{R}^n$ such that every point in $\mathcal{M}$ has a neighborhood in $\mathbb{R}^n$ whose intersection $\mathcal{N}$ with $\mathcal{M}$ is diffeomorphic to $\mathbb{R}^d$ (or equivalently, to the open unit ball in $\mathbb{R}^d$). Here, diffeomorphic means there is a map $\phi : \mathcal{N} \to \mathbb{R}^d$ such that both ϕ and ϕ^{-1} are smooth (i.e., members of the class C^∞ of infinitely differentiable functions).

The Exterior Algebra on $\mathbb{R}^d$

To elucidate the discussion of differential forms, let us first distinguish this from the other common notion of integration. A smooth map $\gamma : [0, 1] \to \mathbb{R}^n$ may be thought of as a curve, embedded or immersed in Euclidean space. There are two somewhat different notions of integration on γ. First, there is a natural measure on the range of γ, namely one-dimensional Hausdorff measure $\boldsymbol{m}$. The integral $\int_\gamma f \, d\boldsymbol{m}$ is then defined for any measurable $f : \mathbb{R}^n \to \mathbb{R}$ as a Lebesgue integral. This is not the kind of integration we are interested in, so there is no need to proceed further in defining Hausdorff measure and Lebesgue integration.

The notion we are interested is more akin to a line integral from physics: a vector-like integrand $f\,dx + g\,dy$ is integrated along the curve $\{(\gamma_1(t), \gamma_2(t)) : 0 \le t \le 1\}$ by taking the inner product of the vector (f, g) with the tangent vector $d\gamma := (\frac{d\gamma_1}{dt}, \frac{d\gamma_2}{dt})$ to obtain

$$\int_\gamma f\,dx + g\,dy := \int_0^1 \left[f(\gamma(t))\frac{d\gamma_1}{dt} + g(\gamma(t))\frac{d\gamma_2}{dt} \right] dt.$$

This is an oriented notion of integration, in the sense that parametrizing the curve γ in the reverse direction would change the sign on the dot product and hence the whole integral. One may observe, however, using the chain rule, that a different parametrization γ in the same orientation produces the same integral. Based on this concept, we now define general real differential forms and their integrals.

Let $\mathcal{M}$ be a real d-manifold in $\mathbb{R}^n$. A collection of k-dimensional vector spaces $\{V_{\boldsymbol{p}} : \boldsymbol{p} \in \mathcal{M}\}$, one for each point of $\mathcal{M}$, is a ***bundle*** over $\mathcal{M}$ if it is locally coordinatized by $\mathcal{M} \times \mathbb{R}^k$. A (smooth) ***section*** of a bundle is a smooth map $\boldsymbol{p} \mapsto u(\boldsymbol{p})$ with $u(\boldsymbol{p}) \in V_{\boldsymbol{p}}$. The bundle with $V_{\boldsymbol{p}} = T_{\boldsymbol{p}}(\mathcal{M})$, the tangent space to $\mathcal{M}$ at $\boldsymbol{p}$, is called the ***tangent bundle***, denoted by $T(\mathcal{M})$. Let $T^*(\mathcal{M})$ denote the ***cotangent bundle*** defined by letting $V_{\boldsymbol{p}} = T_{\boldsymbol{p}}(\mathcal{M})^*$ be the dual space to $T_{\boldsymbol{p}}(\mathcal{M})$. To see why we need this, note that the form $f\,dx + g\,dy$ from the line integral example was an element of the cotangent bundle. For any k-dimensional real vector space V and any $p \le k$, an ***alternating p-linear function*** on V is a linear map from p-tuples of elements of V to $\mathbb{R}$, which is antisymmetric in each pair of arguments. These form a vector space, $\Lambda_p(V)$. Let $\Lambda(V)$ denote the direct sum over p of $\Lambda_p(V)$. The following facts about $\Lambda(V)$ for any k-dimensional vector space, V, may be found in Warner (1983, Chapter 2).

(i) $\Lambda(V)$ is a graded vector space and the dimension of $\Lambda_p(V)$ is $\binom{k}{p}$.
(ii) An associative antisymmetric product may be defined taking p elements of $\Lambda_1(V)$ into $\Lambda_p(V)$, as follows:

$$\alpha_1 \wedge \cdots \wedge \alpha_p(v_1, \ldots, v_p) = \det\left(\alpha_i(v_j)\right) .$$

(iii) Any vector space basis for $\Lambda_1(V)$ generates $\Lambda(V)$ as a ring.
(iv) Let $\{v_1, \ldots, v_k\}$ be any basis for V. For any $\Phi := \{i_1 < \cdots < i_p\}$ of integers from 1 to k, let $v_\Phi^* := v_{i_1}^* \wedge \cdots \wedge v_{i_p}^*$. A basis for $\Lambda_p(V)$ is given by $\{v_\Phi^* : \Phi \text{ a subset of size } p\}$.

The bundle for which $V_p = \Lambda(T_p(\mathcal{M})^*)$ is called the ***exterior algebra bundle***, $\Lambda(\mathcal{M})$, and it is graded by the decomposition into the direct sum of $\Lambda_p(\mathcal{M})$. A section of $\Lambda_p(\mathcal{M})$ is called a ***differential p-form***. The set of these is denoted $E^p(\mathcal{M})$ and the union over p is denoted $E^*(\mathcal{M})$.

Functoriality

All of the preceding definitions take place in the category of smooth manifolds, where the arrows are smooth maps. This means that the bundles are independent of the local parametrization and choice of basis, and that a smooth map between manifolds induces a map, in the appropriate direction, between bundles. To make this more concrete, consider for example a smooth map $f : \mathbb{R}^d \to \mathbb{R}^n$. Suppose this is a chart map for a d-manifold $\mathcal{M} \subseteq \mathbb{R}^n$, so f is a diffeomorphism between a ball in $\mathbb{R}^d$ and neighborhood in $\mathcal{M}$. Let us see how f maps tangent vectors. A tangent vector may be thought of in several ways. The classical way to think of it is as a direction in which a derivative of a real function may be taken. More pictorially, we may think of a tangent vector in $T_p(\mathcal{M})$ as a limit of vectors $\vec{px}$. The tangent vector $\boldsymbol{v}$ in this depiction is the limit of $\vec{px_t}$ where $x_t = p + t\boldsymbol{v}$. Under f, this maps to $Df(p)(\boldsymbol{v})$, the image of $\boldsymbol{v}$ under the differential of the map f at p. For example, the tangent vector known as ∂x_1 maps to the tangent vector $\sum_{j=1}^n (\partial f_j / \partial x_1) \partial x_j$. More generally,

$$f_* \left(\sum_{i=1}^d a_j \partial x_i \right) := \sum_{i=1}^d \sum_{j=1}^n a_i \frac{\partial f_j}{\partial x_i} \partial x_j .$$

Let us next see what map f induces on the cotangent bundle. By definition, for a cotangent vector $u \in T_p^*(\mathcal{M})$ and for $\boldsymbol{v} \in T_p(\mathbb{R}^d)$, $f^*(u)(\boldsymbol{v}) = u(f_*(\boldsymbol{v}))$. Take

$u = dx_j$, the linear function mapping ∂x_j to 1 and ∂x_m to 0 for $m \neq j$. Unraveling the definitions, we see that

$$f^*(dx_j)(\partial x_i) = \frac{\partial f_j}{\partial x_i}$$

and hence

$$f^*(dx_j) = \sum_{i=1}^{d} \frac{\partial f_j}{\partial x_i} dx_i \,.$$

We may now use the naturality of the wedge product to see that

$$f^*(dx_{j_1} \wedge \cdots \wedge dx_{j_d}) = \det\left(\frac{\partial f_{j_k}}{\partial x_i}\right) dx_1 \wedge \cdots \wedge dx_d = J\,dV,$$

where dV denotes $dx_1 \wedge \cdots \wedge dx_d$ and J is the Jacobian of the $(j_1, \ldots, j_d)$-coordinates of f with respect to the coordinates $x_1, \ldots, x_d$.

The Differential Operator

The ring structure on $\Lambda(\mathcal{M})$ is natural: a smooth map $\psi : \mathcal{M} \to \mathcal{N}$ induces a map $\psi^* : \Lambda(\mathcal{N}) \to \Lambda(\mathcal{M})$, and any commuting diagram of smooth maps induces a commuting diagram of the exterior algebras. Let $\{dx_1, \ldots, dx_n\}$ denote the standard basis for $T_p^*(\mathbb{R}^n)$, where dx_i maps the jth standard basis vector to δ_{ij}. The standard basis for p-forms is $\{dx_\Phi\}_\Phi$.

If $\mathcal{M} \subseteq \mathbb{R}^n$ is a d-manifold, then the inclusion $\iota : \mathcal{M} \to \mathbb{R}^n$ induces a pullback $\iota^* : \Lambda(\mathbb{R}^n) \to \Lambda(\mathcal{M})$. The pullback $\{\iota^*(dx_\Phi)\}_\Phi$ of the standard basis in each fiber $T_p^*(\mathbb{R}_n)$ is also denoted dx_Φ.

Let $\mathcal{M}$ be a d-manifold in $\mathbb{R}^n$. We may define a unary operation d on $E^*(\mathcal{M})$ of degree 1 by defining d on $E^p(\mathcal{M})$ as follows. Define

$$d(f\,dx_\Phi) := \sum_{1 \le i \le d} \frac{\partial f}{\partial x_i} dx_i \wedge dx_\Phi \,.$$

Only summands with $i \notin \Phi$ will be nonzero. Extend this by linearity to all of $E(\mathcal{M})$. Then d has the following properties:[1]

(i) $d^2 = 0$.

(ii) If f is a zero form (a smooth function), then $df = \sum_{i=1}^{d} \frac{\partial f}{\partial x_i} dx_i$.

(iii) If $\omega \in E^p(\mathcal{M})$ and $\eta \in E^q(\mathcal{M})$, then

$$d(\omega \wedge \eta) = d(\omega) \wedge \eta + (-1)^p \omega \wedge d(\eta).$$

The operator d is natural in the sense that $d(\psi^*(\omega)) = \psi^*(d\omega)$.

[1] It may appear that d depends on the choice of local coordinatization, because, for example, $f_1 dx_1 + f_2 dx_2$ in one basis is a constant dy_1 in a basis chosen to have this as the first component in each $T_p^*(\mathcal{M})$. In fact, the basis for $T_p^*(\mathcal{M})$ is required to be dual to the basis $\{\partial/\partial x_i\}$ of $T_p(\mathcal{M})$, which carries with it the connection between bases in different fibers.

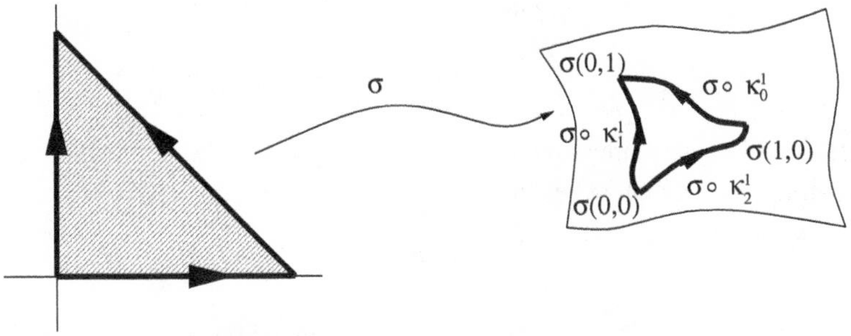

Figure A.1 A singular 2-simplex and its boundary.

Integration of Forms

For each $p \geq 1$, let Δ^p denote the p-simplex in $\mathbb{R}^p$ defined by

$$\{(x_1, \ldots, x_p) \in \mathbb{R}^p : x_i \geq 0 \text{ for all } i \text{ and } \sum_{i=1}^{p} x_i \leq 1\}.$$

We may identify Δ^p with the standard simplex in $\mathbb{R}^{p+1}$ via $(x_1, \ldots, x_p) \leftrightarrow (x_1, \ldots, x_p, 1 - \sum_{j=1}^{p} x_j)$.

When $p = 0$, take Δ^p to be a single point. Associated with the standard simplices are $p+2$ ways of embedding Δ^p as a face of Δ^{p+1}: for $0 \leq i \leq p$, let κ_i^p embed by inserting a zero in the ith position

$$\kappa_i^p(x_1, \ldots, x_p) = (x_1, \ldots, x_{i-1}, 0, x_{i+1}, \ldots, x_p),$$

and let κ_0^p embed into the diagonal face

$$\kappa_i^p(x_1, \ldots, x_p) = \left(1 - \sum_{i=1}^{p} x_i, x_1, \ldots, x_p\right).$$

Let $\mathcal{M}$ be a d-manifold in $\mathbb{R}^n$. For $0 \leq p \leq d$, a ***singular p-simplex*** in $\mathcal{M}$ is defined to be a smooth map $\sigma : \Delta^p \to \mathcal{M}$. Define the space $\mathbb{C}^p(\mathcal{M})$ of ***p-chains*** on $\mathcal{M}$ to be the space of finite formal linear combinations $\sum c_i \sigma_i$ of p-simplices in $\mathcal{M}$. Define the boundary $\partial\sigma$ of a p-simplex σ by

$$\partial\sigma := \sum_{i=0}^{p} (-1)^i \sigma \circ \kappa_i^{p-1}.$$

See Figure A.1

Extend this linearly to the chains of $\mathcal{M}$. It is easy to verify (see Exercise A.2) that

$$\partial^2 = 0. \tag{A.1}$$

For a domain $A \subseteq \mathbb{R}^n$, we define

$$\int_A f\,dx_1 \wedge \cdots \wedge dx_n := \int_A f\,dV,$$

where dV is Lebesgue measure in $\mathbb{R}^n$. Now for any p-simplex $\sigma \in \mathcal{M}$, integration of a p-form ω over σ may be defined by

$$\int_\sigma \omega := \int_{\Delta^p} \sigma^*(\omega),$$

where σ^* is the pullback by σ of ω to the standard p-simplex. We may write $\sigma^*(dx_\Phi)$ more explicitly as $J\,dV$, where J is the Jacobian of the map $(x_1, \ldots, x_p) \mapsto (x_{i_1}, \ldots, x_{i_p})$. The integral $\int_C \omega$ may be defined for any p-chain by extending linearly in C. This allows us to integrate a p-form over any triangulable region of dimension p.

Stokes' Theorem now follows from definitions and some elementary computations (see Warner, 1983, Theorem 4.7):

Theorem A.1.1 (Stokes) *Let ω be a $(p-1)$-form $(p \geq 1)$ on a manifold $\mathcal{M}$ of dimension at least p and let C be a p-chain on $\mathcal{M}$. Then*

$$\int_{\partial C} \omega = \int_C d\omega\,.$$

Some elementary properties of the integral and its relation to more elementary notions may be verified. The proofs of the following iterated integration properties are left as exercises, as is the exact statement of part (ii).

Proposition A.1.2

(i) Define a chain C corresponding to "the unit cube in $\mathbb{R}^n$" and prove that

$$\int_C f\,dx_1 \wedge \cdots \wedge dx_n = \int_0^1 \cdots \left(\int_0^1 f(x_1, \ldots, x_n)\,dx_n \right) \cdots dx_1\,.$$

(ii) Let $\mathcal{M}$ and $\mathcal{N}$ be respectively a p-manifold in $\mathbb{R}^m$ and a q-manifold in $\mathbb{R}^n$. Denote points in $\mathcal{M} \times \mathcal{N}$ by $(\boldsymbol{x}, \boldsymbol{y})$ and denote the projections $(\boldsymbol{x}, \boldsymbol{y}) \mapsto \boldsymbol{x}$ and $(\boldsymbol{x}, \boldsymbol{y}) \mapsto \boldsymbol{y}$ by π and ρ, respectively. Say that an element of the exterior algebra $\Lambda_k(\mathcal{M} \times \mathcal{N})$ is pure of degree (p, q) if it is the wedge $\pi^*\eta \wedge \rho^*\xi$ of forms of respective degrees p and q. The dimension of the space $\Lambda_k(p, q)$ of pure elements of degree (p, q) at a point $(\boldsymbol{x}, \boldsymbol{y})$ is ??

(iii) Suppose $\omega = f(\boldsymbol{x}, \boldsymbol{y})\pi^*\eta \wedge \rho^*\xi$ is a section of $\Lambda_k(p, q)$. If A is a p-chain in $\mathcal{M}$ and B is a q-chain in $\mathcal{N}$, then

$$\int_{A \times B} \omega = \int_A g(\boldsymbol{x}) \cdot \eta,$$

where

$$g(\boldsymbol{x}) = \int_B f(\boldsymbol{x}, \boldsymbol{y}) \cdot \xi.$$

A.2 Differential Forms in $\mathbb{C}^n$

The complex numbers may be identified with $\mathbb{R}^2$. Similarly, $\mathbb{C}^n$ may be identified with $\mathbb{R}^{2n}$; fix the identification that maps $z = (z_1, \ldots, z_j) = (x_1 + iy_1, \ldots, x_j + iy_j)$ to $(x_1, y_1, \ldots, x_j, y_j)$. In this section, points of $\mathbb{R}^{2n}$ are referred to by $(x_1, y_1, \ldots, x_n, y_n)$ rather than $(x_1, \ldots, x_{2n})$.

Let u be a map from $\mathbb{C}^n = \mathbb{R}^{2n}$ to $\mathbb{C}$ (we are not going to view the range as $\mathbb{R}^2$). The following are formal definitions only, although the symbols suggest the origins.

Definition A.2.1 *Define*

$$\frac{\partial u}{\partial z_j} := \frac{1}{2}\left(\frac{\partial u}{\partial x_j} - i\frac{\partial u}{\partial y_j}\right);$$

$$\frac{\partial u}{\partial \overline{z}_j} := \frac{1}{2}\left(\frac{\partial u}{\partial x_j} + i\frac{\partial u}{\partial y_j}\right);$$

$$dz := dx + i\,dy$$

$$d\overline{z} := dx - i\,dy\,,$$

$$\partial u := \sum_{j=1}^{n} \frac{\partial u}{\partial z_j}\, dz_j\,,$$

$$\overline{\partial} u := \sum_{j=1}^{n} \frac{\partial u}{\partial \overline{z}_j}\, d\overline{z}_j\,.$$

In terms of these notations, the d operator may be written

$$du = \partial u + \overline{\partial} u\,.$$

There is an intuitive notion of what it means for the map u to obey the complex structure: polynomials (and convergent power series) in the coordinates z_j should obey the complex structure, but functions such as $|z_1|$ that require decomposing into real and imaginary components $x_1^2 + y_1^2$ should not. This is formalized by the Cauchy-Riemann equations.

Definition A.2.2 *Say that a p-form* ω *is holomorphic if* $\overline{\partial}\omega = 0$*. In particular, taking* $p = 0$*, this defines the notion of a holomorphic function from* $\mathbb{C}^n$ *to* $\mathbb{C}$*.*

One easily verifies that the coordinate functions are holomorphic and that holomorphic functions are closed under sums, products, limits in C^1, and applications of the implicit function theorem. This implies holomorphicity of rational functions, the exponential, the logarithm, and so forth.

In particular, the holomorphic functions form a subring of $C^\infty(\mathbb{R}^{2n})$. A basis for $\Lambda_1(\mathbb{R}^{2n})$ is given by $\{dz_1, \ldots, dz_n, d\overline{z}_1, \ldots, d\overline{z}_n\}$. The ring generated by the subcollection $\{dz_1, \ldots, dz_n\}$ over the ring of holomorphic functions on $\mathcal{M}$ is called the ring of holomorphic forms on $\mathcal{M}$. It is easy to check that the operator d

preserves holomorphicity – indeed if $\overline{\partial}\omega = 0$, then $d\omega = \partial\omega$, which is evidently holomorphic. The notation $dz := dz_1 \wedge \cdots \wedge dz_n$ denotes the ***holomorphic volume form*** in $\mathbb{C}^n$. It is an n-form in $\mathbb{R}^{2n}$, thus ***middle-dimensional***, but is the highest dimensional holomorphic form in $\mathbb{R}^{2n}$. This leads to:

Theorem A.2.3 *If C is any $(d+1)$-chain on a domain U with boundary ∂C, and if ω is any holomorphic d-form on U, then*

$$\int_{\partial C} \omega = 0.$$

Proof The d operator preserves holomorphicity, whereby $\partial\omega$ is holomorphic. There are no holomorphic forms above rank d and $\partial\omega$ is a $(d+1)$-form, and hence $\partial\omega = 0$. Stokes' Theorem then gives

$$\int_{\partial C} \omega = \int_C d\omega = 0.$$

□

If u is holomorphic in a centered polydisk, then it may be represented by an absolutely convergent power series and the terms of the power series may be extracted, leading to (see Hörmander, 1990, 2.2.3 and Theorem 2.2.1):

Theorem A.2.4 (multivariate Cauchy formula) *Let $F(z) = \sum_r a_r z^r$ be a d-variate power series holomorphic in an open polydisk* **D** *containing the origin. Let $T = \prod_{i=1}^d \gamma_i$ be a product of circles γ_i that bound disks D_i of radii b_i, such that the polydisk $\prod_{i=1}^d D_i$ is a subset of* **D**. *Then the multivariate Cauchy formula* (1.3.1) *holds:*

$$a_r = \left(\frac{1}{2\pi i}\right)^d \int_T z^{-r-1} F(z)\, dz,$$

where $dz := dz_1 \wedge \cdots \wedge dz_d$ is the holomorphic volume d-form.

The foregoing definitions for complex forms are based on the definitions for real forms in twice the dimension. However, when dealing with holomorphic forms, one often works in the category of complex analytic spaces and holomorphic maps. In this case, the same results that hold for differential forms on real n-manifolds often hold for holomorphic forms on complex manifolds of the same dimension, n. See, for example, Exercises A.3 and A.4.

A.3 Algebraic Topology

Chain Complexes and Homology Theory

Our motivation for studying homology theory is as follows. A differential form ω is said to be a ***closed form*** if $d\omega = 0$. Many of the forms we care about are closed. For example, if ω is any holomorphic d-form in $\mathbb{C}^d$, then $\overline{\partial}\omega$ vanishes by holomorphicity and $\partial\omega$ vanishes because there are no holomorphic $(d+1)$-forms, and hence ω is closed.

A chain C is a ***cycle*** if $\partial C = 0$. A chain C is a ***boundary*** if $C = \partial \mathcal{D}$ for some chain, $\mathcal{D}$. The boundaries form a subset (in fact a subvector space) of the cycles because $\partial^2 = 0$. Suppose we wish to integrate a closed p-form ω over a boundary $C = \partial \mathcal{D}$. By Stokes' Theorem,

$$\int_C \omega = \int_{\mathcal{D}} d\omega = \int_{\mathcal{D}} 0 = 0 .$$

By linearity, therefore, $\int_C \omega$ depends only on the equivalence class of C in the quotient space of cycles modulo boundaries. Homology theory is the study of this quotient space. It turns out there is a benefit to maintaining generality. Although we care ultimately about the space of singular chains on a manifold, we define chain complexes and the homology functor in a purely algebraic way.

Definitions A.3.1

*(i) A **chain complex** C is a collection $\{C_n : n = 0, 1, 2, \ldots\}$ of complex vector spaces, not necessarily finite dimensional, together with a boundary operator ∂, which is a linear map ∂_n from the n-*chains *C_n to C_{n-1} satisfying $\partial^2 = 0$ (meaning that $\partial_n \circ \partial_{n+1} = 0$ for every n). By definition, $\partial = 0$ on C_0.*

(ii) The group of cycles $Z_n \subseteq C_n$ is the kernel of ∂_n and the group B_n of boundaries is the image of ∂_{n+1}.

(iii) The nth homology group of C is defined by

$$H_n(C) = \frac{Z_n}{B_n} .$$

The notation $H_*(X)$ is used to refer collectively to $H_n(X)$ for all n.

Remark This is sometimes called ***homology with coefficients in*** $\mathbb{C}$, to distinguish it from the analogous construction with coefficients in $\mathbb{Z}$. Although the theory with coefficients in $\mathbb{Z}$ is richer, taking coefficients in a field better suits the purposes of computing integrals (also, see Remark A.3.3).

When C is the complex of singular chains of a d-manifold (or indeed any topological space) $\mathcal{M}$, then $H_k(C)$ is denoted $H_k(\mathcal{M})$ and is called the kth (singular) homology group of $\mathcal{M}$. One thinks of the rank of the homology group $H_n(X)$ as indicating how many cycles there are that don't bound anything; for example, the rank of $H_1(X)$ should be the number of inequivalent circles that can be drawn on X and do not bound a disk in X; the rank of $H_1(X)$ for a connected space X should be zero if and only if X is simply connected.

A map between topological spaces (maps in this category are continuous) induces a natural map on the singular chain complexes (maps in this category commute with ∂). A map between chain complexes induces a natural map on the homology groups. Composing with the inverse, we see that a homeomorphism induces an isomorphism between the homology groups. Thus the homology groups of a topological space are topological invariants.

A homotopy is a map $H : X \times [0,1] \to Y$. When $H_0 := H(\cdot, 0)$ is a homeomorphism, we say that H is a homotopy between the image Y_0 of H_0 and the image Y_1 of $H_1 := H(\cdot, 1)$. Within the space Y, the spaces Y_0 and Y_1 are topologically equivalent. For example, if Y_1 is a single point, then Y_0 can be shrunk to a point inside Y and is a topologically trivial subspace. One way to see why this is true is to examine the homotopy at the chain level. Let C be a chain supported on X and for $j = 0, 1$ let C_j be the image of C under H_j. Let C_H denote the $(d+1)$-chain on the space $X \times [0,1]$, which is the product of C with the standard 1-simplex, σ. Then

$$\partial C_H = C_1 - C_0 + \partial C \times \sigma \,. \tag{A.1}$$

When C is a cycle, this shows C_0 and C_1 to differ by a boundary, meaning that they are in the same homology class.

More general than a homeomorphism is a homotopy equivalence. Say that a map $f : X \to Y$ is a ***homotopy equivalence*** if there is a map $g : Y \to X$ such that $f \circ g$ is homotopic to the identity in Y and $g \circ f$ is homotopic to the identity in X. Homotopic maps induce equal maps on homology, hence homotopy equivalent spaces have naturally identical homology. To see this, one proves, on the chain level, that a homotopy equivalence between topological spaces induces a ***chain homotopy equivalence*** between the singular chain complexes, which induces again an isomorphism between the homology groups (see, e.g., Munkres, 1984, Theorems 12.4, 30.7).

Although the singular chain complex is infinite dimensional, one may also isolate certain subcomplexes whose inclusion into the singular chain complex induces an isomorphism on homology. For example, a topological space is a ***cell complex*** if it may be built from cells homeomorphic to closed simplices of various dimensions, by identifying the boundaries in certain prescribed ways. The corresponding ***cellular chain complex*** $\mathcal{D}$ has a vector space of k-chains of dimension n_k equal to the number of k-cells, and the boundary map is given by the identifications. The singular homology is equal to the homology of $\mathcal{D}$ (Munkres, 1984, Theorem 39.4), and hence it is easy to compute homology groups of a space expressed as a cell complex. Another consequence is that a space built from cells of dimension at most d has vanishing homology above dimension d. We will be interested in cell complexes as spaces over which we integrate differential forms. Let X be a cell complex of dimension d and chain complex $\mathcal{D}$. Each d-simplex corresponds naturally (up to sign) with a generator for the d-chains of $\mathcal{D}$. A sum of all the generators, with any signs, is called a ***representing chain*** for X.

An ***exact sequence*** of abelian groups is a sequence of maps

$$\cdots \to X_{n-1} \to X_n \to X_{n+1} \to \cdots \,,$$

where the image of each map is equal to the kernel of the next. A ***short exact sequence*** is a sequence

$$0 \to X \to Y \to Z \to 0,$$

i.e., the first map is injective, the last is surjective, and $Z \cong Y/\mathrm{Image}(X)$. A short exact sequence of chain complexes is a map of chain complexes, which is a short exact sequence on the n-chains for each n. The useful fact about short exact sequences of chain complexes is that they give rise to long exact homology sequences.

Theorem A.3.2 (the long exact homology sequence) *Let* $0 \to A \to B \to C \to 0$ *be a short exact sequence of chain complexes. Then there is a long exact sequence*

$$\cdots H_{n+1}(B) \to H_{n+1}(C) \to H_n(A) \to H_n(B) \to H_n(C) \to H_{n-1}(A) \to \cdots,$$

where the maps $H_n(A) \to H_n(B) \to H_n(C)$ *are induced by the maps on chains. The maps* $\partial_* : H_*(C) \to H_{*-1}(A)$ *(in speech, the "boundary-star operators") have an explicit natural definition as well.*

The proof is a "diagram chase" and is left as an exercise (Exercise A.7 or see Munkres, 1984, Theorem 23.3).

Remark A.3.3 When X, Y, Z are finite dimensional complex vector spaces of dimensions k, l, m, respectively, then the short exact sequence $0 \to X \to Y \to Z \to 0$ means that $l = k + m$. One may therefore write Y as a direct sum $X \oplus Z$. However, this ***splitting*** is not natural: X embeds naturally in Y, but there is no canonical choice of coset representatives for Y/Z.

Relative Homology and Attachments

Let $Y \subseteq X$ be topological spaces. The inclusion of $Y \hookrightarrow X$ induces an inclusion of chain complexes $C(Y) \hookrightarrow C(X)$. Let $C(X/Y)$ denote the quotient complex whose n-chains are the quotient group $C_n(X)/C_n(Y)$.

Definition A.3.4 (relative homology) *The relative homology of a pair* (X, Y) *is defined to be* $H_*(C(X/Y))$.

The long exact homology sequence for the short exact sequence $0 \to X \to Y \to X/Y \to 0$ is called the long exact sequence for the pair (X, Y). One may think of relative homology roughly as the homology of X if the space Y were to be shrunk to a point – we are looking for cycles that do not bound, but are willing to count a chain as a cycle if its boundary is in Y. In fact, if Y is nicely embedded in X (Y is a deformation retract of an open neighborhood of Y in X), then

$$\pi : (X, Y) \twoheadrightarrow (X/Y, Y/Y) \text{ is a topological quotient } \Rightarrow \pi_* \text{ is an isomorphism} \tag{A.2}$$

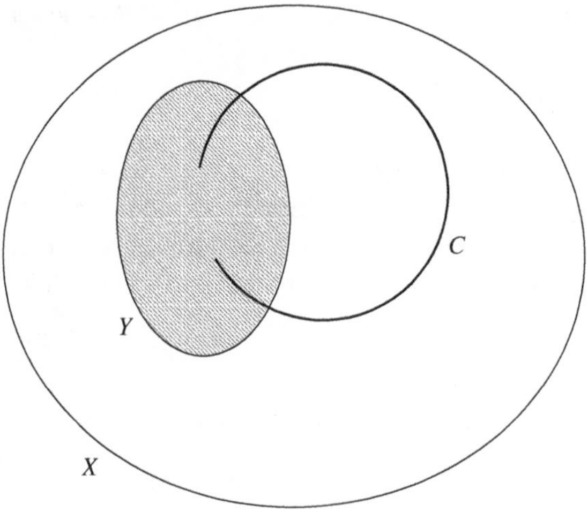

Figure A.2 C is a relative cycle in $C(X, Y)$.

where π_* is the map induced by π on homology (see Munkres, 1984, Exercise 39.3).

Figure A.2 shows a relative cycle in (X, Y). An important feature of relative homology is the ***excision property***:

$$H_n(X, Y) = H_n(X \setminus U, Y \setminus U)$$

if U is in the interior of Y. Informally, the relative homology of (X, Y) "cannot see" the interior of Y.

Definition A.3.5 (attachment) *The attachment of a space Y to a space X along a closed subset $Y_0 \subseteq Y$ by the map $\phi : Y_0 \to X$ is the topological quotient $(X \uplus Y)/\phi$ obtained from the disjoint union of X and Y by identifying each $\mathbf{y} \in Y_0$ with $\phi(\mathbf{y}) \in X$. The triple (Y, Y_0, ϕ) is known as the* ***attachment data***.

Relative homology may be used to compute the homology of an attachment when the homology of the components is known. Let B be the attachment of Y to X by ϕ. In B, the set $X \setminus \phi(Y_0)$ is in the interior of X, so there are isomorphisms

$$H_*(Y, Y_0) \cong H_*(Y/Y_0, Y_0/Y_0) \cong (Y/\phi, Y_0/\phi) \cong H_*(B, X)\,, \tag{A.3}$$

the first two isomorphisms following from (A.2) and the last by excision. There is a lemma, known as the "five lemma," which states that in the following commutative diagram, if the horizontal rows are exact and all but the middle vertical arrow are known to be isomorphisms, then the middle vertical arrow is also an isomorphism.

$$\begin{array}{ccccccccc} A & \longrightarrow & B & \longrightarrow & C & \longrightarrow & D & \longrightarrow & E \\ \downarrow & & \downarrow & & \downarrow & & \downarrow & & \downarrow \\ A' & \longrightarrow & B' & \longrightarrow & C' & \longrightarrow & D' & \longrightarrow & E' \end{array}$$

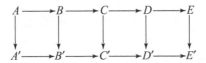

The ∂_* operator in the sequence for the pair (B, X) may be written in terms of the ∂_* operator for the pair (Y, Y_0) as $\phi_* \circ \partial_* \circ \iota_*^{-1}$, where ι_* is the isomorphism obtained in (A.3).

$$\begin{array}{ccc} H_{n+1}(B,X) & \longrightarrow & H_n(X) \\ \iota_* \uparrow \cong & & \uparrow \phi_* \\ H_{n+1}(Y,Y_0) & \xrightarrow{\partial_*} & H_n(Y_0) \end{array}$$

The five lemma then shows that the homology group $H_n(B)$ is determined by the diagram

$$H_{n+1}(B, X) \overset{\phi_*\partial_*\iota_*}{\rightarrow} H_n(X) \rightarrow H_n(B) \rightarrow H_n(B, X) \overset{\phi_*\partial_*\iota_*}{\rightarrow} H_{n-1}(X) .$$

In other words, if we understand the homology groups $H_*(Y, Y_0)$ and $H_*(X)$ and the maps $\phi_* : H_*(Y_0) \rightarrow H_*(X)$, then we may compute the homology of the whole attachment. For this reason, the pair (Y, Y_0) in the attachment data should be thought of as a topological pair.

Products

The product $C = C' \otimes C''$ of chain complexes is defined by letting $C_n = \bigoplus_{k=0}^{n} C'_k \otimes C''_{n-k}$, where a basis for the tensor product $C'_k \otimes C''_{n-k}$ is given by $\bigcup_{k=0}^{n} \sigma \times \tau$, as σ ranges over a basis for C'_k and τ ranges over a basis for C''_{n-k}. The boundary operator is defined by $\partial(\sigma \times \tau) = (\partial\sigma) \times \tau + \sigma \times (\partial\tau)$. In this way, the singular chain complex of a product space is just the product of the chain complexes. The product of the category of pairs of topological spaces is defined by

$$(X', Y') \times (X'', Y'') = \left(X' \times X'', X' \times Y'' \cup Y' \times X''\right) .$$

The singular chain complex C_X/C_Y for a pair (X, Y), which is the product of (X', Y') and (X'', Y''), is then equal to the product of the singular chain complexes for the pairs (X', Y') and (X'', Y'').

The homology of a product is given by the Künneth formula. With coefficients in $\mathbb{C}$, the formula is relatively simple.

Theorem A.3.6 (Künneth product formula) *There is a natural isomorphism*

$$\bigoplus_{p+q=n} H_p(C') \times H_q(C'') \to H_n(C' \otimes C'') .$$

Applying this to the singular chain complexes C' and C'' for two spaces X' and X'' gives an identical-looking formula for $H_n(X' \times X'')$. If C' and C'' are the relative chain complexes for pairs (X', Y') and (X'', Y''), then one obtains

Corollary A.3.7 (Künneth formula for pairs)

$$\begin{aligned} H_n(X' \times X'', X' \times Y'' \cup Y' \times X'') &= H_n((X', Y') \times (X'', Y'')) \\ &\cong \bigoplus_{p+q=n} H_p(X', Y') \times H_q(X'', Y'') . \end{aligned}$$

Cohomology

Given a chain complex C, one may replace each vector space C_n by its dual C^n, the elements of which are called ***cochains***. The maps ∂_n induce maps δ^n in the other direction. Thus we have the ***cochain complex***

$$\cdots \to C^{n-1} \stackrel{\delta^{n-1}}{\to} C^n \stackrel{\delta^n}{\to} C^{n+1} \to \cdots$$

in which $\delta^n \circ \delta^{n-1} = 0$ for all n. The quotient of the kernel of δ^n (the cocycles) by the image of δ^{n-1} (the coboundaries) is called the nth ***cohomology group*** of C and is denoted $H^N(C)$. It is easy to verify that value of a cocycle ν evaluated at a cycle σ depends only on the cohomology class $[\nu]$ of ν and the homology class $[\sigma]$ of σ. This defines a product $\langle \omega, \eta \rangle$ for $\omega \in H^n(C)$ and $\eta \in H_n(C)$. In fact, if X is a cell complex, then this is a pairing and $H^n(X)$ is naturally the dual space of $H_n(X)$.

Any p-form ω may be naturally identified with the cochain of degree p defined by $C \mapsto \int_C \omega$. Using the definition of δ and Stokes' Theorem,

$$\delta\omega(C) := \omega(\partial C) := \int_{\partial C} \omega = \int_C d\omega = d\omega(C) ,$$

in other words, $\delta\omega = d\omega$. The cocycles are thus the closed forms, and the observation at the beginning of this section may be stated as the following theorem.

Theorem A.3.8 (integral depends only on homology class) *Let ω be a closed p-form holomorphic on a domain $\mathcal{M} \subseteq \mathbb{C}^n$. If $p = n$, then ω is always closed. Let C be a singular p-cycle on $\mathcal{M}$. Then $\int_C \omega$ depends on C only via the homology class $[C]$ of C in $H_p(\mathcal{M})$ and on ω only via the cohomology class $[\omega]$ of ω in $H^p(\mathcal{M})$.*

The identification map from p-forms to cochains is in general not a bijection. Nevertheless, this map induces an isomorphism on cohomology with coefficients in $\mathbb{R}$. The complex of p-forms is called the ***de Rham complex*** and its cohomology is known as ***de Rham cohomology***. The theorem asserting its isomorphism to the singular cohomology of a manifold is called the de Rham Theorem. This also works for holomorphic forms on complex manifolds.

Theorem A.3.9 (de Rham Theorem) *Let X be a real manifold. The identification of p-forms with dual chains induces an isomorphism*

$$H^*_{\mathrm{DR}} \cong H^*(X).$$

If, furthermore, X is a complex manifold, then the inclusion of the subcomplex of holomorphic forms into the de Rham complex induces a cohomology isomorphism as well:

$$H^*_{\mathbb{C}-\mathrm{DR}} \cong H^*_{DR}(X) \cong H^*(X).$$

Often we will be integrating a form of a specific type,

$$\omega = \exp(\lambda\, \phi(z))\, \eta\,,$$

where η is a holomorphic k-form and the real part of ϕ is bounded above by c on Y. Let C and C' be chains representing the same relative cycle in $H_k(X, Y)$.

Proposition A.3.10 (asymptotics depend only on relative homology class) *When $C \equiv C'$ in $H_k(X, Y)$ with $\phi \le c$ on Y, then as $\lambda \to \infty$,*

$$\int_C \omega = \int_{C'} \omega + O\left(e^{\lambda c'}\right)$$

for any $c' > c$.

Proof By definition, the difference between C and C' is a relative boundary:

$$C - C' = \partial \mathcal{D} + C''$$

with C'' supported on Y. Thus

$$\int_C \omega - \int_{C'} \omega = \int_{\mathcal{D}} d\omega + \int_{C''} \omega$$
$$\le e^{\lambda c} \int_{C''} |\eta|$$

because $d\omega = 0$. □

A.4 Intersection Classes

Let $\mathcal{V}$ be a smooth complex algebraic hypersurface in $(\mathbb{C}^*)^d$ defined by the vanishing of a square-free polynomial H. Denote $\mathcal{V}_* := \mathcal{V} \cap (\mathbb{C}^*)^d$. Near $\mathcal{V}_*$, there is a natural product structure on $(\mathbb{C}^*)^d$. To see this, first note that H defines a map from $(\mathbb{C}^*)^d$ to $\mathbb{C}$ that vanishes precisely on $\mathcal{V}_*$. To define the product structure, we will define a map $\Phi : U \to \mathcal{V}_*$ on a neighborhood U of $\mathcal{V}_*$ in $(\mathbb{C}^*)^d$ such that Ψ defined by $\boldsymbol{x} \mapsto (\Phi(\boldsymbol{x}), H(\boldsymbol{x}))$ is a homeomorphism from U onto its image in $\mathcal{V}_* \times \mathbb{C}$. Because ∇H is nonvanishing on $\mathcal{V}_*$, the gradient ∇H is nonzero in a neighborhood of $\mathcal{V}$ and hence defines a complex line bundle. The integral surfaces of this bundle have real dimension 2 and define a natural map Φ where $\Phi(\boldsymbol{x})$ is the point of $\mathcal{V}_*$ on whose integral curve the point $\boldsymbol{x}$ lies.

It follows that for any compact $K \subseteq \mathcal{V}_*$, there is an ε such that a neighborhood of $\mathcal{V}_*$ in $(\mathbb{C}^*)^d$ is homeomorphic to $K \times B_\varepsilon$, where B_ε is the centered disk of radius ε. Let U be a sufficiently small neighborhood of $\mathcal{V}_*$ in $(\mathbb{C}^*)^d$. Then $\mathcal{M}' := U \setminus \mathcal{V}$ is a product of $\mathcal{V}_*$ and $\mathbb{C}^*$. The factor $\mathbb{C}^*$ is homotopy equivalent to a circle. By Künneth's product formula,

$$\begin{aligned} H_d(\mathcal{M}') &\cong H_{d-1}(\mathcal{V}_*) \times H_1(\mathbb{C}^*) \oplus H_d(\mathcal{V}_*) \times H_0(\mathbb{C}^*) \\ &\cong H_{d-1}(\mathcal{V}_*) \end{aligned}$$

because $\mathcal{V}_*$ has no homology above dimension $d-1$. The isomorphism between $H_d(\mathcal{M}')$ and $H_{d-1}(\sigma)$ is natural: if σ is any $(d-1)$-chain in $\mathcal{V}_*$ then the homology class $[\sigma]$ on the right-hand side maps to the class of $\sigma \times \partial B_\varepsilon$ on the left-hand side, under the natural product structure. The inclusion $\mathcal{M}' \hookrightarrow \mathcal{M} := (\mathbb{C}^*)^d \setminus \mathcal{V}$ induces an isomorphism on H_d because in the long exact sequence

$$H_{d+1}(\mathcal{M}, M') \to H_d(\mathcal{M}') \to H_d(\mathcal{M}),$$

the fact that the homology dimension of $\mathcal{M}$ and $\mathcal{M}'$ is d causes the initial term to vanish. The inclusion is natural as well, and hence we arrive at this result.

Proposition A.4.1 *$H_{d-1}(\mathcal{V}_*)$ is naturally isomorphic to $H_d(\mathcal{M})$ via a product with a suitably small circle, then mapping by Ψ^{-1}, and then including $\mathcal{M}'$ into $\mathcal{M}$.*

Next, we consider a $(d+1)$-chain $\mathbf{H} \in (\mathbb{C}^*)^d$ that intersects $\mathcal{V}$ transversely. The intersection of $\mathbf{H}$ with $\mathcal{V}$ is a $(d-1)$-chain, σ.

Proposition A.4.2 *The intersection of $\mathbf{H}$ with a small neighborhood U of $\mathcal{V}_*$ in $(\mathbb{C}^*)^d$ is homotopic to $\Psi^{-1}(\sigma \times B_\varepsilon)$ for some $\varepsilon > 0$.*

Proof Cover the range of σ with finitely many chart neighborhoods in which a neighborhood of σ in $\mathcal{V}_*$ is a product of σ with $(2d-2)$-dimensional ball. In each such neighborhood, $\Psi \circ \mathbf{H}$ is a $(d+1)$-chain in the product $\mathcal{V} \times \mathbb{C}$ intersecting $\mathcal{V}$ transversely. Composing with the map that multiplies the non-σ coordinates by a factor of $(1-t)$, we obtain a homotopy to $\sigma \times \mathbb{C}$ and reducing U we obtain a homotopy to $\sigma \times B_\varepsilon$. □

We now apply this in the case where $\mathbf{H}$ is a cobordism, i.e., a $(d+1)$-chain in $(\mathbb{C}^*)^d$ with boundary $\alpha - \beta$ where α and β are two cycles in $\mathcal{M}$. A special case of a cobordism is a homotopy $\mathbf{H} : Z \times [0, 1]$ taking α to β: Z is a simplicial complex, $\mathbf{H}(Z, 0) = \alpha$ and $\mathbf{H}(Z, 1) = \beta$. Note that in general the range of $\mathbf{H}$ is not a subset of $\mathcal{M}$ because it will intersect $\mathcal{V}$.

Let α' and β' be two more d-cycles in $\mathcal{M}$ with $[\alpha'] = [\alpha]$ and $[\beta'] = [\beta]$ in $H_d(\mathcal{M})$, and let $\mathbf{H}'$ be a $(d+1)$-chain in $(\mathbb{C}^*)^d$ with $\partial\,\mathbf{H}' = \alpha' - \beta'$ and having transverse intersection with $\mathcal{V}$. Let $\sigma := \mathbf{H} \cap \mathcal{V}$ and $\sigma' := \mathbf{H}' \cap \mathcal{V}$ be the corresonding intersections.

Theorem A.4.3 *Let α and β be d-cycles in $\mathcal{M}$. Let σ be the intersection of $\mathcal{V}_*$ with a cobordism of α to β that intersects $\mathcal{V}$ transversely. Then the homology class of σ in $H_{d-1}(\mathcal{V}_*)$ is independent of the choice of cycles α and β in the respective homology classes in $H_d(\mathcal{M})$ and independent of the choice of cobordism* $\mathbf{H}$ *between them as long as it intersects $\mathcal{V}$ transversely.*

Proof Let $\alpha', \beta', \mathbf{H}'$ be another choice of representing cycles and connecting homotopy. Let N denote the intersection of $\mathbf{H}$ with a small neighborhood of $\mathcal{V}_*$, let $\Theta := \mathbf{H} - N$ as $(d+1)$-chains, and let $\mathbb{N}'$ and Θ' be defined similarly with $\mathbf{H}'$ in place of $\mathbf{H}$. Then

$$\begin{aligned}\partial(\Theta - \Theta') &= \partial(\mathbf{H}) - \partial(\mathbf{H}') - (\partial N - \partial N') \\ &= (\alpha - \beta') - (\alpha' - \beta') - (\partial N - \partial N').\end{aligned}$$

Projecting to $H_{d+1}(\mathcal{M})$ gives $[\alpha] = [\alpha']$ and $[\beta] = [\beta']$; any boundary becomes zero, hence in $H_{d+1}(\mathcal{M})$,

$$\begin{aligned}0 &= [\partial N] - [\partial N'] \\ &= [\partial \Psi^{-1}(\sigma \times B_\varepsilon)] - [\partial \Psi^{-1}(\sigma' \times B_\varepsilon)] \qquad \text{(by Proposition A.4.2)} \\ &= [\Phi^{-1}(\sigma) \times \partial B_\varepsilon] - [\Phi^{-1}(\sigma') \times \partial B_\varepsilon].\end{aligned}$$

By the isomorphism in Proposition A.4.1, we then see that $[\sigma] = [\sigma']$ in $H_{d-1}(\mathcal{V}_*)$. □

Remarks

1. The class σ depends on the classes $[\alpha]$ and $[\beta]$ only through $[\alpha] - [\beta]$.
2. Existence of $\mathbf{H}$ such that $\partial\,\mathbf{H} = [\alpha] - [\beta]$ is equivalent to the vanishing of $[\alpha] - [\beta]$ in $H_d((\mathbb{C}^*)^d)$. A generator for this cyclic group is any torus $\mathbf{T}(\boldsymbol{x})$. In particular, there is a cobordism between any two d-cycles homologous to a centered torus.

Definition A.4.4 *Given cycles $\alpha, \beta \in \mathcal{M}$ homologous in $(\mathbb{C}^*)^d$ and given $\mathcal{V}$ as above, let*

$$\mathtt{INT}[\alpha, \beta; \mathcal{V}] := [\sigma] \in H_{d-1}(\mathcal{V}_*),$$

where σ is the intersection of $\mathcal{V}$ with a cobordism from α to β. From the preceding discussion we see that $\texttt{INT}[\alpha, \beta; \mathcal{V}]$ *is well defined and depends on α and β only through the class of $\alpha - \beta$ in $H_d(\mathcal{M})$.*

There is a version of this construction in relative homology as well. This will be useful to us because we would like to take β to be a sufficiently larger torus. All of these intersect $\mathcal{V}$, but if they do so in Y, we are still able to define the intersection class.

Theorem A.4.5 *Let Y be a closed subspace of $(\mathbb{C}^*)^d$ and let α and β be relative cycles in in the pair $(\mathcal{M}, \mathcal{M} \cap Y)$ that are homologous in $((\mathbb{C}^*)^d, Y)$. There is a well-defined intersection class* $\texttt{INT}[\alpha, \beta; \mathcal{V}]_Y \in H_{d-1}(\mathcal{V}_*, Y)$ *with the following properties. If $\mathbf{H}$ is any $(d+1)$-chain in $(\mathbb{C}^*)^d$ with $\partial \mathbf{H} = \alpha' - \beta' + \gamma$ with $[\alpha'] = [\alpha]$ and $[\beta'] = [\beta]$ in $H_d(\mathcal{M}, \mathcal{M} \cap Y)$ and $\gamma \in Y$, and if $\mathbf{H}$ intersects $\mathcal{V}$ transversely, then $\mathbf{H} \cap \mathcal{V}$ is a relative cycle in the class* $\texttt{INT}[\alpha, \beta; \mathcal{V}]_Y$.

Proof The proof of the following theorem is exactly the same as the proof of Theorem A.4.3 except that $\partial \Psi^{-1}(\sigma \times B_\varepsilon)$ now has the extra piece $\Psi^{-1}(\partial \sigma \times B_\varepsilon)$. By definition, $\partial \sigma$ is a compact set in the interior of Y, and hence, taking ε small enough, the extra piece is contained in the interior of Y. The rest of the proof is the same. □

By excision, the pair $(\mathcal{M}, \mathcal{M} \cap Y)$ is homotopy equivalent to the pair $(\mathcal{M} \setminus Y^\circ, \partial Y)$. This allows us to extend the previous result to one in which α and β are allowed to intersect $\mathcal{V}$ but only in the interior of Y.

Corollary A.4.6 *Let Y be a closed subspace of $(\mathbb{C}^*)^d$, let $\mathcal{V}_*$ be a smooth complex algebraic hypersurface in $(\mathbb{C}^*)^d$, and let $\mathcal{M} := (\mathbb{C}^*)^d \setminus \mathcal{V}$. Let α and β be relative cycles homologous in $(\mathbb{C}^d, Y)$ intersecting $\mathcal{V}$ only in the interior of Y. There is a well-defined homology class* $\texttt{INT}[\alpha, \beta; \mathcal{V}]_Y \in H_{d-1}(\mathcal{V}_*, \mathcal{V}_* \cap Y)$, *depending only on the class of $\alpha - \beta$ in $H_d(\mathcal{M}, \mathcal{M} \cap Y)$, with the following property. Let $\mathbf{H}$ be a $(d+1)$-chain in $(\mathbb{C}^*)^d$ such that $\partial \mathbf{H} = \alpha - \beta + \gamma$ with γ supported on the interior of Y. If the intersection of $\mathbf{H}$ with $\mathcal{V}$ is transverse away from the interior of Y, then $\mathbf{H} \cap \mathcal{V}$ is a relative cycle representing the class* $\texttt{INT}[\alpha, \beta; \mathcal{V}]_Y$. *In the special case where $\beta = 0$ and Y is the set of points at height c or less, we denote the relative intersection class by*

$$\texttt{INT}[\alpha; \mathcal{V}]_{\leq c}.$$

A.5 Residue Forms and the Residue Integral Theorem

Proposition A.5.1 *Let ξ be a meromorphic form, written as $(G/H)\,d\mathbf{z}$ on a domain $\mathcal{D}$. Let $\mathcal{V}_H$ be the zero set of H and suppose that H has a simple zero everywhere on $D := \mathcal{D} \cap \mathcal{V}_H$. Then the equation*

$$dH \wedge \theta = G\, d\mathbf{z} \tag{A.1}$$

always has a holomorphic solution, and the following uniqueness holds: for any representation of ξ *as* $(G/H)\,dz$ *and any holomorphic solution* θ *to* (A.1), *the restriction* $\eta := \iota^*\theta$ *induced by the inclusion* $D \overset{\iota}{\to} V_H$ *is always the same. We define the residue of* ξ *on* D *to be the form* η, *denoted*

$$\mathrm{Res}(\xi; D) := \eta\,.$$

Proof Uniqueness follows from Exercise A.4: if θ_1 and θ_2 are two solutions, then $dH \wedge (\theta_1 - \theta_2) = 0$ and hence $\iota^*\theta_1 = \iota^*\theta_2$.

To prove existence, suppose first that $H(z) = z_1$. Then a solution to (A.1) is simply $\theta = G(z)\,dz_2 \wedge \cdots \wedge dz_d$. In the general case, proceeding as in Exercise A.4, use the complex implicit function theorem (Hörmander, 1990, Theorem 2.1.2) to find a bi-holomorphic map ψ from a neighborhood of $\boldsymbol{p}$ to $\mathbb{C}^d$ with first coordinate H. Use the special case to solve $dz_1 \wedge \theta_0 = G \circ \psi^{-1}\,dz$; then $\theta := \psi^*(\theta_0)$ solves (A.1). □

Theorem A.5.2 (residue integral theorem) *Let* ω *be holomorphic on* $\mathcal{M}$ *with a simple pole on* $\mathcal{V}$. *Let* α *and* β *be any* d*-cycles in* $\mathcal{M}$ *whose projections to* $H_d((\mathbb{C}^*)^d)$ *are equal. Then*

$$\int_\alpha \omega - \int_\beta \omega = 2\pi i \int_{\mathtt{INT}[\alpha,\beta;\mathcal{V}]} \mathrm{Res}(\omega)\,.$$

Proof As remarked after the proof of Theorem A.4.3 in Appendix A, vanishing of $[\alpha - \beta]$ in $H_d((\mathbb{C}^*)^d)$ by definition implies the existence of a $(d+1)$-chain $\mathbf{H}$ on $(\mathbb{C}^*)^d$ with boundary $\alpha - \beta$. Perturbing generically if necessary, we can assume without loss of generality that $\mathbf{H}$ intersects $\mathcal{V}$ transversely. Letting N denote the intersection of $\mathbf{H}$ with a small neighborhood of $\mathcal{V}$ and $\Theta := \mathbf{H} - N$, the vanishing of holomorphic integrals over d-boundaries (Theorem A.2.3) implies that the integral of the holomorphic d form ω over $\partial\Theta$ vanishes. In other words,

$$\int_\alpha \omega - \int_\beta \omega = \int_{\partial N} \omega\,.$$

By Proposition A.4.2, N is homotopic to a product neighborhood $\sigma \times B_\varepsilon$, where $\sigma := \mathbf{H} \cap \mathcal{V}$. Thus ∂N is homotopic to $\partial(\sigma \times B_\varepsilon)$, which is equal to $\sigma \times \partial B_\varepsilon$ (because σ is a cycle), whence

$$\int_{\sigma \times \partial B_\varepsilon} \omega = \int_\sigma \left(\int_{\partial B_\varepsilon} \omega \right)\,.$$

Using functoriality of the residue, we may change coordinates so that $\mathcal{V}$ is the complex hyperplane where $z_1 = 0$. Thus we need only check it in this case, namely where $H(z) = z_1$. Writing $\omega = (G/z_1)dz_1 \wedge (dz_2 \wedge \cdots \wedge dz_d)$, the iterated integral is

$$\int_\sigma \left[\int_{\partial B_\varepsilon} \frac{G(z)}{z_1}\,dz_1 \right] dz_2 \wedge \cdots \wedge dz_d\,.$$

By standard univariate complex analysis, the inner integral at a point $(z_2, \ldots, z_d)$ is the residue of the meromorphic function $G(\cdot, z_2, \ldots, z_d)/z$ at the pole $(0, z_2, \ldots, z_d)$. This is equal to

$$(2\pi i) \int_\sigma G(0, z_2, \ldots, z_d)\,. \tag{A.2}$$

In this special case, we see that (A.2) is precisely $\int_\sigma \mathrm{Res}(\omega)$. Functoriality then implies the result in general. □

There is a relative version of this result.

Theorem A.5.3 (relative residue integral theorem) *Let Y be any closed subspace of $(\mathbb{C}^*)^d$ such that $H_d((\mathbb{C}^*)^d, Y)$ vanishes. Let ω be holomorphic on $\mathcal{M}$ with a simple pole on $\mathcal{V}$. Let α be a d-cycle in $\mathcal{M}$. Then*

$$\int_\alpha \omega = 2\pi i \int_{\mathtt{INT}[\alpha;\mathcal{V}]} \mathrm{Res}(\omega) + \int_{C'} \omega \tag{A.3}$$

for some chain C' supported on the interior of Y. In particular, if $\omega = \mathbf{z}^{-\mathbf{r}}\eta$ for some holomorphic form η and if Y is the set where the real part of $h_{\hat{\mathbf{r}}} := -\hat{\mathbf{r}} \cdot \log \mathbf{z}$ is at most c, then

$$\int_\alpha \omega = 2\pi i \int_{\mathtt{INT}[\alpha;\mathcal{V}]} \mathrm{Res}(\omega) + O\left(e^{\lambda c'}\right) \tag{A.4}$$

for any $c' > c$ as $\lambda \to \infty$.

Proof By the vanishing of $H_d((\mathbb{C}^*)^d, Y)$ there is a $(d+1)$-chain $\mathbf{H}$ with $\partial\,\mathbf{H} = \alpha + \gamma$ and γ supported on the interior of Y. Let N denote the intersection of $\mathbf{H}$ with a neighborhood of $\mathcal{V}$. As before,

$$\int_\alpha \omega = \int_\gamma \omega + \int_{\partial N} \omega\,.$$

Letting $\sigma := \mathbf{H} \cap \mathcal{V}$, we recall that ∂N is homotopic to $\sigma \times B_\varepsilon$ plus a piece γ' in the interior of Y. Taking $C' := \gamma + \gamma'$, the rest of the proof of (A.3) is the same as that of Theorem A.5.2. The asymptotic estimate follows because $|\int_{C'} \omega| \le e^{\lambda c} \int_{C'} |\eta|$, as in the proof of Proposition A.3.10. □

Notes

As mentioned at the start of Section A.1, the more modern and general treatment of differentiable manifolds is to define the underlying space to be an arbitrary abstract set, together with a set of parametrizations of subsets by open balls in $\mathbb{R}^d$, such that compositions $\phi^{-1} \circ \psi$ of parametrizations ϕ and ψ are smooth maps on their domain in $\mathbb{R}^d$. This is undoubtedly more natural, because the embedding of the manifold in $\mathbb{R}^n$ plays no role in its properties. The use of this review in briefly informing readers who do not already know the material dictates, however, that a shorter path be taken. The embedding in $\mathbb{R}^n$ allows tangent vectors to be defined geometrically rather than as derivations. This seems to me the only way that they can be digested on the first pass. For a comparison of the geometric and abstract definitions, see Lee (2003, Chapter 3).

The material in the first two sections of this chapter is standard graduate level calculus. The organization owes a debt to the text Warner (1983).

Exercises

A.1 (iterated integrals)
Prove Proposition A.1.2.

A.2 ($\partial^2 = 0$)
Verify that $\partial^2 = 0$ in equation (A.1) by proving that $\kappa_j^{p+1} \circ \kappa_j^p = \kappa_{j+1}^{p+1} \circ \kappa_i^p$.

A.3 (kernel of the inclusion map)
(a) Let $f : X \to \mathbb{R}$ be a smooth map on a d-manifold X for which df is everywhere nonvanishing. Let $\mathcal{M}$ be the zero set of f and let $\iota : \mathcal{M} \to X$ denote the inclusion map. Prove that for any $(d-1)$-form η, $\iota^*(\eta) = 0$ if and only if $\eta \wedge df$ vanishes on $\mathcal{M}$. Hint: Use the implicit function theorem to coordinatize X with first coordinate f and use functoriality of $\wedge$ to reduce to the case $f = x_1$. (b) Repeat this for $k \leq d$ functions $f_1, \ldots, f_k$, whose transverse intersection defines a smooth surface $\mathcal{M}$ of codimension k.

A.4 (complexification)
Do Exercise A.3 replacing $\mathbb{R}$ by $\mathbb{C}$; i.e., X is a complex d-manifold and $f : X \to \mathbb{C}$ is analytic. Hint: You can copy the proof, only you need the complex form of the implicit function theorem to be sure your coordinates are holomorphic; see Hörmander (1990, Theorem 2.1.2).

A.5 (lower dimensional integrals vanish)
Let C be a d-chain supported on a submanifold of $\mathbb{C}^d$ of dimension less than d. Show that $\int_C \omega$ vanishes for any holomorphic d-chain ω.

A.6 (an example on the sphere)
Define a 2-form ω in $\mathbb{R}^3$ by $\omega = x\,dy \wedge dz + y\,dz \wedge dx + z\,dx \wedge dy$. Define a 2-chain C i.e., "the unit sphere" and compute $\int_C \omega$ directly from the definitions. Now figure out a shortcut to the same computation using Stokes' Theorem.

A.7 (long exact sequence)
Define the ∂_* operator in Theorem A.3.2 and give a proof of the theorem.

A.8 (univariate residues)
The familiar residue theorem in one variable states that

$$\int_\gamma f = 2\pi i \sum \text{Res}(f; a),$$

where the sum is over poles of the meromorphic f inside the closed contour γ. Derive this from Theorem A.5.2. What are α, β, ω $\mathcal{V}$ and $\texttt{INT}[\alpha, \beta; \mathcal{V}]$?

Appendix B
Morse Theory

Appendix A developed the mathematics (calculus on manifolds) needed to understand the statement and proof of the identity

$$a_r = \left(\frac{1}{2\pi i}\right)^r \int_C \omega, \tag{B.1}$$

where

$$\omega := z^{-r-1} F(z)\, dz$$

and C is any cycle homologous to T in the domain

$$\mathcal{M} := \{z \in \mathbb{C}^d : H(z) \prod_{j=1}^{d} z_j \neq 0\}$$

of holomorphy of the integrand. Proceeding from this point requires Morse theory. The present appendix presents classical Morse theory, along the lines of Milnor (1963). Although we use this only for the special height function $h_{\hat{r}}$, the properties of this function do not enter until the end, for which reason we present the theory for general height functions. To complete the topological analysis, what is actually required is stratified Morse theory. The classical Morse theory in this appendix is a logical prerequisite for the stratified version, which is presented in Appendix C.

B.1 Classical Morse Theory

Morse theory attempts to completely describe the topology of a space X by means of the geometry of X near critical points of a smooth function $h : X \to \mathbb{R}$. Our destination is Theorem B.2.3. This tells us that we may find a basis for each $H_k(X)$ consisting of quasi-local cycles at critical points: for each critical point, $\boldsymbol{p}$, there will be a cycle with height bounded by $h(\boldsymbol{p}) - \varepsilon$ except in an arbitrarily small neighborhood of $\boldsymbol{p}$. This will be accomplished by studying $X^a := \{x \in X : h(x) \leq a\}$ as a increases and showing that the homotopy type of X does not change (the Morse Lemma B.1.2) except at critical points, where a cell is attached (Theorem B.1.3). Along the way, a description of X as a cell complex is given (Theorem B.1.7). A description of the attachments in terms of relative homology is also given, which will be an important simplification in the last section of this appendix.

In the present section, we study classical, or smooth, Morse theory, which pertains to the case where X is a manifold. More general spaces and their complements are handled in subsequent sections.

Homotopy Equivalence Except at Critical Points

Let X be a manifold and $h : X \to \mathbb{R}$ a smooth function, which we think of as height. The ***critical points*** of h are the points $\boldsymbol{p} \in X$ for which $dh = 0$ on the tangent space $T_{\boldsymbol{p}}(X)$. The values $h(\boldsymbol{p})$ at critical points $\boldsymbol{p}$ are called ***critical values*** of h. A critical point $\boldsymbol{p}$ is ***nondegenerate*** for h if the quadratic form given by the quadratic terms in the Taylor approximation for h at $\boldsymbol{p}$ has no null eigenvalues. In coordinates, this means that the determinant of the Hessian matrix $\left[\dfrac{\partial^2 h}{\partial x_i \partial x_j}\right]$ is nonzero when X is locally coordinatized by $x_1, \ldots, x_d$. Although it is traditional to require Morse functions to have distinct critical values, we use a definition that does not require this:

Definition B.1.1 (Morse function) *A proper smooth function $h : X \to \mathbb{R}$ is said to be a **Morse function** if the critical points of h are isolated and nondegenerate. If the critical values are distinct, then h is a **Morse function with distinct critical values**.*

Let X be a manifold with Morse function h with distinct critical values. If a is a real number, let X^a denote the topological subspace $\{\boldsymbol{x} \in X : h(\boldsymbol{x}) \leq a\}$. The fundamental Morse Lemma states that the topology of X^a changes only when a is a critical value of h.

Lemma B.1.2 (Morse Lemma) *Let $a < b$ be real numbers and suppose the interval $[a, b]$ contains no critical values of h. Suppose also that $h^{-1}[a, b]$ is compact. Then the inclusion $X^a \hookrightarrow X^b$ is a homotopy equivalence.*

This is proved in Milnor (1963, Theorem 3.1) by constructing a homotopy on X^b that follows the downward gradient of h.

Attachment at Critical Points

Suppose there is precisely one critical point $\boldsymbol{p}$ with $h(\boldsymbol{p}) \in [a, b]$. The Hessian of h at $\boldsymbol{p}$ is a real symmetric matrix and therefore has real eigenvalues. Define the ***index of** h **at** $\boldsymbol{p}$* to be the number of negative eigenvalues of the Hessian. The index ranges from 0 at a local minimum to the dimension d of X at a local maximum. The following theorem (Milnor, 1963, Theorem 3.2) describes X^b as an attachment of X^a.

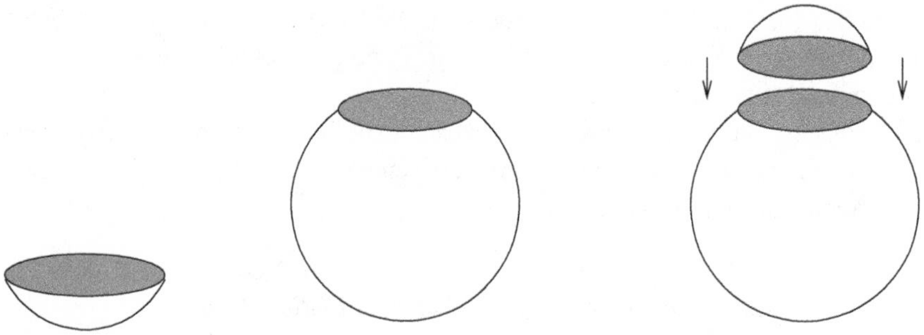

Figure B.1 A sphere, viewed at $a > -1$, $a < 1$, and $a > 1$.

Theorem B.1.3 *Suppose that $h^{-1}[a, b]$ is compact and contains precisely one critical point $\boldsymbol{p}$, with critical value $h(\boldsymbol{p})$ strictly between a and b. Then the space X^b has the homotopy type of X^a with a λ-cell attached along its boundary, where λ is the index of the critical point $\boldsymbol{p}$; when $\lambda = 0$, the boundary is taken to be empty.*

Example B.1.4 Suppose X is the unit sphere in $\mathbb{R}^3$ with height function $h(x, y, z) = z$. (When working in $\mathbb{R}^3$, we often try to arrange for $h = z$ so that "height" is actually height.) There are only two critical points of the height function, namely its minimum and its maximum.

Let us follow X^a as a increases from $-\infty$ to $+\infty$ (see Figure B.1.) For $a < -1$, X^a is empty. As a passes -1, the theorem tells us a zero-cell is added with no identification. Thus X^a is, homotopically, a point. Geometrically, X^a is a small dish, which is contractible to a point. The only other attachment occurs at the end. For $a < 1 \leq b$, the set $X^b \setminus X^a$ is a polar cap. Thus, geometrically as well as homotopically, a 2-cell is attached along its bounding circle. Because the space it is attached to is contractible, all attaching maps are equivalent to mapping the entire boundary to a point, arriving thereby at a sphere. ∎

Example B.1.5 Let X be the torus in $\mathbb{R}^3$ obtained by rotating the circle $(x - 2)^2 + (y - 2)^2 = 1$ about the y-axis. Let $h(x, y, z) = z$. The function h has four critical points, all on the z-axis: a maximum (index 2) at $(0, 0, 3)$, a minimum (index 0) at $(0, 0, -3)$, and saddle points (index 1) at $(0, 0, 1)$ and $(0, 0, -1)$. See Figure B.2.

For $-3 \leq a < -1$, we see as in the previous example that X^a is homotopic to a point (geometrically, a dish). As a passes -1, the theorem tells us to add a 1-cell along its boundary. The only way of attaching a 1-cell to a point is to map both endpoints to the point, leaving a circle. Geometrically, if $a < -1 < b$, one may notice that $X^b \setminus X^a$ is a patch in the shape of a 2-disk with two disjoint segments

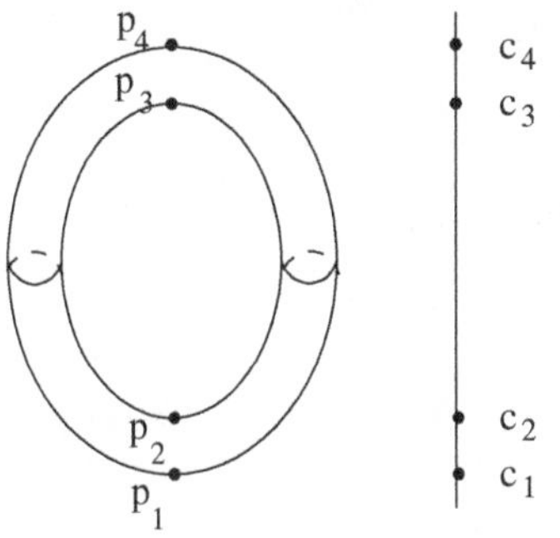

Figure B.2 The critical points on a torus for the standard height function.

of its boundary attached to two disjoint segments of the geometric boundary of X^a.

In general, the space $X^b \setminus X^a$ will be a d-disk, which decomposes into the product of a λ-disk and a $(d-\lambda)$-disk; the attachment is along the boundary of the λ-disk, and the $(d-\lambda)$-disk may be ignored because it is contractible. In the homotopy equivalence category:

$$(X^b, X^a) = (D^\lambda, \partial D^\lambda) \times (D^{d-\lambda}, \emptyset) = (D^\lambda, \partial D^\lambda)$$

which is how the $(D^\lambda, \partial D^\lambda)$ in the theorem arises (see Figure B.3); this is further explained in Milnor (1963).

Returning to the torus, the critical point at height 1 adds another 1-cell modulo its boundary, bringing the homotopy type of X^a up to the wedg of two circles when $1 \leq a < 3$. Finally, for $a < 3 \leq b$, a 2-cell is added modulo its boundary. There is more than one choice for the homotopy type of the attachment, and one must look to see that the attaching map takes the bounding circle to a path homotopic to $\alpha\beta\alpha^{-1}\beta^{-1}$, where α and β are the two 1-cycles that were added in the attachments at heights -1 and 1. At the level of homology, this is zero. ■

Although the theorem specifies the attachment pair, one sees from this example that the computation of the attaching map is not automatic. It will help to have some results that narrow down this computation to certain constructions local to the critical point. The homotopy in the Morse Lemma may be improved so that outside of a neighborhood of $\boldsymbol{p}$, every point is pushed down to a level $c - \varepsilon$. This gives rise to the following definitions.

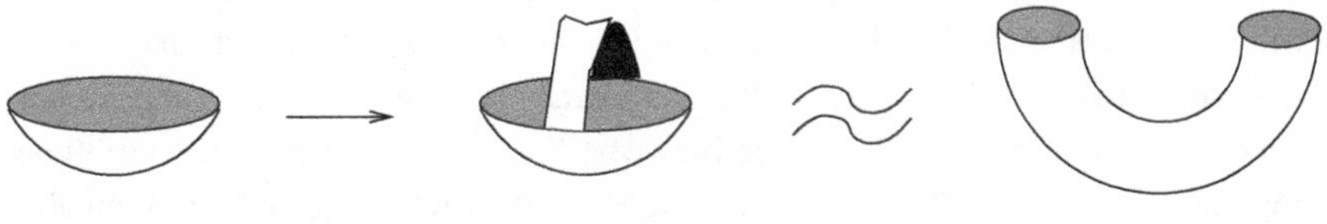

Figure B.3 Crossing the critical value c_2.

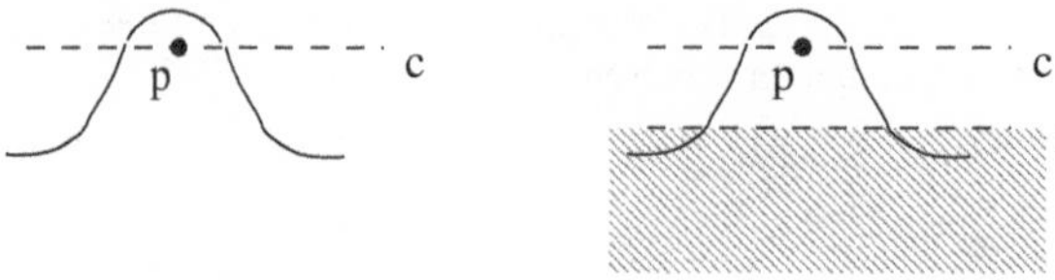

Figure B.4 The space $X^{c,p}$ (dotted line is $h = c$).

Definitions B.1.6 (local pair at a critical point) *Let $\boldsymbol{p}$ be a critical point and let $a < c := h(\boldsymbol{p}) < b$ with no other critical point having height in $[a, b]$. Given $\varepsilon > 0$, let $N_\varepsilon(\boldsymbol{p})$ denote the ε-neighborhood of $\boldsymbol{p}$. By the improved Morse Lemma, the homotopy type of X^b is the same as the homotopy type of $X^{c-\varepsilon} \cup N_\varepsilon(\boldsymbol{p})$ for sufficiently small ε. We denote this space (for any sufficiently small ε) by $X^{c,\boldsymbol{p}}$ (see Figure B.4). Let $X^{\boldsymbol{p},\text{loc}}$ denote the pair $(X^{c,\boldsymbol{p}}, X^{c-\varepsilon})$ for ε sufficiently small.*

The foregoing analysis implies that the attachment pair (X^b, X^a) is homotopy equivalent to $X^{\boldsymbol{p},\text{loc}}$. Suppose now that h is a Morse function whose critical values need not be distinct. If $[a, b]$ contains the unique critical value $c \in (a, b)$, then the homotopy pushes points down to $X^{c-\varepsilon}$ except in a neighborhood of the set of critical points whose value is $\boldsymbol{p}$; because this set is discrete, we see that for sufficiently small ε,

$$(X^b, X^a) \simeq \bigoplus_{\boldsymbol{p}:h(\boldsymbol{p})=c} X^{\boldsymbol{p},\text{loc}} \tag{B.1}$$

that is (X^b, X^a) is homotopy equivalent to the disjoint union of the local pairs at all critical points $\boldsymbol{p}$ with value c. By the excision property, a further description holds as well:

$$(X^{\boldsymbol{p},\text{loc}}) \simeq (X, X \setminus N_\varepsilon(\boldsymbol{p}))$$

for sufficiently small ε.

The last step is to put all this information together to produce the topology of X. At the level of homotopy type, the result is that X has the topology of cell complex, about which certain information is known.

Theorem B.1.7 (Milnor, 1963, Theorem 3.5) *Let X be a manifold and $h : X \to \mathbb{R}$ be a differentiable function with no degenerate critical points. Suppose each X^a is compact. Then X^a has the homotopy type of a cell complex with one cell of dimension λ for each critical point of index λ in X^a.*

Sketch of Proof The proof for the case of finitely many critical points with distinct critical values involves showing inductively that for critical values, c, the homotopy equivalence between $X^{c-\varepsilon}$ and a cell complex may be extended via the attachment of a cell to a homotopy equivalence between $X^{c+\varepsilon}$ and a cell complex with one more cell. The restrictions on critical values are then removed by homotopically perturbing h so as to satisfy the conditions; a limiting argument then removes the finiteness condition. □

Example (sphere and torus continued) The 2-sphere from example B.1.4 is a cell complex with one 2-cell and one 0-cell. There is only one choice for the attachment map. The 2-torus from Example B.1.5 is a cell complex with one 0-cell, two 1-cells, and one 2-cell. There is only one choice for the maps attaching the 1-cells. For the 2-cell, one must look to determine the attaching map.

B.2 Description at the Level of Homology

It is useful to follow the induction through on the level of homology. Let us suppose that we understand the attachment at the level of homology. To be specific, let (B, A) be a pair of the homotopy type of the attachment $(X^{c+\varepsilon}, X^{c-\varepsilon})$. Suppose we understand the ∂^* maps from $H_n(B, A)$ to $H_{n-1}(A)$ in the long exact sequence for the inclusion $A \hookrightarrow B$, in the sense of knowing bases for $H_n(B, A)$ and $H_{n-1}(A)$ and a matrix for ∂^* in these bases. The long exact sequence has a portion

$$H_{n+1}(B, A) \overset{\partial_*}{\to} H_n(A) \overset{\iota_*}{\to} H_n(B) \overset{\pi_*}{\to} H_n(B, A) \overset{\partial_*}{\to} H_{n-1}(A),$$

which may be written as

$$0 \to \frac{H_n(A)}{\mathrm{Image}(\partial_*)} \to H_n(B) \to \ker(\partial_*) \to 0 \,.$$

Thus $H_n(B)$ decomposes as a direct sum (see Remark A.3.3) of the kernel of ∂_*^{n-1} and the ***cokernel*** of ∂_*^n.

This decomposition allows us to construct a basis for the homology of B: for each n, starting with a basis for $H_n(A)$, delete basis elements differing by elements of the image of ∂_* and then add new basis elements indexed by a basis for the kernel (null space) of ∂_*. These new basis elements have an explicit geometric description. The group $H_n(B, A)$ consists of equivalence classes of chains in B whose boundaries lie in A. Let C be such a chain. The image $\partial_*([C])$ is the class of ∂C in $H_{n-1}(A)$, which vanishes if it bounds some n-chain D in A. The pullback of the class $[C]$ by π_*, is the class of the chain $C - D$, which is a cycle because $\partial C = \partial D$ by construction of D. Heuristically,

$$\pi_*^{-1}([C]) = C - \partial_A^{-1}(\partial C). \tag{B.1}$$

One may think of this as the relative cycle C in $Z_n(B, A)$, completed to an actual cycle in a way that stays within A.

Remark The choice of D in this construction is not natural (see Remark A.3.3). A particular composition of a space B as a subspace A to which is attached a space $C := \overline{B \setminus A}$ comes with an explicit inclusion map from ∂C to A, and this induces the ∂_* operator. There may be, however, more than one way to reassemble A, C and ∂_* into B, giving homotopy equivalent spaces but with different homology bases. This is illustrated in the continuation of Example B.1.5.

One further remark on notation: when attaching a space Y along Y_0, the pair (Y, Y_0) is commonly referred to as the attachment data, or in the case of Morse theory, the ***Morse data*** for the attachment. These data should really include the homotopy type of the map ϕ, or else the homotopy type of X together with the attachment data do not determine the homotopy type of the new space. On the level of homology, as we have seen, what we need to know is the relative homology of the pair (Y, Y_0), which is (by excision) the homology of the new space relative to the old space, ***together with the*** ∂_* ***map***.

Filtered Spaces

Suppose that $X_0 \subseteq X_1 \subseteq \cdots \subseteq X_n$ are topological spaces. We call X_n a ***filtered space***. The Morse-theoretic use of filtered spaces is when $X_j = X^{\leq a_j}$. The following lemma, valid for any filtered space, is called the "Pushing Down Lemma" because in the Morse-theoretic situation, we think of the cycle C as being successively pushed toward lower heights.

Lemma B.2.1 (Pushing Down Lemma) *Let $X_0 \subseteq \cdots \subseteq X_n$ be a filtered space and let C be a nonzero homology class in $H_k(X_n, X_0)$ for some k. Then there is a unique positive $j \leq n$ such that for some $C_* \in H_k(X_j, X_{j-1})$,*

$$\iota(C_*) = \pi(C) \neq 0 \tag{B.2}$$

in $H_k(X_n, X_{j-1})$, where ι is the map induced by the inclusion of pairs $(X_j, X_{j-1}) \to (X_n, X_{j-1})$ and π is map induced by the projection of pairs $(X_n, X_0) \to (X_n, X_{j-1})$. If ι is an injection then C_* is unique as well.

Proof For uniqueness, suppose that (B.2) is satisfied for some j and C_*. Let $r > j$ be an integer less than or equal to n. The composition of the two maps

$$(X_j, X_{j-1}) \to (X_n, X_{j-1}) \to (X_n, X_{r-1})$$

induces the zero mapping on homology because any class in the image of the first map has a cycle representative in X_j. Letting π_{r-1} denote projection of (X_n, X_0) to (X_n, X_{r-1}), we have $\pi'(C) = \pi'(\pi(C)) = \pi'(\iota(C_*)) = 0$ and therefore (B.2) cannot hold for $r > j$.

For existence we argue by induction on n. The case $n = 1$ is trivial because then $j = 1$ and $C_* = C$. Assume the result for $n - 1$, and let C be a nonzero class in $H_k(X_n, X_0)$. If the image of C under the projection of (X_n, X_0) to (X_n, X_{n-1}) is nonzero, then we may take C_* to be this image and j to be n. Assume therefore that C_* projects to zero. The short exact sequence of chain complexes for the pairs

$$0 \to (X_{n-1}, X_0) \to (X_n, X_0) \to (X_n, X_{n-1}) \to 0$$

induces the exact mappings

$$H_k(X_{n-1}, X_0) \to H_k(X_n, X_o) \to H_k(X_n, X_{n-1} .$$

By assumption C is in the kernel of the second map, hence is the image under the first map of some nonzero class C'. Applying the inductive hypothesis to C' yields a $j \le n-1$ and a cycle $C_* \in H_k(X_j, X_{j-1})$ satisfying (B.2) with C' in place of C. The commuting diagram

$$\begin{array}{ccccc} & & \begin{array}{c} C' \in \\ (X_{n-1}, X_0) \end{array} & \xrightarrow{\iota_3} & \begin{array}{c} C \in \\ (X_n, X_0) \end{array} \\ & & \pi_1 \big\downarrow & & \pi \big\downarrow \\ \begin{array}{c} C_* \in \\ (X_j, X_{j-1}) \end{array} & \xrightarrow{\iota_1} & (X_{n-1}, X_{j-1}) & \xrightarrow{\iota_2} & (X_n, X_{j-1}) \end{array}$$

allows us to conclude that

$$\pi(C) = \pi(\iota_3(C')) = \iota_2(\pi_1(C')) = \iota_2(\iota_1(C_*)) = \iota(C_*)$$

satisfying (B.2). □

Building up by Succssive Attachments

Suppose that for each $X_{k-1} \subseteq X_k$ we understand the topology of the pair (X_k, X_{k-1}) in the sense that we understand the ∂_* maps for a pair (B, A) homotopy equivalent to (X_k, X_{k-1}). If we also understand the homology groups of X_0, then, inductively on k, we understand the homology groups of all X_k. Apply this to a manifold X filtered by $X^{a_1} \subseteq X^{a_2} \subseteq \cdots$ where there is only one critical point with value between a_{i-1} and a_i for each i and no critical value less than a_1 or equal to any a_i. Theorem B.1.3 allows us to compute the homology group of X^c by induction on the values of $a_i < c$ because we have understand the topology of each attachment.

Let d be the greatest index of any critical point in X^c. If (B, A) is a λ-cell and its boundary, then the rank $H_k(B, A)$ is $\delta_{\lambda,k}$. Therefore, a homology basis for $H_k(X^{c+\varepsilon})$ is composed of a generator for each homology generator of $H_k(X^{c-\varepsilon})$ along with one new generator if the index of the critical point with critical value c is k and the image of the attaching map bounds a k-cell in $X^{c-\varepsilon}$ (equivalently, if the boundary $(k-1)$-sphere is in the kernel of ∂_*); one then must delete a generator if the index of the attachment was $k+1$ and the image of the boundary is nonzero in $H_k(X^{c-\varepsilon})$. If $k = d$ then things are slightly simpler: there is no deletion, because there are no $d+1$-cells to fill in holes of dimension d. In this case the inclusion ι in the Pushing Down Lemma will be injective and C_* will be unique.

Definition B.2.2 (Morse filtration) *If h has distinct critical values, then any filtration as above is called a* **Morse filtration**. *Up to homotopy equivalence, a Morse filtration is one whose pairs (X^{j+1}, X^j) are homotopy equivalent to $X^{\mathbf{p}_j,\text{loc}}$, where $\mathbf{p}_j$ are the critical points listed in order of increasing height. Extend this to*

Morse functions with nondistinct critical values by defining a Morse filtration to be any filtration whose successive pairs are homotopy equivalent to $\bigoplus_{h(\mathbf{p})=c_j} X^{\mathbf{p},\text{loc}}$ *as* c_j *increases through all critical values.*

Putting this all together:

Theorem B.2.3 *If* X *has Morse function* h*, then there is a basis for each* $H_k(X^c)$ *consisting of a single generator for some, but not all points of index* k*. The generator associated with the critical point* $\mathbf{p}$ *with critical value* c *is a cycle in* $X^{c,\mathbf{p}}$*, which projects to the generator for the relative homology group* $X^{\mathbf{p},\text{loc}}$*.*

A running list of these generators may be kept, changing as c *increases past each critical point by the addition of a generator if the index is* k *and* $\partial_* = 0$ *there, and by deleting a generator when the index is* $k+1$ *and* ∂_* *is nonzero there.*

Example (torus continued) In Example B.1.5, there were four critical points: one of index 0, two of index 1, and one of index 2. All ∂_* maps vanished. Therefore, the dimensions of the homology groups are 1, 2 and 1 for $H_0(X)$, $H_1(X)$ and $H_2(X)$, respectively.

As an example of the non-naturality of the homology basis in (B.1), consider the second 1-cell to be added. Let α be the homology class in $H_1(X^0)$ of the first 1-cell. Then the second 1-cell, which is a well-defined homology class β in $H_1(X^2, X^0)$, may be completed to a class in $H_1(X^2)$ in many different ways, resulting in cycles differing by multiples of α. Geometrically, one may for example complete β to the circle $x^2 + z^2 = 1$, or one may instead wrap around the torus any integer number of times.

B.3 Morse Theory for Complex Manifolds

Suppose X is a complex d-manifold. It is well known that the distance to a fixed point $\boldsymbol{p}$ is, for a generic choice of $\boldsymbol{p}$, a Morse function on X. The complex structure implies that the index of this fixed point is at most d (see Goresky and MacPherson, 1988, Section 0.1.5). Thus:

Proposition B.3.1 *A complex* d*-manifold, even though its real dimension is* $2d$*, is homotopy equivalent to a cell complex built of cells of real dimension at most* d*.*

This is proved later as a corollary of a result for complex stratified spaces, which are more general than complex manifolds. A consequence of all this is that for complex d-manifolds, the middle-dimensional homology H_d is the top-dimensional homology of the cell complex, so there are no deletions of generators. The Morse filtration gives a homology basis for $H_d(X)$. Each is local to some critical point $\boldsymbol{p}$ and consists of a λ-cell, $\lambda \le d$, "draped over $\boldsymbol{p}$" so that its boundary lies entirely in $X^{h(p)-\varepsilon}$. One may either work with this as a relative cycle or extend it to an honest cycle via (B.1).

A further special case arises when the height function h is harmonic (the real part of a complex analytic function). Then for each eigenvector $\boldsymbol{v} = (x_1, y_1, \ldots, x_d, y_d)$ of the Hessian, the vector $i\boldsymbol{v} = (-y_1, x_1, \ldots, -y_d, x_d)$ is an eigenvector whose eigenvalue is the negative of the eigenvalue for $\boldsymbol{v}$, and hence the index of h is precisely d at all critical points. Because all attachments have index d, X is homotopy equivalent to a complex built entirely from d-cells. This means that $H_k(X) = 0$ for all $0 < k < d$, which implies that the ∂_* map is always zero on H_d. Thus there is always a generator added and never one deleted, and there is a homology basis for $H_d(X^b)$ with exactly one generator for each critical point.

Remark Topological duality theorems may be used to extend this result to complements of complex d-manifolds. Such a result will be useful to us (think of applying it to $\mathcal{M}$), but instead of deriving it from duality, we will derive it in the more general setting of complex algebraic varieties and their complements (more general because algebraic varieties need not be smooth). See Theorem C.4.1 in Section C.3.

Notes

The classical text on Morse theory is Milnor (1963), from which excellent exposition we have taken ideas for organization and illustration. One departure from the classical presentation is our attitude toward the requirement that h be nondegenerate. All that is required for the Morse Lemma is isolated critical points. Furthermore, in applications to complex integration, the height function is usually harmonic, which in two dimensions limits the possibilities for degeneracy to local behaviors similar to that of the real part of z^k. These topologies are not much more difficult for the degenerate cases ($k \geq 3$) than they are for $k = 2$. We therefore allow degenerate Morse functions.

Exercises

B.1 (lumpy sphere)

Let X be a sphere with a lump, i.e., a patch on the northern hemisphere where the surface is raised to produce a local, but not global, maximum of the height function. List the critical points of the lumpy sphere and determine the homotopy types of the attachments. This gives a description of the lumpy sphere as a cell complex different from the complex with just two cells. Use this to compute the homology and verify it is the same as for the nonlumpy sphere.

B.2 (bouquet of spheres)

Let X be a smooth complex algebraic variety of complex dimension d in $\mathbb{C}^{2d}$ and let h be a Morse function on X, which is the real part of a complex analytic function. Suppose X has five critical points. Can you determine the homotopy type of the pair $(X^{\leq b}, X^{\leq a})$ when $a \to -\infty$ and $b \to +\infty$?

B.3 (any downward patch generates)

Let $\mathcal{M}$ be a manifold with Morse function h having distinct critical values and let x be a critical point of index k. Let P be any submanifold of $\mathcal{M}$ diffeomorphic to an open k-ball about x and such that h is strictly maximized on P at x. Prove that P is a homology generator for the local homology group $H_k(\mathcal{M}^{h(x)+\varepsilon}, \mathcal{M}^{h(x)-\varepsilon})$. (Hint: This is true of any embedded k-disk through x in $\mathcal{M}^{h(x)}$ that intersects the ascending $(n-k)$-disk transversely.)

Appendix C
Stratification and Stratified Morse Theory

In this chapter we give a number of results from Goresky and MacPherson (1988) that give the topology of a stratified space X in terms of changes in topology in the spaces $X^{\leq c}$ as c passes through values $h(p)$, where p is a critical point in the stratified sense. Some of the relevant *defintions* and results were given in Section 5.4, and others appear in Section 8.2. In particular, we develop from scratch the notion of a Whitney stratified space and of Morse functions and critical points in the stratified sense. We discuss the nonproper extensions of this and then summarize a number of basic results of Goresky and MacPherson (1988). The final sections deal with specific properties enjoyed by complex algebraic varieties.

C.1 Whitney Stratification

Let I be a finite partially ordered set and define an *I-decomposition* of a topological space Z to be a partition of Z into a disjoint union of sets $\{S_\alpha : \alpha \in I\}$ such that

$$S_\alpha \cap \overline{S_\beta} \neq \emptyset \iff S_\alpha \subseteq \overline{S_\beta} \iff \alpha \leq \beta \, .$$

Definition C.1.1 (Whitney stratification) *Let Z be a closed subset of a smooth manifold $\mathcal{M}$. A* ***Whitney stratification*** *of Z is an I-decomposition such that*

(i) Each S_α is a manifold in $\mathbb{R}^n$.
(ii) If $\alpha < \beta$, if the sequences $\{x_i \in S_\beta\}$ and $\{y_i \in S_\alpha\}$ both converge to $y \in S_\alpha$, if the lines $l_i = \overline{x_i \, y_i}$ converge to a line l and the tangent planes $T_{x_i}(S_\beta)$ converge to a plane T of some dimension, then both l and $T_y(S_\alpha)$ are contained in T.

The definition is well crafted. The conditions are easy to fulfill; for example, every algebraic variety admits a Whitney stratification, which we stated earlier as Proposition 5.4.3. This is proved, for example, in Hironaka (1973, Theorem 4.8). The conditions have strong consequences; for example, they are strong enough to

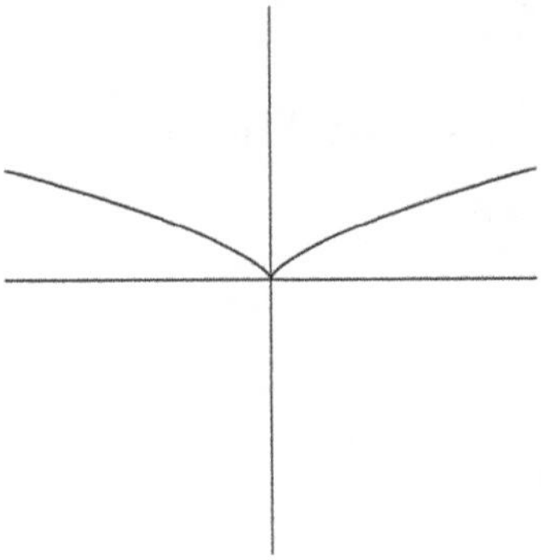

Figure C.1 One stratum is the singular point and one stratum is everything else.

make the proofs of stratified Morse theorems work. Actual stratifications, when needed, are usually quite natural. For algebraic varieties, stratifications are computable; for recent work on the complexity, see Rannou (1998) and Mostowski and Rannou (1991).

Example C.1.2 A smooth manifold is a stratified space with a single stratum. ■

Example C.1.3 Let Z be a union of affine subspaces of $\mathbb{R}^n$. Let $\{A_\alpha : \alpha \in I\}$ be the ***lattice of flats***, i.e., all intersections of subfamilies of the affine spaces, ordered by inclusion. Then the sets $S_\alpha = A_\alpha \setminus \bigcup_{\eta < \alpha} S_\eta$ are a Whitney stratification of Z.

An interesting special case is when $\mathbb{R}^{2d}$ is viewed as $\mathbb{C}^d$. The simplest nontrivial example is when Z is the union of two complex lines in $\mathbb{C}^2$. The intersection point σ is a zero-dimensional stratum, and removing σ from each of the two complex lines gives two 2-dimensional strata. ■

Example C.1.4 Let Z be a real algebraic curve $\{f(x, y) = 0\}$ in $\mathbb{R}^2$ with f irreducible. Let $Y = \{(x, y) : \nabla f(x, y) = 0\}$ be the set of singular points of Z. Taking $Z \setminus Y$ to be one stratum and each singleton $\{y\}$ to be another produces a Whitney stratification of Z. Figures C.1 and C.2 show two examples of this, the first curve $x^2 - y^3$ having a cusp at the origin and the second curve

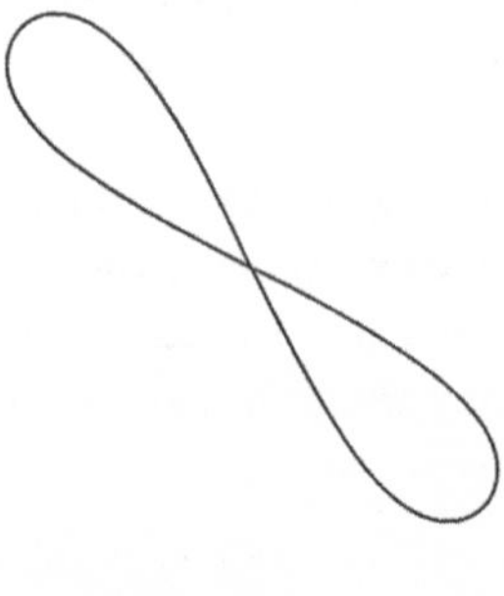

Figure C.2 Again, one stratum is the singular point and one stratum is everything else.

$19 - 20x - 20y + 5x^2 + 14xy + 5y^2 - 2x^2y - 2xy^2 + x^2y^2$ having a self-intersection at (1, 1). ■

Example C.1.5 Let Z be a complex algebraic hypersurface in $\mathbb{C}^3$ defined by $f(x, y, z) = 0$ and suppose ∇f vanishes along an algebraic curve γ. It is possible that $\{\gamma, Z \setminus \gamma\}$ is a Whitney stratification for Z. However, if γ is not smooth, then any Whitney stratification of Z must have at least three strata, one containing singularities of γ, one containing the rest of γ, and one containing $Z \setminus \gamma$. ■

Singular Loci and Effective Computation

Let H be a polynomial in d variables vanishing on the variety $\mathcal{V}$. We would like to use effective algebraic techniques to understand the singularities of $\mathcal{V}$ and to compute a stratification of $\mathcal{V}$. Recall that an ideal $\mathcal{I} \subseteq \mathbb{C}[\mathbf{z}]$ is said to be ***radical*** if $f^2 \in \mathcal{I} \Rightarrow f \in \mathcal{I}$ for all f. Each ideal $\mathcal{I}$ has a smallest radical ideal $\sqrt{\mathcal{I}}$ that contains it; these define the same varieties, and the correspondence between ideals and varieties is one-to-one on radical ideals. Specifically, Hilbert's Nullstellensatz asserts (see Eisenbud, 1995, Theorem 1.1) that if $\mathcal{I}(\mathcal{V})$ denotes the annihilator of $\mathcal{V}$ and $\mathcal{V}_{\mathcal{I}}$ denotes the variety on which all functions in $\mathcal{I}$ vanish, then $\mathcal{I}(\mathcal{V}_{\mathcal{I}}) = \sqrt{\mathcal{I}}$ and $\mathcal{V}_{\mathcal{I}(\mathcal{V})} = \mathcal{V}$.

Before computing a Whitney stratification of $\mathcal{V}$, let us look for any way to decompose $\mathcal{V}$ into smooth manifolds, not necessarily satisfying the Whitney tangency conditions. One such stratification, namely the ***singular locus stratification***, is given as follows. Let $\mathcal{V}_0$ be all the smooth points of $\mathcal{V}$, i.e., all the points $\mathbf{z}$ such that $\mathcal{V}$ is a manifold in a neighborhood of $\mathbf{z}$. The space $\mathcal{V}_0$ is, by definition, locally a $(2d - 2)$-manifold, hence it is globally and we may take $\mathcal{V}_0$ to be one stratum. The complement $\mathcal{L}_0 := \mathcal{V} \setminus \mathcal{V}_0$ is known as the ***singular locus***. It may have components of different dimension, but the highest dimesion, as a real manifold, is at most $2d - 4$. Decompose this into components. For each component S, do the following. If S is smooth, take it to be a stratum, and if not, take the smooth points of S to be a stratum. Doing this for all components of $\mathcal{L}_0$ gives a decomposition into manifolds of everything except the set $\mathcal{L}_1$ of singular points of components of $\mathcal{L}_0$. Continue inductively in this way, each time lowering the maximum dimension of the components and therefore ending after at most d steps. This yields the singular locus decomposition.

It is worth a brief discussion of how one actually computes this. It is easiest always to work with radical ideals. Computing the radical of an ideal is easy, and Maple's `PolynomialIdeals` package has a command to do this, namely `Radical`. For principal ideals, such as $\langle H \rangle$, computing the radical is simply a matter of removing repeated factors so as to make H square free.

Thus the first step is to pass to the radical $\sqrt{H}$. Conveniently, there is a Maple command `Radical` in the `PolynomialIdeals` package (it requires conversion from a list to an ideal via the command `PolynomialIdeals`). The singular locus

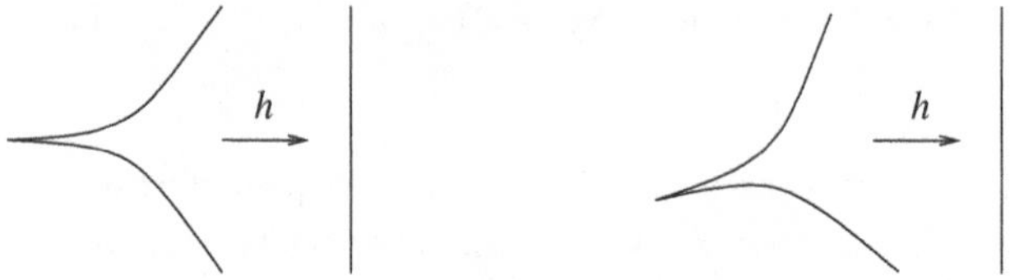

Figure C.3 A non-Morse function (left) and a Morse function (right).

of $\mathcal{V}$ is just the set where the gradient of $\sqrt{H}$ vanishes. This may be computed via a generic Maple computation:

```
H := PolynomialIdeals[Radical] (H);
I_1 := Basis([H, diff(H,z1), ..., diff(H,zd)],
       tdeg (z1, ..., zd);
```

The stratum $\mathcal{V}_0$ is the complement of this. To continue, if $\mathcal{V}_0$ is not all of $\mathcal{V}$, take the radical of I_1 and iterate with this in place of the principal ideal $\langle H \rangle$.

C.2 Critical Points and the Fundamental Lemma

Critical Points for Stratified Spaces

Fix a Whitney stratification $\{S_\alpha : \alpha \in I\}$ of a closed subset Z of a smooth manifold $\mathcal{M} \subseteq \mathbb{R}^n$. The following generalization of the notion of a Morse function is given in Goresky and MacPherson (1988, Section I.2.1).

Definition C.2.1 (Morse function on a stratification) *A **Morse function with distinct critical values** on Z is the restriction to Z of a smooth function $h : \mathcal{M} \to \mathbb{R}$ such that*

(i) $h|_Z$ is proper and has distinct critical values.
(ii) For each $\alpha \in I$, $h|_{S_\alpha}$ is a Morse function, meaning that its critical points are nondegenerate; that is, its Hessian is nonsingular at each critical point.
(iii) If $\mathbf{p}$ is a critical point of $h|_{S_\alpha}$ and T is a limit of tangent planes $T_{\mathbf{p}_i}(S_\beta)$ as $\mathbf{p}_i \to \mathbf{p}$ in a stratum S_β with $\beta > \alpha$, then either $T = T_{\mathbf{p}}(S_\alpha)$ or T contains a tangent vector on which $dh(\mathbf{p})$ does not vanish.

Dropping the requirement of distinct critical values, one has simply a ***Morse function***. A standard perturbation argument shows that coinciding critical values do not affect the topology. An example is shown in Figure C.3 of two height functions, one failing the third condition and one satisfying it: on the left, the limit of tangent lines at the cusp is horizontal, so annihilated by dh.

The stratified version of the Fundamental Morse Lemma (Lemma B.1.2) is as follows.

Lemma C.2.2 (Stratified Morse Lemma [Goresky and MacPherson, 1988, page 6]) *Let X be a stratified space with Morse function h. Let $a < b$ be real* *numbers and suppose the interval* $[a, b]$ *contains no critical values of* h. *If also* $h^{-1}[a, b]$ *is compact, then the inclusion* $X^a \hookrightarrow X^b$ *is a homotopy equivalence.*

The result usually quoted as the description of the attachment in the stratified case (a stratified version of Theorem B.1.3) is Goresky and MacPherson (1988, Theorem, page 69). This computes the change in topology of a stratified space X on which the function h is proper. When h is a continuous function on $(\mathbb{C}^*)^d$, this requires the subset X to be closed. We are chiefly interested in the space $X = \mathcal{V}^c$, which is not closed. Dealing with nonproper height functions requires two extra developmental steps. The first is to develop a system for keeping track of the change in topology of the complement of a closed space up to a varying height cutoff. This computation is similar to the one for the space itself. Goresky and MacPherson stated the two results together in a later version of the main theorem (Goresky and MacPherson, 1988, Theorem, page 122), and we follow their example, stating the results together in Theorem C.3.3 below. Before stating this, however, we need to address a second way in which the function h can fail to be proper.

Compactifications

Even on a closed stratified space there may be points of finite height at infinity, or more generally, the set $\{x \in X : h(x) = c\}$ may be unbounded for a finite value of c. This issue is trickier than the issue of changing between a space and its complement because it requires that h extend continuously to some compactification of $\mathcal{V}$. In Pemantle (2010, Conjecture 2.11), the following conjecture appears.

Conjecture C.1 *Let* $\mathcal{V}$ *be an algebraic hypersurface in* $(\mathbb{C}^*)^d$ *and let* $h = h_{\hat{\mathbf{r}}}$ *for some positive vector* $\hat{\mathbf{r}}$. *Then there exists a compact space* $\mathcal{V}^\dagger$ *such that*

(i) $\mathcal{V}$ *embeds as a dense set of* $\mathcal{V}^\dagger$;
(ii) h *extends to a continuous function mapping* $\mathcal{V}^\dagger$ *to the extended real line* $[-\infty, \infty]$.

The importance of the conjecture is that the main results of stratified Morse theory are known to hold for $\mathcal{M} := (\mathbb{C}^*)^d \setminus \mathcal{V}$ as well as for $\mathcal{V}$ when the conjecture holds. Corresponding to the fundmental lemma, for instance, we have Theorem SMT-A of Goresky and MacPherson (1988, Section 1.2).

Lemma C.2.3 (Stratified Morse Lemma for complements) *If* (i) *and* (ii) *of Conjecture C.1 hold for* $\mathcal{V}$ *and if the interval* $[a, b]$ *contains no critical values for* h *on* $\mathcal{V}^\dagger$, *then the inclusion* $\mathcal{M}^{\leq a} \hookrightarrow \mathcal{M}^{\leq b}$ *is a homotopy equivalence.*

In dimension 2, for generic $\hat{\mathbf{r}}$, the height function $h_{\hat{\mathbf{r}}}$ is proper because the algebraic curve $\mathcal{V}$ avoids points at infinity of finite height. Once $d \geq 3$, there will be points at infinity of finite height. This means an ad hoc argument is needed to verify the conclusions of the conjecture to be assured that the only changes in topology of $\mathcal{M}^c$ occur at critical values (in the stratified sense) for h on $\mathcal{V}$. This problem is not solved in this appendix or elsewhere in the book. Rather, we continue with what we can; namely, we describe the attachment that yields the change in topology in $\mathcal{V}$ and its complement locally at a critical point of $\mathcal{V}$.

C.3 Description of the Attachments

We return to our plan for using Morse theory to find generators for $H_d(\mathcal{M})$. Athough our principal aim is to describe $\mathcal{M}$, not $\mathcal{V}$, results in the literature are always stated in two parts, so as to cover both cases, and we continue to adhere to this. Let X be a stratified space with Morse function h, and let $\mathbf{p}$ be a critical point in a stratum S.

Definition C.3.1 (tangential Morse data) *Define the* ***tangential Morse data*** *at* $\mathbf{p}$ *to be the Morse data at* $\mathbf{p}$ *for the height function* $h|_S$ *on the smooth manifold* S. *By Theorem B.1.3, this is the pair* $(B^\lambda, \partial B^\lambda)$, *where* λ *is the index of* $h|_S$ *at* $\mathbf{p}$ *and* B^λ *denotes the ball of dimension* λ. *This data does not include the map* ϕ.

One property enjoyed by Whitney stratifications is that near S, the space X is locally a product. Specifically, the ***normal slice*** $N_S(\mathbf{p})$ at any point $\mathbf{p} \in S$, defined to be the intersection of X with a small closed disk D containing $\mathbf{p}$ and transverse to the stratum S, is always of the same topological type as $\mathbf{p}$ varies along S. The normal slice $N(\mathbf{p})$ has boundary $L(\mathbf{p}) := \partial D(\mathbf{p}) \cap X$, which is called the ***normal link*** at $\mathbf{p}$, and $N(\mathbf{p})$ is a cone over $L(\mathbf{p})$ with vertex $\mathbf{p}$.

Definition C.3.2 (normal Morse data)

(i) Suppose $\mathbf{p} \in X$ *is a critical point with critical value* c. *Define the* ***normal Morse data*** *at* $\mathbf{p}$ *to be the pair* $(N(\mathbf{p}) \setminus X, N(\mathbf{p})^{c-\varepsilon} \setminus X)$.

(ii) Suppose $\mathbf{p} \notin X$ *is a critical point with critical value* c. *Define the normal Morse data to be* $(N(\mathbf{p}) \cap X, N(\mathbf{p})^{c-\varepsilon} \cap X)$. *It is shown in Goresky and MacPherson (1988) that* $(l^+(\mathbf{p}), \partial l^+(\mathbf{p}))$ *and* $(l^+(\mathbf{p}), l^0(\mathbf{p}))$ *are also of the same homotopy type, where* $l^+(\mathbf{p}) = L(\mathbf{p}) \cap h^{-1}[c, \infty)$ *and* $l^0(\mathbf{p}) = L(\mathbf{p}) \cap h^{-1}[c]$.

The following theorem justifies the terminology by characterizing the Morse data at $\mathbf{p}$.

Theorem C.3.3 (Goresky and MacPherson, 1988, Theorem SMT B, page 8) *Let* X *be a stratified space with Morse function* h. *Then the attachment pair at the critical point* $\mathbf{p}$ *in a stratum* S *is the product (in the category of pairs) of the normal and the tangential Morse data. Specifically, if* $\mathbf{p} \in X$ *then the attachment pair at* $\mathbf{p}$ *is*

$$(B^\lambda \times (N(\mathbf{p}) \setminus X), B^\lambda \times (N(\mathbf{p})^{c-\varepsilon} \setminus X) \cup \partial B^\lambda \times (N(\mathbf{p}) \setminus X)),$$

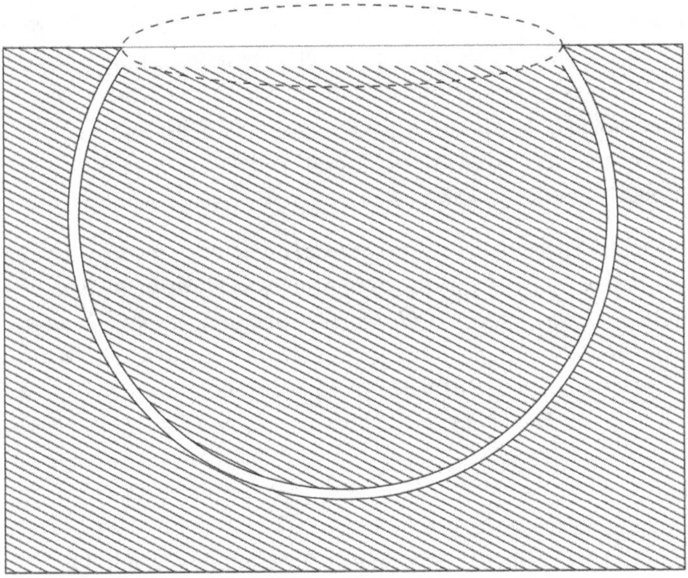

Figure C.4 The complement of the unit sphere up to height $+1/2$.

whereas if $\boldsymbol{p} \notin X$, *then the attachment pair at* $\boldsymbol{p}$ *is*

$$(B^\lambda \times l^+(\boldsymbol{p}), B^\lambda \times l^0(\boldsymbol{p}) \cup \partial B^\lambda \times l^+(\boldsymbol{p})).$$

Remark Goresky and MacPherson have this to say about the statement and proof of this result (Goresky and MacPherson, 1988, page 9): "Theorem SMT Part B, although very natural and geometrically evident in examples, takes 100 pages to prove rigorously in this book."

Example C.3.4 (complement of S^2 in $\mathbb{R}^3$) Let X be the complement of the unit sphere $S \subseteq \mathbb{R}^3$. The function $h(x, y, z) = z$ extends to a proper height function on $\mathbb{R}^3$, which is Morse with respect to the stratification $\{S, X\}$.

There are no critical points in X but there are two in S: the South pole and the North pole. In each case the normal slice is an interval minus a point, so the normal data is homotopy equivalent to (S^0, S^0_-), where S^0 is two points, one higher than the other, and S^0_- is the lower of the two points. For the South pole, which has index 0, the tangential data is a point, so the attachment is (S^0, S^0_-), which is the addition of a disconnected point. Figure C.4 illustrates that for $-1 < a < 1$, the space X^a is in fact the union of two contractible components. The North pole has index 2, so the tangential data at the North pole is $(D^2, \partial D^2)$, a polar cap modulo its boundary. Taking the product with the normal data gives two polar caps modulo all of the lower one and the boundary of the upper one. This is just the upper polar cap sewn down along its boundary, the boundary being a point in one of the components. Thus one component becomes a sphere and the other remains contractible. We may write such a space as $S^2 \vee S^0$. ∎

Suppose we have a closed space $Y \subseteq \mathbb{R}^d$, whose complement X we view as a stratified space with Morse function h. Let $\mathbf{p}$ be a critical point for h in some stratum, S. There is a local coordinatization of Y as $S \times B_{\mathbf{p}}$, where $B_{\mathbf{p}}$ is a small ball of dimension $d - k$, where k is the dimension of S. The set $B_{\mathbf{p}} \setminus Y$ is just the ball minus the origin, so it is a cone over $L(\mathbf{p})$ with vertex $\mathbf{p}$. Any chain in $B_{\mathbf{p}} \setminus Y$ may be brought arbitrarily close to $\mathbf{p}$.

Definition C.3.5 (quasi-local cycles) *A **local** absolute or relative cycle at a point $\mathbf{p}$ is a cycle that may be deformed so as to be in an arbitrarily small neighborhood of $\mathbf{p}$. Given a stratified space, with Morse function h, a **quasi-local cycle** at a critical point $\mathbf{p}$ of the stratification is a cycle $C_\perp \times C_\|$, where $C_\|$ is a disk in S on which h is strictly maximized at $\mathbf{p}$, $C^\perp$ is a local cycle in $(B_{\mathbf{p}} \setminus Y, (B_{\mathbf{p}} \setminus Y)^{h(\mathbf{p})-\varepsilon})$, and the product is taken in any local coordinatization of a neighborhood of $\mathbf{p}$ by $B_{\mathbf{p}} \times S$.*

C.4 Stratified Morse Theory for Complex Manifolds

Suppose that X is a complex variety. It turns out that the Morse data has an alternate description obeying the complex structure. Let S be a stratum containing a critical point $\mathbf{p}$. Let $N(\mathbf{p})$ be a small ball in the normal to S at $\mathbf{p}$ and define the ***complex link***, $\mathcal{L}(S)$, to be the intersection of a X with a generic hyperplane (linear space of complex codimension 1) $A \subseteq N(\mathbf{p})$ that comes close to $\mathbf{p}$ but does not contain it. It is shown in Goresky and MacPherson (1988, page 16) that the normal Morse data at $\mathbf{p} \in X$ is given in terms of $\mathcal{L}(S)$ by

$$(\mathrm{Cone}^{\mathbb{R}}(\mathcal{L}(S)), \mathcal{L}(S)) . \tag{C.1}$$

Suppose that X has dimension d, S has dimension k, and the ambient space has dimension n (all dimensions are complex). Then $N(\mathbf{p})$ is a complex space of dimension $n - k$, its intersection with the generic hyperplane has dimension $n - k - 1$, whence

$$\dim_{\mathbb{C}} \mathcal{L}(S) = d - k - 1 .$$

In fact, the homeomorphism type of the complex link depends on X and S but not on the individual choice of $p \in S$, nor the ambient space, nor the choice of Morse function h on the stratified space X (see Goresky and MacPherson, 1988, Section II:2.3).

Suppose next that X is the complement of a d-dimensional variety in $\mathbb{C}^{d+1}$. A formula for the Morse data at a point $p \notin X$ in a stratum S is given (see Goresky and MacPherson, 1988, page 18) by

$$(\mathcal{L}(S), \partial\mathcal{L}(S)) \times (B^1, \partial B^1), \tag{C.2}$$

where B^1 is a 1-ball, that is, a real interval.

Theorem C.4.1

(i) If X is a complex variety of dimension d, then X has the homotopy type of a cell complex of dimension at most d.

(ii) If X is the complement in a domain in $\mathbb{C}^n$ of a complex variety of dimension d, then X has the homotopy type of a cell complex of dimension at most $2n - d - 1$.

Remark The proof of this result in Goresky and MacPherson (1988) appears somewhat difficult, mostly due to the necessity to establish the invariance properties of the complex link. The result, however, is very useful. For example, suppose that X is the complement of the zero set of a polynomial in n variables. Then $d = n - 1$ and the homotopy of dimension of X is at most n. Note that X may have strata of any complex dimension $j \leq d$, and that the complement of a j-dimensional complex space in $\mathbb{C}^n$ has homotopy dimension $2n - 2j$. The theorem asserts that the complex structure prevents the dimensions of contributions at strata of dimensions $j < d$ from exceeding the dimension of the contributions from d-dimensional strata.

Proof (i) Let us assume that this variety is embedded in some $\mathbb{C}^n$ and that the height function has been chosen to be the square of the distance from a generic point. We examine the homotopy type of the attachment at a point $\boldsymbol{p}$ in a stratum of dimension k. It suffices, as in the proof of Proposition B.3.1, to show that each attachment has the homotopy type of a cell complex of dimension at most d.

First, if $k = d$ ($\boldsymbol{p}$ is a smooth point), then, as was observed before stating Proposition B.3.1, the index of h is at most d. The attachment is $(B^i, \partial B^i)$ where i is the index of h, so in this case the homotopy type of the attachment is clearly at most d.

When $k < d$, we proceed by induction on d. The tangential Morse data has the homotopy type of a cell complex of dimension at most k.

The space $\mathcal{L}(S)$ is a complex analytic space, with complex dimension 1 less than the dimension of the normal slice $N(\boldsymbol{p})$, i.e., of dimension $d - k - 1$. The induction hypothesis shows that the homotopy dimension of $\mathcal{L}(S)$ is at most $d - k - 1$. Taking the cone brings the dimension to at most $d - k$ and adding the dimension of the tangential data brings this up to at most d, completing the induction.

(ii) When X is the complement of a variety, V, still assuming h to be square distance to a generic point, all critical points are in V, not in X. Again, it suffices to show that the attachments all have homotopy dimension at most $2n - d - 1$, and again we start with the case $k = d$. Here, $\boldsymbol{p}$ is a smooth point of V so the normal data is the same as for the complement of a point in $\mathbb{C}^{n-d}$, which is $S^{2(n-d)-1}$. The tangential data has homotopy dimension at most d, so the attachment has dimension at most $2n - d - 1$.

When $k < d$, again proceed by induction on d. The link $\mathcal{L}(S)$ is the complement of $V \cap A$ in A. We have $\dim_{\mathbb{C}} N(\mathbf{p}) = n - k$, $\dim_{\mathbb{C}}(A) = n - k - 1$, and $\dim_{\mathbb{C}}(V \cap A) = d - k - 1$ because V has codimension $n - d$ and interects A generically, and because $k \leq d - 1$. The induction hypothesis applied to the complement of $V \cap A$ in A shows that $\mathcal{L}(S)$ has the homotopy type of a cell complex of dimension at most $2(n - k - 1) - (d - k - 1) - 1 = 2n - d - k - 2$. The normal Morse data is the product of this with a 1-complex and hence has homotopy dimension at most $2n - d - k - 1$, and taking the product with the tangential Morse data brings the dimension up to at most $2n - d - 1$, completing the induction. □

It will be useful to summarize the results from this section for complements of manifolds, applying the Künneth formula to obtain

Theorem C.4.2 *Let X be the complement of a complex variety of dimension d in $\mathbb{C}^{d+1}$. Then X may be built by attaching spaces that are homotopy equivalent to cell complexes of dimension at most $d + 1$. Consequently, $H_d(X)$ has a basis of quasi-local cycles that may be described as*

$$\mathcal{B} = \{\sigma_{\mathbf{p},i}\}_{\mathbf{p},i},$$

where $\mathbf{p}$ ranges over critical points of strata, $\sigma_{\mathbf{p},i} \in X^{c,\mathbf{p}}$; for each fixed $\mathbf{p}$, the projection $\pi_* : X^{c,\mathbf{p}} \to (X^{c,\mathbf{p}}, X^{c-\varepsilon}) = X^{\mathbf{p},\mathrm{loc}}$ maps the set $\{\sigma_{\mathbf{p},i}\}$ to a basis for the relative homology group $H_d(X^{\mathbf{p},\mathrm{loc}})$.

Notes

The idea to use Morse theory to evaluate integrals was not one of the original purposes of Morse theory. Nevertheless, the utility of Morse theory for this purpose has been known for forty years. Much of the history appears difficult to trace. The first author learned it from Yuliy Baryshnikov, who related it as mathematical folklore from V. I. Arnol′d's seminar. The smooth Morse theory in this chapter (and some of the pictures) is borrowed from Milnor's classic text (1963). Stratified Morse theory is a relatively new field, in which the seminal text is Goresky and MacPherson (1988). Most of our understanding derives from this text.

The technical, geometric nature of the arguments coupled with the statements presented in increasing iterations from purely informal and intuitive to purportedly completely rigorous leave us sometimes unable to be certain what has been proved. We don't believe that Goresky and MacPherson (1988) is at fault for this any more than is the norm for publication in geometric subfields of mathematics. In some cases, arguments presented here rely on results from Goresky and MacPherson (1988) that fall in this category. The most notable is Theorem C.4.1, which relies on the complex link apparatus. For us, there is some uncertainty about whether dimensions and cones are quoted over $\mathbb{R}$ or $\mathbb{C}$ in various cases, and whether and when X needs to be a complex hypersurface as opposed to a variety of greater codimension.

Exercises

C.1 (singular locus stratification)
Let $f(x, y, z) := x^2 + y^2 z$ be the ***Whitney Umbrella***. Compute the singular locus stratification of f. Prove that this is also a Whitney stratification or prove that it is not and find a refinement i.e.

C.2 (computing with the help of Morse functions)
Let X be the complement in $\mathbb{C}^2$ of the smooth curve $x^2 + y^2 = 1$. Define a Morse function and use it to compute the homology of X.

References

Adamczewski, B. and J. P. Bell (May 2012). "Diagonalization and rationalization of algebraic Laurent series." *ArXiv e-prints*. arXiv:math.NT/1205.4090 [math.NT] (cit. on p. 314).

Aĭzenberg, I. A. and A. P. Yuzhakov (1983). *Integral representations and residues in multidimensional complex analysis*. Vol. 58. Translations of Mathematical Monographs. Translated from the Russian by H. H. McFaden, Translation edited by Lev J. Leifman. Providence, RI: American Mathematical Society (cit. on pp. 8, 251).

Aldous, D. (2013). "Power laws and killed branching random walls." Webpage http://www.stat.be-keley.edu/user/aldous/Research/OP/brw.html. Accessed 2013-01-24 (cit. on p. 61).

Ambainis, A. et al. (2001). "One-dimensional quantum walks." In: *Proceedings of the 33rd Annual ACM Symposium on Theory of Computing*. New York: ACM Press, pp. 37–49 (cit. on p. 190).

Askey, R. and G. Gasper (1972). "Certain rational functions whose power series have positive coefficients." *Amer. Math. Monthly* **79**, pp. 327–341.

Atiyah, M., R. Bott, and L. Gårding (1970). "Lacunas for hyperbolic differential operators with constant coefficients, I." *Acta Mathematica* **124**, pp. 109–189 (cit. on pp. 12, 253, 257, 261, 262, 267, 276).

Banderier, C., P. Flajolet, et al. (2001). "Random maps, coalescing saddles, singularity analysis, and Airy phenomena." *Random Structures Algorithms* **19**. Analysis of algorithms (Krynica Morska, 2000), pp. 194–246 (cit. on p. 313).

Banderier, C. and P. Hitczenko (2012). "Enumeration and asymptotics of restricted compositions having the same number of parts." *Discrete Appl. Math.* **160**, pp. 2542–2554 (cit. on p. 315).

Baryshnikov, Y., W. Brady, et al. (2010). "Two-dimensional quantum random walk." *J. Stat. Phys.* **142**, pp. 78–107 (cit. on pp. 159, 164, 186, 190, 207, 297, 298).

Baryshnikov, Y. and R. Pemantle (2011). "Asymptotics of multivariate sequences, part III: quadratic points." *Adv. Math.* **228**, pp. 3127–3206 (cit. on pp. 8, 12, 149, 159, 207, 253, 260–263, 265, 267, 268, 275, 276).

Baryshnikov, Y. and R. Pemantle (2013). "Morse theory of the complement of the complexification of a real hyperplane arrangement." *In preparation* (cit. on p. 236).

Bena, I. et al. (May 2012). "Scaling BPS Solutions and pure-Higgs States." *ArXiv e-prints*. arXiv:hep-th/1205.5023 [hep-th] (cit. on p. 207).

Bender, E. A. (1973). "Central and local limit theorems applied to asymptotic enumeration." *J. Combinatorial Theory Ser. A* **15**, pp. 91–111 (cit. on pp. 6, 287).

Bender, E. A. (1974). "Asymptotic methods in enumeration." *SIAM Rev.* **16**, pp. 485–515 (cit. on p. 6).

Bender, E. A. and J. R. Goldman (1970/1971). "Enumerative uses of generating functions." *Indiana Univ. Math. J.* **20**, pp. 753–765 (cit. on p. 43).

Bender, E. A. and L. B. Richmond (1983). "Central and local limit theorems applied to asymptotic enumeration. II. Multivariate generating functions." *J. Combin. Theory Ser. A* **34**, pp. 255–265 (cit. on pp. xi, 6, 7, 163, 207, 286, 287).

Bender, E. and L. B. Richmond (1996). "Admissible functions and asymptotics for labelled structures by number of components." *Elec. J. Combin.* **3**, Research Paper 34, 23 pp. (electronic).

Bender, E. A. and L. B. Richmond (1999). "Multivariate asymptotics for products of large powers with applications to Lagrange inversion." *Electron. J. Combin.* **6**, Research Paper 8, 21 pp. (electronic) (cit. on pp. 7, 287).

Bender, E. A., L. B. Richmond, and S. G. Williamson (1983). "Central and local limit theorems applied to asymptotic enumeration. III. Matrix recursions." *J. Combin. Theory Ser. A* **35**, pp. 263–278 (cit. on p. 7).

Bender, E. and S. G. Williamson (1991). *Foundations of Applied Combinatorics*. Redwood City, CA: Addison Wesley.

Berenstein, C. A. and R. Gay (1991). *Complex variables*. Vol. 125. Graduate Texts in Mathematics. An introduction. New York: Springer-Verlag (cit. on p. 6).

Bertozzi, A. and J. McKenna (1993). "Multidimensional residues, generating functions, and their application to queueing networks." *SIAM Rev.* **35**, pp. 239–268 (cit. on pp. 8, 221, 251).

Bierstone, E. and P. Milman (1997). "Canonical desingularization in characteristic zero by blowing up the maximum strata of a local invariant." *Inventiones Mathematicae* **128**, pp. 207–302 (cit. on p. 276).

Björner, A. et al. (1999). *Oriented matroids*. Second ed. Vol. 46. Encyclopedia of Mathematics and its Applications. Cambridge: Cambridge University Press (cit. on p. 213).

Bleistein, N. and R. A. Handelsman (1986). *Asymptotic expansions of integrals*. Second ed. New York: Dover Publications (cit. on pp. 11, 78, 87, 104).

Borcea, J., P. Branden, and T. Liggett (2009). "Negative dependence and the geometry of polynomials." *J. AMS* **22**, pp. 521–567 (cit. on p. 276).

Bostan, A. et al. (2007). "Differential equations for algebraic functions." In: *ISSAC 2007*. New York: ACM, pp. 25–32 (cit. on p. 44).

Bousquet-Mélou, M. and M. Petkovšek (2000). "Linear recurrences with constant coefficients: the multivariate case." *Discrete Math.* **225**. Formal power series and algebraic combinatorics (Toronto, ON, 1998), pp. 51–75 (cit. on pp. 26, 28, 30, 32, 43).

Boyd, S. and L. Vandenberghe (2004). *Convex optimization*. Cambridge: Cambridge University Press (cit. on p. 204).

Bressler, A. and R. Pemantle (2007). "Quantum random walks in one dimension via generating functions." In: *2007 Conference on Analysis of Algorithms, AofA 07*. Discrete Math. Theor. Comput. Sci. Proc., AH. Assoc. Discrete Math. Theor. Comput. Sci., Nancy, pp. 403–412 (cit. on p. 284).

Briand, E. (2010). "Covariants vanishing on totally decomposable forms." In: *Liaison, Schottky problem and invariant theory*. Vol. 280. Progr. Math. Basel: Birkhäuser Verlag, pp. 237–256 (cit. on p. 219).

Canfield, E. R. and B. McKay (2009). "The asymptotic volume of the Birkhoff polytope." *Online J. Anal. Comb.* **4**, p. 4 (cit. on p. 252).

Carteret, H. A., M. E. H. Ismail, and B. Richmond (2003). "Three routes to the exact asymptotics for the one-dimensional quantum walk." *J. Phys. A* **36** (33), pp. 8775–8795 (cit. on p. 284).

Castro, F. (1984). "Théorème de division pour les opérateurs diffèrentiels et calcul des multiplicités." PhD thesis. Thèse de 3ème cycle, Université de Paris 7 (cit. on p. 117).

Chabaud, C. (2002). "Séries génératrices algébriques: asymptotique et applications combinatoires." PhD thesis. Université Paris VI (cit. on p. 292).

Chayes, J. T. and L. Chayes (1986). "Ornstein-Zernike behavior for self-avoiding walks at all non-critical temperatures." *Comm. Math. Phys.* **105**, pp. 221–238 (cit. on p. 200).

Chen, W. Y. C., E. Deutsch, and S. Elizalde (2006). "Old and young leaves on plane trees." *Eur. J. Combin.* **27**, pp. 414–427 (cit. on p. 316).

Chyzak, F., M. Mishna, and B. Salvy (2005). "Effective scalar products of D-finite symmetric functions." *J. Combin. Theory Ser. A* **112**, pp. 1–43 (cit. on p. 44).

Chyzak, F. and B. Salvy (1998). "Non-commutative elimination in Ore algebras proves multivariate identities." *J. Symbolic Comput.* **26**, pp. 187–227 (cit. on pp. 39, 117, 118).

Comtet, L. (1964). "Calcul pratique des coefficients de Taylor d'une fonction algébrique." *Enseignement Math. (2)* **10**, pp. 267–270 (cit. on p. 35).

Comtet, L. (1974). *Advanced combinatorics*. enlarged. The art of finite and infinite expansions. Dordrecht: D. Reidel Publishing (cit. on pp. 20, 194, 281).

Conway, J. B. (1978). *Functions of one complex variable*. Second ed. Vol. 11. Graduate Texts in Mathematics. New York: Springer-Verlag (cit. on pp. 6, 49, 126, 258).

Corteel, S., G. Louchard, and R. Pemantle (2004). "Common intervals of permutations." In: *Mathematics and computer science. III*. Trends Math. Basel: Birkhäuser, pp. 3–14 (cit. on p. 20).

Coutinho, S. (1995). *A primer of algebraic D-modules*. Vol. 33. London Mathematical Society Student Texts. Cambridge: Cambridge University Press (cit. on p. 117).

Cox, D. A., J. Little, and D. O'Shea (2005). *Using algebraic geometry*. Second ed. Vol. 185. Graduate Texts in Mathematics. New York: Springer (cit. on pp. 11, 107, 118).

Cox, D., J. Little, and D. O'Shea (2007). *Ideals, varieties, and algorithms*. Third ed. Undergraduate Texts in Mathematics. An introduction to computational algebraic geometry and commutative algebra. New York: Springer (cit. on pp. 108, 118).

De Bruijn, N. G. (1981). *Asymptotic methods in analysis*. Third ed. New York: Dover Publications (cit. on p. 87).

Delabaere, E. and C. J. Howls (2002). "Global asymptotics for multiple integrals with boundaries." *Duke Math. J.* **112**, pp. 199–264 (cit. on p. 315).

De Loera, J. A. and B. Sturmfels (2003). "Algebraic unimodular counting." *Math. Program.* **96**. Algebraic and geometric methods in discrete optimization, pp. 183–203 (cit. on pp. 231, 251, 252).

Denef, J. and L. Lipshitz (1987). "Algebraic power series and diagonals." *J. Number Theory* **26**, pp. 46–67 (cit. on p. 314).

DeVore, R. A. and G. G. Lorentz (1993). *Constructive approximation*. Vol. 303. Grundlehren der Mathematischen Wissenschaften. Berlin: Springer-Verlag (cit. on pp. 242, 243).

DeVries, T. (2011). "Algorithms for bivariate singularity analysis." PhD thesis. University of Pennsylvania (cit. on pp. 143, 183, 280).

DeVries, T. (2010). "A case study in bivariate singularity analysis." In: *Algorithmic probability and combinatorics*. Vol. 520. Contemp. Math. Providence, RI: Amer. Math. Soc., pp. 61–81 (cit. on pp. 8, 183, 184).

DeVries, T., J. van der Hoeven, and R. Pemantle (2012). "Effective asymptotics for smooth bivariate generating functions." *Online J. Anal. Comb.* **7**, to appear (cit. on pp. 8, 178–180, 183).

Dobrushkin, V. (2010). *Methods in algorithmic analysis*. Boca Raton, CRC Press. (Cit. on p. 63).

Drmota, M., B. Gittenberger, and T. Klausner (2005). "Extended admissible functions and Gaussian limiting distributions." *Math. Comp.* **74**. no. 252 (electronic) (cit. on p. 159).

Du, P. et al. (2011). "The Aztec Diamond edge-probability generating function." *Preprint* (cit. on p. 273).

Durrett, R. (2004). *Probability: theory and examples*. Third ed. Belmont, CA: Duxbury Press, p. 497 (cit. on p. 288).

Eisenbud, D. (1995). *Commutative algebra*. Vol. 150. Graduate Texts in Mathematics. With a view toward algebraic geometry. New York: Springer-Verlag (cit. on p. 353).

Fayolle, G., R. Iasnogorodski, and V. Malyshev (1999). *Random walks in the quarter-plane*. Vol. 40. Applications of Mathematics (New York). Algebraic methods, boundary value problems and applications. Berlin: Springer-Verlag (cit. on pp. xi, 43).

Flajolet, P. and A. Odlyzko (1990). "Singularity analysis of generating functions." *SIAM J. Discrete Math.* **3**, pp. 216–240 (cit. on pp. 58, 63).

Flajolet, P. and R. Sedgewick (2009). *Analytic combinatorics*. Cambridge University Press, p. 824 (cit. on pp. xi, 10, 15, 63, 292, 293, 295, 304, 306).

Flatto, L. and S. Hahn (1984). "Two parallel queues created by arrivals with two demands. I." *SIAM J. Appl. Math.* **44**, pp. 1041–1053 (cit. on p. 43).

Flatto, L. and H. P. McKean (1977). "Two queues in parallel." *Comm. Pure Appl. Math.* **30**, pp. 255–263 (cit. on p. 43).

Flaxman, A., A. W. Harrow, and G. B. Sorkin (2004). "Strings with maximally many distinct subsequences and substrings." *Electron. J. Combin.* **11**, Research Paper 8, 10 pp. (electronic) (cit. on pp. 113, 290).

Foata, D. and M.-P. Schützenberger (1970). *Théorie géométrique des polynômes eulériens*. Lecture Notes in Mathematics, Vol. 138. Berlin: Springer-Verlag (cit. on p. 43).

Forsberg, M., M. Passare, and A. Tsikh (2000). "Laurent determinants and arrangements of hyperplane amoebas." *Adv. Math.* **151**, pp. 45–70 (cit. on p. 127).

Furstenberg, H. (1967). "Algebraic functions over finite fields." *J. Algebra* **7**, pp. 271–277 (cit. on pp. 38, 314).

Galligo, A. (1985). "Some algorithmic questions on ideals of differential operators." In: *EUROCAL '85, Vol. 2 (Linz, 1985)*. Vol. 204. Lecture Notes in Comput. Sci. Berlin: Springer, pp. 413–421 (cit. on p. 117).

Gao, Z. and L. B. Richmond (1992). "Central and local limit theorems applied to asymptotic enumeration. IV. Multivariate generating functions." *J. Comput. Appl. Math.* **41**. Asymptotic methods in analysis and combinatorics, pp. 177–186 (cit. on pp. 7, 287).

Gårding, L. (1950). "Linear hyperbolic partial differential equations with constant coefficients." *Acta Math.* **85**, pp. 1–62 (cit. on pp. 149, 261).

Gel'fand, I. M., M. M. Kapranov, and A. V. Zelevinsky (1994). *Discriminants, resultants, and multidimensional determinants*. Mathematics: Theory & Applications. Boston, MA: Birkhäuser Boston (cit. on pp. xii, 11, 118, 124, 127–129, 131).

Gel'fand, I. and G. Shilov (1964). *Generalized Functions, Volume I: Properties and Operations*. Trans. by E. Saletan. New York: Academic Press (cit. on p. 268).

Gessel, I. M. (1981). "Two theorems of rational power series." *Utilitas Math.* **19**, pp. 247–254 (cit. on p. 44).

Gittenberger, B. and Mandlburger (2006). "Hayman admissible functions in several variables." *Elec. J. Combin.* **13**, Research Paper 106, 9 pp. (electronic) (cit. on p. 159).

Goresky, M. and R. MacPherson (1988). *Stratified Morse theory*. Vol. 14. Ergebnisse der Mathematik und ihrer Grenzgebiete (3) [Results in Mathematics and Related Areas (3)]. Berlin: Springer-Verlag (cit. on pp. 11, 99, 142, 348, 351, 354–360).

Goulden, I. P. and D. M. Jackson (2004). *Combinatorial enumeration*. With a foreword by Gian-Carlo Rota, Reprint of the 1983 original. Mineola, NY: Dover Publications (cit. on pp. 15, 22, 43, 281, 292).

Gourdon, X. and B. Salvy (1996). "Effective asymptotics of linear recurrences with rational coefficients." In: *Proceedings of the 5th Conference on Formal Power Series and Algebraic Combinatorics (Florence, 1993)*. Vol. 153, pp. 145–163 (cit. on pp. 47, 183).

Gülen, O. (1997). "Hyperbolic polynomials and interior point methods for convex programming." *Math. Oper. Res.* **22**, pp. 350–377 (cit. on p. 257).

Hardy, G. H. and S. Ramanujan (2000a). "Asymptotic formulæ for the distribution of integers of various types [Proc. London Math. Soc. (2) **16** (1917), 112–132]." In: *Collected papers of Srinivasa Ramanujan*. AMS Chelsea Publ., Providence, RI, pp. 245–261 (cit. on p. 62).

Hardy, G. H. and S. Ramanujan (2000b). "Une formule asymptotique pour le nombre des partitions de n [Comptes Rendus, 2 Jan. 1917]." In: *Collected papers of Srinivasa Ramanujan*. AMS Chelsea Publ., Providence, RI, pp. 239–241 (cit. on p. 62).

Hautus, M. L. J. and D. A. Klarner (1971). "The diagonal of a double power series." *Duke Math. J.* **38**, pp. 229–235 (cit. on pp. 38, 283).

Hayman, W. K. (1956). "A generalisation of Stirling's formula." *J. Reine Angew. Math.* **196**, pp. 67–95 (cit. on pp. xi, 51, 63, 137).

Henrici, P. (1988). *Applied and computational complex analysis. Vol. 1*. Wiley Classics Library. Power series – integration – conformal mapping – location of zeros, Reprint of the 1974 original, A Wiley-Interscience Publication. New York: John Wiley & Sons (cit. on p. 6).

Henrici, P. (1991). *Applied and computational complex analysis. Vol. 2*. Wiley Classics Library. Special functions – integral transforms – asymptotics – continued fractions, Reprint of the 1977 original, A Wiley-Interscience Publication. New York: John Wiley & Sons (cit. on pp. 6, 56, 63, 87).

Hironaka, H. (1973). "Subanalytic sets." In: *Number theory, algebraic geometry and commutative algebra, in honor of Yasuo Akizuki*. Tokyo: Kinokuniya, pp. 453–493 (cit. on p. 351).

Hörmander, L. (1983). *The analysis of linear partial differential operators. I*. Vol. 256. Grundlehren der Mathematischen Wissenschaften. Distribution theory and Fourier analysis. Berlin: Springer-Verlag (cit. on p. 308).

Hörmander, L. (1990). *An introduction to complex analysis in several variables*. Third ed. Vol. 7. North-Holland Mathematical Library. Amsterdam: North-Holland Publishing (cit. on pp. 204, 326, 337, 339).

Hwang, H.-K. (1996). "Large deviations for combinatorial distributions. I. Central limit theorems." *Ann. Appl. Probab.* **6**, pp. 297–319 (cit. on p. 287).

Hwang, H.-K. (1998a). "Large deviations of combinatorial distributions. II. Local limit theorems." *Ann. Appl. Probab.* **8**, pp. 163–181 (cit. on p. 287).

Hwang, H.-K. (1998b). "On convergence rates in the central limit theorems for combinatorial structures." *Eur. J. Combin.* **19**, pp. 329–343 (cit. on p. 287).

Isaacson, E. and H. B. Keller (1994). *Analysis of numerical methods*. Corrected reprint of the 1966 original [Wiley, New York; MR0201039 (34 #924)]. New York: Dover Publications (cit. on p. 30).

Jockusch, W., J. Propp, and P. Shor (Jan. 1998). "Random Domino Tilings and the Arctic Circle Theorem." *ArXiv Mathematics e-prints*. arXiv:math/9801068 (cit. on p. 13).

Kaloshin, V. Y. (2005). "A geometric proof of the existence of Whitney stratifications." *Mosc. Math. J.* **5**, pp. 125–133 (cit. on p. 144).

Kandri-Rody, A. and V. Weispfenning (1990). "Noncommutative Gröbner bases in algebras of solvable type." *J. Symbolic Comput.* **9**, pp. 1–26 (cit. on p. 117).

Kashiwara, M. (1978). "On the holonomic systems of linear differential equations. II." *Invent. Math.* **49**, pp. 121–135 (cit. on p. 118).

Kauers, M. and P. Paule (2011). *The Concrete Tetrahedron*. Leipzig: Springer-Wien (cit. on p. 15).

Kauers, M. and D. Zeilberger (2011). "The computational challenge of enumerating high-dimensional rook walks." *Advances in Applied Mathematics* **47**, pp. 813–819 (cit. on p. 315).

Kenyon, R. and A. Okounkov (2007). "Limit shapes and the complex Burgers equation." *Acta Math.* **199**, pp. 263–302 (cit. on p. 274).

Kesten, H. (1978). "Branching Brownian motion with absorption." *Stochastic Processes Appl.* **7**, pp. 9–47 (cit. on p. 24).

Knuth, D. (2006). *The Art of Computer Programming*. Vol. I–IV. Upper Saddle River, NJ: Addison-Wesley (cit. on p. xi).

Kogan, Y. (2002). "Asymptotic expansions for large closed and loss queueing networks." *Math. Probl. Eng.* **8**, pp. 323–348 (cit. on p. 221).

Krantz, S. G. (2001). *Function theory of several complex variables*. Reprint of the 1992 edition. AMS Chelsea Publishing, Providence, RI (cit. on p. 131).

Kredel, H. (1993). *Solvable Polynomial Rings*. Aachen, Germany: Verlag Shaker (cit. on p. 117).

Larsen, M. and R. Lyons (1999). "Coalescing particles on an interval." *J. Theoret. Probab.* **12**, pp. 201–205 (cit. on pp. 13, 27).

Lee, J. M. (2003). *Introduction to smooth manifolds*. Vol. 218. Graduate Texts in Mathematics. New York: Springer-Verlag (cit. on p. 339).

Leray, J. (1959). "Le calcul différentiel et intégral sur une variété analytique complexe. (Problème de Cauchy. III)." *Bull. Soc. Math. France* **87**, pp. 81–180 (cit. on p. 8).

Lichtin, B. (1991). "The asymptotics of a lattice point problem associated to a finite number of polynomials. I." *Duke Math. J.* **63**, pp. 139–192 (cit. on p. 8).

Limic, V. and R. Pemantle (2004). "More rigorous results on the Kauffman-Levin model of evolution." *Ann. Probab.* **32**, pp. 2149–2178 (cit. on p. 45).

Lin, S. and R. Pemantle (2013). "Computation of second order asymptotics in Laplace-type integral for marginal distributions." *In preparation* (cit. on p. 104).

Lipshitz, L. (1988). "The diagonal of a D-finite power series is D-finite." *J. Algebra* **113**, pp. 373–378 (cit. on p. 38).

Lipshitz, L. (1989). "D-finite power series." *J. Algebra* **122**, pp. 353–373 (cit. on pp. 36, 37, 44).

Lladser, M. (2003). "Asymptotic enumeration via singularity analysis." PhD thesis. The Ohio State University (cit. on pp. 8, 313).

Lladser, M. (2006). "Uniform formulae for coefficients of meromorphic functions in two variables. I." *SIAM J. Discrete Math.* **20**, 811–828 (electronic) (cit. on pp. 8, 313).

Mikhalkin, G. (2000). "Real algebraic curves, the moment map and amoebas." *Ann. of Math. (2)* **151**, pp. 309–326 (cit. on p. 131).

Mikhalkin, G. (2004). "Amoebas of algebraic varieties and tropical geometry." In: *Different faces of geometry*. Vol. 3. Int. Math. Ser. (N.Y.) Kluwer/Plenum, New York, pp. 257–300 (cit. on p. 131).

Milnor, J. (1963). *Morse theory*. Based on lecture notes by M. Spivak and R. Wells. Annals of Mathematics Studies, No. 51. Princeton, NJ: Princeton University Press (cit. on pp. 11, 340, 341, 343, 344, 349, 360).

Milnor, J. (1968). *Singular points of complex hypersurfaces*. Annals of Mathematics Studies, No. 61. Princeton, NJ: Princeton University Press (cit. on p. 140).

Mostowski, T. and E. Rannou (1991). "Complexity of the computation of the canonical Whitney stratification of an algebraic set in $\mathbf{C}^n$." In: *Applied algebra, algebraic algorithms and error-correcting codes (New Orleans, LA, 1991)*. Vol. 539. Lecture Notes in Comput. Sci. Berlin: Springer, pp. 281–291 (cit. on p. 352).

Munarini, E. and N. Z. Salvi (2002/04). "Binary strings without zigzags." *Sém. Lothar. Combin.* **49**, Art. B49h, 15 pp. (electronic).

Munkres, J. R. (1984). *Elements of algebraic topology*. Menlo Park, CA: Addison-Wesley Publishing (cit. on pp. 328–330).

Noble, R. (2010). "Asymptotics of a family of binomial sums." *J. Number Theory* **130**, pp. 2561–2585 (cit. on p. 299).

Odlyzko, A. M. (1995). "Asymptotic enumeration methods." In: *Handbook of combinatorics, Vol. 1, 2*. Amsterdam: Elsevier, pp. 1063–1229 (cit. on pp. xi, 7, 278).

Orlik, P. and H. Terao (1992). *Arrangements of hyperplanes*. Vol. 300. Grundlehren der Mathematischen Wissenschaften [Fundamental Principles of Mathematical Sciences]. Berlin: Springer-Verlag (cit. on p. 228).

Paris, R. B. (2011). *Hadamard expansions and hyperasymptotic evaluation*. Vol. 141. Encyclopedia of Mathematics and its Applications. An extension of the method of steepest descents. Cambridge: Cambridge University Press (cit. on p. 315).

Paris, R. B. and D. Kaminski (2001). *Asymptotics and Mellin-Barnes integrals*. Vol. 85. Encyclopedia of Mathematics and its Applications. Cambridge: Cambridge University Press (cit. on p. 315).

Passare, M., D. Pochekutov, and A. Tsikh (2011). "Amoebas of complex hypersurfaces in statistical thermodynamics." *ArXiv e-prints*. arXiv:math-ph/1112.4332 [math-ph] (cit. on p. 131).

Pemantle, R. (2000). "Generating functions with high-order poles are nearly polynomial." In: *Mathematics and computer science (Versailles, 2000)*. Trends Math. Basel: Birkhäuser, pp. 305–321 (cit. on p. 252).

Pemantle, R. (2010). "Analytic combinatorics in d variables: an overview." In: *Algorithmic probability and combinatorics*. Vol. 520. Contemp. Math. Providence, RI: Amer. Math. Soc., pp. 195–220 (cit. on pp. 138, 143, 144, 159, 355).

Pemantle, R. and M. C. Wilson (2002). "Asymptotics of multivariate sequences. I. Smooth points of the singular variety." *J. Combin. Theory Ser. A* **97**, pp. 129–161 (cit. on pp. 8, 12, 13, 148, 159, 163, 164, 169, 192, 207, 251, 290).

Pemantle, R. and M. C. Wilson (2004). "Asymptotics of multivariate sequences. II. Multiple points of the singular variety." *Combin. Probab. Comput.* **13**, pp. 735–761 (cit. on pp. 8, 12, 13, 104, 148, 159, 222, 251).

Pemantle, R. and M. C. Wilson (2008). "Twenty combinatorial examples of asymptotics derived from multivariate generating functions." *SIAM Rev.* **50**, pp. 199–272 (cit. on pp. 8, 137, 221, 278, 289, 295, 299).

Pemantle, R. and M. C. Wilson (2010). "Asymptotic expansions of oscillatory integrals with complex phase." In: *Algorithmic probability and combinatorics*. Vol. 520. Contemp. Math. Providence, RI: Amer. Math. Soc., pp. 221–240 (cit. on pp. 8, 104, 105, 162, 316).

Petersen, T. K. and D. Speyer (2005). "An arctic circle theorem for Groves." *J. Combin. Theory Ser. A* **111**, pp. 137–164 (cit. on p. 271).

Petkovšek, M., H. S. Wilf, and D. Zeilberger (1996). $A = B$. Wellesley, MA: A K Peters (cit. on p. 44).

Pólya, G. (1969). "On the number of certain lattice polygons." *J. Combin. Theory* **6**, pp. 102–105 (cit. on p. 278).

Pólya, G. and G. Szegő (1998). *Problems and theorems in analysis. II*. Classics in Mathematics. Theory of functions, zeros, polynomials, determinants, number theory, geometry, Translated from the German by C. E. Billigheimer, Reprint of the 1976 English translation. Berlin: Springer-Verlag (cit. on p. 30).

Raichev, A. and M. C. Wilson (2007). "A new method for computing asymptotics of diagonal coefficients of multivariate generating functions." In: *2007 Conference on Analysis of Algorithms, AofA 07*. Discrete Math. Theor. Comput. Sci. Proc., AH. Assoc. Discrete Math. Theor. Comput. Sci., Nancy, pp. 439–449 (cit. on p. 314).

Raichev, A. and M. C. Wilson (2008). "Asymptotics of coefficients of multivariate generating functions: improvements for smooth points." *Electron. J. Combin.* **15**, Research Paper 89, 17 (cit. on pp. 8, 314).

Raichev, A. and M. C. Wilson (Feb. 2012a). "A new approach to asymptotics of Maclaurin coefficients of algebraic functions." *ArXiv e-prints*. arXiv:1202.3826 [math.CO] (cit. on pp. 183, 314).

Raichev, A. and M. C. Wilson (2012b). "Asymptotics of coefficients of multivariate generating functions: improvements for multiple points." *Online J. Anal. Combin.* **7**, to appear (cit. on pp. 8, 309, 314).

Rannou, E. (1998). "The complexity of stratification computation." *Discrete Comput. Geom.* **19**, pp. 47–78 (cit. on p. 352).

Riddell Jr., R. J. and G. E. Uhlenbeck (1953). "On the theory of the virial development of the equation of state of mono-atomic gases." *J. Chem. Phys.* **21**, pp. 2056–2064 (cit. on p. 43).

Riesz, M. (1949). "L'intégrale de Riemann-Liouville et le problème de Cauchy." *Acta Mathematica* **81**, pp. 1–223 (cit. on p. 267).

Rockefellar, R. T. (1966). *Convex analysis*. Princeton: Princeton University Press (cit. on p. 130).

Rogers, D. G. (1978). "Pascal triangles, Catalan numbers and renewal arrays." *Discrete Math.* **22**, pp. 301–310 (cit. on p. 288).

Rubel, L. A. (1983). "Some research problems about algebraic differential equations." *Trans. Amer. Math. Soc.* **280**, pp. 43–52 (cit. on p. 44).

Rubel, L. A. (1992). "Some research problems about algebraic differential equations. II." *Illinois J. Math.* **36**, pp. 659–680 (cit. on p. 44).

Safonov, K. V. (1987). "On conditions for the sum of a power series to be algebraic and rational." *Mat. Zametki* **41**, pp. 325–332, 457 (cit. on p. 305).

Safonov, K. V. (2000). "On power series of algebraic and rational functions in $\mathbf{C}^n$." *J. Math. Anal. Appl.* **243**, pp. 261–277 (cit. on pp. 305, 307, 308).

Saito, M., B. Sturmfels, and N. Takayama (2000). *Gröbner deformations of hypergeometric differential equations*. Vol. 6. Algorithms and Computation in Mathematics. Berlin: Springer-Verlag (cit. on pp. 44, 117).

Shapiro, L. W. et al. (1991). "The Riordan group." *Discrete Appl. Math.* **34**. Combinatorics and theoretical computer science (Washington, DC, 1989), pp. 229–239 (cit. on p. 288).

Sprugnoli, R. (1994). "Riordan arrays and combinatorial sums." *Discrete Math.* **132**, pp. 267–290 (cit. on p. 288).

Stanley, R. P. (1997). *Enumerative combinatorics. Vol. 1*. Vol. 49. Cambridge Studies in Advanced Mathematics. With a foreword by Gian-Carlo Rota, Corrected reprint of the 1986 original. Cambridge: Cambridge University Press (cit. on pp. 13, 15, 43, 231, 278).

Stanley, R. P. (1999). *Enumerative combinatorics. Vol. 2*. Vol. 62. Cambridge Studies in Advanced Mathematics. With a foreword by Gian-Carlo Rota and appendix 1 by Sergey Fomin. Cambridge: Cambridge University Press (cit. on pp. 25, 33–35, 38, 43, 302, 304).

Stanley, R. P. (1980). "Differentiably finite power series." *Euro. J. Combin.* **1**, pp. 175–188 (cit. on p. 44).

Stein, E. M. (1993). *Harmonic analysis: real-variable methods, orthogonality, and oscillatory integrals*. Vol. 43. Princeton Mathematical Series. With the assistance of Timothy S. Murphy, Monographs in Harmonic Analysis, III. Princeton, NJ: Princeton University Press (cit. on pp. 81, 87, 94, 104, 105).

Sturmfels, B. (2002). *Solving systems of polynomial equations*. Vol. 97. CBMS regional conference series in mathematics. Providence RI: American Mathematical Society, pp. viii+152 (cit. on p. 118).

Szegö, G. (1933). "Über gewisse Potenzreihen mit lauter positiven Koeffizienten." *Math. Z.* **37**, pp. 674–688 (cit. on p. 270).

Theobald, T. (2002). "Computing amoebas." *Experiment. Math.* **11**, 513–526 (2003) (cit. on pp. 127, 131).

Van der Hoeven, J. (2009). "Ball arithmetic." *HAL preprints*. HAL: `00432152` (cit. on p. 183).

Van Lint, J. H. and R. M. Wilson (2001). *A course in combinatorics*. Second ed. Cambridge: Cambridge University Press (cit. on pp. 10, 15).

Varchenko, A. N. (1977). "Newton polyhedra and estimation of oscillating integrals." *Functional Anal. Appl.* **10**, pp. 175–196 (cit. on pp. 161, 250).

Vince, A. and M. Bona (Apr. 2012). "The number of ways to assemble a graph." *ArXiv e-prints*. arXiv:1204.3842 `[math.CO]` (cit. on p. 316).

Wagner, D. G. (2011). "Multivariate stable polynomials: theory and application." *Bull. AMS* **48**, pp. 53–84 (cit. on p. 276).

Ward, M. (2010). "Asymptotic rational approximation to Pi: Solution of an unsolved problem posed by Herbert Wilf." *Disc. Math. Theor. Comp. Sci.* **AM**, pp. 591–602 (cit. on p. 63).

Warner, F. W. (1983). *Foundations of differentiable manifolds and Lie groups*. Vol. 94. Graduate Texts in Mathematics. Corrected reprint of the 1971 edition. New York: Springer-Verlag (cit. on pp. 321, 324, 339).

Whitney, H. (1965). "Tangents to an analytic variety." *Annals Math.* **81**, pp. 496–549 (cit. on p. 98).

Wilf, H. S. (1994). *Generatingfunctionology*. Second ed. Boston, MA: Academic Press (cit. on p. 278).

Wilf, H. S. (2006). *Generatingfunctionology*. Third ed. Available at `http://www.math.upenn.edu/~wilf/DownldGF.html`. Wellesley, MA: A K Peters (cit. on pp. 10, 15, 43).

Wilson, M. C. (2005). "Asymptotics for generalized Riordan arrays." In: *2005 International Conference on Analysis of Algorithms*. Discrete Math. Theor. Comput. Sci. Proc., AD. Assoc. Discrete Math. Theor. Comput. Sci., Nancy, 323–333 (electronic) (cit. on pp. 8, 288, 293).

Wimp, J. and D. Zeilberger (1985). "Resurrecting the asymptotics of linear recurrences." *J. Math. Anal. Appl.* **111**, pp. 162–176 (cit. on p. 314).

Wong, R. (2001). *Asymptotic approximations of integrals*. Vol. 34. Classics in Applied Mathematics. Corrected reprint of the 1989 original. Philadelphia, PA: Society for Industrial and Applied Mathematics (SIAM) (cit. on pp. 87, 104).

Zeilberger, D. (1982). "Sister Celine's technique and its generalizations." *J. Math. Anal. Appl.* **85**, pp. 114–145 (cit. on p. 44).

Author Index

Subject Index

Note: A boldface page number refers to the page on which the term is defined